THE
PARTNERSHIP

A  HISTORY OF THE
APOLLO-SOYUZ TEST PROJECT

EDWARD CLINTON EZELL & LINDA NEUMAN EZELL

INTRODUCTION TO THE
DOVER EDITION BY

PAUL DICKSON

DOVER PUBLICATIONS, INC.
MINEOLA, NEW YORK

Bibliographical Note

This Dover edition, first published in 2010, is an unabridged republication of the work originally published in the NASA History Series as NASA SP-4209, Washington, D.C., in 1978. The frontispiece from the original edition has been moved to the inside front cover. A new Introduction by Paul Dickson has been added to this edition.

Library of Congress Cataloging-in-Publication Data

Ezell, Edward Clinton.
    The partnership : a history of the Apollo-Soyuz test project / Edward Clinton Ezell, Linda Neuman Ezell ; introduction to the Dover edition by Paul Dickson. — Dover ed.
        p. cm.
    Originally published: Washington : National Aeronautics and Space Administration, 1978, inseries: NASA SP ; 4209.
    Includes index.
    ISBN-13: 978-0-486-47889-0
    ISBN-10: 0-486-47889-0
    1. Apollo Soyuz Test Project. 2. Astronautics—International cooperation. I. Ezell, Linda Neuman. II. Title.

TL788.4.E95 2010
629.45'4—dc22

                                                                    2010021689

Manufactured in the United States by Courier Corporation
47889001
www.doverpublications.com

To

Hugh Latimer Dryden
1898-1965

and

Anatoliy Arkadyevich Blagonravov
1894-1975

Introduction to the Dover Edition

In July 2010, surviving members of the 1975 Apollo-Soyuz mission met in New York City to celebrate the thirty-fifth anniversary of the rendezvous and docking of two spacecraft: the American Apollo (often referred to as Apollo 18, although there was a previously scheduled moon landing mission numbered Apollo 18 which was cancelled for budgetary reasons) and the Soviet Soyuz 19. The Apollo-Soyuz Test Project (ASTP) was the first human spaceflight joint venture between the United States and what was then the Soviet Union.

Astronaut Vance Brand, who served as the Apollo command module pilot, was at the reunion and said "Apollo-Soyuz started a big thing. It has evolved into other cooperative programs—Mir and the International Space Station (ISS). The International Space Station has something like fifteen countries involved. There is a lot of cooperation that I think was never expected back in the 1970s and it has been a very good thing." By the time of the ASTP reunion the United States had become somewhat dependent of the Russian space program. NASA was by then under contract with Russia to carry astronauts and cargo to the International Space Station (ISS) until the U.S. was again ready to deploy a commercial spacecraft or a U.S. successor to the Space Shuttle, probably around 2015.

A third of a century earlier, on July 17, 1975, the two Cold War rivals linked spacecraft that were built halfway around the world from one another. The mission of the Apollo-Soyuz was conceived as a symbolic gesture to end the Space Race between the two countries, a competition that began when the Soviets were the first to launch an earth-orbiting satellite—Sputnik 1—into orbit on October 4, 1957. The Apollo-Soyuz project was symbolic of the increasing political detente between the two countries. At that moment the older Cold War competition yielded to increasing cooperation in the realm of space.

Symbolism aside, the architects of the plan realized there was a practical reason for cooperating—the safety of astronauts and cosmonauts. They wanted compatible docking equipment on space stations to permit rescue operations in space emergencies. This was an issue which had been raised in the public mind as a result of the 1969 space disaster movie *Marooned* and subsequently rendered real during the voyage of Apollo 13 in 1970. The spaceship experienced a major failure but the crew was saved by their use of the onboard Lunar Module as a "lifeboat" during the return trip to earth.

After rendezvous and docking, the Apollo and the Soyuz spacecraft remained docked together for forty-four hours. During that time, they conducted experiments, exchanged flags and signatures, ate together and proposed non-alcoholic toasts—actually borscht in a tube labeled as vodka—to their two nations and the future of space exploration.

The Partnership was originally published in 1978 when the value of the mission was still very much a matter of controversy. There were vociferous critics in congress in the years leading up to the project who objected to it on several levels, beginning with its cost ("a costly space circus" said one editorial) and ending with a strong distrust of the Soviet's ability to make it work. Others saw it as a giveaway of American technology. Robert B. Hotz, editor-in-chief of *Aviation Week and Space Technology,* saw it as pure propaganda and editorialized: "the real tragedy for this country was the decision to put its scarce space dollars into the political fanfare of Apollo-Soyuz."

The book is a fascinating and deftly written narrative. It underscores the conditions under which this complex mission was accomplished. The challenges were not only technical but cultural, ideological, and political. Some challenges were especially fascinating in light of the progress of the intervening years. Consider the matter of communications between the two countries. This is how the Ezells describe an early attempt to iron out some preliminary questions about the docking mechanism: "Because the overseas circuits were busy, nearly forty minutes passed before the NASA party reached the [Soviet] Academy of Sciences. For seventy-five minutes, the two sides struggled with the initially awkward process of talking through an interpreter over a not-too-perfect overseas telephone connection. On the Soviet side, the process was complicated by the fact that they were using two telephone handsets, but in Houston a conference arrangement circumvented the necessity of passing the phone from one person to the other." Communication was made even more cumbersome because each side had to transcribe the conversation, type it up and mail it half way around the world.

The authors not only show us how it was all put together but bring out the human side of the story, which includes the distinct personalities of the five spacefarers—three Americans and two Soviets.

Despite fears that Soviet engineering and science might not be up to the project, and that the Soviet spacecraft might not be safe, eventually the project was regarded by both nations as an exercise in perfection in which differences in everything from hardware to language to units of measurement were overcome. The voices of the critics who saw the mission as a boondoggle or a technological giveaway faded as cooperation between the two countries became the norm.

The 2010 ASTP reunion was notable in the sense that the participants got to relive the historic rendezvous and docking for perhaps the last time. They also got to honor their fallen comrade Donald Kent "Deke" Slayton, one of the original seven Mercury astronauts, whose first and last mission was ASTP.

"I opened the hatch and I saw the face of Tom Stafford," said Alexei Leonov at the reunion press conference. In addition to serving as the Soviet commander for the ASTP mission, Leonov was, ten years before ASTP, the first to conduct a spacewalk. "I said, 'Hello Tom! Hello Deke!' and at this moment we shook hands."

PAUL DICKSON
September 2010

Foreword

In the early days of the space age, when costs for exploration were projected, members of government and the scientific community often suggested that those nations with the greatest experience in space flight band together in joint programs. The United States and the Union of Soviet Socialist Republics, both heavily committed to space travel, were usually identified as the countries that should cooperate rather than compete. But, as long as the machines to accomplish such feats were little past the concept and drawing board stages, cooperative efforts would have been possible only with great difficulty, if at all.

By the end of the 1960s, some form of cooperation in manned space flight made more sense from a technical standpoint. Both nations had achieved some space goals and both had mission-proven spacecraft. Joint development of a new spacecraft would have been no easier at this stage than in the early years. But if each nation furnished a craft and together the nations figured out how to use them in a cooperative orbital flight, a useful step toward learning to work together in other fields would be taken. Even this, however, was a monumental task.

Communication was a bigger problem than technology in developing the joint program—and it was not necessarily a language problem. The philosophies of spacecraft design, development, and operations were so widely separated that a great chasm of differences had to be bridged before the technical work could begin. Several Soviet and American Working Groups, as this book relates, spent long hours, over many months, negotiating and reconciling the differences to produce a successful Apollo-Soyuz Test Project mission.

I had some concerns at the beginning of the cooperative program. We in NASA rely on redundant components—if an instrument fails during flight, our crews switch to another in an attempt to continue the mission. Each Soyuz component, however, is designed for a specific function; if one fails, the cosmonauts land as soon as possible. The Apollo vehicle also relied on astronaut piloting to a much greater extent than did the Soyuz machine. Moreover, both of these spacecraft, in their earlier histories, suffered tragic failures. By the time of the mission, all aspects of the two programs (hardware as well as procedures) that would be needed in the joint venture had been discussed frankly.

The exchange of people was perhaps a more significant gain than coming to some mutual understanding on how programs are conducted in the two countries and working out a joint flight project. Only about a hundred American and no more than two hundred Soviet managers, engineers, pilots, and technicians ever came into direct contact with each other, but millions of their countrymen watched with interest and discussed the activities, the families, and the ways of life (their similarities as well as their differences).

During the Apollo-Soyuz Test Project, and even afterwards, there were charges that the program was an American technological giveaway. These charges were unfounded. NASA's conduct of its space programs has been covered by the media in great detail and descriptions of its systems can be found in many technical journals in libraries and bookstores. However, no one can build an Apollo or a Soyuz merely by reading a book or visiting a factory. These craft are the products of many, many incremental steps, lasting for years, and of the development of a personnel reservoir capable of managing a space program from concept through operations. Both sides did gain some new knowledge, but the benefits accrued by working together probably outweigh any potential threat. Apollo-Soyuz was the product of an evolutionary process of nearly 20 years. This book traces the events that led to this cooperative flight and then introduces the reader to five men, from two nations, as they worked together in the vastness of space.

Christopher C. Kraft, Jr.
Director, Lyndon B. Johnson Space Center
November 1977

Preface

Apollo and Soyuz docked in space on 17 July 1975. The American and Soviet space teams met in orbit to test an international docking system and joint flight procedures. Sometimes lost in the extensive coverage given the event by the media was the fact that the Apollo-Soyuz Test Project (ASTP) was only a *first* step—an experiment. Implicit in the preparations for the first international rendezvous and docking was the idea that in the future manned space flight—both routine flights and rescue missions—could use the hardware concepts and mission procedures developed by the National Aeronautics and Space Administration and the Soviet Academy of Sciences. As a first step, ASTP was a success. The hardware was sound, and specialists from the two nations worked truly as a team. This history of ASTP is also a first step.

Apollo and Soyuz were still 16 months away from their rendezvous when we began this history in April 1974. But interest in an official record of the joint effort goes back to at least the summer of 1972, when ASTP emerged as a full-scale project after the Nixon-Kosygin summit agreement on cooperation in space. Throughout NASA, individuals who were preparing for the mission were aware that they were involved in a unique experience. Nearly all these people had originally come to the space agency during the Cold War to help ensure American preeminence in space. But with ASTP, they were asked to cooperate with their rival. Indeed, they were expected to build and test hardware that would permit a joint flight by mid-1975. Not everyone in NASA was sympathetic with this goal, but nearly all were intrigued by the challenge.

NASA employees have thrived on challenges. As members of a brand new agency, they had dared to overcome the risks involved in putting a man into orbit. Project Mercury had been the answer to that first bold challenge. They mastered the difficulties of space rendezvous in the second manned program, Gemini. And in the boldest of all challenges in the span of a single decade, they worked together to send men to the moon and return them safely. During the Skylab missions, they broke new barriers as man learned to live for extended periods of time in the zero-gravity environment of space. But flying a joint mission with the Soviet Union would be more than just a technological feat; it would require diplomacy, hardheaded perseverance, and good humor. NASA accepted the new challenge, despite pessimistic voices inside and outside the agency.

There is an infectious spirit of optimism at NASA. Individuals do not go about saying they are optimists; they just act in ways that indicate they are. Contracting for a history of ASTP before the hardware was finished and before the mission was flown was one example of this positive frame of mind. The ASTP team at the Johnson Space Center (JSC) in Houston knew that Apollo and Soyuz would rendezvous and dock in space.

This history is an official history only because it was sponsored by NASA. The authors were invited through a contract to record their version of the events that led to, shaped, and emerged from the joint flight. When we first met with Glynn S. Lunney, the American Technical Director for ASTP, we asked, "Why do you want to have a history written?" Lunney responded that he had never asked himself precisely that question but that he did desire to see preserved the subtlety of human interaction that he had observed during the first four years of the project. Lunney went on to suggest that the technical aspects of ASTP were not nearly as interesting, or perhaps as significant, as the working relationships that had emerged among the technical specialists of the two nations. Written documents tend often to be dry and distilled, he thought. Lunney wanted a historian to see firsthand some of the personal interplay so that the flavor of the working sessions could be preserved along with the story that could be found in more conventional documents.

Our history is to a large extent based upon oral records. Sometimes dubbed "combat historians," or less favorably, "instant historians," we stalked the halls of the joint meetings in Houston with tape recorders in hand. Although never quite a part of the furniture, we were not an apparent disturbance to any of the negotiations we witnessed. And although we never traveled to the Soviet Union, those who did gave freely of their time, recollecting their experiences or answering our questions. Sometimes we cornered them in the halls between negotiating sessions, at other times by telephone. But whether it was over a quick cup of coffee while they waited for Xerox copies of a document or during a hamburger break, these men and women went out of their way to help, to explain, and to re-explain.

In addition to this firsthand observation of ASTP activities and interviews with participants, we had the typical "embarrassment of riches" that has faced all those who have written history for NASA.* Several early participants had already retired their "desk archives" to the JSC history office by the spring of 1974, when the authors began receiving all

---

*Barton C. Hacker and James M. Grimwood, *On the Shoulders of Titans: A History of Project Gemini,* NASA SP-4203 (Washington, 1977), Preface.

correspondence originating from the Apollo Spacecraft Program Office relating to ASTP as part of the daily distribution of such materials. In the future, when researchers look at the correspondence files that we have left behind at the Johnson Space Center and see "BE4/EZELL," they will know that the historians were reading everyone's mail. Much of the material we sifted through was extremely detailed. We could learn how many electrical connectors for the VHF/AM transceiver were being shipped to Moscow or what the latest revisions were to the "Joint Crew Activities Plan"; so we spent many days separating the nitty-gritty telexes and test data from the material that would permit us to tell the larger story.

The book that emerged from these efforts has both strengths and weaknesses. First, we have told essentially the NASA side of the story. We had free access to American materials and members of the NASA team. In addition, NASA has an ongoing history program, which makes the historian's task an easier one. Most of the information on earlier programs is readily at hand in published histories or works in progress. The Soviet space program by contrast is shrouded in mystery. The Soviets have not produced any comparable historical studies of their programs, and when we requested Soviet assistance with this history we were informed politely, but firmly, that they did not wish to discuss history. As a consequence, we had only limited opportunities to speak with members of the Soviet ASTP team. Where possible, to balance our presentation, we have cited Russian language sources, but our story remains one told from the American perspective.

Second, history written as events are unfolding can be neither entirely objective nor complete. But we have attempted to be fair in our judgments as we explained what the project meant to the participants through their personal recollections—recollections that otherwise might not have been preserved. We have tried to write an interesting narrative, sufficient in technical detail for the intelligent reader to grasp the mechanical elements of ASTP, but simple enough so that pages do not become bogged down by complex description. Those who worked on ASTP know that for every page of description in this history there are often hundreds of pages of technical documents, thousands of feet of computer tape, and seemingly endless hours of work. Some will be dismayed that their efforts were passed over or given only a line or two, but our goal has been to preserve some of the spirit of ASTP with the hope that some historians in the future will evaluate the project's significance more fully. Years will pass before we know if the partnership of so many engineers, spacemen, negotiators, and diplomats represents a stepping stone, plateau, or pinnacle in the history of international cooperation. Only time will determine the true perspective of their performance.

Third, there are topics that we chose not to discuss in detail because they will be recorded in other NASA publications. For example, we did not describe in depth the manufacturing history of the Apollo spacecraft, since that is covered in the fourth volume of *The Apollo Spacecraft: A Chronology* (NASA SP-4009) and is the subject of the forthcoming history "Chariots for Apollo." We may be accused of slighting certain groups—the State Department, the Department of Defense, or Rockwell International, the spacecraft contractor. But we think that our treatment of these organizations in this history reflects adequately their participation in ASTP. More than any single manned space flight before, ASTP was a Johnson Space Center enterprise. Technical negotiations were conducted almost exclusively by personnel from Houston. Even NASA Headquarters typically assumed an advisory and supportive role, with the notable exception of Deputy Administrator George M. Low, who played a central part in planning and directing the program. When it came to the design of the docking system and the docking module, the JSC engineers took the lead and basically told the contractor in detail what they wanted. Again, this was a departure from earlier programs and does not reflect the manner in which the Space Shuttle was to be developed. We hope our book adequately reflects the unique nature of Apollo-Soyuz.

Because the flight of Apollo and Soyuz can be understood only in the international context from which it emerged, we have presented two introductory chapters that describe the early years of Cold War competition (chap. I) and the first efforts at cooperation (chap. II). The next chapter describes the evolution of manned spacecraft in the U.S. and U.S.S.R. (chap. III), while "Mission to Moscow" (chap. IV) outlines the experiences of the American technical specialists during their first visit to the U.S.S.R. in October 1970. In January 1971, discussion about cooperation in space flight turned from general talk of the "future" to specific proposals for a test mission using existing hardware (chap. V). During the ensuing 16 months, NASA and Soviet Academy engineers began to learn to work with one another, and by May 1972 the two sides were confident that they could design and build the necessary hardware by mid-1975 (chap. VI). Once given the official seal of approval at the Nixon-Kosygin summit in May 1972, work began in earnest toward the creation of a test project (chap. VII). As the hardware evolved, the United States and the Soviet Union monitored progress with reviews, planned public release of ASTP information (chap. VIII), and selected their crews, who began their technical and linguistic training for the flight (chap. IX). Final reviews of the project were held in the spring of 1975, while critics questioned the wisdom and safety of the joint mission (chap. X). All the efforts culminated in a nearly flawless flight in July 1975 (chap. XI), and the only unanswered question concerned what

PREFACE

the future would hold for cooperation in space between two nations that
had dared to break down old rivalries.

As for accolades to those who helped us with this history, their names
are best preserved in our essay on sources, which describes the materials we
used, where they came from, and how they are arranged for future use.

On 24 July 1975 after Apollo had splashed down and the crew was
aboard the U.S.S. *New Orleans,* we chanced to encounter Glynn Lunney as
he left the Mission Operations Control Room. Suit coat over his shoulder, he
smiled and said, "Now you have a story to tell." He was right.

Edward Clinton Ezell
Linda Neuman Ezell
Houston
July 1976

Notes

Controversy often surrounds the conversion of the Cyrillic characters used in the Russian language into the Roman letters used in English. In the absence of a universally accepted standard, we have transliterated Russian personal and place names after the pattern established by the Soviets themselves and as recorded in the English language version of the ASTP documents.

The metric system poses an equally thorny problem. NASA is committed to the national goal of metrication, and in 1973 the space agency prohibited the use of English weights and measures in all publications, including its historical series. But NASA engineers were not thinking metric all the time. ASTP documents often recorded specifications in metric form, but they almost always added the English equivalent. Furthermore, the actual production of the American components for the joint mission was done on manufacturing tooling calibrated in standard American engineering units. As some passages in the text indicate, this English/metric schizophrenia caused occasional troublesome moments when the switch from metric drawings to English tooling required extra care to determine that all the critical dimensions were correct. In this history we have followed the trend to metrication, and we have used the *systeme internationale d'unites* (SI) with one major exception. We have chosen millimeters of mercury to designate cabin pressures. The Soviet space community has commonly used this measurement, while their American counterparts have used pounds per square inch (psi). Neither side talked in pascals (newtons per square meter), the approved SI measurement, which would have puzzled nearly everyone who worked on ASTP.

Another metric unit requires explanation. In the English system, "pound" is used to describe both mass and force. The creators of the metric system devised two distinct units, the familiar gram for mass and the less familiar newton for force—thus pounds of weight become grams, but pounds of thrust become newtons. Where we wrote about thrust, we used both newtons and the equivalent in pounds. Elsewhere, we have given only the metric units to keep from cluttering the pages with endless conversions in parentheses.

Contents

List of Illustrations

CONTENTS

Table

Prologue
The Paine-Keldysh File

Air Force One, the President's airplane, was flying westward across the Pacific in late July 1969 toward the anticipated splashdown site of *Apollo 11*. As man's first visit to an extraterrestrial body neared its conclusion, four men in the plane informally discussed the future of manned exploration in space. President Richard M. Nixon, Secretary of State William P. Rogers, National Security Adviser Henry A. Kissinger, and Administrator Thomas O. Paine of the National Aeronautics and Space Administration all knew that the Apollo program was a watershed, making the first lunar landing and those that would follow the end of an initial phase of space exploration. The age of Mercury, Gemini, and Apollo had been one of national adventure and single-flight spacecraft. The next step into space would call for reusable spacecraft and space stations. One question in particular remained to be answered: Would the character of space exploration change from costly and duplicative competition to cooperation among nations?

The concern for future cooperative space ventures was uppermost in Paine's thoughts. He directed his companions' attention to the desirability of greater substantive international cooperation in space projects, especially with the Soviet Union. Paine argued convincingly for NASA's plans to seek increased multinational space ventures. The President and his advisers agreed that this was a laudable goal, and they encouraged Paine to pursue his contacts with the Soviets.[1]

Tom Paine, the third administrator of NASA, brought to the agency an abiding belief that the Soviet Union and the United States eventually would have to consider working together, abandoning the competitive nature of space flight. His beliefs concerning the necessity for closer working relationships between the two superpowers went back many years. When he returned to college after World War II, "learning the Russian language was one of the two fields [he] selected for its long-range implications (the other was nuclear energy)."[2] As he studied the future of manned space flight and other aspects of man's investigations of the cosmos, Paine became convinced "that the conquest of space [was] a job of such enormity that a new partnership of major nations should be organized with the U.S./U.S.S.R.

leaders demonstrating the way. This required, of course, a complete reversal of our previous rationale of U.S./U.S.S.R. competition as the justification for NASA's bold programs."[3] Such an approach had been fine for the 1960s. Paine later reflected on this decision:

> ... I decided—and I hope I made the right decision—that although Jim Webb certainly had done a tremendous job of building up NASA and the program on the basis of the Russian threat, that times had changed. The time had come for NASA to stop waving the Russian flag and to begin to justify our programs on a more fundamental basis than competition with the Soviets.[4]

Thus, throughout his time with NASA, Paine tried to tone down the competitive aspects of Soviet-American space relations. He concentrated on developing a rapprochement with the Soviets that might spread into other parts of society. He also believed that elimination of the "Russian threat" rationale would force NASA to develop a space program based upon new foundations. This would not mean that competition with the Soviet Union would be eliminated; Paine saw that as a natural aspect of space exploration. However, he thought that it should be a more open, friendly contest. He also expressed the belief that NASA should not "scare the American public with such a competition but ... do it as a matter of national pride."[5] Paine's efforts to establish a new posture with the Soviets began two months before the flight of *Apollo 11.*

Following his appointment as Administrator on 5 March 1969, Paine renewed proposals made by his predecessors by calling once again for international cooperation in the scientific study of outer space.* The efforts to establish a foundation for cooperative space enterprises during the post-Sputnik years, 1957-69, had been filled with recurring frustrations and dashed hopes (see chaps. I and II). Despite skepticism on the part of some of his staff, at the end of April Paine began official correspondence with the Soviet Academy of Sciences. With his letter to Anatoliy Arkadyevich Blagonravov, Chairman of the Academy's Commission on Exploration and Use of Space, Paine forwarded a copy of the NASA management handbook sent to all potential participants in space scientific studies, *Opportunities for Participation in Space Flight Investigations.*[6]

Administrator Paine urged Academician Blagonravov to solicit from his scientific community proposals for experiments to be flown on American spacecraft, with complete assurance that those experiments would be given full consideration based upon their scientific merit. Paine told his Soviet correspondent that "the close collaboration which would be required to

---

*Paine had served as Deputy Administrator from 5 Feb. to 7 Oct. 1968, at which time he became Acting Administrator, effective with the resignation of James E. Webb.

integrate Soviet experiments into American spacecraft should engender closer working relationships than we have been able to achieve and establish a basis for still further commonality of purpose and program." Paine hoped that the Soviet scientists would be interested in NASA's plans to place a laser-ranging retroreflector on the moon during the *Apollo 11* lunar landing, because this reflector would permit precise measurement of lunar orbital phenomena. Paine concluded by saying, "The participation of Soviet scientists in this and other opportunities will be warmly welcomed. Of course, if the Soviet Academy should find itself in a position to extend similar opportunities to American scientists, this too would be welcomed."[7]

Later in May, Paine tried to find a suitable time and place for a conversation with Blagonravov. In a letter dated 29 May, he suggested that "it would be useful if we attempted at an early date to arrange a meeting and informal discussion which could further our mutual interests in cooperative space projects." Such talks had not been possible during an earlier visit by Blagonravov to New York, nor had Paine's own travel plans for Europe during the summer of 1969 afforded a suitable occasion. "However, another opportunity will be presented by the launching of *Apollo 11* from Cape Kennedy, now scheduled for July 16. I would be very pleased if you could be there." Sensitive to possible concerns on the part of the Soviet Academy of Sciences, Paine continued, "I appreciate the questions which arise in connection with such an invitation. I assure you that my invitation is offered in all sincerity and entirely for the purpose of permitting you to view an event which is of interest to all of us who are engaged in space programs, and to provide an opportunity for private discussions on the subject of cooperation." While there was the almost certain possibility that such a meeting would be in the public eye, Paine stressed that "steps could be taken to avoid publicity attached to such a visit by you." Therefore, he asked if Blagonravov could accept the invitation.[8] Blagonravov declined.[9] Undeterred, Paine waited for a more auspicious moment to continue his efforts.

The successful lunar landing became an important element in the course of subsequent discussions of space cooperation between the Soviets and the Americans. Following the landing of *Eagle* and the pioneering moon walks of Neil A. Armstrong and Edwin E. Aldrin, Jr., on 20 July 1969, the Soviet Union joined the ranks of official well-wishers congratulating the United States. On the following day, Soviet Premier Alexsey Nikolayevich Kosygin took the opportunity afforded by a farewell conversation with former Vice President Hubert H. Humphrey to compliment the Americans on their accomplishment and to express his interest in widening talks with United States officials on the topic of space cooperation.[10]

The news coverage in the Soviet Union of the *Apollo 11* flight was equally warm. Scientist Cosmonaut Konstantin Petrovich Feoktistov typified the public comments in his press and television statements. Hailing the flight as a landmark, he reflected in an *Izvestiya* article, "This without a doubt is a major development of cosmonautics. . . . The very fact of the first landing of human beings on another celestial body cannot but stimulate the imagination. What recently had been pure fantasy is now a reality."[11] Georigy Ivanovich Petrov, Director of the Institute of Cosmic Research, called the mission an "outstanding achievement," while suggesting that more information for each ruble could have been obtained through the use of unmanned, automated spacecraft, a sentiment that still has its supporters in the American scientific community as well.[12] The race for the moon had ended.

The first steps toward closer cooperation grew out of a formal exchange of letters between Administrator Paine and the President of the Soviet Academy of Sciences, Mstislav Vsevolodovich Keldysh. A distinguished physicist who had specialized in space mechanics, Keldysh had been among the well-wishers following the return of *Apollo 11*. He told Paine that he "warmly" congratulated the United States on the successful lunar landing and return, as "this achievement is a great contribution to the opening up of the cosmos in further progress of world science."[13] Paine responded with the suggestion that Keldysh might wish to select a delegation of Soviet scientists to attend the briefings at NASA Headquarters in Washington on 11-12 September to discuss the proposed experiments to be carried on the Viking mission to Mars, then scheduled for 1973. The presentations were to include findings of the 1969 Mariner investigations and also a description of the current status of the spacecraft design and planning for the mission. The Administrator was confident that the Soviet scientists would find the briefings informative. Dr. Paine suggested that this occasion could also serve as an opportunity for an informal discussion between "your scientists and a small group of NASA personnel." As before, the Paine rationale for this proposal was to maximize the scientific benefits for the manpower and money expended.

> We have just completed a very extensive and detailed planning activity, and have outlined possible courses of action for NASA over the next decades. We would be pleased to discuss these and hope that your scientists would be able to discuss some of the future plans for the Soviet program.

To keep the talks manageable, Paine suggested that they be limited initially to planetary exploration.[14]

The Soviets did not receive the Paine letter until 3 September; thus, they were unable to take proper advantage of it. Keldysh was nevertheless "very grateful" to Paine for the "courteous" invitation, but he regretted that he could not "gather together a group of Soviet scientists in such a short time to participate in this meeting." Keldysh suggested that the doors not be closed on expanded cooperation and asked for copies of the materials to be distributed at the Viking briefings, "in order that Soviet scientists could develop possible proposals from our side. Later it would be possible to exchange opinions on this question."[15]

Paine responded in a letter on 15 September with the materials requested by Keldysh. Speaking to the problem of timing, Paine regretted that he had not given the Soviets more advanced notice, "but I believe that this circumstance need not thwart the purpose of my invitation." Paine went further and said, "In order to compensate for your inability to attend the Viking briefing this week, we are prepared to provide a meeting for your people as soon as you can arrange for them to get to Washington." Returning to the theme of his 21 August letter, Paine suggested that such a briefing could also be accompanied with a broader discussion of the respective plans that the Soviet Academy and NASA had for planetary exploration.[16]

The Academy of Sciences in its subsequent decision not to participate in the Mars landing program in no way rejected the possibility of future cooperative efforts. After a study of the Viking materials, Keldysh responded that immediate Soviet participation in the Viking program was not feasible from their point of view. This response reflected a difference in scientific philosophy and not a put-off for political reasons. Keldysh pointed out that "the investigation of planets by automatic spacecraft requires a complex program of measurement, which determines the flight plan and actual design of the spacecraft. The installation of individual instruments, which in essence would duplicate the measurements planned by your scientists, would hardly be worthwhile."[17]

As the correspondence between Keldysh and Paine developed, the Space Task Group* presented a report to the President: *The Post-Apollo Space Program: Directions for the Future.* When President Nixon requested this study on 13 February 1969, the lunar landing of *Apollo 11* was a foregone conclusion. Once man had reached the moon, a new set of goals would have to be developed. In the ensuing eight months, the Task Group provided a forum for discussions with governmental agencies, the Congress, and participants from industry, universities, professional societies, and the public. The completed report provided the basis for an informed discussion of the future direction of the American space effort.[18] By the time the Space Task Group had completed its deliberations and produced its

---

* The Space Task Group consisted of Vice President Spiro T. Agnew, Chairman; Secretary of the Air Force, Robert C. Seamans, Jr.; Administrator of NASA, Thomas O. Paine; Science Adviser to the President, Lee A. Dubridge; and the following observers: U. Alexis Johnson, Under Secretary of State for Political Affairs; Glenn T. Seaborg, Chairman, Atomic Energy Commission; and Robert P. Mayo, Director, Bureau of the Budget.

report, the first moon landing had passed into history; the perspective of the report reflected a new era.

Assessing the international aspects of the *Apollo 11* flight, the Task Group stated that the "Achievement of the Apollo goal resulted in a new feeling of 'oneness' among men everywhere. It inspired a common sense of victory that can provide the basis of new initiatives for international cooperation." Looking back on the preceding twelve years of space flight, the report declared that the United States and the Soviet Union had been portrayed widely "as in a 'race to the Moon' or as vying over leadership in space." Candidly, the Task Group reported that "this has been an accurate reflection of one of the several strong motivations for U.S. space program decisions over the previous decade."[19] In looking for new goals for the space program, the Space Task Group suggested that international cooperation was one of the themes emerging from the Apollo experience that should be an essential element of future programs:

> The landing on the Moon has captured the imagination of the world. It is now abundantly clear to the man in the street, as well as to the political leaders of the world, that mankind now has at his service a new technological capability, an important characteristic of which is that its applicability transcends national boundaries. If we retain the identification of the world with our space program, we have an opportunity for significant political effects on nations and peoples and on their relationships to each other, which in the long run may be quite profound.[20]

In keeping with the spirit of the Space Task Group's report, Paine transmitted copies of it, together with NASA's more detailed report *America's Next Decades in Space,* to the Soviet Academy of Sciences. In his cover letter of 10 October 1969, Paine told Keldysh that these documents might "suggest to you as they do to me, possibilities for moving beyond our present very limited cooperation to space undertakings in which the Soviet Union and the United States could undertake major complementary tasks to the benefit of both our countries." Paine added that he would be pleased to initiate discussions should Keldysh feel that "there may now be some reasonable chance for progress." In closing, the Administrator welcomed a visit from Keldysh to the United States, or he was prepared to travel to the Soviet Union. Tom Paine saw the glimmer of hope for a mutual space effort, and he intended to pursue that opportunity.[21]

The Keldysh response supported Paine's belief that cooperation was possible. Keldysh said that he fully shared Paine's "point of view concerning the advantages of international cooperation and the coordination of plans for scientific investigations which are conducted in space." The Soviet scientist also agreed with Paine that this was an area in which Soviet-

American cooperation was of a "limited character" and that there was "a need for its further development." Perhaps a meeting between representatives of the Soviet Academy and NASA would be beneficial, but the preparation for such a meeting would require time. Keldysh expected to be able to address this matter more fully in three or four months. Then "we could return to this matter and reach an understanding on the time and place for our meeting and the schedule. . . ."[22] Now that the Soviets seemed to be planning for substantive talks, American government agencies began an internal discussion on what it would mean to engage in such talks. Following the informal conversation aboard Air Force One, the President formed an interagency committee to study the ramifications—positive and negative—that would arise relative to cooperative space ventures with the Soviet Union. The committee was then to present policy alternatives to the White House.* With the exception of the Department of Defense representatives, the members of this committee favored broader efforts toward cooperation. One suggestion for joint work concerned those areas of manned space activity affecting safety and common flight operations procedures—for example, the development of compatible docking hardware and the standardization of flight control and rendezvous systems to permit the creation of a reciprocal space rescue capability. In such a project, both countries stood to benefit; but clearly both sides would have to exchange much more information if a rendezvous and docking system were to get beyond the talking stage. The candid opinion in Washington, including the State Department, was that there would be no early progress in obtaining such discussions with the Soviets.[23]

While the interagency committee deliberated, Dr. Paine responded to Academician Keldysh's December letter. The Administrator had hoped for an earlier encounter; now he looked forward to receiving word in the early spring concerning the Soviet preference for a time and place for an initial conference.[24] A key step toward a meeting between officials from NASA and the Soviet Academy was an informal dinner in New York City at the Lotus Club, when a serious cooperative proposal was discussed for the first time.

Since Academician Blagonravov was in New York, Paine thought that this was an appropriate occasion for them to become acquainted. It also seemed to be the right time for a "discreet discussion" on joint space

---

*This committee, formed in the latter part of 1969, consisted of representatives from the Department of State, the Department of Defense, the Office of Science and Technology, the Space Council, and NASA. State coordinated the activities of the committee, even though the department had basically played an advisory role in the earlier NASA discussions with the Soviets.

7

ventures.[25] The amiable conversation touched on many subjects. Paine mentioned to his guest that Neil Armstrong planned to deliver a paper at the COSPAR* meetings scheduled for 20-29 May 1970 in Leningrad, and Paine said he hoped that Armstrong would have an opportunity to visit some of the Soviet scientific facilities. Blagonravov responded that the cosmonauts would be pleased to show their American counterpart their facilities and some of the other space-related institutes. Paine then summarized for Blagonravov the substance of his testimony earlier that day on the problems encountered during the unsuccessful lunar flight of *Apollo 13*. Paine also described NASA's efforts to develop increased foreign participation in the United States space program. During the course of the evening, Paine asked Blagonravov for his views on the possibility of developing joint programs for planetary exploration and for work toward astronaut/cosmonaut safety. Along this line, Paine suggested that it might be worthwhile to discuss incorporating compatible docking mechanisms on future spacecraft, such as space stations and shuttles. The latter concern reflected the proposals of the President's interagency committee.[26]

While Blagonravov did not respond directly, both the Administrator and his Assistant for International Affairs, Arnold W. Frutkin, felt that their Soviet guest could be relied upon to transmit a favorable report on the meeting to the U.S.S.R. policy makers. As Tom Paine was later to reflect, "We had no reasons to expect a favorable reaction" from Moscow, but there was no reason not to try.[27] Frutkin, judging from his previous contacts with Blagonravov, felt that some "new signal" was in the works and that it would likely come in response to the Paine-Keldysh correspondence. Frutkin also noted that Blagonravov was not likely to play a prominent role in later discussions. The elder Soviet space statesman had referred several times to his upcoming 76th birthday.[28]

Closer cooperation took a step forward at the 13th annual meeting of COSPAR in Leningrad. Soviet Premier Kosygin sent a message that seemed to signify a new trend—"International cooperation in space exploration and in the use of outer space for peaceful purposes must be based upon the development of mutual understanding and trust among the peoples." Kosygin saw that there was "growing cooperation on an international scale in space research," and he noted, "further progress in this field can open up still greater prospects for mankind."[29] While Neil Armstrong received an exceptionally warm reception from the predominantly Soviet audience,

*The International Committee on Space Research, or, as it is probably more widely known, COSPAR, is a subdivision of the International Council of Scientific Unions (ICSU) to which the United States belongs through the National Academy of Sciences.

PROLOGUE

George M. Low, the Deputy Administrator of NASA, had significant private talks with Soviet officials.

On the second morning of the COSPAR sessions, 21 May 1970, Low met with President Keldysh. The two men began their conversation with an exchange of books. Low presented a new book of photographs taken by Lunar Orbiter, while Keldysh reciprocated with a book on the Soviet space program. Low then told the President of the Academy that NASA officials were still eager to hear of possible proposals for cooperation and that Dr. Paine was prepared to meet him at any time and place. Keldysh said that he had waited until the Academy had something positive to offer. He then indicated to Low that such a proposal likely would be made in the near future. Low assured Keldysh that NASA would give positive consideration to any proposal, underscoring the fact that NASA was "most anxious" to start cooperative efforts with the Soviet Union in space. Summarizing his impressions for the record, Low concluded, "The meeting was pleasant, and communications between us appeared to be good."[30] A less formal discussion of this same topic had been undertaken ten days earlier by Dr. Philip Handler of the U.S. National Academy of Sciences during his visit to the U.S.S.R.

Handler later recounted how he became involved in the Soviet-American space dialogue. "My personal introduction to the possibility that I might play a useful role with respect to Soviet-American cooperation began when I accompanied Tom Paine and Jim Webb to President Johnson's ranch" on 2 November 1968 for the presentation of NASA awards to outgoing Administrator Webb and the *Apollo 7* crew. On the flight to Johnson City, Texas, conversation turned to the need for greater international cooperation. Handler recalled, "I pointed out that among my other goals as the new President of this Academy was the development of closer scientific ties between our Academy and that of the Soviet Union." Both Paine and Webb gave him encouragement but warned him not to become discouraged if he did not meet with early success. These men were aware of the long and unfruitful efforts in which NASA had been engaged with the Soviets.[31]

Before he had an opportunity to talk with the Soviets, Handler saw a movie that influenced his thinking concerning manned space flight.

In the early spring of 1970, . . . I saw a special showing of the film *Marooned* in which . . . an American astronaut is marooned in orbit, unable to return to earth, and has a relatively limited oxygen supply remaining. While preparations are made on earth for rescue by NASA, a Soviet spacecraft is caused to change its course so as to closely approach the helpless American craft. A Soviet cosmonaut then undertakes a space walk and delivers some tanks of oxygen to the marooned American permitting him to survive until the American rescue is possible.*

About a week before Handler's departure for the Soviet Union, he saw Tom Paine; *Marooned* was still in the back of his mind. During their conversation, Paine and Handler reviewed various possibilities for cooperation with the Soviets. Paine told him of his correspondence with Keldysh and urged Handler to press the discussion of this subject with the Soviets. Handler later reflected, "it was my clear intention to catalyze the process knowing full well that if I could secure agreement with the Soviet Academy to begin cooperative ventures seriously, from then on the negotiations would have to be directly with NASA."[32]

The two days that Handler spent in Moscow, 11-12 May 1970, were filled with talks on a broad range of topics relating to the whole realm of cooperation between the two scientific communities. At one point, Handler found an opportunity to discuss the question of space cooperation with President Keldysh, Dzhermen Mikhaylovich Gvishiani (Premier Kosygin's son-in-law and Deputy Minister for Science and Technology), and a group of younger Soviet scientists. Handler's approach was less tactful than that which had been pursued by NASA officials; "I confronted them with copies of a recent article in the *New York Times* and in *Science* magazine recounting the rather disgraceful history of their failure to react to the many initiatives offered by NASA." Handler urged closer cooperation by describing the basic scenario of the film *Marooned*. The fact that "an American film should portray a Soviet cosmonaut as the hero who saves an American's life came to them as a visible and distinct shock."

In response to Handler's general comments that surely the time had come for joint space ventures "for reasons of economy, for reasons of the symbolism it might offer humanity, and to accelerate the pace of space exploration," the Soviets said they were preparing a set of replies to Dr. Paine. Handler understood that the proposals would center on three specific areas. First, the Soviets would suggest a more vigorous program for the exchange of scientific data from space experiments. Second, they would recommend a unified system of communication with spacecraft and ground stations. Finally, they would suggest wider exploitation of both nations' meteorological satellites.[33]

According to Handler, the suggestion that the two nations work toward the development of a "mutually acceptable single docking mechanism on

---

*The motion picture was based upon a novel of the same title by Martin Caidin published by E. P. Dutton, 1964. The adventure story was set in the era of Project Mercury, while the 1969 screenplay by Mayo Simons was set in the Apollo period with a crew of three, not one as Handler recollected.

space stations planned by both groups" caused considerable discussion. After some private conversation in Russian in which some of the young scientists appeared to urge favorable consideration of this idea, Gvishiani and Keldysh quietly told Handler that they were not in a position to give a definitive reply at the moment; they were sympathetic, but would have to refer the matter to higher authorities. The two Soviet officials asked Handler if he could wait for a reply and further if he planned to discuss this proposal with the American press upon his return home; Handler indicated that he would remain silent until he had their reply. The Soviets promised to direct a response to either Paine or Handler at an early date.[34]

Neither Tom Paine nor Philip Handler could have known then how close they were to a dramatic offer on the part of the Soviet Academy of Sciences. On 11 July, Anatoliy Fedorovich Dobrynin, the Soviet Ambassador to the United States, called Handler at the National Academy of Sciences. Ambassador Dobrynin asked him to receive Ye. A. Belov, the newly appointed Scientific Attache at the Soviet Embassy, who had a message from Academician Keldysh. At the subsequent meeting, Belov, having just arrived from Moscow and reading from his own handwritten notes, discussed a number of the questions that had been left open after the May talks with Handler. He also brought specific word from Keldysh that the Presidium of the Soviet Academy, in consultation with other appropriate groups, was prepared to discuss common docking mechanisms for space stations.[35]

The message from Keldysh indicated that the Soviet Academy would be pleased to respond favorably if the National Academy issued an official request for a discussion of cooperation in space. The Soviet message to Handler could be interpreted as an indication that the Soviet space scientists thought that the National Aeronautics and Space Administration was subordinate to the National Academy of Sciences, just as their Institute of Space Research is a subdivision of the Soviet Academy.* However, Handler perceived the Keldysh request differently. The National Academy provided a "comfortable channel" of communication through which the Soviets could

---

*The National Academy of Sciences, established 3 Mar. 1863 by congressional charter, has enjoyed a close relationship with the Federal government, but it is not an official body. Instead, it is an organization of distinguished scientists who act in an advisory capacity to governmental agencies. The Academy does not have laboratories of its own, but seeks to stimulate scientific research for the public welfare through existing university and government facilities. The Academy of Sciences of the U.S.S.R. is, on the other hand, an official government institution. The Soviet Academy, which traces its beginnings back to 1725, performs a number of significant roles. Among them is a direct involvement in higher education, and many of the Academy's institutes grant academic titles and graduate degrees.

indicate their interest in cooperative discussions. If the American government was serious in its suggestions, then the proper agency, NASA, would address the matter formally. Handler subsequently wrote an explanatory letter on behalf of the National Academy of Sciences to the effect that further discussions should be conducted between the Soviet Academy and NASA.[36] Meanwhile, Administrator Paine sent the official response for the United States, clarifying the role of the space agency: "As the government agency responsible for civil space activities, NASA has direct responsibility for any discussions with Soviet officials regarding actions we might take together to assure compatible docking systems in our respective manned space flight programs."[37]

Should the Soviet Academy agree to discuss this subject, Paine continued, the Manned Spacecraft Center in Houston, as a preparatory measure, would welcome, in the near future, two Soviet engineers; these visitors would have the opportunity to examine NASA's current designs for docking mechanisms and to discuss future docking concepts. The next step would be joint talks between responsible officials from NASA and the Soviet Academy. Paine saw important benefits from such discussions. "If we can agree on common systems, and I foresee no particular technical difficulty, we will have made an important step toward increased safety and additional cooperative activities in future space operations." The Administrator then referred to his recent decision to resign from that post for personal reasons. He assured Keldysh that his decision would in no way alter NASA policies concerning space cooperation. "Thus, you should understand our past and current correspondence as official rather than personal, although this matter has my wholehearted support."[38]

Paine followed his 31 July letter with another on 4 September 1970, in which he told the President of the Soviet Academy that NASA was still interested in common docking equipment. The Administrator restated his invitation for a visit to Houston by Soviet technical experts and suggested that the Academy officials might wish to consider the idea of a test flight in which a Soviet spacecraft would rendezvous and dock with the American space laboratory Skylab, then scheduled for launch in 1973. Paine said that NASA felt it would be feasible to install a Soviet docking fixture in the multiple docking adapter on Skylab. Explaining subsequently the motivation for this suggestion, Paine commented, "The Skylab docking proposal was made so that we could convince the Soviets of the reality of our proposal. We made this specific to avoid initiating prolonged general discussions in which everyone agreed to 'cooperate' but nothing actually happened." [39] While Paine did not expect the Soviets to accept this particular proposal, he did hope that it would elicit workable counter-proposals and discussions. To give Keldysh and his associates a better idea of the nature of Skylab, Paine enclosed a summary description of the space station in his letter.[40]

Paine and Keldysh were moving rapidly toward the same goal. Paine's letter of 4 September crossed in the mail with a letter of the 11th from Keldysh. Keldysh indicated that the "leadership of the U.S.S.R. Academy of Sciences understands the entire importance and timeliness" of discussing a compatible rendezvous and docking system. "There is no doubt that a positive solution of this question would constitute an important contribution by Soviet and American scientists to the cause of space exploration in the interests of world science and the progress of all mankind." To get the talks under way, the Soviet Academy proposed preliminary discussions in Moscow scheduled for either October or the latter half of November—which is to say, the Soviets wanted to meet either before or after the "October" Revolution holidays in early November.[41]

Turning to specific items to be discussed at a joint meeting, Keldysh listed four topics for consideration. First, there were questions associated with the alternative spacecraft configurations for a rendezvous and docking mission. Second, it was necessary to enumerate the flight procedures to be standardized for such a mission. Third, a decision was needed on the type and number of technical groups to work out the hardware requirements. And finally, time should be set aside to consider plans for future working sessions. "I hope, my dear Mr. Paine, that the National Aeronautics and Space Administration will find our proposal completely acceptable and will promptly inform us of the precise date for the beginnings of the talks."[42]

Paine's resignation became effective on 15 September, and the task of responding to the Keldysh letter fell on the Acting Administrator, George Low. On 25 September, Low reaffirmed the continuing official desire to hold talks with the Soviets. "As Acting Administrator, I shall be continuing Paine's efforts to find ways in which we can develop cooperation between our two countries in space research beyond its present limited extent." In accepting the Soviet invitation to send NASA personnel to visit Moscow, Low suggested that the 26th and 27th of October would be satisfactory.[43]

Turning to the agenda proposed by Keldysh, which was acceptable to NASA, Low defined the approach the Americans would like to follow in discussing those subjects. Under the first item, the Americans would expect to exchange views on possible mission profiles, the types of spacecraft to be employed, and the kinds of docking systems that might be developed. Within the scope of the second topic, Low said NASA would be prepared to share background on operating procedures, docking hardware, communications links, interconnecting ground systems, spacecraft atmosphere, and crew transfer techniques. The third subject for discussion, working groups, would afford the two sides an opportunity to consider the best way to approach the technical areas listed in the second agenda item. Under the final topic, plans for future work, Low thought it would be appropriate to arrange for an early review of the working group findings. While waiting for the Soviet reply, the Americans prepared for a journey to Moscow. Five men were selected to make the trip: from NASA Headquarters, Arnold Frutkin; from the Manned Spacecraft Center in Houston, Director Robert R. Gilruth; Glynn S. Lunney, Chief, Flight Directors Office; and Caldwell C. Johnson, Chief, Spacecraft Design Division; and from the Marshall Space Flight Center, George B. Hardy, Skylab Program Office. Keldysh answered Low's letter with a telegram confirming the acceptability of the 26th and 27th of October for a meeting.[44] The next step was a flight to Moscow.

I
The Years Before

The predominant theme underlying the joint flight of Apollo and Soyuz was international cooperation in space exploration. After conducting separate and competitive programs for several years, the two major spacefaring nations embarked upon a collaborative effort to rendezvous and dock manned spacecraft in earth orbit. To understand why cooperation came slowly, the point and counterpoint of Soviet-American relations in the space age must be considered, because international relations and foreign policy decidedly influenced space programs.

For the study of geophysical questions of common international interest, man-made satellites had initially been promoted as valuable scientific instruments. But it soon became apparent that scientific endeavors could not easily cross national boundaries nor could science policy be separated from the realities of international politics. The technology that launched satellites could also deliver warheads. Thus, early proposals made in the name of scientific knowledge were frustrated by national interests and the demands for military security. From the beginning, the barriers to truly cooperative space projects seemed insurmountable. Before Apollo and Soyuz could fly together, the Americans and the Soviets had to seek out a rationale for cooperation.

Initial efforts to explore the new ocean of space developed as a result of the International Geophysical Year (IGY), a cooperative international program established to study a broad spectrum of scientific questions. The idea for an IGY, first suggested by a group of scientists gathered at the Silver Spring, Maryland, home of James Van Allen in the spring of 1950, grew rapidly in scope. Early discussions on the best way to obtain simultaneous measurements and observations of the earth and the upper atmosphere from a point above the earth had prompted Lloyd V. Berkner, head of the Brookhaven National Laboratory, to propose a re-creation of the International Polar Years (1882 and 1932), in which the scientists of many nations had studied a common topic—the nature of the polar regions. Berkner proposed shortening the interval between such programs to 25 years, to coincide with a period of maximum solar activity. The European scientific community endorsed the concept through the International Council of Scientific Unions, but expanded the project to study the whole planet and renamed it the International Geophysical Year, which embodied an 18-month period of study from 1 July 1957 through 1958. Ultimately, scientists from 67 nations took part.[1]

Several participants believed that the IGY would be enhanced by using artificial satellites to gather geophysical and astrophysical data from above the atmosphere. In September 1954, Berkner, as President of the International Scientific Union and Vice President of the *Comité speciale de l'année géophysique internationale* (CSAGI), set up two informal committees to study the utility of a scientific program. These committees were chaired respectively by S. Fred Singer of the University of Maryland and Homer E. Newell, Jr., of the Naval Research Laboratory. From these deliberations came resolutions favoring the use of such satellites. Berkner then sought endorsement by CSAGI.

The *Comité speciale* included members of the Soviet Academy of Sciences. At first, the Soviets had not responded to the invitations, and when the May 1954 deadline for submitting proposals passed without a word from the Academy, there was concern that the Cold War climate would prevent any significant cooperation. Then on the eve of the IGY meetings in Rome, the Soviet embassy there announced that U.S.S.R. scientists would attend. But during the meetings that followed, the Soviet representatives were remarkably silent. They sat without comment through the discussion and approval of an American proposal for orbiting an artificial satellite.[2]

The resolution drafted by the Americans at the IGY meeting presented a bold challenge:

> In view of the great importance of observations during extended periods of time of extra-terrestrial radiations and geophysical phenomena in the upper atmosphere, and in view of the advanced state of present rocket techniques, CSAGI recommends that thought be given to the launching of small satellite vehicles, to their scientific instrumentation, and to the new problems associated with satellite experiments, such as power supply, telemetering, and orientation of the vehicle.[3]

Two nations had the wealth and the technology to respond to this challenge, the United States and the Soviet Union. Berkner and his colleagues knew that more than scientific riches would result from the first successful flight of a man-made moon. Political and psychological prestige would also proceed from such an accomplishment.

The competition between the United States and the Soviet Union for international prestige was part of the Cold War between those countries. Their alliance to defeat the Axis powers in World War II had been in many ways an uneasy one. With victory over the common enemy, they began to view each other with increasing apprehension and mistrust. Many in both countries decided that their respective ideologies were fundamentally incompatible and that, sooner or later, their countries would clash. This attitude fueled the flames of mistrust, as each side perceived hostility and threat in the other's behavior and responded in such a way as to reinforce the initial suspicions.[4]

In the resultant rivalry, technology, as translated into both industrial capacity and military hardware, became a major indicator of national prestige and power. Both the United States and the Soviet Union had emerged as victors from World War II because the industrial sector of their societies could provide troops in the field with the machines of war in quantities that German industry proved incapable of sustaining. Among the new weapons devised during that war, two would become critical in the postwar world. One was the atomic bomb developed by the United States; the other was the V-2 rocket created by Germany. The significance of the first atomic weapons was immediately apparent after Hiroshima and Nagasaki. The implications of ballistic rockets were less clearly seen immediately following the war, since the V-2s had been less than perfect as military weapons. Nevertheless, both the United States and the Soviet Union developed rockets and nuclear weapons.

By the early months of 1955, the CSAGI proposal for IGY satellites was a topic of serious consideration by scientific and military leaders in America. Alan T. Waterman, director of the National Science Foundation, spearheaded the effort to convince President Dwight D. Eisenhower that the IGY satellite project should be pursued. The military services hesitated to engage in purely scientific investigations because of the expense; however, enthusiasm over the opportunity to participate did exist. A Department of Defense study supported the scientific satellite proposal as long as it did not hinder the development of military satellites or impede other military programs. Further, a Defense spokesman said, "the satellite itself and much of the information as to its orbit would be public information; the means of launching would be classified."[5]

While the discussion of an American satellite developed, the Soviets announced on 15 April 1955 that they had created a "permanent high-level, interdepartmental commission" within the U.S.S.R. Academy of Sciences "for interplanetary communications." Moscow Radio announced on 26 April that the Soviet Academy of Sciences planned not only to launch a satellite but also to explore the moon by means of a remote-controlled vehicle. These statements fueled a growing belief within the Eisenhower administration that the Soviet Union was about to announce plans for an IGY satellite. At least one man in the administration, Nelson A. Rockefeller, was concerned over the propaganda potential of such an announcement. Rockefeller, the President's special assistant, had reviewed the military comments on the proposed scientific satellite. He concluded that the project should be approved and announced before the Soviets made their statement:

> I am impressed by the costly consequences of allowing the Russian initiative to outrun ours through an achievement that will symbolize scientific and technological advancement to people everywhere. The stake of prestige that is involved makes this a race we cannot afford to lose.[6]

The military comments, somewhat more cautious, noted that the "unmistakable relationship" of the IGY satellite "to intercontinental ballistic missile technology might have important repercussions on the political determination of free world countries to resist Communist threats." The Central Intelligence Agency reportedly was convinced in the spring of 1955 that the Soviet Union intended to be the first nation to orbit an IGY satellite. Implicit in these attitudes and statements is acceptance of competition between the United States and the Soviet Union in space.[7] On 29 July 1955, Presidential News Secretary James C. Hagerty officially announced that the United States would launch "small earth-circling satellites" as part of its participation in the IGY.

The announcement elicited an interesting response from the Soviets observing the sessions of the International Astronautical Congress in Copenhagen. Leonid Ivanovich Sedov, who headed the Commission on Interplanetary Communications, in a press conference held at the Soviet Legation in Copenhagen made the following comments on 2 August:

> Recently in the U.S.S.R. much consideration has been given to research problems connected with the realization of interplanetary communications, particularly the problems of creating an artificial earth satellite. The practicability of technological artificial satellite projects is already well known to engineers, designers, and scientific workers engaged in or interested in rocket technology. In my opinion, it will be possible to launch an artificial earth satellite within the next two years, and there is a technological possibility of creating artificial satellites of various sizes and weights.
>
> From a technical point of view, it is possible to create a satellite of larger dimensions than that reported in the newspapers which we had the opportunity of scanning today. The realization of the Soviet project can be expected in the comparatively near future. I won't take it upon myself to name the date more precisely.[8]

While this statement was reported in various ways in the American press, there was general agreement that this was an official announcement that the Soviets would indeed launch a satellite. The edited official version of Sedov's statement that appeared in *Pravda* was certainly more circumspect than the reports in the Western press. Reaction among American scientists was mixed. Some were alarmed, others were disdainful, but the majority were more curious about Soviet plans than they were concerned that the first satellite would not be launched by the United States.[9]

Against this backdrop of ideological differences and technological competition, the orbiting of *Sputnik I* by Soviet technicians on 4 October 1957, followed a month later by *Sputnik II* with its canine passenger Laika—and its implications for manned space flight—assumed great significance. The Soviets had obtained a visible and indisputable technological first and had apparently developed a rocket technology that also could be used for military purposes. Americans not only perceived the technological challenge of this accomplishment but also saw the obvious meaning of this first earth satellite for prestige and military power. As their Soviet counterparts reaped political, military, and scientific returns from their new star, American leaders embarked upon a period of deep, worried self-examination. The obvious response to the Soviet feat was an intensification of the American program to launch a satellite and an increase in the tempo of military rocket research. Declared or not, a bilateral technological competition had begun in space exploration and military rocketry.[10]

At the beginning it was impossible to separate the military and civilian aspects of the new competition—a circumstance that would complicate later attempts to cooperate in space. Soviet satellites were launched on military rockets, as was the first American satellite. Before it was transformed into NASA and entrusted with the civilian portion of the American space

At the Soviet Legation in Copenhagen, August 1955, interpreter Sannikov relays news from Professor K. F. Ogorodnikov and Academician L. I. Sedov who are seated next to him that the Soviet Union intends to launch an artificial earth satellite during the IGY (Associated Press photo).

program, the National Advisory Committee for Aeronautics (NACA) showed a tendency to lump the scientific and military aspects of space into the single package of Cold War competition. NACA's Special Committee on Space Technology surveyed the problem in the spring of 1958 and recommended an integrated program of development for long-range missiles and space vehicles, saying:

> One of the prime objectives established in preparing this report was that of accomplishing a manned lunar landing in advance of the Soviets. Such an accomplishment would firmly establish Western technological supremacy and be of great psychological value. Due to the strategic location of the moon for space travel and warfare, an even greater and more permanent value would be derived by such a landing—that of claiming the moon for the United Nations of the Western World.

Clearly, the dominant theme was "to catch up with and ultimately surpass the Soviets in the race for leadership on this planet and for scientific and military supremacy in space."[11]

Ironically, the cooperative spirit of the IGY that had spawned projects to orbit satellites became overshadowed by the urge to either maintain the lead or surpass the leader in this new technological arena. Two conflicting goals thus emerged. First was the desire to establish national pre-eminence in science and technology, as an adjunct to the broader Cold War rivalry. Second was the wish to develop international ties through cooperative studies of the cosmos, as reflected by the aims of the IGY. To meet the Soviet challenge, the American government created a separate space agency, and the conflicting themes of competition and cooperation were present in the discussions that led to the creation of the National Aeronautics and Space Administration. While the establishment of a space agency was in large measure a response to the Soviet achievement in launching the first satellite, the fact that the new organization was under civilian leadership testified to the desire of President Eisenhower to avoid, if at all possible, an extension of the military aspects of Cold War into outer space. From the very beginnings of the American satellite project, Eisenhower had supported the position that space exploration should be undertaken for peaceful purposes only.[12]

Through the months of work by various executive and congressional groups, the drafting and redrafting of bills, and the inevitable compromising on and off the floor of Congress, the two potentially conflicting themes survived.[13] The National Aeronautics and Space Act of 1958 opens with a declaration of policy that includes two specific purposes:

Sec. 102.
(c) (5) The preservation of the role of the United States as a leader in aeronautical and space science and technology . . . ;

(7) Cooperation by the United States with other nations and groups of nations. . . .[14]

Arnold Frutkin, who was given the responsibility of directing the International Programs office of NASA* in 1959, later commented on the dual challenge placed before the new agency:

> While facing up to the grim reality of competition between the great powers, the Congress nevertheless elected to place some hope, if not faith, in the simultaneous practice of cooperation. . . . both courses of action—the competitive and the cooperative—were pursued simultaneously in the early years of the space age.

This parallel approach was entirely conscious. NASA's second Administrator, James E. Webb, said on more than one occasion that "space, like Janus, looks in two directions." As Frutkin perceived this complex process, "This was only part and parcel of the age old strategy of pursuing the battle vigorously while seeking and preparing for an armistice."[15] NASA's Office of International Programs faced a unique and difficult task.

ORIGINS OF THE OFFICE FOR INTERNATIONAL PROGRAMS

It was not altogether clear at first exactly what role the Office of International Programs was to play in the overall mission of NASA. The Space Act of 1958 was signed into law on 29 July, and T. Keith Glennan and Hugh L. Dryden were sworn in as Administrator and Deputy Administrator on 19 August. NASA officially came into existence on 1 October. In the whirlwind rush, the question of international programs was just one of a host of pressing concerns.

As early as May, draft organization charts had shown a position for an Assistant for International Activities.[16] The idea for this staff office reflected the view of the National Advisory Committee for Aeronautics on organization. When Glennan was appointed, he asked the management consultant firm McKinsey and Company to study the various proposals for NASA managerial structure. McKinsey suggested the creation of an office devoted solely to international questions. First, it would provide a central point of coordination and assistance for the Administrator and other officials in the development of a cooperative international program of "space research and development," and, second, the office would provide staff support to the State Department on matters that concerned foreign policy and space affairs. The International Office was also to serve as a

*See appendix A for the 29 Jan. 1959 NASA organization chart.

clearinghouse and coordinating body for exchange of scientific and technical information, arrangement of cooperative facilities in other countries, and coordination of a host of scientific activities, such as weather observation.[17]

Glennan accepted the recommendation and appointed a Director of the Office of International Cooperation, who, within nine months was replaced by Arnold Frutkin.[18] The forty-one-year old Frutkin brought with him a sober realism born of his experiences during the IGY. In May 1957, Frutkin had joined the staff of the National Academy of Sciences as Director of the Office of Public Affairs of the U.S. Committee for the IGY. Concurrently, he served as Deputy to the Executive Director of that committee. As a consequence, Frutkin had witnessed firsthand many of the frustrations of working with other national committees, especially the difficulties encountered with the Soviet committee.

Frutkin reflected on the IGY and its meaning for the exploration of space in his book, *International Cooperation in Space.* Looking at the day-to-day efforts of the IGY, he held that the idea of "shoulder-to-shoulder cooperation" was "a substantially misleading picture." In short, Frutkin saw the IGY as "a collection of national programs, independently working toward purely scientific objectives loosely coordinated by a nongovernmental mechanism." While the IGY did construct "scientific bridges across political chasms," he argued that "the bridges had no effect on the chasms; these remained and no traffic other than scientific crossed them."[19]

From Frutkin's vantage point, the broad success that characterized many cooperative scientific endeavors did not extend into space research. Scientific representatives of the Soviet Union "stubbornly restricted IGY agreements for the exchange of information in this area. . . . attempts to improve the situation . . . were unavailing." Frutkin summarized: "Extensive efforts to apply the usual IGY data exchange formulas to space came to naught. . . . Clearly, the cold war had reached into the IGY and frostbitten one of its major arms, the space program."[20]

But what did the experiences of the IGY say to the man who would be responsible for government-to-government considerations of collaboration in space activities? First, "it remains most important to recognize that those who molded the IGY were probably far freer from disabling political considerations than would have been the case if governmental representatives had attempted to frame a similar program." Second, the IGY "was a notable element among the forces that gave the U.S. national space program its peculiar shape" when NASA was created in 1958. Clearly, Frutkin perceived that the difficulties experienced by his non-government colleagues in the IGY would be magnified within NASA should that agency negotiate for international cooperation with the representatives of other governments. His earlier experiences with the IGY and his concern for realism in international negotiations were to temper his approach to cooperative ventures in the years that followed.[21]

FIRST EFFORTS TO ESTABLISH A BASIS FOR COOPERATION

By the fall of 1959, NASA had the mandate to cooperate, and it had set up the administrative machinery to formulate policy concerning international programs; but what did *cooperation* and *international programs* mean? How and with whom would NASA cooperate? What would be the subject matter for international agreements? There were, of course, those areas in which NASA needed the assistance of other nations, notably to establish tracking stations for both manned and unmanned spacecraft. Also, NASA hoped to encourage other nations to join in scientific experiments involving American spacecraft. And there was a third category of possible cooperation—the Soviet Union. Skillful negotiation would be required in this pursuit, as the Soviet Union was a coequal, perhaps the technological leader, in space flight. Thus, while it was difficult enough to deal with nations nominally friendly, negotiations with the Soviets were always to be a special case. How and for what reasons would cooperative programs be developed between the Americans and the Soviets?[22]

Before Frutkin arrived at NASA, Deputy Administrator Hugh Dryden had made several important contacts with other nations. Homer E. Newell, Jr., Assistant Director for Space Sciences, had taken the lead to organize the international community interested in space flight by convening the first organizational meeting of the International Committee for Space Research (COSPAR) in November 1958. COSPAR had been created to perpetuate the cooperative aspects of space investigation that had been part of the IGY,* but the international body quickly became a victim of Cold War politics.[23]

A debating society environment plagued the United Nations discussions of cooperation on the new frontier; nuclear disarmament was the stumbling block. Following *Sputnik I,* much had been said about preventing the introduction of weapons into space. Indicative of the divergence of opinion between the Americans and the Soviets on this subject were the letters exchanged during the spring of 1958 between Eisenhower and Soviet Premier Nikolai Bulganin. Eisenhower asserted that the peaceful use of

---

*Over the years, COSPAR has grown in stature, but it still remains a non-governmental body, hence an unofficial point of contact at which scientists can exchange views. While delegates from the Soviet Academy of Sciences are official spokesmen for their country, representatives of the National Academy of Sciences do not speak for the U.S. government.

space—prohibiting the use of space for military gain—was "the most important problem which faces the world today. . . . We face a decisive moment in history. . . ." Addressing the problem of developing rockets for military applications, Eisenhower raised the question of learning from past failures:

> . . . a decade ago, when the United States had a monopoly of atomic weapons and of atomic experience, we offered to renounce the making of atomic weapons and to make the use of atomic energy an international asset for peaceful purposes only. . . . The nations of the world face today another choice perhaps even more momentous than that of 1948. That relates to the use of outer space. Let us this time, and in time, make the right choice, the peaceful choice.
>
> There are about to be perfected and produced powerful new weapons which, availing of outer space, will greatly increase the capacity of the human race to destroy itself. . . . can we not stop the production of such weapons which would use or, more accurately, misuse, outer space, now for the first time opening up as a field for man's exploration? Should not outer space be dedicated to the peaceful uses of mankind and denied to the purposes of war? That is my proposal.[24]

Premier Bulganin responded that reserving space for peaceful purposes depended on prior solution of the problem of disarmament in general:

> We, of course, do not deny the importance of the question of using outer space for peaceful purposes exclusively, i.e., first of all, of the question of the prohibition of intercontinental ballistic missiles with nuclear warheads. I hope, however, Mr. President, that you will agree that this question can be considered only as a part of the general problem of the prohibition of nuclear and rocket weapons. It is for that very reason that the Soviet Union, in the interest of strengthening peace and reaching agreement on questions of disarmament, is also prepared to discuss the question of intercontinental missiles, provided the Western powers are prepared to agree on the prohibition of nuclear and hydrogen weapons, the cessation of tests of such weapons and the liquidation of foreign military bases in the territories of other states. . . .[25]

In the succeeding exchange of letters between the two states and in the debates in the U.N., the discussions bogged down over the relation of space to questions of national security and disarmament. The two space powers, who also were the two nuclear powers, defined differently the problem at hand. American leaders sought to ban the militarization of outer space; this seemed a logical step and an opportunity that should not be lost. The Soviets, however, saw sinister motives behind the American proposals. The Russians saw themselves surrounded by American and allied military power.

In addition to their bases in the continental United States, the Americans had installations in the U.K., Western Europe, the Middle East, and the Far East. With such facilities, outer space was not needed to launch an attack. The Soviets, lacking such advanced bases, relied upon the development of the intercontinental ballistic missile (ICBM)—a strategic weapon whose parabolic trajectory arced into space. America's proposal to neutralize space was thus seen as an attempt to deprive the Soviet Union of her only defense against the nuclear strike capabilities being developed by the Americans. Both nations sought to neutralize outer space, but only on terms that would be advantageous to themselves.[26]

Debate in the U.N. divided along ideological lines, and NASA's desire to use that body as the foundation for developing a program of space cooperation foundered.* Glennan and his colleagues came to believe that negotiations with the Soviets would have to be direct, bilateral, and more private than the open forum of either COSPAR or the U.N. As a consequence, the NASA leadership sought to engage the Soviets in less formal talks. Typical of these early contacts were the discussions between representatives of the Soviet Academy of Sciences and NASA during the annual meetings of the American Rocket Society. At the mid-November 1959 meeting of the Society in Washington, for example, Soviet space scientists Sedov, Blagonravov, and V. I. Krassovsky presented papers on the nature of Soviet space research.[27] Dryden met privately with the Soviets to exchange views. They agreed that their countries should cooperate more closely in space science, and Dryden made it clear that NASA was ready to talk about issues of mutual interest. The Soviets warned that such an undertaking should proceed "step by step." However, Frutkin reported that "when pressed, they were not prepared to identify the first possible step."[28]

In an effort to demonstrate American willingness for closer relations, George Low gave the Soviet guests a tour of the Langley Research Center in Virginia, where among other things he showed them a model of a Mercury spacecraft. The Soviets were polite but noncommittal, and the hoped-for invitation to see Soviet space-flight facilities never materialized.[29]

The Soviets continued to insist that the proper forum for discussing space cooperation was the United Nations; and the Americans remained acutely aware that discussions in that arena, as long as the Soviets enjoyed the technological lead, could only result in a Soviet propaganda advantage.

*In Jan. 1960, NASA created an ad hoc Office for the U.N. Conference that was to address the issues raised by the General Assembly call for an international conference on the peaceful uses of outer space. This office was headed by John Hagen. When the conference failed to materialize, the office was disbanded in Sept. 1961. Rosholt, *Administrative History of NASA*, pp. 127-128.

COMPETITION VERSUS COOPERATION: 1959-1962

For NASA personnel interested in fostering cooperative projects with the Soviet Union, the political climate of 1959-1962 was frustrating. These were the years of Soviet Premier Nikita Sergeyevich Khrushchev's foreign policy that on the one hand sought detente with the West while on the other exploited "every major trouble spot, every embarrassment" to damage Western influence and prestige. To quote one assessment:

> There appeared to be two Khrushchevs: one, a "coexistentialist" eager for enhanced intercourse between the U.S. and the U.S.S.R.; dropping hints (to be sure so obscure as to remain at the time undecipherable) about the necessity for a virtual alliance of the two powers; the other, a militant Communist and bully ready to cash in on each and every weakness and hesitation of the West, threatening nuclear obliteration if his opponent would not submit.

Khrushchev did not want a crisis that would lead inexorably to nuclear disaster, but he was a skillful poker player who successfully bluffed the leaders of the country that had originated the game, until the confrontation over missiles in Cuba.[30]

Nineteen fifty-nine was a year of political maneuvering. Vice President Richard Nixon and Premier Khrushchev held their "kitchen debate" at an American exhibition in Moscow's Sokolniki Park,[31] and Khrushchev later made his ostentatious, but largely ceremonial, visit to the U.S. It was also the year of the first Soviet lunar probes. *Luna I,* launched in January, was the first spacecraft to penetrate interplanetary space; *Luna II,* launched during the Premier's visit to the U.S., was the first spacecraft to hit the moon. Then in October, *Luna III* swung around the moon and photographed its back side. But the debates and visits did nothing to solve international problems; successful moon probes certainly did not enhance the chances for cooperation between the two nations—especially when contrasted with the high number of U.S. launch failures in 1959.

In the next year, however, Soviet and American heads of state had to deal with realities of international politics that could not be brushed aside. Khrushchev had wanted a summit meeting for several years; now such a meeting seemed less than desirable. Following his visit to the United States, Khrushchev had visited Peking. From the Soviet standpoint, discussions with the Chinese were unsatisfactory, causing the ideological split between the two nations to widen and heading the Chinese on an increasingly independent course. This problem, together with the hardening positions of the American, British, and French on the question of two Germanys, made a summit meeting with the Americans undesirable. Just as the potentially embarrassing get-together approached, American pilot Francis Gary Powers became an unintentional celebrity when his Lockheed U-2 high-altitude reconnaissance aircraft was downed deep in Soviet territory.

The U-2 incident had three immediate consequences. First, it solved Khrushchev's dilemma. He could now avoid the summit meeting without accepting the responsibility for wrecking it. Second, the United States suffered a serious international embarrassment when President Eisenhower took personal responsibility for the U-2 flight.[32] Third, the credibility of the National Aeronautics and Space Administration was questioned because it had served as a cover for this clandestine, intelligence-gathering overflight.

On 5 May 1960, on orders from the White House, NASA stated that one of its U-2 research planes used "to study gust-meteorological conditions found at high altitude" had been missing since 1 May. Then six days later, Eisenhower admitted publicly that the flight actually had been part of a military reconnaissance program conducted with his permission. While the administration had to cope with the impact of the U-2 mission at the abortive Paris summit conference and later during Khrushchev's visit to the United Nations in September, NASA had to fight the notion that there was more to the civilian program than was being admitted in public.

An immediate issue was Soviet participation in the Tiros weather satellite program. "It's part of our national policy that space research is for peaceful purposes," Arnold Frutkin told a *Wall Street Journal* reporter. "We want to have an open program. And the best way to prove this to other countries is to have them participate in our experiments."[33] NASA had long planned to solicit the cooperation of other nations, including the U.S.S.R., in studying cloud photographs taken by the Tiros satellite. Soviet participation would have gone a long way to allay fears that Tiros was looking at more than the weather patterns, but the Soviets saw—or purported to see—the satellite as another U-2. A year later NASA Administrator James E. Webb labeled as "political opportunism" their attacks on the Tiros program and their refusal to participate.[34]

Even without the U-2 incident, 1960 was not a propitious time to talk about cooperative ventures in space. The American public was watching a very close political contest between John F. Kennedy and Richard Nixon; a key campaign topic was the state of the nation's defenses against nuclear attack by the Soviet Union. During the campaign, the trade journal *Missiles and Rockets* invited the candidates to respond to a series of statements on space and defense. The first proposition asked if they would "recognize as national policy that we are in a strategic space race with Russia." Kennedy's response was published first:

> We are in a strategic space race with the Russians, and we have been losing. The first man-made satellite to orbit the earth was named *Sputnik*. The first living creature in space was Laika. The first rocket to the moon carried a Red

flag. The first photograph of the far side of the moon was made with a Soviet camera. If a man orbits earth his year his name will be Ivan. These are unpleasant facts that the Republican candidate would prefer us to forget.

Control of space will be divided in the next decade. If the Soviets control space they can control earth, as in past centuries the nation that controlled the seas dominated the continents. This does not mean that the United States desires more rights in space than any other nation. But we cannot run second in this vital race. To insure peace and freedom, we must be first.[35]

Nixon responded later in a manner that was uncharacteristic of the Eisenhower administration, which had played down the idea of a space race. Candidate Nixon argued:

If the Eisenhower Administration had not long ago recognized that we were in a strategic race with Russia, our space record would not be as creditable as it is today.

Twenty-six satellites and 2 space probes have been launched successfully by the United States.

Six satellites and 2 space probes have been launched successfully by the Soviet Union.

Today 13 United States satellites are in orbit; only 1 Russian satellite remains in orbit.

Eight United States satellites in orbit are still transmitting; the sole Russian satellite in orbit is not transmitting.

The United States has recovered 2 satellite payloads from orbit while the U.S.S.R. claims to have recovered one.

Despite the greater weight of U.S.S.R. space vehicles, the United States has gathered far more scientific information from space. In instrumentation, communications, electronics, reliability, and guidance, United States space vehicles have made gigantic strides.

In short, the United States is not losing the space race or any other race with the Soviet Union. Today we are ahead of the U.S.S.R. From a standing start in 1953, we have forged ahead to overcome an 8-year Russian lead. And we will continue to maintain a clear cut lead in the race for space.[36]

While the candidates debated, NASA and the Eisenhower administration attempted to keep a line open with the Soviets on space cooperation. Frutkin had talked informally with Academician Anatoliy Arkadyevich Blagonravov about the possibility of using *Echo I,* the balloon-like passive communications satellite, for communications experiments between the United States and the Soviet Union. *Echo I* had been launched on 12 August 1960, three days before the International Astronautical Congress convened in Stockholm, and the delegates had heard a message recorded by President Eisenhower, transmitted part of the way by the satellite.[37] On 22 September, the President in an address to the United Nations suggested a four-point proposal for the peaceful exploration of space, using as his precedent the 1959 Antarctic Treaty, which had prompted scientific research and barred military activity from that continent.[38] However, the future of Eisenhower's hope for an agreement on the peaceful uses of outer space would depend upon the efforts of the new President and the individuals within NASA.

Kennedy's election in November 1960 portended a number of changes for defense and space programs. Subsequently, Kennedy asked his Vice President-elect to serve as his senior adviser on space policy and as chairman of the National Aeronautics and Space Council. Lyndon B. Johnson's first task was to recommend a new Administrator for NASA, Glennan having resigned effective the last day of the Eisenhower administration. As Johnson began the search, Kennedy announced on 11 January 1961 the appointment of Jerome B. Wiesner of MIT to be his assistant for science and technology. The same month appeared the "Wiesner Report," prepared by a committee of science advisers who had worked with the Kennedy campaign.

Expanding upon campaign themes, this document criticized the space program under the Eisenhower administration. But while belaboring some aspects, especially the manned space-flight project, the report foresaw "exciting possibilities for international cooperation" in space exploration and communications. Such projects would prosper if "carried out in an atmosphere of cooperation as projects of all mankind instead of in the present atmosphere of national competition."[39] Kennedy pursued the same theme in his inaugural address.

Kennedy's speech was notable because of its hopeful and skillful rhetoric, expressing the desire for new beginnings in foreign policy, including a reduction in the level of conflict between the United States and the Soviet Union. To that end, he appealed to the Soviets: "Let both sides seek to invoke the wonders of science instead of its terrors. Together let us explore the stars, conquer the deserts, eradicate disease, tap the ocean depths and encourage the arts and commerce. . . ." And Kennedy continued to espouse the cooperative theme in his State of the Union address on 30 January 1961. The President invited all nations, including the U.S.S.R., "to join with us in developing a weather prediction program, in a new communications satellite program and in preparation for probing the distant planets of Mars and Venus, probes which may someday unlock the deepest secrets of the universe." He repeated the hopes of his science advisers that the arms race could be kept from spreading into space. "Both nations would help themselves as well as other nations by removing these endeavors from the bitter and wasteful competition of the Cold War." This was to be a recurring theme in Kennedy's public comments.[40]

At the time of these pronouncements, and to this day, debate has existed over the depth of the new President's initial understanding of the space issue relative to the realities of international power politics.[41] Missiles and space had been a warm issue during the campaign; Kennedy had insisted that the previous administration had allowed national defense to slip in relation to Soviet strength. After Kennedy assumed the Presidency, the "missile gap" proved to have been a myth; but the problem remained to fit the national space program into the power equation by which American military and political leaders would evaluate the "strength" of their nation versus that of the Soviet Union.

Ten days after his inauguration, Kennedy followed the recommendation of his Vice President and nominated James E. Webb to be Administrator of the space agency. At first hesitant to accept the position, which he felt would have been more satisfactorily filled by a scientist or engineer, Webb had agreed once he understood that Kennedy was seeking a policy maker who could manage scientists and engineers. Upon accepting the assignment, Webb announced that Hugh Dryden, the other presidential appointee in NASA, would continue as Deputy Administrator. With directions from the President to make a comprehensive review of NASA programs, Webb went before the Senate for hearings on his confirmation. He was confirmed on 9 February and sworn in on the 14th.[42]

As the first months of 1961 slipped away, Kennedy and Webb became convinced that second place in space exploration would carry the negative impression that the United States was second rate in military strength as well. This conclusion once again pointed to the dilemma of competition versus cooperation in space exploitation. On the one hand, Kennedy genuinely wanted to cooperate in this arena with the Soviets; on the other hand, military and technical superiority had to remain with the United States. Events during the spring of 1961 swiftly determined his choice between these conflicting goals.

The successful one-orbit flight of Yuri Alekseyevich Gagarin on 12 April 1961 was a significant element in the subsequent American deliberations. While this event was anticipated by the Kennedy administration, the Soviet feat was still another blow to the American image at home and abroad. The Soviet Union constantly stressed three themes in exploiting the first manned space flight:

1. the Gagarin flight was evidence of the virtues of "victorious socialism";

2. the flight was evidence of the global superiority of the Soviet Union in all aspects of science and technology;

3. the Soviet Union, despite the ability to translate this superiority into powerful military weapons, wanted world peace and general disarmament.[43]

Such a challenge could not go unanswered. Theodore Sorenson later commented, overdramatically perhaps, that "As the Soviet Union capitalized on its historic feat in all corners of the globe, Kennedy congratulated Khrushchev and Gagarin and set to work."[44]

Even as John Kennedy was rolling up his sleeves and consulting his advisers, other events were unfolding that would complicate the political scene. None too secretly, a band of approximately 1500 Cuban refugees was preparing to launch an invasion of Fidel Castro's Cuba. The exact impact of this military and political fiasco on the subsequent decision to go to the moon has been repeatedly argued by many of those associated with the Kennedy administration. John Logsdon concludes in his study of the events:

> The fiasco of the Bay of Pigs reinforced Kennedy's determination, already strong, to approve a program aimed at placing the United States ahead of the Soviet Union in the competition for firsts in space. It was one of the many pressures that converged on the president at the time, and thus its exact influence cannot be isolated. As president, Kennedy could treat few issues in isolation anyway, and there seems to be little doubt that the Bay of Pigs was in the front of his mind as he called Lyndon Johnson to his office on April 19 and asked him to find a "space program which promises dramatic results in which we could win."[45]

By the end of April 1961, Kennedy had decided that the dramatic program would be a manned lunar landing. The suborbital flight of Alan B. Shepard in his *Freedom 7* spacecraft on 5 May was a much needed positive accomplishment, which brought favorable public response. On 8 May, Vice President Johnson presented to the President a memorandum prepared by NASA Administrator Webb and Defense Secretary Robert S. McNamara— "Recommendations for our National Space Program: Changes, Policies, Goals." The Webb-McNamara memorandum suggested that manned space flight could be an effective means of enhancing national prestige:

> Major successes, such as orbiting a man as the Soviets have just done, lend national prestige even though the scientific, commercial or military value of the undertaking may by ordinary standards be marginal or economically unjustified. . . . The non-military, non-commercial, non-scientific but "civilian" projects such as lunar and planetary exploration are, in this sense, part of the battle along the fluid front of the cold war.[46]

John Kennedy agreed.

On 25 May in a speech on "Urgent National Needs," the President reminded the Congress that "these are extraordinary times. We face an extraordinary challenge." After addressing himself to a number of other important issues, Kennedy turned to the subject of space. This new frontier was just another aspect of the "battle that is going on around the world between freedom and tyranny. . . ." Therefore, "Now it is time to take longer strides—time for a great new American enterprise—time for this nation to take a clearly leading role in space achievement, which in many ways may hold the key to our future on earth." One of those "longer strides" Kennedy proposed was the landing of an American on the moon. The President believed "that the Nation should commit itself to achieving the goal, before this decade is out, of landing a man on the moon and returning him safely to earth." This goal was that bold type of challenge that had peculiar appeal to the young President. "No single space project in this period will be more impressive to mankind, or more important for the long-range exploration of space; and none will be so difficult or expensive to accomplish."[47]

Thus, space competition between the United States and the Soviet Union was reaffirmed by Kennedy's speech. What did this mean to NASA, and particularly what did it mean for NASA's mandate to cooperate? During 1961, the NASA position on the prospects of Soviet-American space cooperation was one of basic skepticism. Administrator Webb was committed by the Webb-McNamara memorandum of 8 May to support a program of American technological pre-eminence in space. Any program of cooperation would have to occur within a framework that would not jeopardize America's chances of establishing that position.

In June 1961, in response to questioning, NASA submitted a series of formal statements to the Senate Committee on Aeronautical and Space Sciences. "In general, how cooperative have the Soviets been in sharing the results of their space experiments?" NASA responded that the difference between the attitude of the U.S. and that of the U.S.S.R. was one of degree. The Soviets were judged to have been quite active in international meetings.

In a 25 May 1961 address to joint session of the U.S. Congress, President John F. Kennedy establishes the goal "of landing a man on the moon and returning him safely to earth" before the decade is out.

They had presented papers and discussed problems of mutual interest with their international colleagues, but it was the NASA opinion that they had not operated with an openness comparable to that of scientists from other nations.[48] Throughout 1961, NASA spokesmen told Congress and the American public that while NASA still sought space cooperation with the U.S.S.R., the attitude and actions of the Soviets left little hope for success.

Public remarks by Soviet officials in 1961 on space cooperation were equally ambivalent. On 13 February, Kennedy congratulated Khrushchev on the launch of a space probe to Venus.[49] In his reply two days later, Khrushchev thanked Kennedy for his "high appraisal to this outstanding achievement of peaceful science." The Soviet leader, in referring to Kennedy's inaugural and State of the Union invitations to the Soviets, said that "such an approach . . . impresses us and we welcome these utterances of yours." But the Soviet Premier still saw disarmament as the key to the problem: "We consider that favorable conditions for the most speedy solution of these noble tasks facing humanity would be created through the settlement of the problem of disarmament."[50]

With Gagarin's *Vostok I* April flight, the tone of the Soviet statements on cooperation in space changed. Clearly the Soviets enjoyed their sense of technological superiority, but still they did not totally abandon the thought of cooperation with the U.S. Academician Sedov,* in his public congratulations to Alan Shepard for suborbital flight, was careful to point out that the Gagarin flight was of greater significance. He also restated the Soviet position on the relationship of international cooperation in space flight to the question of disarmament:

> Soviet scientists and scientists of other countries, who are occupied with scientific research in space, are participating in mutual discussions on the results achieved, and we can speak on the beginning of fruitful cooperation. Nonetheless, the problem of international scientific cooperation on space flights in general is still not resolved. It is evident that such cooperation will be successful only upon the favorable development of international relations and the realistic solution of the problem of disarmament.[51]

Later at the Washington meetings of the International Astronautical Federation during October, Sedov was asked if the U.S. and the U.S.S.R. would be able to collaborate in launching large payloads. Sedov replied, "I think it will be possible in the future, not only between the Russians and Americans but with other countries as well." Deputy Administrator Dryden

*Sedov was Chairman of the Commission for the Promotion of Interplanetary Flights, U.S.S.R. Academy of Sciences, as well as President of the International Astronautical Federation.

observed at the time that "Sedov and I have discussed this possibility many times. If the decision were ours alone, there would be no problem."[52]

Coming at a time when East-West tensions had worsened, optimistic statements about cooperation in space hardly seemed realistic. The two-day confrontation between Kennedy and Khrushchev during the June 1961 Vienna summit was from Kennedy's perspective a disaster. But in one of the rare moments of amicability, Kennedy suggested that the two nations pool their space efforts and "go to the moon together." Khrushchev's immediate response was "all right," but upon reflection the mercurial Soviet leader decided that such a venture would not be practical. The boosters used for manned space flight had military implications. That triggered considerations of disarmament, and that brought the discussions back to the Cold War. There the proposed joint trip to the moon died.[53]

The unsuccessful Vienna summit was followed by the crisis over the Berlin Wall. With that physical barrier between East and West Berlin erected on 13 August 1961, Khrushchev once again raised the question of the divided status of Germany. For the second time in three years, Khrushchev threatened to sign a separate peace treaty with the East German Government, thus forcing the Americans to deal with a separate communist state. On 25 July, Kennedy told the nation in a somber television address that the United States would go to war should that become necessary to defend a free Berlin. Khrushchev reacted strongly to what he perceived to be an ultimatum from the President of the United States, and while the two sides negotiated the Berlin issue, the Soviet Union dramatically broke the three-year old moratorium on atmospheric nuclear weapons tests. Beginning on 1 September 1961, the tests continued for two months. They were culminated with a 58-megaton explosion, the most powerful hydrogen device to have been tested at that time by either nation.[54] While events such as these would seem to pose insurmountable barriers to cooperation in space, Russian and American scientists managed to keep the discussions alive.

Threats to world peace posed by the succession of summer and autumn crises, while not unnoticed, seemed far distant from the pleasant atmosphere of the lodge at Smugglers Notch, Vermont. For four days, 5-8 September 1961, scientists from ten countries, including the U.S.S.R., gathered for the Seventh International Conference on Science and World Affairs.* Included

*Americans present included E. Rabinowitch, Professor of Biophysics, University of Illinois; H. Brown, California Institute of Technology; P. Doty, Harvard University; and I. I. Rabi, Professor of Physics, Columbia University. The Soviets included A. A. Blagonravov; A. V. Topchiev, Vice President, Soviet Academy of Sciences; I. Y. Tamm, physicist; and N. N. Bogolubov, physicist. British representatives included Professor P. M. S. Blackett, physicist, London University; Sir John Cockcroft, nuclear physicist, Cambridge University; and the Rt. Hon. Philip Noel-Baker. Henry Kissinger, Harvard, and George Kistiakowsky, former science adviser to President Eisenhower, attended the sessions on disarmament.

in a broad spectrum of proposals relating to greater cooperation among the world's scientists were suggestions for a program of space cooperation between the U.S. and the U.S.S.R. Four areas in which the scientists felt that cooperation was possible were (1) a worldwide system of weather satellites and forecasting; (2) an international program of communications satellites; (3) an international exchange of data relating to space biology; and (4) a joint program for the scientific exploration of the moon and the planets.[55] Despite the international debate engendered by the Soviet resumption of nuclear arms tests, there was an atmosphere of good will at Smugglers Notch.[56] The fragility of such scientist-to-scientist efforts was clearly demonstrated two months later.

In November 1961, NASA and the U.S. Department of Commerce sponsored an International Satellite Workshop in Washington. American representatives explained their plans for the further exploitation of weather satellites and encouraged other nations to participate in the gathering and use of satellite data. The Americans expected delegates from the U.S.S.R., Poland, and Czechoslovakia, since visas had been sought by representatives of those countries. On the second day of the workshop, it became apparent that the Soviets would not attend. To most contemporary observers the lesson was clear: cooperation in space matters was a political consideration that could be understood only in the broader context of East-West relations.[57] Nineteen sixty-one, the fifth year of the space age and NASA's third, had not been a good year for space cooperation. Indeed, as one commentator has reflected: "For all the style and excitement of the new team, and all the great promise, 1961 was a terrible year for the Kennedy Administration."[58] International tensions would not lessen during 1962, but the opportunity for cooperation in space would seem more real. Two men would work hard to give that opportunity a chance to mature—Hugh Dryden of NASA and Anatoliy Blagonravov of the Soviet Academy of Sciences.

II

Dryden and Blagonravov

Soviet-American discussion about cooperation in space received new impetus in the spring of 1962. Following the successful three-orbit flight of John H. Glenn, Jr., on 20 February, Premier Khrushchev sent a congratulatory message to President Kennedy. This letter, which called for closer cooperation in space activities, might first have appeared disingenuous when viewed against the tense political background of the preceding year. But there had been considerable planning in the Soviet Union for just such an overture.

At its Twenty-second Congress on 17 October 1961, the Communist party of the Soviet Union considered closer cooperation with other nations and urged the Soviet government to pursue such a policy "in the fields of trade, cultural relations, science, and technology."[1] An early step toward the implementation of this goal came in December 1961 when the Soviet delegation to the United Nations ended its boycott of the Committee on the Peaceful Uses of Outer Space and other international organizations, such as the World Meteorological Organization.[2] These actions were but a prelude.

KHRUSHCHEV-KENNEDY LETTERS: FEBRUARY-MARCH 1962

John Glenn's flight in his Mercury spacecraft *Friendship 7* was good for NASA, good for the United States, and excellent for international relations. Previously, the news media and public figures in the U.S.S.R. had spoken disparagingly of the American suborbital missions flown by Alan Shepard and Virgil I. Grissom. For example, at a session of the Twenty-second Party Congress, Cosmonaut Gherman Stepanovich Titov made a typical critique of the American space program. "We fly in orbit around the earth, and they jump up in ballistic curves. . . . We should like to wish them success in making orbital flights." Adding a touch of comparative politics, he commented further, "if they do want to emerge into orbital flights let them build a reliable launching pad, let them build socialism."[3]

After *Friendship 7*'s 4-hour and 55-minute flight, the Soviet attitude changed. Although quick to point out that this achievement was simply a repeat, and a briefer one at that, of Titov's day-long mission, the Soviet news media did give extensive coverage to the American flight.[4] More significantly, newspapers that reported the details of the flight also carried the text of a congratulatory letter to Kennedy from Khrushchev.

Khrushchev congratulated the American people and their President for "the successful launching of a spaceship with a man on board." The Premier saw this to be one more step "toward mastering the cosmos"; this time an American had been "added to the family of astronauts." Khrushchev hoped that:

> ... the genius of man, penetrating the depth of the universe, will be able to find ways of lasting peace and insure the prosperity of all peoples on our planet Earth which, in the space age, though it does not seem so large, is still dear to all of its inhabitants.
>
> If our countries pooled their efforts—scientific, technical, and material—to master the universe, this would be very beneficial for the advance of science and would be joyfully acclaimed by all peoples who would like to see scientific achievements benefit man and not be used for "cold war" purposes and the arms race.[5]

While the words of the Soviet leader could have been dismissed as a propaganda ploy, President Kennedy and his White House advisers decided to take the Soviet message at its face value and respond positively.

Kennedy's reply was direct and immediate. "I welcome your statement that our countries should cooperate in the exploration of space." Moreover, he told Khrushchev that he had "long held this same belief" and that he had championed such cooperation in his speeches to the American public. While supporting the supervisory role of the U.N. in the field of space cooperation,

President Kennedy rides with John H. Glenn and General Leighton I. Davis following Glenn's orbital flight aboard Friendship 7.

the President saw that the U.S. and the Soviet Union had a peculiar responsibility to lead the way toward international cooperation. As a consequence, Kennedy said that he had asked certain members of his administration to prepare "new and concrete proposals for immediate projects of common action" that he hoped would be discussed by representatives from the two countries at an early date "in a spirit of practical cooperation."[6]

In a news conference on 21 February, the President reported that he found Khrushchev's proposal "most encouraging" and "beneficial to the advance of science." The President also indicated, "It is increasingly clear that the impact of Colonel Glenn's magnificent achievement yesterday goes far beyond our own times and our own country," or, as Kennedy phrased it later in his press conference, now we "have more chips on the table than we did some time ago."[7] When asked by reporters how far the U.S. would go in cooperating with the Soviet Union, Kennedy responded that it would be "premature" for him to say, but he added that "we all know from long experience that it's more difficult to transform these general expressions into specific agreements." Only time would tell if practical results would follow, and the President promised to withhold judgment until "we see whether the rain follows the warm wind in this case."[8]

At NASA, the Kennedy response to the Khrushchev suggestion for closer scientific and technological collaboration was a surprise.* The White House staff had prepared a reply to Khrushchev after an inquiry to Arnold Frutkin's NASA International Programs Office concerning the possibility of developing a list of "concrete" proposals.[9] Following the dispatch of the Kennedy letter to Khrushchev, representatives from the White House and the State Department worked with a list of possible joint activities drawn up by the space agency for inclusion in a more detailed letter to the Soviet Premier. During the work on these proposals, neither NASA Deputy Administrator Hugh Dryden nor Frutkin had any direct contact with the President or his White House staff. NASA worked at a distance with the Department of State acting as an intermediary.†[10] They knew the President wanted to cooperate with the Soviets on space projects if possible. But what was possible? Was the President willing to sacrifice other aspects of NASA's programs to obtain

---

*Dryden and Frutkin indicated that the initiative for the Kennedy response of 21 Feb. came from the White House, although NASA received the message through the State Department. Dryden felt that Presidential Science Adviser Jerome Wiesner might have been the source of this particular response, but he was not certain.

†The NASA contacts in the State Department were George C. McGhee, Under Secretary of State for Political Affairs; Philip J. Farley, Special Assistant to the Secretary of State for Atomic Energy and Outer Space; and Robert F. Packard, Farley's assistant.

a closer cooperative relationship with the Soviets? In the absence of a clear mandate from the President, Frutkin's conservative approach toward cooperation prevailed. While not the dramatic stand desired by some Kennedy staff members, the NASA efforts were based upon previous experience with the Soviets in space negotiations.

The 7 March 1962 letter that Kennedy sent to the Soviet Union was based on a conscious strategy aimed at enhancing the possibility of obtaining a cooperative relationship.*[11] Negotiations would be conducted at the technical level, not at the head of state level where politics might intrude. Such discussions would involve coordination of efforts in space research without calling for the integration of experiments of one nation into the spacecraft or ground equipment of the other. This parallel effort would be coupled with the reciprocal exchange of data.

Arnold Frutkin has summarized the key topics proposed in Kennedy's letter to Khrushchev:

> (1) the establishment of an operational world weather satellite system through the coordinated launching by the US and the USSR of weather satellites in complementary orbits, the resulting data to be distributed globally through existing meteorological channels;
>
> (2) the exchange of spacecraft tracking services, each side providing equipment suited to its own requirements to be erected and operated on the other's territory by the other's technicians;
>
> (3) mapping of the earth's magnetic field in space, a matter "central to many scientific problems," by satellites which the countries would launch, one each, in complementary orbits;
>
> (4) an invitation to the Soviet Union to join in programs already under way with other countries for the joint testing of intercontinental communications satellites (each country providing a ground terminal suitable for working with US communications satellites and participating in an international ground station coordinating committee).[12]

Beyond these four points, Kennedy briefly touched on the possibility of pooling and exchanging data gathered in space medicine and of exploring plans for future manned and automated space flight. This effort on the part of the White House staff to keep broader topics open for discussion was

---

*There have been some charges that the Kennedy proposals represented nothing new. Former Kennedy White House science aide Eugene Skolnikoff also charged NASA with selecting "only those projects which it thought would be technically and politically desirable." Accordingly, NASA was interested only in the exchange of information and not "intimate cooperation that would have involved joint research and development programs." Arnold Frutkin would not disagree with the specifics, but he would take exception with the interpretation. He felt that NASA should deal with those projects that were possible, not with those that were desirable simply because they were idealistic and dramatic.

indicative of a desire to let the Soviets know that dialogue could evolve into something larger. Kennedy therefore stressed that the points raised in his letter were not intended "to limit our mutual consideration of desirable cooperative activities."[13]

As the work on the Kennedy letter progressed, NASA, the State Department, and the President's Science Adviser decided to go ahead and appoint a technical negotiator in anticipation of a positive response from Khrushchev.* Dryden, NASA's Deputy Administrator, was the unanimous choice, and President Kennedy approved the appointment on 19 March. The following day the President received a reply from the Soviets. In Dryden's words, "Now events moved very rapidly."[14]

Chairman Khrushchev's 20 March response to the Kennedy proposal contained a lengthy preamble restating a desire to preserve space for peaceful exploration and exploitation of those studies that would benefit all nations. Khrushchev's shopping list of proposals contained some that were nearly identical to those suggested by Kennedy, plus two new ones. Suggestions that were similar centered on cooperation in communications and weather satellites, data collection relating to the earth's magnetic field, exchange of space medicine information, and organization of a system for observing and tracking vehicles launched to the moon or the planets. The new topics dealt with the rescue of spacecraft and with space law.[15]

Khrushchev was agreeable to drafting an international pact providing "for aid in searching for and rescuing spaceships, satellites and capsules that have accidentally fallen." This agreement seemed particularly important "since it might involve saving the lives of cosmonauts...." Rescue operations and returning space hardware pointed also to the further necessity of attending to the "important legal problems" of space that confronted the spacefaring nations.[16]

To begin the dialogue, Khrushchev told Kennedy that the Soviet representatives to the U.N. Committee on the Peaceful Uses of Outer Space were being instructed to meet with their American counterparts. Further, Khrushchev seemed to indicate a relaxation of one of the barriers that had been hindering concrete discussions: disarmament no longer was held to be the basic prerequisite to such talks, though it was a conditioning factor. It seemed obvious to the Soviet leader "that the scale of our . . . cooperation in the peaceful conquest of space . . . is to a certain extent related to the solution of the disarmament problem." Therefore, Khrushchev felt that "until an agreement on general and complete disarmament is achieved, both

---

*Administrator James E. Webb represented NASA in this discussion, with George McGhee of the State Department and Science Adviser Jerome B. Wiesner.

our countries will . . . be limited in their abilities to cooperate in . . . space."
If the question of disarmament could be satisfactorily resolved, "Considerably broader prospects for cooperation and uniting our scientific-technological achievements, up to and including joint construction of spacecraft for reaching other planets—the moon, Venus, Mars—will arise. . . ."[17]

In a news conference on 21 March, President Kennedy announced that he was gratified by the Khrushchev reply, and that steps would be taken to initiate an early discussion with the Soviets, with Dryden as his technical representative. Kennedy said that the U.S. would make "all possible efforts to carry forward the exploration and use of space in a spirit of cooperation for the benefit of all mankind."[18] The rhetoric sounded promising, but the work remained. As Kennedy said, "an agreement to negotiate does not always mean a negotiated agreement."[19]

THE FIRST DRYDEN-BLAGONRAVOV AGREEMENT–1962

As Soviet and American reporters analyzed the exchange between their political leaders, NASA officials prepared for discussions with the Soviets.[20] With State Department help, the NASA Office of International Programs drafted three informal position papers expanding the major points of Kennedy's 7 March letter.[21] Dryden and Frutkin then traveled to New York City to meet with Academician Blagonravov on 27 March for their first exploratory talks; the exchanges were informal and preliminary.* Both parties had agreed in advance that formal negotiations would begin later. The Kennedy-Khrushchev letters were discussed, but to Dryden "It became obvious as the talks proceeded that Academician Blagonravov had left Moscow [either] before the exchange of letters between Chairman Khrushchev and President Kennedy, or so soon thereafter that he had not discussed the several proposals in any detail with other scientists, and that he had received few instructions from Moscow."[22] Blagonravov promised to study the NASA position papers and respond with formal position statements at a subsequent meeting.

Dryden believed that these first conversations were "generally free of cold-war propaganda. On one or two occasions there were remarks that cooperation could be on a much larger scale if the disarmament negotiations were successful, but the main interest seemed to be . . . finding possible

---

*The American delegation also included D. F. Hornig, J. W. Townsend, Jr., P. S. Thacher, R. W. Porter, and L. Bowdin. The other Soviet participants were Y. A. Barinov, G. S. Stashevsky, R. M. Timberbaev, and V. A. Zaitzev.

A. A. Blagonravov and H. L. Dryden have an informal chat in the lobby of the U.S. Mission to the United Nations before beginning their talks on space cooperation, March 1962 (New York Times photo).

beginning steps for cooperation." At one juncture, Blagonravov raised the issue of American nuclear tests in the atmosphere, and subsequently at a meeting in the Soviet mission in New York City, he briefly mentioned spy satellites. Dryden replied politely but firmly that his authority was limited to technical matters; political and legal issues were outside his authority.[23] Frutkin later reported that "Blagonravov accepted this position philosophically, not raising such issues again."[24]

As Frutkin saw it, the Soviets seemed hesitant to discuss the possibilities of cooperative efforts in space medicine, even though this topic had been proposed by Khrushchev, and Blagonravov quickly dismissed the American proposal to conduct experiments with high-altitude balloons. He said that his country disliked balloons, an obvious reference to American programs to disseminate propaganda leaflets from balloons over Eastern Europe.[25] On the question that had been raised by Khrushchev's letter concerning outer-space pollution, Blagonravov "expressed concern" regarding the negative impact of one nation's experiments on the scientific work of another. Specifically, he was referring to Project West Ford, a target for Soviet criticism.* Frutkin also perceived that the Soviets were not eager to become immediately involved in joint space flight. "Blagonravov stated that

---

*Project West Ford, a USAF program conceived at MIT's Lincoln Laboratory, involved launching into earth orbit 350 million copper threads (17.78 millimeters long and 0.254 millimeters in diameter), which would serve as reflector antennas for short wavelength communications (8000 megahertz). The experiment promised to make global radio coverage invulnerable to jamming. Project West Ford, approved on 4 Oct. 1961 by the White House, met with mixed international scientific reactions, being criticized by many scientists as a possible threat to the study of radio astronomy or as an alteration to the environment of space, but the project was praised by NATO politicians as a significant deterrent defense system. On 10 May 1963, a second attempt to orbit the disputed payload was successful; the dipoles ejected and formed a compact cloud, circling the earth every 166 minutes in a near-polar orbit at a height of 3704 kilometers. *Science* on 16 Dec. reported that nearly all of Project West Ford's dipoles had reentered the atmosphere.

current programs were too far along to permit coordination at this date. The coordination of future programs . . . seemed possible."[26]

The guarded sense of optimism felt by Dryden and Frutkin was expressed only in private.[27] In a brief joint statement from Blagonravov and Dryden on 30 March, the press was told that the representatives of the two nations "have now concluded their preliminary discussions." They also announced that they intended to meet again during either the COSPAR sessions scheduled for 30 April-10 May in Washington or the meeting of the Scientific-Technical and Juridical Subcommittee of the U.N. Outer Space Committee. Additional scientists from both nations would join in these technical discussions. This was not hard news, but the statement indicated that both parties realized their work had just begun.[28]

Soviet public reaction to the proposed cooperation was favorable. On 12 April 1962 at the government-sponsored Cosmonautics Day celebrations, both Gagarin and Titov were quoted in the Soviet press as favoring cooperation between the two countries, especially if it led to a reduction in armaments.*[29] Mstislav Vsevolodovich Keldysh, President of the Soviet Academy of Sciences, declared that he favored Soviet-American space cooperation as a route toward the solution of many scientific concerns.[30] This basic theme was repeated in an interview with Khrushchev by Gardner Cowles, editor of *Look* magazine. Khrushchev saw a joint expedition to the moon as technically and scientifically possible; only the political problem of the military character of space rockets stood in the way.[31]

Reaction in the U.S. to space cooperation with the Soviets was mixed. Glenn's flight had reassured many Americans who had been worried about the nation's position in the space race. Most public figures were still committed to establishing American pre-eminence in space. Senator Margaret Chase Smith, the ranking Republican member of the Aeronautical and Space Sciences Committee, felt that the United States had little to gain from cooperation, especially since the nation was committed to "superiority over Russia on really important space development."[32] However, Representative George P. Miller, Chairman of the House Committee on Science and Astronautics, approached the possibility of cooperation in a more positive fashion. In welcoming the Khrushchev overture to cooperate, Congressman Miller said, "This is something we must do. We must accept their offer in good faith unless, and until, proven otherwise. The world expects this of

---

*Cosmonautics Day, 12 Apr., was created by the Presidium of the Supreme Soviet not only to celebrate the anniversary of Gagarin's first space flight but also to remind the Soviet public and the world of the accomplishments and goals of the Soviet space program. It has become annually customary for *Pravda* and *Izvestiya* to feature articles at this time written by the cosmonauts on different aspects of space flight. Gagarin, until his death in an aircraft crash in Mar. 1968, and Titov were frequent authors of items promoting international peace and cooperation.

us."[33] The wider public reaction seemed to mildly favor cooperation so long as it did not have a negative impact on the American goal in space—the Kennedy-inspired goal to reach the moon during this decade.[34]

Vice President Lyndon Johnson, on 10 May 1962, summed up the feelings of many American politicians in a speech dedicating the NASA Space Exhibit at the Seattle World's Fair. Cooperation in space could be the route to greater understanding between the United States and the Soviet Union. Joint scientific efforts might make other political areas easier to discuss, but the burden of cooperative programs was a mutual one. The Vice President, "with a spirit of cautious optimism," was able to tell his audience "that the Soviet Union appears to realize that—in outer space, at least—there may be something to be gained by cooperating with the rest of humanity."[35]

Meanwhile, Dryden was preparing for the next round of discussions with the Soviets, to be held at the end of May in Geneva.[36] Dryden was concerned about the political considerations behind the Kennedy administration desire to discuss collaboration; thus, he sought to determine the President's position. Unfortunately, Dryden never had the opportunity to discuss the matter directly with Kennedy or his top White House advisers. His closest contact to the President was George C. McGhee at the State Department.

Dryden, a scientist turned administrator called upon to be an international negotiator, sat down with McGhee on 18 May and asked him how the President wanted the negotiations conducted. Were these discussions intended to arrive at true cooperation, or were they only propaganda? Was it a sincere effort to get negotiations going or merely something for public display? As Dryden told McGhee, the nature of the goal "would make some difference in the approach." McGhee assured Dryden that "the President had in mind real cooperation, that he was as anxious to go just as far as the Soviets would go." With the nature of his mission somewhat more clear, Dryden made ready for his trip to Europe.[37]

Dryden and Blagonravov met in Geneva on 27 May 1962. Both men had traveled to the Swiss city for the first meeting of the Technical Subcommittee of the U.N. Committee on the Peaceful Uses of Outer Space. While there was no direct connection between the bilateral Soviet-American talks and the U.N. meeting, the negotiators found such an occasion convenient to pursue their private discussions. The two men, despite their obvious political constraints, worked well together. In 12 days, they succeeded in hammering out agreement on three points.*[38]

---

*Frutkin in *International Cooperation in Space* gives a detailed account of the negotiations and some of the difficulties encountered by Dryden and Blagonravov.

As reported by Frutkin, "this first agreement embraced three projects, following the US proposals on meteorology and geomagnetism very closely and reflecting Blagonravov's new interest in the Echo experiment in satellite communications." The two principal negotiators were satisfied with their progress. Dryden commented to reporters that approval of the agreements by the American and Soviet governments would mark an "important step" in space cooperation. At the joint news conference on 8 June, Blagonravov added that they would have been wasting their time if they had not "believed the work to be of major significance."[39] The two men departed to their respective capitals to secure the necessary government approvals for their proposals.

The Dryden-Blagonravov agreement provided for a two-month study period, during which either party could suggest changes to the proposals. As it developed, neither country sought amendments, and Soviet Academy President Keldysh and NASA Administrator Webb exchanged letters on 18 and 30 October 1962 that formalized the agreements.[40] Much political and technical work lay ahead—work that was hindered by the grave situation created by the discovery of Soviet Intermediate Range Ballistic Missiles in Cuba.*[41]

When the joint announcement of the bilateral space agreement was made to the U.N. on 5 December 1962, the somber and tense days of October lingered in the minds of many American and Soviet political figures. Indeed, the joint announcement had been postponed until December because of a Presidential order during the Cuban crisis decreeing "that there be no further action on the U.S.-U.S.S.R. outer space bilateral until the Cuban situation has been settled."[42] An atmosphere of restraint accompanied the official announcements when they were made. Administrator Webb indicated simply:

> This is an important step toward cooperation among nations of the world to increase man's knowledge and use of his special environment. The careful preparation for such a joint cooperative effort made by Academician A. A. Blagonravov and Dr. Hugh L. Dryden is a sound basis on which to proceed. The United States will make every effort to facilitate this undertaking.[43]

The official Soviet news agency, Tass, stated briefly: "There is no doubt that this agreement will make a great contribution to the conquest of the universe

*Kennedy publicly announced on 22 Oct. 1962 the presence in Cuba of Soviet missiles capable of striking a large part of the U.S. A naval blockade was imposed to intercept further shipments, and the President bluntly demanded that the Soviets withdraw their missiles. By early November, aerial reconnaissance showed that the Cuban bases were being dismantled and the missiles crated for return to the U.S.S.R.

and to the further advance of international cooperation between scientists."[44]

The next step in implementing the agreements called for creating joint working groups. To facilitate the establishment of those technical parties, Dryden and Blagonravov met in Rome on 11-20 March 1963 and again in Geneva during the following May. The result of these two meetings was a document—the "First Memorandum of Understanding to Implement the Bilateral Space Agreement of June 8, 1962."[45] The details for the weather satellite launching and the data exchange project were concluded with relative ease. But the agreement on the communications satellite experiments with *Echo II* was more difficult to arrange because of technical complexities. Proposals for a coordinated launch of geophysical satellites to study the earth's magnetic field were finalized at the May meeting.*[46]

The process for conducting the negotiations followed an unofficial protocol, which established a precedent for subsequent discussions. In Rome, the first two days were essentially ceremonial. Following the formalities held first at the American Embassy and then at the Soviet Embassy, the working sessions began. Generally, the pattern of the meetings called for the discussion of draft documents, during which the two negotiating teams compared points and argued matters of substance and wording until an agreed document was assembled in both English and Russian.[47]

In testimony before the Senate Committee on Aeronautics and Space Sciences, Dryden reflected on the possible motivations that underlay the Soviet decision to subscribe to these cooperative agreements. It was Dryden's personal belief that "this group of scientists who are interested in collaboration have been given a hand to see what they can come up with." Both groups of negotiators had decided that they "could not agree on anything which did not show a benefit to both countries." Looking at the nature of the joint discussions, Dryden felt that there was a "possibility that the political elements in Russia may at some point shut this off." Dryden was assuming, as did other American scientists, that Blagonravov and his associates in the Soviet Academy represented "what you might call a liberal group in Russia," which sought to begin "limited cooperation within the

*For details of the discussions, see Frutkin's *International Cooperation in Space,* pp. 97-105. The co-chairmen of each of the three Working Groups were as follows: Working Group 1 (weather)—M. Tepper, Director, Program of Weather Satellite Applications, NASA, and V. A. Bugayev, Director, Central Institute of Weather Forecasting, U.S.S.R.; Working Group 2 (communications)—L. Jaffee, Director, Communications Systems, NASA, and I. V. Klokov, Deputy Minister of Communications, U.S.S.R.; and Working Group 3 (geomagnetic study)—L. Cahill, Director of Physics, Office of Space Sciences, NASA, and Yu. D. Kalinin, Deputy Director of the Institute of Terrestrial Magnetism, U.S.S.R.

political climate of their own country and of the times."[48] Frutkin, however, challenged the notion expressed by Dryden and others that "technical cooperation does not involve a party political line."[49]

The concept that scientists have a unique position in the scheme of things, as a result of the international character of their work, has a long history. Equally strong is "the notion that the scientist can play a special role and effective role in establishing and cementing improved relations among nations. . . ."[50] In the post-World War II era, there has been a strong feeling of internationalism within the community of science and technology, especially in the U.S. where a number of scientists urged their fellows to lead the way toward greater scientific cooperation among nations. But among scientists, as among all peoples, there are both internationalists and nationalists. Frutkin contends:

> The evidence appears to be overwhelming that scientists confronted with the exigencies of national need have reacted much as other patriotic citizens, professional and nonprofessional. In part, this follows from an interaction between science and government which produces a rough alignment even in democratic countries. International ties, real or fancied, have not weighed in the balance in any significant way. . . . When we say that science is international we mean that it is international where scientific matters of essentially professional character are concerned, and not really where political matters are concerned.[51]

Thus, Dryden was correct in his report that both teams of negotiators could only agree to those activities that were of mutual benefit, but he may have been too generous in his analysis when he said that politics did not influence technical cooperation. Simply, in some areas of negotiations, politics were less obtrusive than in other areas. Indeed, the passage of time would show that there were political considerations behind all the technical agreements.

American public response to the 16 August 1963 announcement of the Soviet-American bilateral space agreement was conditioned by the successful conclusion of the nuclear test ban treaty and speculation over rumors of a joint manned space flight. On 25 July 1963, representatives of the U.S., the Soviet Union, and the United Kingdom initialed a treaty prohibiting nuclear weapons tests in the atmosphere and space and under water; relaxation of nuclear tension made the space agreement between the Americans and Soviets seem all the more promising. A *New York Times* article on the Dryden-Blagonravov "Memorandum of Understanding" termed the idea of cooperative manned space flights "a logical outgrowth of the present agreement."[52] Rumors circulated that there might possibly be a joint lunar mission in the planning, speculation developed partly as a result of the visit of British astronomer Sir Bernard Lovell. In the latter half of 1963, according to Frutkin, there ensued in the story of U.S.-U.S.S.R. space relationships "by all odds the strangest chapter. . . ."

THE KENNEDY PROPOSAL FOR A JOINT MOON FLIGHT

Sir Bernard Lovell, a professor at the University of Manchester and Director of the Jodrell Bank radio telescope facility, had been active in the international astronautics community for many years. The Jodrell Bank observatory was scheduled to play a key role in the Soviet-American communications satellite experiments agreed to in Rome. During June and July 1963, Sir Bernard was the guest of the Soviet Academy of Sciences on an unprecedented tour, for a Western scientist, of the major optical and radio observatories. In a letter to Dryden dated 23 July 1963, Lovell described his visit:

> During this time I was taken to the major Soviet optical and radio observatories and to the deep space tracking network, a station which has not so far been seen by Western eyes or by many Soviet scientists so I was told, I mention this at the beginning of this letter because it does seem to underline the apparently genuine desire of the Academy to extend its cooperation with the West.[53]

After describing the "cooperative programs" that he had negotiated with the Soviets, he reported on conversations in which his hosts had discussed the plans for future Soviet efforts in space. Included in Soviet comments was an apparent postponement of a manned program of lunar exploration. Lovell told Dryden that President Keldysh of the Soviet Academy had given three reasons for favoring automated unmanned spacecraft for exploring the lunar surface:

> (1) Soviet scientists could see no immediate solution to the problem of protecting the cosmonauts from the lethal effects of intense solar outbursts.
> (2) No economically practical solution could be seen of launching sufficient material on the moon for a useful manned exercise with reasonable guarantee of safe return to earth.
> (3) The Academy is convinced that the scientific problems involved in the lunar exploration can be solved more cheaply and quickly by their unmanned, instrumented lunar program.[54]

Sir Bernard reported that he had argued in favor of a manned lunar expedition, and Keldysh said that a Soviet program to send cosmonauts to

the moon might be revived if the issues raised in the three objections could
be overcome. Furthermore, Keldysh was reported to have suggested:

> ... that the Academy believed that the time was now appropriate for
> scientists to formulate on an international basis (a) the reasons why it is
> desirable to engage in the manned lunar enterprise and (b) to draw up a list of
> scientific tasks which a man on the moon could deal with which could not be
> solved by instruments alone. The Academy regarded this initial step as the
> first and most vital in any plan for proceeding on an international basis.[55]

In concluding his report to Dryden, Lovell said that he had promised
Keldysh to convey the substance of these discussions to the "appropriate
authorities in the United Kingdom and the United States of America." Now
that Lovell had discharged his promise, a major question remained. What did
his conversation with President Keldysh signify?

There were various American interpretations of the Lovell letter. To
some observers, this seemed to be strong, reliable data from a prominent
scientist that the Soviets had dropped out of the race to the moon.
Furthermore, the Soviet Union seemed willing to talk about cooperation in a
joint program of lunar exploration. This would mean a dramatic shift from
the concept of coordinated space ventures to integrated programs, a change
that would require deeper study and extensive discussions between the U.S.
and the Soviet Union. Other commentators on the Soviet space program,
including Dryden, viewed the Keldysh remarks to Lovell simply as a
propaganda ploy that would require the Americans to submit their lunar
program to an international body for scrutiny.[56] Whatever the motivation,
the conversations reported by Lovell were newsworthy, and the press asked
President Kennedy to address the substance of these remarks on 17 July.

"Would we still continue with our moon program" if the Soviets should
drop out of the lunar race, the press asked? The President said that he knew
only what he had heard or read in news reports; therefore, he had to
conclude that only time would tell what the true Soviet intentions were.
Kennedy did see that the Soviets were "carrying on a major [technological]
campaign and diverting greatly needed resources to their space effort. With
that in mind," the President thought, "we should continue" our effort to go
to the moon. Betraying a sense of skepticism, he suggested that "the
prediction in this morning's paper that they are not going to the moon ...
might be wrong a year from now." When pressed to defend Apollo and the
moon landing should the Soviets quit the race, Kennedy touched on the
strategic importance of sending an American to the moon:

> The point of the matter always has been not only of our excitement of
> interest in being on the moon, but the capacity to dominate space, which
> would be demonstrated by a moon flight, I believe is essential to the United

States as a leading free world power. That is why I am interested in it and that is why I think we should continue, and I would be not diverted by a newspaper story.[57]

But two months later on 20 September, President Kennedy in a surprise address before the General Assembly of the United Nations raised the possibility of a "joint expedition to the moon."[58] How are Kennedy's two positions to be reconciled? At one point, the President called for American domination of the space frontier in the 1960s, and at another time he argued that "space offers no problems of sovereignty," so "why, therefore, should man's first flight to the moon be a matter of national competition? Why should the United States and the Soviet Union, in preparing for such expeditions, become involved in immense duplications of research, construction, and expenditure?"[59] The "why" of competition versus cooperation had been a matter of much discussion among the White House staff prior to Kennedy's U.N. address.

Two days before Kennedy's speech, McGeorge Bundy, a Presidential assistant, addressed the question of cooperation and competition in a "Memorandum for the President." NASA Administrator Webb had reported to Bundy that the agency anticipated continued suggestions from the Soviets that the two nations cooperate in space. Indeed, the subject of the Lovell letter and the idea of cooperative lunar exploration had been discussed by Blagonravov and Dryden in a New York luncheon meeting.[60] The dramatic newspaper reports of the meeting raised questions that Bundy passed along to Kennedy.[61] "The obvious choice was whether to press for cooperation or to continue to use the Soviet space effort as a spur to our own." In this same memorandum, which was prepared as background for the President's meeting that same morning with Administrator Webb, Bundy indicated that there was some "low-level disagreement" on this topic within NASA.* He argued that in his own "hasty judgment" a decision was called for on competition or cooperation. If competition was favored, then the U.S. should make every effort to meet the goal of a lunar landing before the end of the 1960s. "*If we cooperate,* the pressure comes off, and we can easily argue that it was our crash effort [in] '61 and '62 which made the Soviets ready to cooperate."[62]

Later on the morning of 18 September, the President met briefly with James Webb. Kennedy told him that he was thinking of pursuing the topic of cooperation with the Soviets as part of a broader effort to bring the two

---

*The "low-level disagreement" Bundy mentions refers to press accounts of a 17 Sept. 1963 speech in which Manned Spacecraft Center Director Robert Gilruth had told the National Rocket Club that a joint American-Russian space flight—especially one to the moon—would present almost insuperable technological difficulties.

countries closer together. He asked Webb, "Are you sufficiently in control to prevent my being undercut in NASA if I do that?" As Webb remembered that meeting, "So in a sense he didn't ask me if he should do it; he told me he thought he should do it and wanted to do it. . . ." What he sought from Webb was the assurance that there would be no further unsolicited comments from within the space agency. Webb told the President that he could keep things under control.[63]

Late on the following day, Bundy called Webb to tell him that the President had decided to include a statement about space cooperation with the Soviets in his U.N. address. Bundy informed Webb that Kennedy wanted "to be sure that you know about it."[64] The new paragraph, drafted by Arthur M. Schlesinger, Jr., another Kennedy aide, had not been included in the earlier drafts of the speech circulated at NASA.[65] Upon receiving the President's message, Webb immediately telephoned directions to the various NASA centers "to make no comment of any kind or description on this matter."[66]

The President's proposal for a joint expedition to the moon was intended to be a step toward improved Soviet-American relations. The impact of the speech was quite the reverse. Moscow and the Soviet press virtually ignored the U.N. address.*[67] Officially, the Soviet government did not comment.[68] In the U.S., the public remarks either strongly supported the idea of a joint flight or equally forcefully opposed it.[69]

Reaction within NASA itself was varied. During a news conference in Houston on the day of the President's address, Associate Administrator Robert C. Seamans, Jr., stated that Kennedy's proposals came as no great surprise. He said that many "large areas" for cooperation existed, such as exchanges of scientific information and space tracking data, but he emphasized that there were no plans for cosmonauts to fly aboard an Apollo spacecraft. Deputy Associate Administrator for Manned Space Flight George E. Mueller shared Seamans's view. He compared future U.S.-U.S.S.R. cooperation in space to joint explorations in Antarctica. Scientists from both nations worked in the same region, but "they got there in different ships." Robert Gilruth, Director of MSC, expressed the concerns of technical specialists about an integrated mission.[70]

Speaking before the National Rocket Club three days before the Kennedy address to the U.N., Gilruth had said that he "would welcome the opportunity to go behind the scenes in the Soviet Union and see what

---

*The paper *Za Rubezhom* saw the Kennedy proposal as a propaganda stunt. A Walter Lippman column reprinted by *Pravda* saw the primary value of Kennedy's speech to be the opportunity it offered the U.S. to escape a unilateral visit to the moon.

they're doing, what they have learned." But then he added that a joint space flight involving the melding of equipment would pose difficulties. "I tremble at the thought of the integration problems." Gilruth emphasized that he was speaking as a working engineer and not as "an international politician." He said that American space engineers had enough difficulties mating the hundreds of electrical, mechanical, and pyrotechnic connections between American launch vehicles and spacecraft. Noting "how difficult these integration problems are" from a technical standpoint within a single agency, he said that the engineering problems inherent in combining the hardware of two nations would be "hard to do in a practical sort of way." At the 20 September MSC news conference, he added that such problems "are very difficult even when [the hardware components] are built by American contractors."[71] Gilruth's fears were unfounded for the time being; there would be no joint missions in the foreseeable future.

Thus the optimism generated by the Lovell report regarding joint flight ventures turned into disillusionment.*[72] The political climate—domestic and international—would not support bold proposals for cooperation. Most Americans believed that the U.S. was firmly committed to be the first nation on the moon; an executive or scientific wish to cooperate should not deter the country from obtaining that goal. The clearest statement of the national

---

*The Lovell letter was disavowed by the Soviets in the winter of 1963. Keldysh repudiated the letter in a radio broadcast on 14 Oct., while Khrushchev indicated that the U.S.S.R. was still part of the race to the moon.

Copyright © 1963 Chicago Sun-Times and reproduced by courtesy of Wil-Jo Associates, Inc., and Bill Mauldin.

"I WAS AFRAID SOMETHING LIKE THIS MIGHT HAPPEN IF YOU KEPT TALKING ABOUT IT."

THE JOINT FLIGHT PLAN

MOON
BUG LANDS
SPACECRAFT LEAVES BUG
JOINED VEHICLES GO TO MOON
BUG JOINS SPACECRAFT
EARTH
RUSSIANS LAUNCH SPACECRAFT
U.S. LAUNCHES BUG
BUG REJOINS SPACECRAFT
JOINED VEHICLES ORBIT MOON
SPACECRAFT RETURNS TO EARTH
SPACECRAFT ORBITS EARTH

RUSSIANS LAUNCH
U.S. LAUNCHES
U.S. BOOSTER FALLS OF

An ingeniou...

HOW WE

President Kennedy's recent proposal to t
U.N. that the United States and Russia c
operate in going to the moon drew dissents fro
many space experts. They said that joini
Soviet and U.S. equipment would present eno
mous technical problems and require an u
thinkable disclosure of secret information. B
last week Thomas Turner, an engineer with R
public Aviation Corp., devised an ingenio
scheme for making the President's plan wo

Thomas Turner of the Republic Aviation Corporation teamed up with Mel Hunter to suggest a way that the Americans and the Soviets could go to the moon together. Drawings for Life *by Mel Hunter (© 1963 Time Inc.).*

JOINT MOON FLIGHT CONTINUED

COSMONAUTS, I PRESUME:
WE LINK UP IN SPACE

It is the orbital docking of the U.S. bug to the Soviet spacecraft that gets around the technical and security problems of a joint flight. The docking mechanism, shown in the drawing at right, is crucial to the plan.

If American engineers wanted to mate their bug with a Russian spacecraft and booster on earth, they would have to know all the Soviet booster's characteristics. Otherwise the combined bug-spacecraft-booster could not withstand the stresses and vibrations of launch. But boosters also propel intercontinental ballistic missiles, and the Russians refuse to reveal the secrets of any of their rockets. With space docking, this disclosure becomes unnecessary. The Russians launch their own spacecraft. And the U.S. launches the bug—with a much smaller booster than our separate moon program calls for. The only hardware requiring cooperative development by the two countries is the docking contrivance—the docking head on the bug and the docking collar of the spacecraft must lock together securely. To facilitate even this cooperation, Turner suggests the U.S. bug manufacturer could build both collar and head and supply the collar to the Russians. They would simply fit it on the face of their spacecraft. The only other exchanges of

information required by the docking scheme would be on communications and on settling on a common atmosphere to be maintained in both bug and capsule.

At a time when both the U.S. and the Soviet Union are concerned about the enormous costs of their moon programs, the plan for this joint venture offers especially attractive economic benefits. Instead of building all the units necessary for a lunar landing, each nation could concentrate only on its parts of the program. The U.S., for example, could stop or slow work on its own three-man spacecraft and huge booster. Russia could avoid development of the sophisticated bug. This division of labor, saving billions, could have the further advantage of advancing, by months or even years, the day men finally get to the moon.

OUR ASTRONAUT JOINS THEIRS. The cutaway drawing at right shows the first vital transfer when the U.S. astronaut leaves the bug through a tunnel and enters the Soviet spacecraft as they orbit earth. A spring-actuated catch locks on the docking head, the cylinder which protrudes from the front of the bug into the capsule's docking collar,

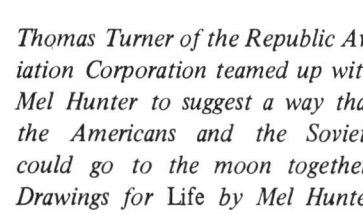

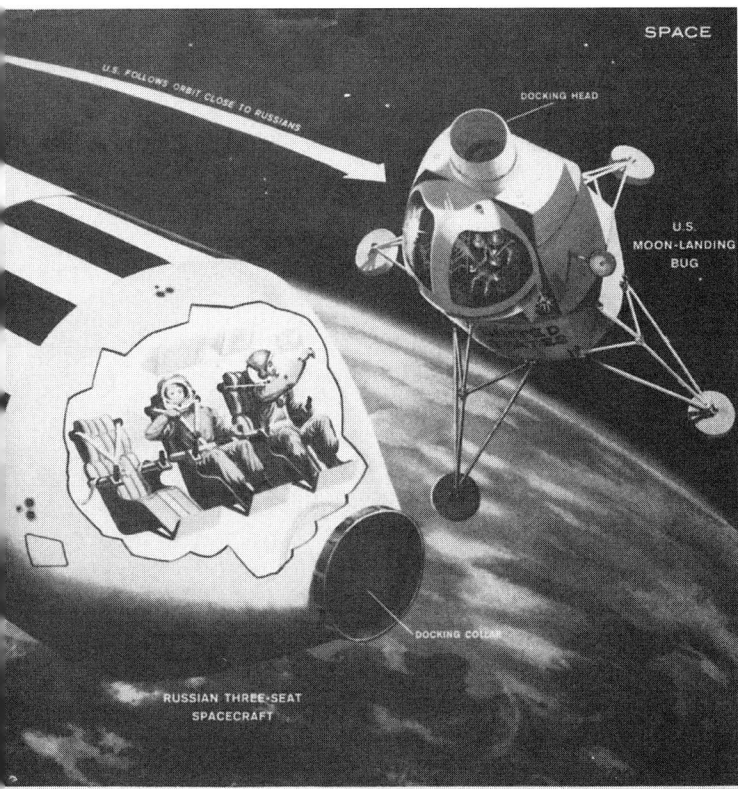

U.S. FOLLOWS ORBIT CLOSE TO RUSSIANS

DOCKING HEAD

U.S.
MOON-LANDING
BUG

DOCKING COLLAR

RUSSIAN THREE-SEAT
SPACECRAFT

scheme for meeting the Russians in space

SPACE RENDEZVOUS. Soviet spacecraft (*center*) and U.S. bug prepare to join, or "dock," in orbit. Docking head on bug fits into spacecraft's collar.

CAN JOIN IN A MOON TRIP

Orbiting the earth in a three-seat Soviet spacecraft, two Russian cosmonauts prepare to receive an American astronaut, arriving in a U.S. lunar landing vehicle nicknamed the "bug" (*see drawing above*). This delicate docking, or joining, of the two nations' vehicles in space would be the key to their joint moon flight. Previous ideas for a cooperative moon flight envisioned mating the necessary vehicles into a single huge booster rocket on earth and launching them together. This produces the security problems discussed on the next page. But Turner's flight plan, diagramed at left, calls for *two* initial launchings, one from the U.S.S.R., one from the U.S. The Russians fire into earth orbit a spacecraft with a three-seat capsule and a rocket to drive it from earth orbit into moon orbit and back to earth. Into this same orbit the U.S. would fire the two-seat bug already being developed by the U.S. to ferry astronauts between the Apollo lunar orbiting vehicle and the moon. After the two dock, the astronaut would join the Russians in their capsule for the flight into lunar orbit. Once there, the astronaut and a cosmonaut would enter the bug, descend to the moon, then fly back to rejoin the remaining cosmonaut in the capsule. They would then leave the bug drifting in space and all three would return to earth in the Russian spacecraft.

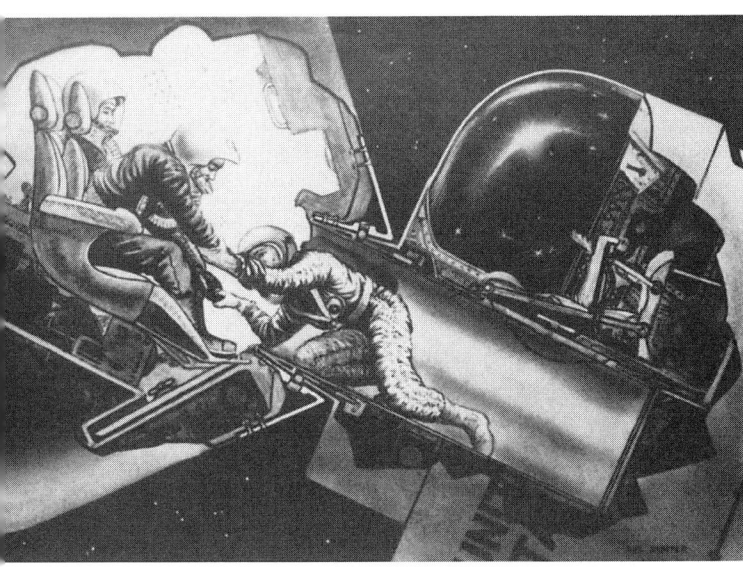

attitude toward the Kennedy proposal of a joint moon venture came in December, when Congress passed an appropriations bill carrying the following stipulation:

> No part of any appropriation made available to the National Aeronautics and Space Administration by this Act shall be used for expenses of partici- pating in a manned lunar landing to be carried out jointly by the United States and any other country without consent of the Congress.[73]

This basic provision was repeated in the NASA appropriations acts for fiscal years 1964-1966. President Lyndon Johnson called this clause an "un- necessary and undesirable restriction."[74]

Johnson attempted throughout the winter of 1963 to keep the door to cooperation open. On 2 December, Ambassador Adlai E. Stevenson told the Political Committee of the U.N. that the President had instructed him to reaffirm the Kennedy proposal for a joint Soviet-American expedition to the moon. Without referring to the political storm in Congress over the idea of any proposals for joint flight ventures, Stevenson said, "If giant steps cannot be taken at once, we hope that shorter steps can. We believe there are areas of work, short of integrating the two national programs, from which all could benefit." Therefore, he suggested that "we should explore the opportunities for practical cooperation. . . ."[75] The task of negotiating these "small steps" fell once more upon the shoulders of Hugh L. Dryden and Anatoliy Arkadyevich Blagonravov.

THE DRYDEN-BLAGONRAVOV TALKS—1964-1965

At the outset of 1964, a tangible result of the initial Dryden- Blagonravov discussions came when NASA launched the communications satellite *Echo II*. Two weeks before the launch, Blagonravov had notified Dryden that the Academy of Sciences would participate in the tracking and communications experiments with *Echo II* as agreed in the Geneva talks of May 1963. In the same message, he informed Dryden that information would be forthcoming shortly detailing their plan for cooperation in meteorological studies. The Americans were cautiously enthused by this step forward.[76]

From Vandenberg Air Force Base on 25 January, the balloon satellite of laminated Mylar plastic and aluminum was placed in near-polar orbit.*[77] Two days later, Academician Blagonravov announced that Soviet ground

---

Echo II, placed into orbit by a Thor-Agena B launch vehicle, weighed 243 kilograms, but when inflated it had a diameter of 41 meters.

stations were tracking *Echo II.* Some of these optical facilities had observed the inflation of the satellite, and three observatories had succeeded in photographing it.[78] On that same date, NASA received raw tracking data, and later the Soviets forwarded photographic materials and a preliminary analysis of orbital data obtained when the satellite was not being observed by U.S. tracking facilities. The second phase of the experiments with the communications satellite, beginning 22 February and continuing into March, consisted of 34 communications exercises between the Manchester University radio telescope at Jodrell Bank in the U.K. and Zimenki Observatory at Gorki University in the Soviet Union.[79]

Dryden discussed with guarded enthusiasm the meaning of the joint U.S.-U.S.S.R. tests with *Echo II* in testimony before the Senate Committee on Aeronautical and Space Sciences in March 1964. At first glance, Dryden thought that the real significance of the tests was that the two teams had taken "advantage of existing programs, approved and executed on their own merits, to provide an opportunity for scientists and engineers of both countries to gain experience in working together for their mutual benefit." This was "a pioneer venture . . . designed as a coordinated rather than joint effort." Dryden thought it interesting that the Soviets had re-christened *Echo II* the "Friendly Sputnik."[80]

A year later in March 1965, Dryden's remarks to the Senate were to be less effusive. He prefaced his comments on cooperation with the U.S.S.R. with the statement: "we engage in cooperative international activities for two reasons—to further the NASA mission and to advance the foreign policy objectives of the United States." He then bluntly presented a final assessment of the *Echo II* test project:

> The Soviet side observed the critical inflation phase of the satellite optically and forwarded the data to us. They did not provide radar data, which would have been most desirable, but they had not committed themselves to do so. The Soviets provided recordings and other data of their reception of the transmissions via ECHO from Jodrell Bank. On the other hand, the communications were carried out in only one direction instead of two, at less

Echo II passive communications satellite undergoes pre-flight inflation tests.

interesting frequencies than we would have liked, and with some technical limitations at the ground terminals used. I do not want to over-emphasize any technical benefits from this project. It was, however, a useful exercise in organizing a joint undertaking with the Soviet Union.[81]

The intervening year had bred some caution and doubt at NASA as to the future of cooperation between the two space powers. At the end of May, Administrator Webb had commented on the twin goals of cooperation and competition. He did not see any inconsistency in pursuing both goals simultaneously:

> I think it makes good sense. The greater our lead in space, the more willing the Soviet Union may become to give up its hopes for world domination and the victory of communism everywhere. The greater our lead in space the more ready the Soviet Union may become to cooperate with us in mutually beneficial ways that will lessen the dangers of nuclear war and advance the cause of freedom.

Webb also cautioned his audience not to expect cooperation overnight.[82]

Dryden and Blagonravov met twice in 1964, but their negotiations were short on concrete results. The first meeting, which coincided with the May COSPAR sessions in Florence, Italy, was limited to discussing an agenda for a second meeting to be held at Geneva during the convocation of a U.N. subcommittee on the Peaceful Uses of Outer Space.[83] During late May and early June, the two negotiators discussed the progress of implementing the details of the 1963 "First Memorandum of Understanding." One major new point centered on an accord to publish several joint volumes of material on space biology and medicine, a field that Dryden indicated "has a considerable bearing on the future of manned space flight, although there was no talk at Geneva of a joint manned flight."*[84]

In reviewing the results of his 1964 meetings with Blagonravov, Dryden told the press that he had discussed cooperation with President Johnson prior to his departure for Geneva and that he had been instructed "to seek to widen the areas of cooperation with the Soviet Union" in space activities. In private conversations with Blagonravov, Dryden conveyed the President's willingness to go as far with cooperative efforts as the Soviet government wished to proceed. As Dryden summarized the American position, "We are always, always have been, prepared to go somewhat farther than they have been willing to do."[85]

Dryden also gave the press his perception of the Soviet attitudes toward

---

*As a result of these negotiations, which were formalized in Oct. 1965, NASA and the Soviet Academy of Sciences jointly published in 1975 and 1976 a three-volume work in four books called *Foundations of Space Biology and Medicine.*

cooperation. He noted "evidence of a very great desire to have cooperative agreements" and an equally strong wish to begin cooperation in space biology and medicine. Counterbalancing this apparent willingness to cooperate was the Soviet concern for secrecy. The "secrecy with regard to engineering and rockets and instruments and spacecraft" had assured a very slow pace and meager results for the two years of negotiations. Dryden felt that as long as the Soviets pursued this course of keeping space data classified, the future of Soviet-American efforts to cooperate would be determined by the pace that the U.S.S.R. wished to follow. Thus, the Deputy Administrator concluded that much patience was called for on the American side, but he also believed that patience was justified since "the prospects are good for a very slow widening of the area of cooperation. . . ."*[86]

Dryden's cautious testimony during the March 1965 congressional hearings indicated that progress had been slow. Data from ground-based magnetic observatories had been exchanged, and the transmission of weather data on the "cold line," a special cable link between Moscow and Suitland, Maryland, had been started in October 1964.†[87] Dryden summarized the status of the joint efforts; "I would describe the situation as a form of limited coordination of programs and exchange of information rather than true cooperation." He continued his report saying, "they have not responded to any proposals which would involve an intimate association and exposure of their hardware to our view." Nor had the Soviets demonstrated "anything in the nature of a joint group working together." When asked if the prospect for the future was one of continued competitiveness, Dryden answered in the affirmative, "As near as we can tell at the moment."[88]

But Dryden's work was coming to an end. Since late 1961, he had been waging a quiet personal battle with an incurable malignancy. He had not yielded to his illness but instead had doubled his work load, as he labored to see Project Apollo and other key NASA programs started toward successful conclusions. In the last four years of his life, he seemed always on his way to attend an in-house conference or to catch a plane for an international meeting. On 16 November 1965, after a series of transcontinental speaking engagements, he entered the National Institutes of Health. Sixteen days later, on 2 December, Dryden was dead at the age of 67.[89]

Pravda carried a Tass communique from Geneva listing the points of the Dryden-Blagonravov talks and noting that the joint efforts in space biology and medicine would be "of great practical value for assuring the life, health, and safety of cosmonauts making orbital flights, as well as for future flights into deep space."

†The "cold line" was so designated to differentiate it from the emergency "hot line," which had been agreed to in 1963 by the Soviets and Americans to reduce the risk of war by miscalculation or accident.

The decade of the 1960s witnessed an increasing tempo of manned space flights; a central theme surrounding these flights was competition between the United States and the Soviet Union. From March 1962 to November 1964, Dryden and Blagonravov had met six times to formulate a basis for cooperation, but the element of competition had prevailed. With Dryden's death, a strong voice for cooperation with the Soviets disappeared. Administrator Webb's primary concern now was the goal of placing a man on the moon ahead of the Soviet Union. As the U.S. and the U.S.S.R. ventured forth on their separate routes to the conquest of space, the idea of cooperation remained, but only as a dream.[90]

III

Routes to Space Flight

By the mid-1950s, the idea of manned space flight emerged from the realm of fantasy to become a topic of serious technical discussion. Frederick C. Durant III, President of the International Astronautical Federation (IAF), told the delegates gathered in 1954 at Innsbruck, Austria, that "the feasibility of space flight is no longer a topic for academic debate, but a matter of time, money and a program."[1] To illustrate his point, Durant showed the Walt Disney Productions motion picture *Man in Space* during the August 1955 Sixth Congress of the IAF in Copenhagen.

After an introductory discussion on the evolution of rockets, three American proponents of "man in space" addressed different aspects of manned space flight. Willy Ley described the prospects for utilizing rockets in space travel and the steps required to build a space station that could orbit 1730 kilometers above the earth. Through the medium of an animated cartoon character, "Homo Sapiens Extra-Terrestrialis," Heinz Haber explained some of the questions raised by "space medicine," illustrating the physiological hazards—acceleration loads, weightlessness, cosmic radiation, meteorites—that the first space travelers would encounter. Finally, Wernher von Braun closed the film with a discussion of his conceptual design for a 55-meter tall, 1280-metric ton, four-stage interplanetary rocket that could carry a crew of six into the cosmos.*[2] The IAF delegates were enthusiastic about this 33-minute movie, especially in the light of President Eisenhower's earlier announcement that the United States would launch artificial satellites during the International Geophysical Year.

Among the viewers of *Man in Space* were Leonid Ivanovich Sedov and Kyril Feodorovich Ogorodnikov, the first Soviets to attend an IAF Congress. They spoke with Durant about borrowing the film for use in the Soviet Union, saying it would be "very good to have here a copy of Walt Disney's

---

*In 1955, Ley was a writer of factual science publications centering on rocketry and space exploration; Haber was a member of the physics department at UCLA, after having worked five years as a research scientist with the Air Force School of Aviation Medicine; and von Braun was Chief of the Guided Missile Development Division at the Army's Redstone Arsenal.

excellent film for private demonstration."[3] It is likely that the Soviets viewed *Man in Space* as proof of growing American interest in solving the basic problems associated with manned space flight. Sedov and Ogorodnikov wanted to use the Disney picture to promote their own nation's efforts in rocketry and space research. To Soviet space enthusiasts, the movie was at once an encouragement and a warning.

Seven years after that Copenhagen meeting, both the U.S.S.R. and the U.S. orbited and returned their first space pilots. Vostok and Mercury were possible within such a short span of time because engineers and scientists had amassed a wealth of basic engineering and scientific data directly applicable to the questions posed by manned space flight. In those early years, much of the work was duplicative, as security restrictions forced Soviet and American researchers to repeat the same fundamental investigations; but if the competitive environment was wasteful, it also spurred the development of space flight technology. Seemingly, man would have crossed the barriers of the space frontier without the element of international competition, but it was precisely that element that did give rise to the space program—and made

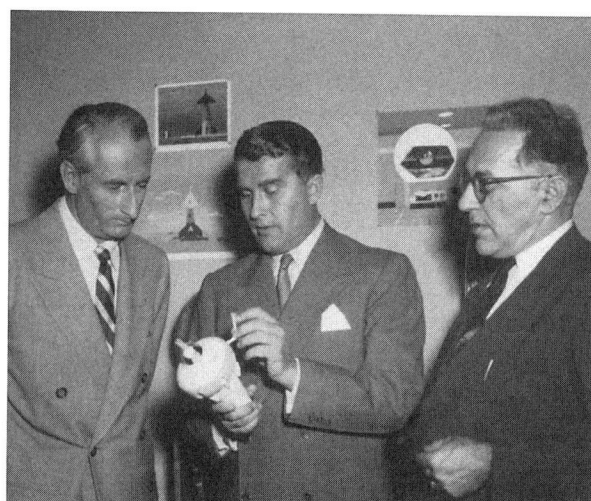

Heinz Haber, Wernher von Braun, and Willy Ley examine a prop from the Disney movie Man in Space *(©Walt Disney Productions).*

Wernher von Braun points to the final stage of the manned spacecraft he described in the movie Man in Space *(© Walt Disney Productions).*

funds available. Fantasy yielded to reality; and that reality was the orbiting hardware.

THE CHALLENGE OF SPACE FLIGHT

Vostok and Mercury were first steps, designed to explore the concept of manned space flight. Maxime A. Faget, chief designer of the Mercury spacecraft, summarized their importance:

> Since these flights were initial efforts, the purpose of the flights was limited to the basic experience of launching the spacecraft and crew into orbit, having them remain there for a period of time, and then having them return safely to earth. These flights were made at low altitude with the spacecraft barely high enough to avoid appreciable drag from the upper fringes of the atmosphere. . . . the amount of energy required for launching was minimized, and the flight was made safer, since the difficulty of making a reentry maneuver was also minimized. . . . these flights . . . proved that it was practical for man to fly in space.[4]

While providing valuable lessons in the design and operation of spacecraft, Vostok and Mercury also demonstrated two different approaches to accomplishing the same tasks.

The rapid onset of multi-gravity forces accompanying the rocket launch was one of the primary concerns that faced the two technical teams. During the powered ascent from earth, crewmembers had to be protected from the increased "g-loads," vibration, and noise. It was known, from aircraft and centrifuge experiments, that human tolerance to increased gravity forces varied with the duration of exposure and the attitude of the body with reference to the force. Soviets and Americans agreed that the reclining position permitted a pilot to absorb heavy acceleration loads more comfortably than in any other posture.* In the U.S., Faget, William M. Bland, Jr., Jack C. Heberlig, and their engineering colleagues decided in favor of a couch contoured to the form of each individual astronaut to protect him from g-loads. Soviet designers also used the form fitting couch, and all Mercury and Vostok pilots rode semi-supine in their own tailor-made seats.[5]

Once a pilot overcame the initial acceleration forces of flight, he would encounter the phenomenon of gravity balanced by centrifugal force, generally called weightlessness or zero g. Flight physicians contended that the absence of gravity might affect man's physical and mental performance, but in the

---

*Experiments with rocket sleds and the centrifuge indicated that pilots could endure forces 20 times that of the earth's gravity, a load well in excess of those anticipated in normal flights and above those expected under emergency conditions.

face of limited information, the effect of zero g was mainly a topic of speculation. Some medical doctors wondered if the human organism, tailored to earth's gravity, would continue to function normally when suddenly deprived of that force. Other physicians worried about the reaction of particular internal organs to the succession of changes imposed by acceleration, weightlessness, and deceleration. Heinz Haber and Otto Gauer, who had studied the question of weightlessness in Germany, had concluded that more experimental data were needed to permit a better analysis of the role of zero g in manned space flight.[6]

Since it was impossible to duplicate weightlessness on earth, scientists conducted tests with animals borne aloft by rockets. In the U.S. in 1947, experimenters began launching live organisms with V-2 rockets. On 20 September 1951, the monkey Yorick and 11 mice were recovered after an Aerobee flight to 72 kilometers. From this and two subsequent Aerobee monkey launches, James P. Henry and David G. Simon concluded that weightlessness and acceleration forces did not adversely affect the animals.[*7]

Soviet rocket engineers and physicians also sent animals to high altitudes, and their canine experiments led them to the same conclusions that the Americans had reached with primates and rodents. At first, the Soviet tests were conducted using pressurized capsules; then they experimented with dogs wearing special space suits and traveling in unpressurized cabins. In one case, Albina and Tsyganka were ejected from the descending launch vehicle at an altitude of 85 kilometers; both dogs rode safely back to earth in their space suits and ejection seats.[†] These experiments convinced the Soviets that acceleration and weightlessness did not pose impossible barriers to manned flight. The significance of this conclusion was made clear to the rest of the world when the Soviets sent Laika into orbit with *Sputnik II* on 23 November 1957. Although she was not returned to earth, Laika ate, barked, and moved about in her space cabin for seven days without apparent ill effects.[8]

LIFE SUPPORT SYSTEMS

But there were other questions raised by the unknowns of space environment. Man in space would be absolutely dependent upon an artificial en-

---

[*]Both physicians played subsequent roles in aerospace medicine. Henry became director of the animal program in Project Mercury. Simon went on to pilot a Project Man High balloon to 31 kilometers for a 32-hour study of man's performance in near space in Aug. 1957.

[†]Albina and Tsyganka were veteran travelers and members of the first group of nine canine cosmonauts. Subsequently, the Soviet scientists trained eight more dogs for experimental flight and landing by parachute.

vironment. One Soviet author described the life support system as "a set of engineering, physical-chemical and medical-biological resources" that will "satisfy the needs of man for oxygen, food and water" in order to create "normal living conditions for man in a flight vehicle."[9] The atmosphere on earth is a mixture of 80 percent nitrogen and 20 percent oxygen, with small quantities of water vapor and carbon dioxide, plus traces of other gases. Since the astronaut would continually breathe oxygen and generate carbon dioxide and water vapor, the spacecraft needed devices to replenish oxygen and to eliminate excess carbon dioxide. While both the Soviet and American engineers removed carbon dioxide and humidity by using lithium hydroxide canisters, they approached the problem of oxygen supply differently.

The Soviets decided upon a cabin pressure equal to about one atmosphere (760 millimeters of mercury [mm Hg]) and an 80/20 nitrogen-oxygen composition, which would be essentially the same as on earth. The Americans adopted a cabin pressure of 258 mm Hg, or the equivalent of approximately 1/3 atmosphere, and elected to use a pure oxygen environment. While the Soviet system had the advantage of simplicity and minimal danger from fire (always present with oxygen), it had the disadvantage of exposing the cosmonaut to potential decompression should he have to switch to his space suit life support system in an emergency. American cabin and suit pressures were similar, so that a switch from cabin to suit system oxygen would not subject the crew to the "bends." Astronauts were required to prebreathe oxygen prior to launch to remove the nitrogen from their blood streams, reducing the possibility of decompression sickness, or aeroembolism. This absence of nitrogen in the atmosphere also generated the requirement for flameproofing all materials used in the cabin.*[10]

Soviet and American technicians also differed in the manner by which they replenished spacecraft oxygen. There are three ways to store oxygen—as a high-pressure gas; as a cryogenic fluid; or as a solid, chemically combined with other elements. Storage as a gas requires strong, high-pressure tanks, which are heavier than the oxygen with which they are filled. Liquid oxygen can be stored in lighter and smaller tanks than those required for gaseous oxygen, but it must be kept very cold, below 90 kelvins (−297° F); this would require special thermal insulation. Chemical systems that release oxygen upon contact with carbon dioxide and water vapor have three drawbacks—weight; volume; and variable performance, based upon a number of

---

*Since life has evolved in the 80/20 nitrogen-oxygen atmosphere, over long periods of time breathing undiluted oxygen at sea level pressure (760 mm Hg) can be toxic. Toxicity diminishes as the pressure is reduced, and when pure oxygen is breathed at a pressure approximating the partial pressure of oxygen at sea level (181 mm Hg), there are no detectable adverse effects. For this reason, American engineers chose 258-mm-Hg pressure for use in Mercury, Gemini, and Apollo spacecraft.

factors, such as the crewman's metabolic rate, cabin temperature, and humidity.[11]

To replenish cabin oxygen, Soviet environmental control system designers selected a "chemical bed" system based upon alkali metal superoxides, which liberate oxygen as they absorb moisture and form more alkali, which in turn absorbs carbon dioxide. Despite the lack of precision control and the amount of space required for the apparatus, the Soviets favored the chemical bed because it eliminated the problems encountered with high-pressure bottles for gas and the precise temperature controls required for liquid oxygen. In the U.S., John F. Yardley, John R. Barton, Richard S. Johnston, and Faget were successful in arguing for pure oxygen atmosphere at a pressure of 258 mm Hg, since it met the weight and volumetric requirements imposed by the design limitations of the Mercury spacecraft. Although the development of spherical pressure bottles for gaseous oxygen was a challenge, the American designers felt that the effort was justified by reliability.[12] A key goal of Project Mercury engineering was reliability, to be established through use of proven concepts, redundant systems, and extensive testing. Soviet and American engineers selected an environmental control system that satisfied their respective design goals and criteria for reliability.

REENTRY VEHICLES: SPHERES VS. BLUNT BODIES

The choice of reentry vehicle configuration reflected additional differences in approach. The central and most visible difference between the Vostok and Mercury spacecraft was their external configuration. Beneath the streamlined launch shroud, the orbital/reentry portion of Vostok was spherical, while the basic shape of Mercury was a truncated cone. The spacecraft designers studied the alternative shapes for reentry vehicles and made their choices based upon standards established within their own programs.

The Soviets, under the leadership of Sergei Pavlovich Korolev, chief designer of spacecraft, reviewed the different possibilities and chose the sphere for their reentry configuration. According to Korolev, among non-lifting shapes the spherical reentry body alone possessed an inherent dynamic stability as it plunged back into the earth's atmosphere. He rejected the conical craft, because its tendency to pitch and yaw would have required an elaborate attitude control system, plus greater reliance upon man as pilot rather than man as passenger.*

---

*The role of man in space flight has been one of the basic and continuing philosophical differences between the Soviet and American space programs. Americans have sought to make the astronaut a central figure in the operation of the spacecraft, especially in his ability to veto automatic systems. The Soviets have preferred to rely upon automated systems on the ground and in the air, with the cosmonaut playing a secondary and more limited role.

The orbital configuration of Vostok consisted of a spherical cabin with an attached equipment cluster.[*][13] Prior to descent, the spacecraft was oriented for reentry by means of a solar sensor located in the equipment compartment. This maneuver aimed the retrorockets so that they fired along the line of flight, slowing the craft as it entered its descent trajectory. Upon termination of retrofire, the cabin separated from the instrument section, which subsequently burned up as it entered the atmosphere. Vostok was then a simple sphere, descending along a ballistic trajectory, protected from the intense reentry temperatures by an ablative coating that shielded the entire craft.[†][14]

Vostok reentered like a bullet, following the path dictated by the retrorocket impulse; there was no attitude control. By placing the sphere's center of gravity behind and below the cosmonaut, the spacecraft designers assured Vostok pilots from Gagarin to Bykovskiy and Tereshkova the proper orientation for ejection from the "lander" when it reached 7000 meters. At that altitude, the bolts securing the pilot's hatch were severed explosively, and the hatch was blown away. Two seconds later the cosmonaut and his couch were ejected from the craft to begin a parachuted descent to 4000 meters.[‡] At that height, the cosmonaut continued his return by means of his own parachute. Also at 4000 meters, a parachute opened to slow the final descent of the spacecraft.[15]

In their study of reentry, the Americans evolved their own theories regarding optimum spacecraft configuration. In June 1952, H. Julian Allen of the NACA Ames Aeronautical Laboratory addressed the problem of structural heating during atmospheric reentry. His research led to the formulation of the "blunt-body principle," a radical departure from the streamlined aircraft of the early fifties. Allen's work indicated that a blunt shape would be most suitable for a body reentering the earth's atmosphere, since 90 percent of the friction heat would be dissipated through the bow shock wave. Tests five years later, in 1957, with a scale model Jupiter-C nosecone demon-

---

[*]K.P. Feoktistov, who had prime responsibility for design details of Vostok, described the two sections as "a recoverable capsule (accommodating the spaceman and his life-support equipment, flight controls, communication, on-board systems controls and landing controls) and an instrument compartment (housing various instruments and units of spaceship systems controlling orbital flights, communications, telemetering measurements, orbit parameters, power supply, etc.); that is, all that contributed to orbital flight alone."

[†]Hartley A. Soulé recalls that in American circles the spherical "shape was specifically criticized because the weight of the material to completely shield the surface from reentry heat would [have precluded] launching with programmed ICBM boosters." The Soviets had the launch vehicle capability that kept this extra weight from being such a serious concern. Some American designers favored the spherical shape to reduce the problems associated with attitude control, but others feared that "the lack of orientation might result in harm to the occupant during the deceleration period."

[‡]According to one source, this delay was incorporated after the loss of a pilot who was testing the ejection seat system during a drop test of the Vostok.

strated that the remaining heat could be dissipated through use of an ablative coating on a heatshield. Although his studies were directed toward resolving the nosecone reentry problem of the ballistic missile, they were later applicable to the Mercury spacecraft. During the ensuing years, heat-resistant materials of the ablative and heat sink types were perfected by government and industry.

Beginning in 1954 and continuing through 1958, Allen and two associates, Alfred J. Eggers, Jr., and Stanford E. Neice, examined the relative merits of three types of hypersonic spacecraft—ballistic, skip, and glide. They prepared in early 1954 a theoretical discussion of the alternative configurations that could be used for manned spacecraft, "A Comparative Analysis of the Performance of Long-Range Hypervelocity Vehicles." For manned satellite missions, any of the three craft could be boosted to orbital velocity by a rocket and then be separated from the launch vehicle for either free flight or earth orbit. The skip vehicle, which would reenter the atmosphere by an intricate series of dips and skips, would require the greatest boost capacity, and would encounter excessive aerodynamic heating during reentry. The glider-type craft, although heavy, would require a smaller boost capacity and would have a greater degree of pilot control during the reentry phase of the mission; the glider was a promising concept, but it would also be a long term project, since it would require extensive engineering and development. The third option was the ballistic shape, which was simply a blunt, non-lifting, high-drag projectile. Although without aerodynamic controls, its blunt configuration would provide superior thermal protection to the pilot, and its lighter weight would permit longer range missions. Moreover, the deceleration forces would be minimized if the vehicle reentered at the correct angle. The Ames researchers concluded that "the ballistic vehicle appears to be a practical man-carrying machine, provided extreme care is exercised in supporting the man during atmospheric entry."[16]

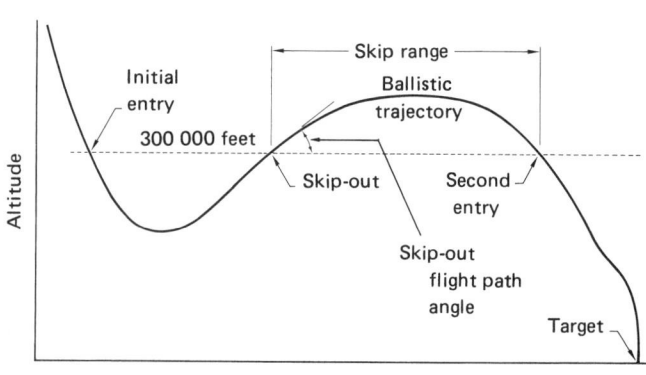

A 1963 sketch illustrating a possible skip reentry trajectory of the Apollo spacecraft.

As time passed, Eggers became convinced of the superiority of the manned satellite glider over the ballistic satellite, but he also knew that the rockets then on the American drawing boards could not put the glider into orbit. He had two concerns when he thought of using the ballistic vehicle—the deceleration loads and the absence of control once the craft entered the atmosphere. The latter problem dictated a large landing area, perhaps as much as several thousand square kilometers. By late 1957 Eggers was proposing a semi-ballistic vehicle in which the best elements of the glider and the ballistic shapes were combined. Further progress on manned spacecraft was influenced by the Air Force and by research in progress at the Langley Memorial Aeronautical Laboratory.[17]

On 29-31 January 1958, the Air Research and Development Command held a closed conference at Wright-Patterson Air Force Base, during which 11 aircraft and missile firms outlined for Air Force and NACA representatives their classified proposals for manned satellites. These variations on the three basic configurations discussed previously ranged in projected weight from 454 to 8165 kilograms and involved mainly the use of multistage launch vehicles. Since there was such a difference in technology among the various proposals, the estimated development time ranged from one to five years. Looking back on this period, Robert R. Gilruth recalls:*

> Because of its great simplicity, the non-lifting, ballistic-type of vehicle was the front runner of all proposed manned satellites, in my judgment. There were many variations of this and other concepts under study by both government and industry groups at that time. The choice involved considerations of weight, launch vehicle, reentry body design, and to be honest, gut feelings. Some people felt that man-in-space was only a stunt. The ballistic approach, in particular, was under fire since it was such a radical departure from the airplane. It was called by its opponents "the man in the can," and the pilot was termed only a "medical specimen." Others thought it was just too undignified a way to fly.[18]

While subject to considerable criticism, the concept of a simple ballistic manned satellite gained important support from a group of NACA engineers who started work on just such a spacecraft, borrowing on the experience and technology available in recent research on nosecones for intercontinental ballistic missiles. Max Faget was one of the key members of the NACA group interested in this effort. In January 1958, he had identified himself as a supporter of the ballistic reentry vehicle when he proposed to NACA Headquarters that a non-lifting spherical capsule be considered for orbital flight.

---

*Robert R. Gilruth had been Assistant Director of the Langley Aeronautical Laboratory since 1952 and was named Manager of the Space Task Group, which was assigned responsibility for Project Mercury on 5 Nov. 1958.

NACA expressed little interest in the idea, but Faget continued his studies of ballistic vehicles and spoke out for adoption of this concept when occasions arose. Less than a week after an Air Force man-in-space conference in March 1958,* Gilruth called Faget and a group of top Langley engineers together to discuss a NACA conference on high speed aerodynamics, scheduled to begin at the Ames laboratory on 18 March. The "Langley position" that emerged from the conference reflected the thinking of Faget and his colleagues on a ballistic spacecraft launched by a ballistic missile booster.[19]

The Ames conference was the last in a series of formal symposia; as such it attracted nearly 500 people from NACA, the military, and the aircraft and missile industry. The 46 papers presented during the three-day meeting summarized the most advanced aerodynamic thinking within the Advisory Committee's laboratories on hypersonic, orbital, and interplanetary flight. Faget presented the first paper, "Preliminary Studies of Manned Satellites—Wingless Configuration: Non-lifting," in which he and his co-authors pointed out the inherent advantages of the ballistic approach. First, ballistic missile research, development, and production experience was directly applicable to this type of spacecraft. Equally significant, the choice of a ballistic flight trajectory minimized the amount of automatic stabilization, guidance, and control equipment required on board the craft, thus saving critical weight and reducing the chance of equipment malfunction. Faget and his associates also demonstrated that their proposed craft could be returned from orbit by a modest-power retrorocket system. The Langley engineers went so far as to propose a specific ballistic configuration—a cone, 3.4 meters long and 2.1 meters in diameter, protected on the blunt end by a heatshield. He concluded that "as far as reentry and recovery is concerned, the state of the art is sufficiently advanced so that it is possible to proceed confidently with a manned satellite project based upon the ballistic reentry type of vehicle."[20]

The Mercury spacecraft grew out of this 1958 conceptual study prepared at Langley. After an additional two months of design studies, preliminary specifications for a manned satellite were drafted during June by Langley personnel under the supervision of Faget and Charles W. Mathews. Following a number of revisions and additions, these specifications were used for the Project Mercury spacecraft contract with McDonnell Aircraft

---

*The Air Force held a working conference on 10-12 Mar. at the Air Force Ballistic Missile Division, Los Angeles, in support of its program "Man in Space Soonest" (MISS). At that time, the Air Force concept consisted of three stages—a high-drag, no-lift, blunt-shaped spacecraft to get man in space soonest, with landing to be by parachute; a more sophisticated approach by possibly employing a lifting vehicle or one with a modified drag; and a long-range program that might end in a space station or a trip to the moon.

Corporation. All this work occurred during the months in which the National Aeronautics and Space Act was being drafted and enacted by Congress. Gilruth remembered working out of the old NACA building in Washington during the summer of 1958; it had been hot, humid, and busy.[21]

In designing the Mercury spacecraft, the key word was simplicity. The goal was a spacecraft that represented "the simplest and most reliable approach—one with a minimum of new developments and using a progressive buildup of tests." Employing these criteria, "It was implicit . . . that we use the drag-type reentry vehicle; an existing ICBM booster; a retrorocket to initiate descent from orbit; a parachute system for final approach and landing; and an escape system to permit the capsule to get away from a malfunctioning launch rocket."[22] Although Vostok and Mercury emerged from the design process with different external configurations, their designers had met the same problems and had made some remarkably similar decisions. Undoubtedly, the key decision was to keep the first step into space a simple one. While the Mercury space vehicle would become more complex and so-

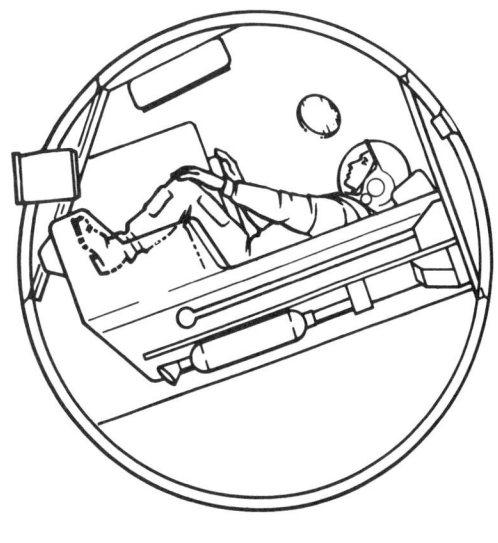

Comparative cutaway views of Mercury and Vostok spacecraft drawn to the same scale. Note ejection seat in the Soviet craft.

Mercury

Vostok

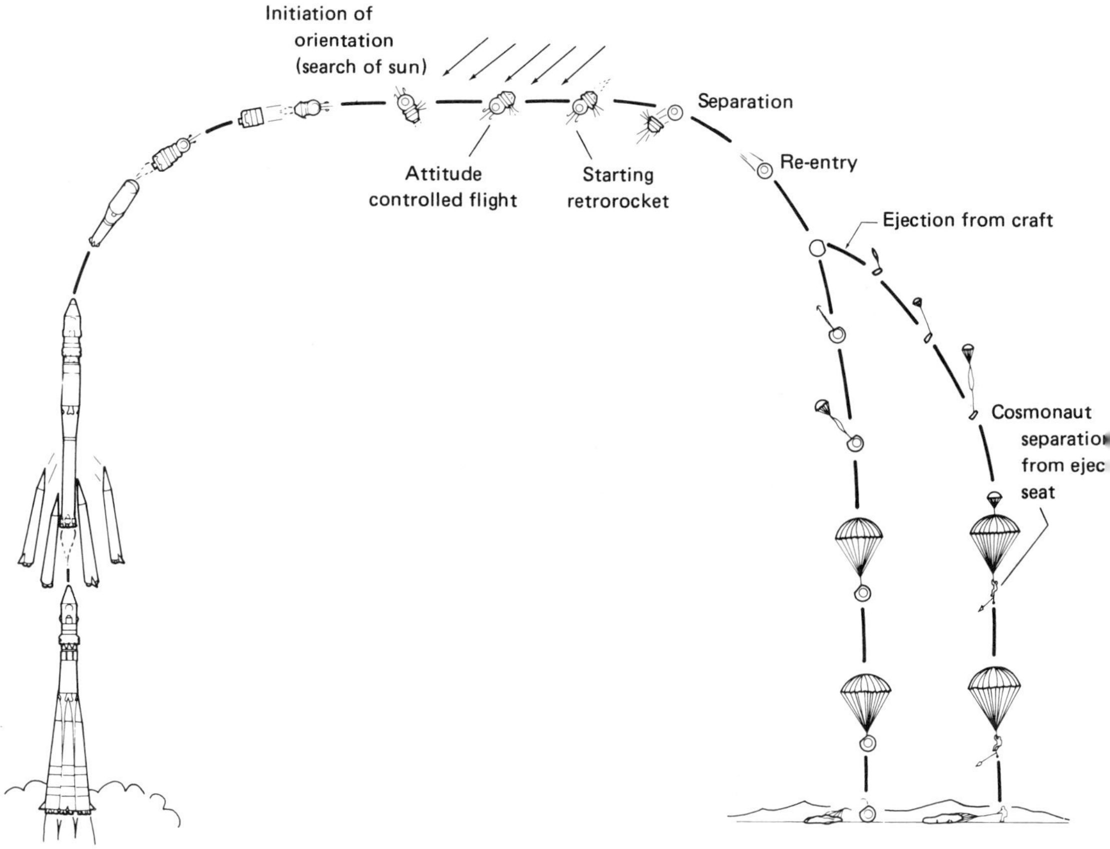

Typical mission profile for Vostok flights.

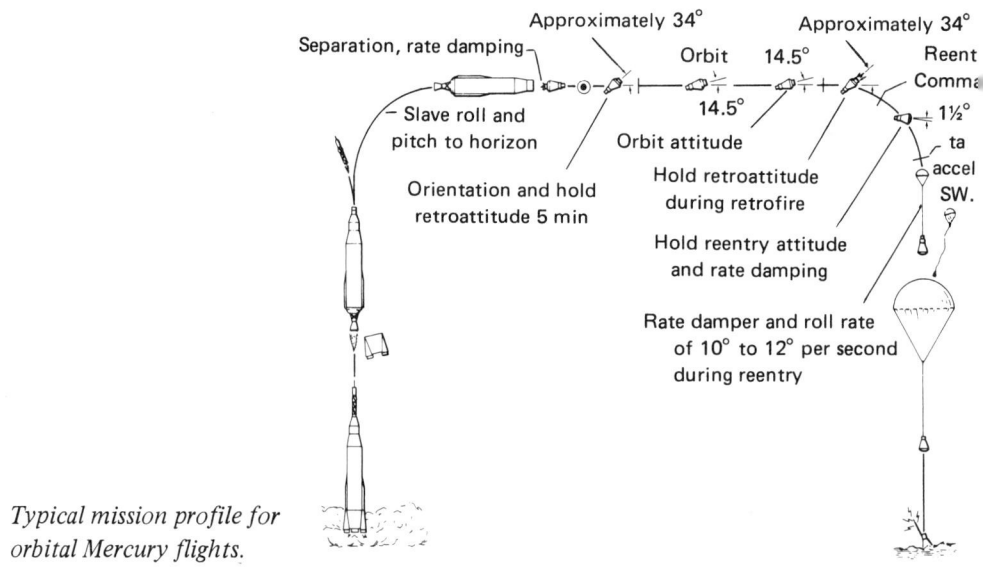

Typical mission profile for orbital Mercury flights.

phisticated during the developmental process, the emphasis on reliability and relative simplicity remained.

VOSTOK AND MERCURY: FIRST FLIGHTS INTO SPACE

The years 1958-1961 were busy ones in both the United States and the Soviet Union for the development of manned space vehicles.* According to Konstantin Petrovich Feoktistov, the details for mockups and breadboard models of Vostok were worked out and then built during 1959. Final developmental work on the "carrier rocket" was being conducted simultaneously at the launch site.[23] By 15 May 1960, the Soviets had progressed sufficiently with the development of their spacecraft and the adaptation of their ICBM boosters as launch vehicles to commence a series of five unmanned test flights. These Vostok precursor flights, *Korabl Sputnik I* through *V,* were designed to collect additional data on the effects of space environment (especially solar radiation) on biological specimens and to test the spacecraft systems. The flights included no unforeseen physiological problems associated with manned space missions, but the first and third spaceships did encounter trouble upon reentry. The problem centered on the proper orientation for retroengine firing, a difficulty that was worked out by the time the fourth and fifth test missions were flown in March 1961. Feoktistov indicated that a round of technical discussions led to major changes in the spacecraft during September-December 1960. "In late 1960-early 1961 the revised technical documentation was used for the manufacture of the spaceships." The Soviets were ready to begin manned space flight operations.[24]

The rationale of the six Vostok flights has been summarized by design engineer and cosmonaut Feoktistov. Yuri Gagarin's flight on 12 April 1961 was a single-orbit checkout of the spacecraft systems. Rather than the ballistic shots used at first by the Americans, the Soviets preferred an orbital mission to collect additional data on weightlessness, a topic of considerable concern to Soviet flight surgeons. Indeed, for the second mission, flown by Gherman Titov on 6 August 1961, the medical specialists had urged that the duration be held to just two or three orbits so they could judge the effects of zero gravity, but the designers and Titov wanted to go for a day-long mission, a goal that coincided with political considerations as well. Feoktistov later hinted that the one year hiatus in manned flight following *Vostok II* may have been related to the motion sickness experienced by Titov. Andriyan Grigoryevich Nikolayev and Pavel Romanovich Popovich in *Vostok III*

---

*Appendix B lists the major Soviet and American developmental (unmanned) and manned flights.

At left, Cosmonaut B. V. Volynov examines radio transmitter, while unidentified comrade reaches inside Vostok spacecraft during winter training exercises (Tass from Sovfoto). At right, Charles J. Donlan, Assistant Director, Project Mercury (left); Robert R. Gilruth, Director, Project Mercury; and Maxime A. Faget, Chief, Flight Systems Division, stand in front of recovered Mercury-Redstone 1A unmanned spacecraft after its recovery 19 December 1960.

and *IV* completed their dual mission in August 1962. Though they did not actually rendezvous, they appear to have been within 5 kilometers of each other, thus giving the Soviet trajectory specialists an opportunity to study the problem of rendezvous and to track two spacecraft simultaneously. In June 1963, Valeriy Fedorovich Bykovskiy's flight aboard *Vostok V* lasted nearly five days, and during the last three days he was accompanied in orbit by *Vostok VI,* piloted by Valentina Vladimorovna Tereshkova, the only woman to fly in space to date. Vostok was the "necessary foundation for . . . further development of manned space vehicles in the Soviet Union."[25]

While the Soviet cosmonauts were monopolizing world headlines, work on Project Mercury continued. NASA had embarked upon a step-by-step program of spacecraft and booster qualification trials in 1959. The test program, divided into two parts, sought first to qualify the Redstone and Atlas missiles as launch vehicles for manned spacecraft, and second to "man rate" the Mercury spacecraft itself. The Mercury-Redstone phase of the program covered a 31-month period, during which six missions were flown. The results were mixed. On the very first launch attempt (MR-1), early separation of an electrical ground line to the booster aborted the mission. On the second flight (MR-1A), all systems worked satisfactorily, but problems again appeared in the primate "Ham" mission (MR-2), when over-acceleration caused a higher trajectory and longer downrange travel than had been anticipated. As a consequence, an extra flight was scheduled before a manned launch was attempted. Then on 5 May 1961, less than a month after the Gagarin mission, Alan Shepard became the first American in space, flying a suborbital trajectory in his spacecraft *Freedom 7.* Gus Grissom in *Liberty Bell 7* made the second suborbital flight on 21 July. The data gathered from these two successful missions justified canceling the remaining Mercury-Redstone flights.

Then came the step to orbital flight for which the Atlas missile had been selected as the launch vehicle. When the program was approved in October 1958, no other booster could have been chosen if the objectives of the program were to be accomplished in a reasonable length of time. So, as had the Soviets, the Americans decided to "man rate" an intercontinental ballistic missile. The 57-month flight phase for Atlas began with the launch of the Big Joe Atlas with a boilerplate model of the Mercury spacecraft on 9 September 1959. The first production spacecraft mounted on an Atlas launch vehicle (MA-1) was launched on 29 July 1960. After about 60 seconds, launch vehicle and adapter failed structurally. Because no spacecraft escape system was used, the spacecraft was destroyed upon impact. Follow-

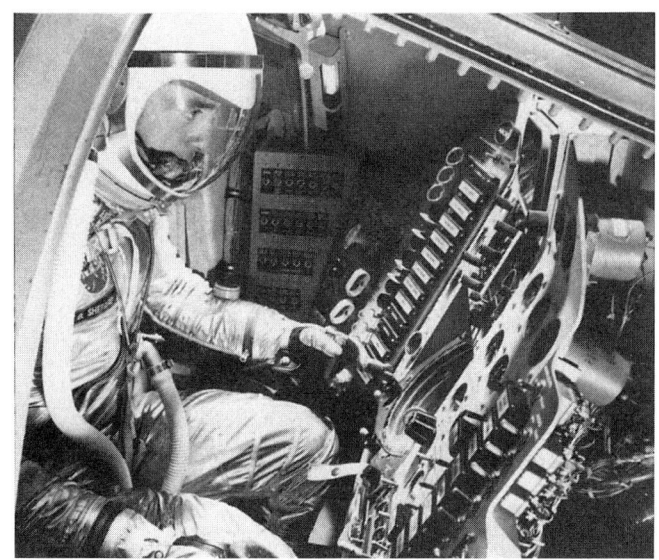

First American into space, Alan Shepard, practices for his suborbital mission in the Mercury procedures simulator.

ing an intensive seven-month investigation, modifications were introduced to stiffen the adapter between the launch vehicle and the spacecraft and to otherwise improve the structural integrity of the entire upper part of the Atlas. An interim version of the alteration was used without difficulty on the flight of MA-2, while the final version was not tested until the unmanned orbital flight of MA-4 on 13 September 1961. More than two months later on 29 November, Enos, a trained chimpanzee, was launched on a planned three-orbit mission. During the flight of MA-5, the attitude control system performed abnormally, and ground control brought the spacecraft down after two orbits. The problem, as demonstrated on later flights, could have been corrected by an astronaut, thus confirming the American judgment favoring manual overrides of automatic control systems.

After a series of frustrating delays caused by unfavorable weather and fuel leaks, John Glenn became the first American to orbit the earth. His flight was followed by a three-orbit mission flown by M. Scott Carpenter, in which the only problem was an attitude misalignment at the time of retrofire, causing a 402-kilometer landing overshoot. As the next step toward a day-long mission, Walter M. Schirra piloted a six-orbit mission on 3 October 1962. By drifting in flight, he conserved critical fuel and demonstrated the feasibility of longer duration missions. The 34-plus-hour mission of L. Gordon Cooper on 15-16 May 1963 was Project Mercury's last flight.[26]

Mercury and Vostok demonstrated the feasibility of placing a human being in orbit, observing his reactions to the space environment, and returning him safely to earth at a known point. While the Soviet designers assigned limited tasks to their cosmonauts, NASA went one step beyond to demonstrate that man could function as "an invaluable part of the space flight systems as pilot, engineer and experimenter." The next stage was the development of more flexible and multi-place spacecraft for the conduct of more intricate missions—the era of Gemini and Voskhod.

VOSKHOD AND GEMINI: INTERMEDIATE STEP

Even as the Vostok and Mercury programs were entering their operational phases, engineers in the U.S. and the U.S.S.R. were undertaking the design of a second generation manned spacecraft. The Americans began with an effort to extend the capabilities of the Mercury craft, the so-called Mark II version, and ended up designing an essentially new two-man vehicle capable of greater maneuverability, rendezvous and docking, and flights of a duration that would equal the period anticipated for the lunar mission of Project Apollo. The Soviets, apparently spurred by the goals set for Project Gemini, decided to modify their Vostok spacecraft for multi-man flights. Where Voskhod was an attempt to exploit more fully a tested design, Gemini became geared to the creation of new systems and to the testing of unproven flight concepts that would be applied to even bolder missions in the future.

By December 1961, Project Gemini received formal approval from Washington as the second major project in NASA's manned space program; however, much of the design work had been done and many of the major decisions had already been made.[27] The character of the new effort was shaped by two converging lines of thought. The most influential consideration was President Kennedy's decision in May that committed the U.S. to a manned lunar expedition before the end of the 1960s. NASA advance planners had been thinking about a mission to the moon, but in the time frame of the 1970s, dependent upon the development of a new, larger launch vehicle called Nova. This rocket would be capable of lifting a spacecraft that could fly directly to the moon, land, and then return to earth. This method of reaching the moon—called direct ascent—was readily accepted because it would almost certainly work. However, within NASA there was a group of engineers who supported the development of an alternative route involving the orbital rendezvous of two or more spacecraft.

John C. Houbolt of Langley and a group of his associates felt that orbital rendezvous promised significant savings in fuel, weight, and time, especially if it were done in lunar orbit rather than earth orbit. A lunar expedition based upon the rendezvous concept might be assembled with much smaller rockets than a direct mission would need, launch vehicles that could be available well before Nova. Orbital rendezvous had the disadvantage, however, of being a new and untested idea. No one could predict how difficult or hazardous a rendezvous and linkup in space might be. As long as there was no pressing deadline for a lunar mission, direct ascent offered the easier and safer approach, but with the Presidential creation of a specific timetable, the supporters of rendezvous could press their case for a quicker and cheaper path to the moon. The idea still had to be tried to determine its feasibility, and "Gemini was first and foremost a project to develop and prove equipment and techniques for rendezvous."[28]

Project Gemini was also influenced by a second important consideration, the desire to make a major jump in the state of spacecraft technology. The engineers who had worked on Mercury had seen a number of possible improvements that could have been used if they had not been held back by a combination of considerations—weight, time, and the desire to keep the first spacecraft simple. While the Mercury designers had justifiably been preoccupied with solving the basic problems of manned space flight, it had taken too long to build and check out the handcrafted spaceship. James A. Chamberlin, chief designer of Gemini, described the difficulties in Mercury brought about by numerous design constraints:

> Most system components were in the pilot's cabin; and often, to pack them in
> this very confined space, they had to be stacked like a layer cake and com-

ponents of one system had to be scattered about the craft to use all available space. This arrangement generated a maze of interconnecting wires, tubing, and mechanical linkages. To replace one malfunctioning system, other systems had to be disturbed; and then, after the trouble had been corrected, the systems had to be checked out again.[29]

Chamberlin saw an opportunity to make Mercury Mark II, which became Gemini, a more easily assembled and serviced vehicle. He began by modularizing all systems and assembling the components of each system into compact packages, which were so placed that any system could be removed without tampering with another. Simultaneously, he sought to arrange most of the packages on the outside walls of the pressurized cabin for easy access; this would also permit several technicians to work on different systems at the same time.

In an effort to eliminate some of the trouble spots identified in Mercury, Chamberlin simplified his systems wherever possible. He reduced the complexity of the relays that controlled the automatic systems on board the craft. The new design relied upon pilot control with automatic backup flight systems. The result was a much simpler machine. Another change was the elimination of the rocket-powered escape tower used in Mercury, cutting hundreds of kilograms of extra weight, numerous relays, and much complex wiring. This in part was made possible by the change from the liquid-fueled Atlas rocket to the less explosive, hypergolic-fueled Titan II.* Whereas safety required an automatic abort system to propel the pilot away from the highly explosive Atlas in a launch emergency, Chamberlin could equip the new spacecraft with pilot-actuated ejection seats.

The year 1961 was a creative one for Gemini. It began with discussions at Langley in January and continued with the March Wallops Island talks regarding post-Mercury possibilities for manned space flight.[†] By mid-1961, the desire for an advanced technology spacecraft and the Presidential decision to press forward with the Apollo lunar program had led to a concrete proposal for a new spacecraft.

> Project Gemini owed its origins to its predecessor—it built on the technology and experience of Project Mercury—and to its successor—it derived its chief justification from Project Apollo's concerns. The new project acquired other objectives as well: testing of the concept of controlled landing, determining the effects of lengthy stays in space, and training ground and flight crews.[30]

*Hypergolic fuel ignites spontaneously upon contact with its oxidizer, thereby eliminating the need for an ignition system, as well as being less dangerous in some emergency situations.

†In attendance were Abe Silverstein, Robert Gilruth, George Low, James Chamberlin, Walter Williams, Paul Purser, Maxime Faget, Charles Mathews, and Charles Donlan.

ROUTES TO SPACE FLIGHT

With the creation of a Gemini Project Office at the Manned Spacecraft Center in Houston, the program moved into its development phase.

Throughout the development period, 1962-1963, Gemini engineers and managers worked to solve technical problems and to meet a tight budget. "Within NASA and without, Apollo and the trip to the moon always held center stage."[31] Toward the end of 1963, the first Gemini launch vehicle and spacecraft were being prepared for qualifying trials. Early April 1964 saw the first of Gemini's 12 flights, an unmanned test of the spacecraft and booster which produced excellent results. Further test flights were postponed as hurricane season arrived on the Florida coast. Meanwhile, the Soviets had launched their first multi-place spacecraft.

When given the assignment to place three cosmonauts into orbit in the same spaceship, designer Korolev set about to redesign Vostok.[32] Apparently, the most important consideration in his decision to modify an existing design rather than to create a new one was the boost capacity of the launch vehicles at his disposal. From the fragmentary details available, it appears that by 1963 Korolev and his colleague Leonid Aleksandrovich Voskresensky were already well along in the design work of an advanced spacecraft capable of long-duration earth orbital missions. This vehicle, which would later publicly emerge as Soyuz, was much heavier than Vostok, and the

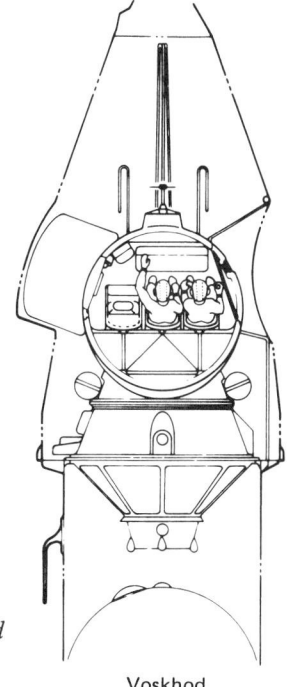
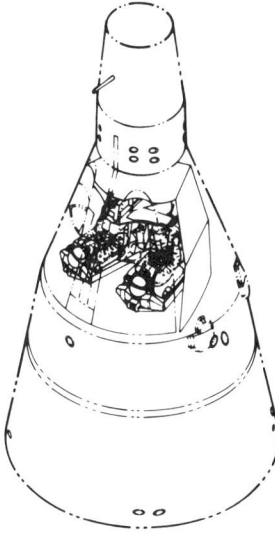

Simplified interior view of Voskhod and Gemini spacecraft.

Voskhod   Gemini

Soviets planned to launch it with the standard Vostok launch vehicle, plus a new and still untested upper stage that would provide the necessary additional thrust.[33] The evidence suggests that as this design work progressed, the Soviet political leadership grew concerned over the possibility that the U.S. would launch a two-man vehicle before the Soviets could.* In particular, Khrushchev wanted the Soviet multi-man space mission to come first to maintain the Soviet lead in space accomplishments.[34] Since Korolev could not hope to perfect his advanced spacecraft and improved launch vehicle in the time remaining before the first Gemini flight, he turned to the task of modifying Vostok to carry a three-man crew.

As he approached the task of altering the Vostok interior, Korolev had two problems of equal magnitude—how to make room for three persons, and how to keep the weight of the completed vehicle as close to that of the original as possible. He first eliminated the ejection seat. This change saved weight and made it possible to accommodate three form-fitting couches. To make room for the crew, Korolev planned to have the Voskhod cosmonauts fly in a "shirt sleeve environment." The Soviet designer could risk eliminating space suits since he and his staff had created a virtually leakproof spaceship.† Removal of the ejection apparatus would force the crew to ride to earth in the spacecraft, thus necessitating the development of a "soft-landing" system. Korolev attacked this problem by adding two pieces of equipment, a second parachute to supplement the one previously used to slow the Vostok reentry sphere and a rocket-powered landing apparatus in the parachute shroud lines that would reduce the craft's velocity to less than one meter per second at touchdown.[35]

There appears to have been a number of unsuccessful trials with the soft landing system, including some tests in which monkeys were killed. According to an official Soviet publication, "At Korolev's instructions, a series of Voskhod-type spacecraft were launched, until he was convinced that the soft-landing system worked impeccably."[36] This series included *Cosmos 47*, launched on 6 October 1964 and identified subsequently as an unmanned precursor to *Voskhod I*, which flew six days later.[37]

The flight of *Voskhod I* was another space spectacular for the U.S.S.R. On board were Command Pilot Vladimir Mikhaylovich Komarov; Dr. Boris

---

*According to the U.S. public announcement of Project Gemini made on 8 Dec. 1961, the first manned flight would occur in 1963-1964. At NASA, Deputy Administrator Dryden had "long expected the U.S.S.R. to make every effort to modify a Vostok, which is large enough to carry more than one man, to obtain an earlier flight" than those scheduled with Gemini.

†The design of the life support system had required a very tightly sealed spacecraft, because the 760-mm-Hg pressure represented the total volume of gas on board. The chemical replenishment system changed only the composition of the gases, not the volume. In the absence of a capacity to repressurize the cabin, the Soviets built and tested their craft to ensure that they were leak free.

Borisovich Yegorov, a medical doctor serving as flight physiologist; and the spacecraft engineer Feoktistov, who acted as an onboard technical scientist. The day-long mission, equivalent to three man-days for the life support system, was completed without reported difficulty. Toward the end of the flight, Komarov expressed the crew's eagerness to continue the flight for another day, but Korolev, quoting Shakespeare, replied, "There are more things in heaven and earth, Horatio," vetoing the request. Longer missions would come, but for the present it was best to adhere to the flight plan. On 13 October, the retrorockets fired, and the craft began its reentry.

> As on the *Vostok* flights, the spacecraft's parachutes opened at an altitude of 7 kilometers. When it came close to the ground, the soft-landing system automatically went into operation. Streams of gases, expelled from nozzles in the direction of the ground, reduced the touchdown velocity to virtually zero. The cosmonauts did not feel the impact.[38]

With the success of this first flight, Korolev and his associates were ready to fly again. Meanwhile, NASA was preparing for a second unmanned Gemini-Titan flight.

Both space teams were fully occupied during 1965. The second Gemini mission was launched from the Kennedy Space Center on 19 January. This suborbital qualification test of the spacecraft's structure, onboard systems, and reentry heat protection was a success, and the spacecraft was recovered two hours after splashdown. Just over a month later on 22 February, the Soviet launch crews sent aloft *Cosmos 57,* a rehearsal for *Voskhod II,* which flew on 18 March.[39] The two-man crew, Command Pilot Pavel Ivanovich Belyayev and Copilot Alexei Arkhipovich Leonov, completed a 26-hour mission, during which Leonov took the first extravehicular steps into space. The Soviets equipped *Voskhod II* with a special inflatable airlock, and Leonov, prior to entering it, prebreathed pure oxygen for over an hour to reduce the amount of nitrogen in his bloodstream and body tissues. After entry, he pressurized his space suit, checked it for leaks, adjusted his helmet, and tested the closed oxygen life support system; Belyayev then closed the hatch between the main cabin and the airlock. Following the gradual depressurization of his narrow compartment, Leonov stepped out into space.[40] This airlock arrangement resulted in a minimum reduction of the original cabin pressure, apparently necessitated by the lack of an onboard repressurization system. The Soviets continued to rely upon a chemical bed for generating oxygen, modified only to the extent required to support a second or third crewman.

Belayayev and Leonov had to land their spacecraft manually when the solar-orientation system malfunctioned. Reentry by means of the automatic-descent sequence and solar-orientation system, the technique used in

Soviet technicians complete checkout of Voskhod II spacecraft. Note the aerodynamic shroud that protects the reentry vehicle during launch (Novosti from Sovfoto).

all previous Soviet manned space flights, had been planned for the 17th orbit. When trouble was discovered, Belyayev asked permission to undertake a manual reentry on the 18th orbit. Korolev counted off the seconds until retrofire, and the command pilot fired the retrograde rockets high over Africa. *Voskhod II* overshot the recovery area and landed in a dense forest on the snow-covered slopes of the Ural Mountains. After hours of searching, helicopters dropped supplies to Belyayev and Leonov, who had to spend that night in the snow. Another day passed before the cosmonauts and their rescuers could be airlifted to safety.[41] While the U.S.S.R. celebrated the rescue of the crew and Leonov's 12-minute sortie into the void of space, the American team was preparing for the first manned Gemini flight.

On 23 March, Gus Grissom and John W. Young flew their spacecraft "Molly Brown" in a four-hour evaluation flight of the craft and launch vehicle.* Grissom and Young established a space-flight first by maneuvering in orbit. They employed the orbit attitude and maneuver system 90 minutes after launch for a precisely timed 75-second burn, which cut the spacecraft speed by 15 meters per second and dropped it into a nearly circular orbit.

> Three quarters of an hour later, during the second revolution, Grissom fired the system again, this time to test the ship's translational capability and shift the plane of its orbit by one-fiftieth of a degree. During the third pass, the pilot completed the fail-safe plan with a two and a half minute burn that dropped the spacecraft's perigee to 72 kilometers (45 miles) and ensured reentry even if the retrorockets failed to work.[42]

The retroengines did work, but there were still some surprises. At first all went well, but then "Molly Brown" seemed to be off course. The Gemini

---

*"Molly Brown," the "unsinkable" heroine of a Broadway stage hit, had seemed a logical choice for Grissom's second ship, as his Mercury *Liberty Bell 7* had sunk shortly after splashdown.

spacecraft produced far less lift than predicted, and as a consequence *Gemini 3* was about 84 kilometers short of its intended splashdown point. After a few nervous minutes, Navy swimmers arrived on the scene via helicopter to attach a flotation collar.

With the basic success of the first Gemini flight, the project gained momentum, permitting a routine launch nearly every other month throughout 1965 and 1966. There were difficulties, to be sure, but the simplified manufacture and checkout procedure permitted holding to this busy schedule. Beginning on 3 June 1965, James A. McDivitt and Edward H. White II conducted a four-day mission aboard *Gemini IV*.* This was the first long-duration flight, best remembered for White's 20-minute space walk, which added a new abbreviation to the public vocabulary—EVA (extravehicular activity). *Gemini IV*'s difficulties with a practice rendezvous meant that the next Gemini crew would be concerned with practicing that capability before the full-dress rendezvous experiment planned for the sixth mission. Andre J. Meyer, Jr., of the Gemini Project Office commented, "There is a good explanation on what went wrong with rendezvous. . . ." The crew and some of the flight planners "just didn't understand or reason out the orbital mechanics involved. . . ."

> Catching a target in orbit is a game played in a different ball park than chasing something down on Earth's essentially two-dimensional surface. Speed and motion in orbit do not conform to Earth-based habit, except at very close ranges. To catch something on the ground, one simply moves as quickly as possible in a straight line to the place where the object will be at the right time. As *Gemini IV* showed, that will not work in orbit. Adding speed also raises altitude, moving the spacecraft into higher orbit than its target. The paradoxical result is that the faster moving spacecraft has actually slowed relative to the target, since its orbital period, which is a direct function of its distance from the center of gravity, has also increased. As the *Gemini IV* crew observed, the target seemed to gradually pull in front of and away from the spacecraft. The proper technique is for the spacecraft to reduce its speed, dropping to a lower and thus shorter orbit, which will allow it to gain on the target. At the correct moment, a burst of speed lifts the spacecraft to the target's orbit close enough to the target to eliminate virtually all relative motion between them. Now on station, the paradoxical effects vanish, and the spacecraft can approach the target directly.[43]

Gemini V's first day in space was a worrisome one, during which a wire to a heater that pressurized the fuel cells was found to be faulty. The lowest

---

*After Gus Grissom had given the quasi-official name of "Molly Brown" to his spacecraft, NASA's top triumvirate, James Webb, Hugh Dryden, and Robert Seamans, Jr., decided that "all Gemini flights should use as official spacecraft nomenclature a single easily remembered and pronounced name. Consequently, the next mission will be called 'Gemini IV' and the code name will be 'Gemini.' "

pressure at which the fuel cell would function was determined after Gordon Cooper powered down the craft and consulted with the ground. But the rendezvous evaluation pod with which *Gemini V* was to maneuver had already been released and had drifted away, so the Gemini crew had to practice its rendezvous with coordinates radioed to them by Houston. Charles "Pete" Conrad, Jr., and Cooper would rendezvous with a phantom vehicle. The success of each "phantom rendezvous" made the Gemini flight planners more confident about the feasibility of bringing two manned spacecraft together. The next step was a rendezvous of Gemini with an Agena target vehicle.

But plans went awry when the Agena target vehicle exploded before going into orbit on 25 October 1965. The flight of *Gemini VI,* ready for launch with Walter M. Schirra, Jr., and Thomas P. Stafford, was postponed. Walter F. Burke and John F. Yardley of McDonnell Aircraft Corporation began to discuss a Gemini-to-Gemini rendezvous within minutes of the Agena failure. Three days of intensive deliberation led to a decision for a *Gemini VII/VIA* rendezvous mission. The two-shot mission was inspired by the concern that the Soviets might be planning similar flights, as well as by the desire to turn a minor defeat into a major accomplishment.

> That a plan of such scope could be suggested, thought about, decided upon, and announced in scarcely three days was a sign of the managerial and technical trust that Gemini had already come to inspire. William D. Moyers, the President's Press Secretary, told the news media about the plan and answered questions from reporters. Moyers said the mission was targeted for January but gave no specific date. Back at MSC, however, everyone from Gilruth on down was working toward an early December flight.[44]

After 38 days of extensive crew training and spacecraft preparation, the dual Gemini mission began on the afternoon of 4 December 1965. For 11 days, Frank Borman and James A. Lovell, Jr., aboard *Gemini VII* carried out their tests on the effects of long duration in space, especially the problems associated with personal hygiene and comfort. On the morning of 15 December, Schirra and Stafford were launched on the fifth manned Gemini flight and the first genuine rendezvous mission.* Their third launch attempt was a success, and *Gemini VIA* was on her way to meet *VII.* During the ensuing six hours, Schirra and Stafford executed a series of maneuvers that brought them closer to the Borman-Lovell spacecraft. After 3 hours and 15 minutes into the mission, the *VIA* crew locked onto *VII*'s radar transponder, 434

---

*An earlier scheduled launch on 12 Dec. did not take place because an electrical umbilical connector separated prematurely; the crew did not eject but waited removal by the ground crew, something that would have been impossible in Mercury. See Hacker and Grimwood, *On the Shoulders of Titans,* pp. 514-517.

Meeting in space: Gemini VII/VIA *rendezvous, 15 December 1965.*

kilometers distant. There followed a series of precise maneuvers that led to the first sighting of the target vehicle at five hours and four minutes into the mission. At 05:56:00 ground elapsed time, the two vehicles met in space with only 37 meters separating them; the first manned rendezvous was a fact.

There was some controversy over the claim by the Americans that they had been the first to rendezvous in space. Nikolayev and Popovich had been given credit for the same feat by *Pravda* when they flew together in *Vostok III* and *IV*. When Popovich was asked by an *Izvestiya* correspondent if it were possible to compare his formation flight with Nikolayev to that of *Gemini VII* and *VIA,* Popovich said:

> I think it is possible. The first formation flight in cosmonautics history at an orbit near earth was made in August, 1962 by Andrian Nikolayev and myself flying the space ships Vostok-3 and Vostok-4. As you remember, at that time our ships came to within five kilometers distance in space. Thus, in principle, the American experiment of an orbit rendezvous repeats in some degree what we did. But of course there are differences too. During the three years which elapsed since our flight the cosmonautical techniques advanced a great deal. This allowed the Gemini-6 Command Pilot, Walter Schirra, to accomplish with exactitude a series of maneuvers to approach Gemini-7. Of course, the skill of Walter Schirra played a great part in it.[45]

Wally Schirra saw more to rendezvous than Popovich claimed:

> Somebody said ... when you come to within three miles [five km], you've rendezvoused. If anybody thinks they've pulled a rendezvous off at three miles, have fun! This is when we started doing our work. I don't think rendezvous is over until you are stopped—completely stopped—with no relative motion between the two vehicles, at a range of approximately 120 feet [37

m]. That's rendezvous! From there on, it's stationkeeping. That's when you can go back and play the game of driving a car or driving an airplane or pushing a skateboard—it's about that simple.[46]

For more than three revolutions of the earth, the two NASA spacecraft flew together, separated by ranges of 0.3 meter to 91 meters, while the crew of *VIA* tested stationkeeping* and flyaround techniques. After a five-hour sleep period during which they had "parked" 16 kilometers away from the other craft, Schirra and Stafford prepared to go home. With a brief transmission, "Really a good job, Frank and Jim," Schirra flipped *VIA* around, blunt-end forward, jettisoned the equipment section, and waited for the automatic retrofire.[47] As the *Gemini VIA* crew went through the process of reentry, recovery, and return to the U.S., Borman and Lovell worked with Mission Control to ensure that the remaining time of their scheduled 14-day mission did not hold any surprises. Two days later, after some anxious moments over the fuel cell, *Gemini VII* returned safely to earth, proving that man could work and survive in space for the length of time that it would take him to travel to the moon and back.

Each of the five remaining Gemini flights strengthened the conviction and technical certainty that an American could land on the lunar surface and return before 1970. On 16 March 1966, Neil A. Armstrong and David R. Scott conducted the first manned docking when they nosed *Gemini VIII* into the docking adapter of an Agena target vehicle. But shortly after the two vehicles had locked together, a spacecraft thruster stuck open, sending the two astronauts into a dizzying ride through space. They undocked from the Agena, but *Gemini VIII* only spun faster. They were forced to use their reentry control thrusters to restore stability, so ground control told the crew to prepare for immediate reentry. While the early termination of the mission at 10 hours and 41 minutes was most exasperating, the crew did return safely. And they had proved that docking two spacecraft in orbit was possible.

Tom Stafford and Eugene A. Cernan rode *Gemini IXA* into orbit on 3 June 1966 to work further on orbital maneuvers, but when they completed their first rendezvous with the target vehicle, the crew discovered a problem with the docking adapter that precluded the docking phase of the flight. They did continue rendezvous exercises, however, simulating the meeting of an Apollo command module with a lunar module in lunar orbit. *Gemini IXA* also provided an important lesson on the difficulties of working outside a spacecraft in zero gravity, as Cernan left the spacecraft to perform some experiments to get the feel of this new environment.[48]

---

*A definition of stationkeeping is "remaining in a particular, precise orbit with a constant velocity, usually at a given distance from a companion body or another vehicle."

The final three Gemini missions in 1966 built upon the experiences of the earlier flights. They were complex missions with multiple maneuvers; they were designed to test rendezvous and docking, to explore more fully the problems of working outside the spacecraft, and to conduct other experiments that would yield valuable information for Project Apollo. *Gemini X* and *XI* reduced the worry about radiation, demonstrating that it could be avoided during trips into deep space. *Gemini XI*'s first-revolution rendezvous with an Agena target vehicle simulated the meeting of an Apollo command module and lunar module. The automatic reentry of these last two flights gave additional proof that man could return from long missions in space with both manual and automatic control over the final approaches to the landing site. Gemini, in accumulating 1940 man-hours in space flight as opposed to 55 in Mercury, had seasoned flight and ground crews for Apollo; had developed the techniques for rendezvous, docking, and EVA; and shown that astronauts could stay in space as long as two weeks without physical damage.

SOYUZ–DEVELOPMENT OF THE SPACE STATION; APOLLO–VOYAGE TO THE MOON

Less than two weeks after the splashdown of *Gemini XII*, on 28 November 1966, the Soviets launched *Cosmos 133*, an unmanned test of their new manned spacecraft–Soyuz. In the 18 months between the last flight of Voskhod and the first unmanned test of Soyuz, the Soviet space program had lost three important advocates. Premier Khrushchev had stepped down from his post on 14 October 1964, the day following the return of *Voskhod I*; L.A. Voskresensky, Korolev's top assistant, had died on 15 December 1965 after preparing *Voskhod II* for flight; and a month later the Chief Designer himself was dead.[49] While the new Soviet leaders reviewed the competitive space program they had inherited from Khrushchev, the space design group continued the development of Soyuz.

Two elements appear to have slowed the initial pace of the Soyuz project. Soviet engineers needed time to perfect a new upper stage for their basic launch vehicle to provide sufficient power to boost the heavier Soyuz into orbit, and the political requirements to launch a multi-manned Voskhod after Vostok had diverted them. By the end of 1966, the Soviets resolved their various design questions and launched a series of four Cosmos precursor flights that led to the 23 April 1967 launch of *Soyuz 1.** That new spacecraft was designed to exploit knowledge gained in earlier flights, permitting extended missions that would allow Soviet specialists to gather additional

---

*The unmanned Cosmos flights are summarized in appendix B.

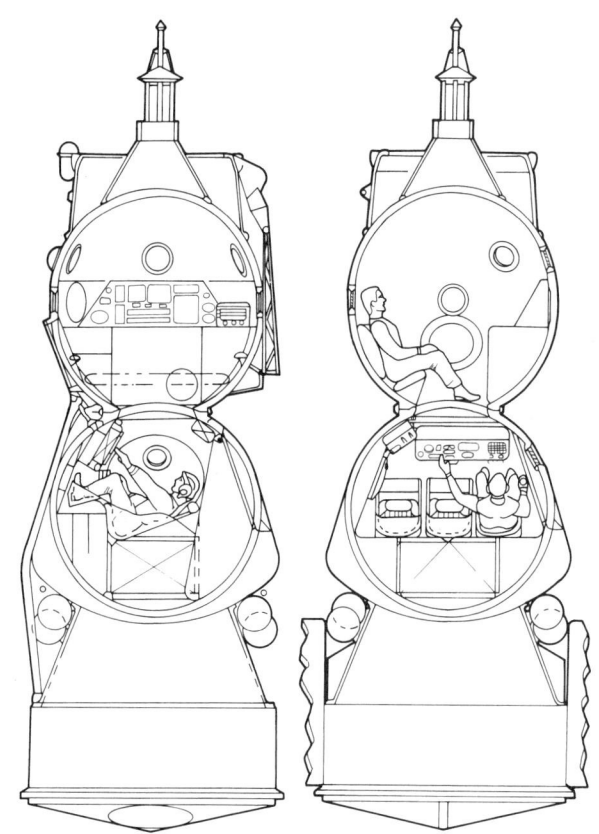

Artist's conception of Soyuz interior as prepared in 1969 by W. M. Taub at the Manned Spacecraft Center.

data on man in space and to investigate the problems of rendezvous and docking. According to the Soviets, the basic purpose of these Soyuz missions was the development of an earth-orbiting space station; others speculated that Soyuz was their entry into the competition to reach the moon.[50]

Work on Soyuz combined elements both old and new. The spacecraft consisted of three major components—the cosmonauts' cabin (descent vehicle), occupied during the launch and reentry phases of the flight; an orbital module, partitioned from the descent vehicle by an airtight hatch; and an instrument assembly module. The descent vehicle had evolved from the earlier Vostok and Voskhod spheres but was fitted with a new heatshield which gave the cabin a bell-shaped external appearance. Unlike its predecessors, Soyuz was designed to have stabilized and controlled reentry.

Various equipment and apparatus for spacecraft control, communication and life support systems are installed in the cosmonauts' cabin. The main and reserve parachute systems are located in special containers. The spacecraft control console, on which are mounted the instruments for monitoring the operation of systems and assemblies, navigation equipment, a television screen and switches for controlling the onboard systems are installed directly in front of him. Lateral auxiliary consoles, for example, the console for medical monitoring of the state of the cosmonauts . . . are arranged alongside the center console. An optical sighting device—a navigation device—is installed in a special porthole.[51]

The Soviet design team retained the form-fitting couches and equipped the descent vehicle with landing rockets located beneath the heatshield, which was jettisoned shortly before touchdown.

Nearly spherical in shape, the orbital module was designed to house equipment for scientific experiments and serve as an airlock for extravehicular activity. The crew would eat, rest, and sleep here. Television, movie, and still photography cameras, along with food, medicine, and personal hygiene gear were stowed in the orbital compartment, which also had an oxygen generation system typical of those used in earlier Soviet spacecraft.[52]

The cylindrical instrument module housed the two 3.9-kilonewton (880-pound-of-thrust) spacecraft engines, the attitude control thrusters, and onboard equipment that otherwise would have cluttered the interior of the spacecraft. In the pressurized portion of this compartment were the temperature controls for the cabins, the radio and telemetry transmitters, and the attitude control system. A set of solar panels attached to the instrument/equipment section provided electrical power during the mission. Protected by a shroud at launch, these panels unfolded once the craft reached orbit. The radio and radar antennas, also folded at launch, deployed subsequently.[53]

Soyuz 1, a test mission, was flown with a crew of one, Vladimir Komarov. This initial mission was fraught with trouble and ended in disaster. The first indication of problems came on the second day of flight, 24 April

Soviet space pioneers Yuri Gagarin and Vladimir Komarov, on the eve of the latter's ill-fated flight aboard Soyuz 1 (Soviet Academy of Sciences photo).

1967, when the spacecraft began to tumble during the 15th and 16th revolutions. Komarov experienced difficulty in bringing his ship under control and found that he was expending far more control fuel than was desirable. As with *Voskhod II,* the automatic orientation system did not function properly, and after communicating with ground control, a process that was impaired by the tumbling, Komarov decided to attempt a manual landing during the 17th orbit. He was unable to obtain the proper orientation for retrofire and went into the next orbit, where he succeeded in bringing his craft under control. He jettisoned the orbital and instrument assembly modules and fired the retroengines at the proper moment, but the Soyuz reentry vehicle continued to revolve about its axis. This motion caused the shroud lines to become entangled when he attempted to deploy the parachute at 70 000 meters. With no parachute, the descent vehicle crashed to earth at a velocity of 450 kilometers per hour. At 6:15 a.m. Vladimir Mikhailovich Komarov was dead.[54]

The loss of a cosmonaut on his return from space struck sorrow in hearts around the globe. President Johnson and Vice President Humphrey expressed their sadness at the loss of "this distinguished space pioneer." Just three months earlier on 27 January 1967, American astronauts Gus Grissom, Edward White, and Roger B. Chaffe had perished when fire swept through their Apollo spacecraft (Apollo 204) as it underwent tests at KSC. NASA Administrator Webb, in voicing his regret at the Soviet loss, suggested that Komarov's death and those of the Apollo astronauts indicated the need for closer cooperation between the two space programs. "Could the lives already lost have been saved if we had known each other's hopes, aspirations and plans? Or could they have been saved if full cooperation had been the order of the day?"[55] But the competitive motivation behind manned space flight still outweighed the desire to cooperate. While a Special State Commission investigated the Soyuz mishap, NASA and American aerospace industries were implementing the recommendations and changes contained in the report of the Apollo 204 Review Board.[56]

Apollo design and development had progressed with reasonable speed since the first consideration of that project in 1959. After 16 months of preliminary study and work, Robert Gilruth on 1 September 1960 called for the creation of an Apollo Projects Office which would have the responsibility of defining the spacecraft configuration. This office became a subordinate part of Max Faget's Flight Systems Division and was headed by Robert O. Piland. Building upon earlier discussions, the initial work began. The command-center module became the crew quarters for all phases of the mission, and the propulsion module held all redundant and orbital maneuvering systems. Willard M. Taub, working for Caldwell Johnson, took all these ground rules and prepared a set of rough sketches of the command module, and by the end of October he had evolved a fairly detailed layout of the crew quarters.[57] All of this work preceded the first manned flights of Project Mercury and the conception of Project Gemini.

Concurrent with in-house design efforts, NASA awarded contracts to three aerospace companies to conduct independent feasibility studies for an advanced manned spacecraft, but it was the work conducted by Taub for Johnson which survived. The General Electric D-2 reentry vehicle proposal bears remarkable external similarity to the Soyuz descent module.

As NASA and industry specialists worked to define the Apollo spacecraft, President Kennedy on 25 May 1961 established manned lunar landing as the primary American goal in space. NASA had not yet issued spacecraft specifications, selected a spacecraft contractor, chosen a family of launch vehicles, or settled the question of direct ascent versus a form of orbital rendezvous for the moon voyage. During the next 18 months, several key decisions gave Apollo more form and direction. On 9 August 1961, NASA selected the Instrument Laboratory of the Massachusetts Institute of Technology to develop the guidance and navigation equipment. At the end of November, following formal presentations by potential spacecraft contractors, North American Aviation, Inc., was selected as prime contractor for the command and service modules. In January 1962, the Saturn C-5 was chosen as the Apollo launch vehicle. Then on 11 July 1962, NASA announced at a press conference in Washington that lunar orbit rendezvous had been approved as the mission mode.* Grumman Aircraft Engineering Corporation had already begun development of the third Apollo craft—the lunar excursion module.[58]

As it evolved through the processes of conceptualization, design, and development, the Apollo spacecraft was composed of two parts, the command and service modules. Called CM for short, the command module was a multipurpose space cabin internally organized to function as a combined cockpit, office, laboratory, communications center, galley, sleeping quarters, and personal hygiene center. It was constructed with an inner pressure shell to provide structural and environmental integrity and an outer wrap-around heatshield for thermal and radiation protection during flight and reentry. This form of construction yielded maximum strength for minimum weight (5450 kilograms). Conical in shape, the CM was 3.23 meters high and 3.91 meters at the base. The service module (SM), which had an overall length of 7.54 meters and a launch weight of 23 950 kilograms, contained the main space-

---

*This decision climaxed one of the most extensive and intensive studies ever conducted by NASA. The final decision was based on the conclusion that lunar orbit rendezvous was more desirable from the standpoint of meeting the proposed schedule, budget, and mission goals.

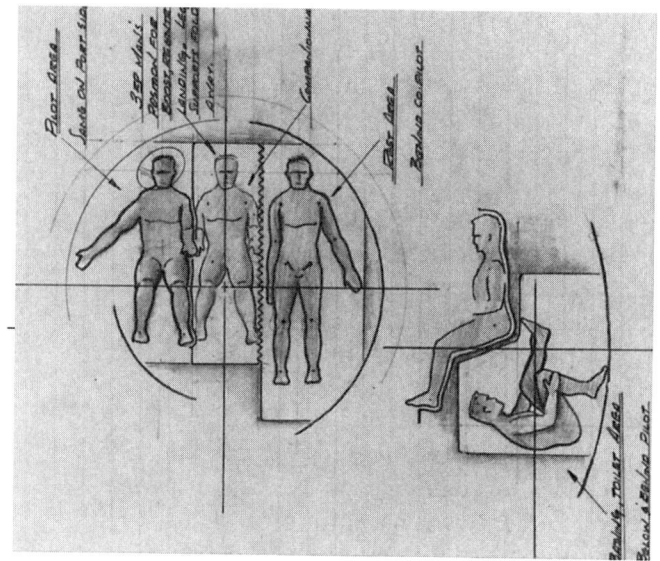

Early conceptual drawing of Apollo cabin interior sketched by C. C. Johnson.

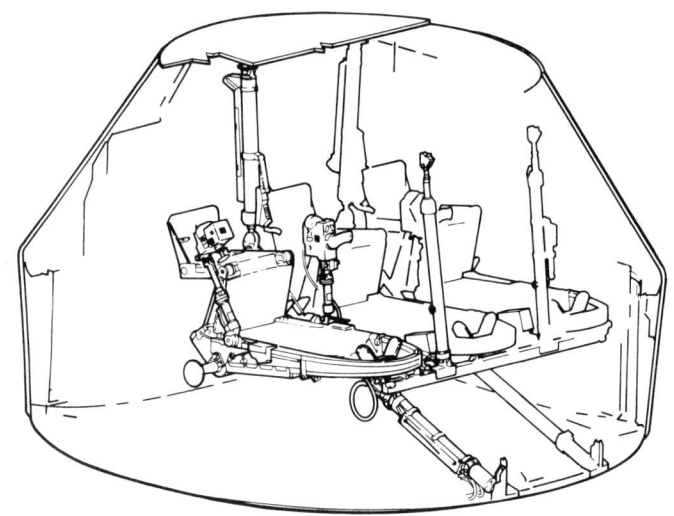

Couch suspension system inside the Apollo command module.

craft propulsion system, reaction control system, and most of the spacecraft consumables (oxygen, water, propellants, and hydrogen). Work on both the spacecraft and the launch vehicle during the years 1962-1966 progressed at a pace that permitted the first manned Apollo flight to be scheduled for 21 February 1967. These plans were altered, however, when the flash fire occurred that year.

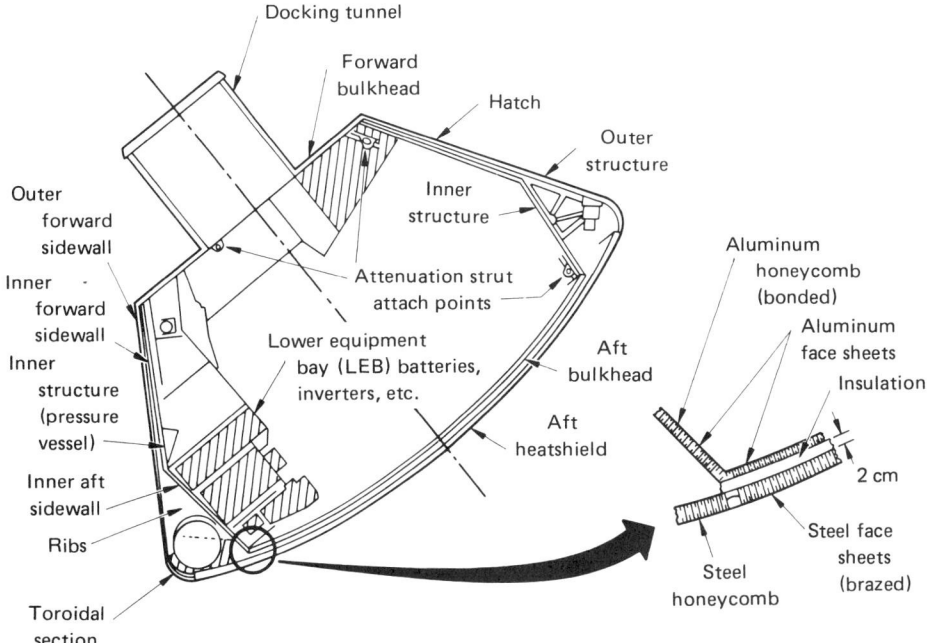

Docking tunnel

Forward bulkhead

Hatch

Outer structure

Inner structure

Outer forward sidewall

Inner forward sidewall

Inner structure (pressure vessel)

Inner aft sidewall

Ribs

Toroidal section

Attenuation strut attach points

Lower equipment bay (LEB) batteries, inverters, etc.

Aft bulkhead

Aft heatshield

Aluminum honeycomb (bonded)

Aluminum face sheets

Insulation

2 cm

Steel face sheets (brazed)

Steel honeycomb

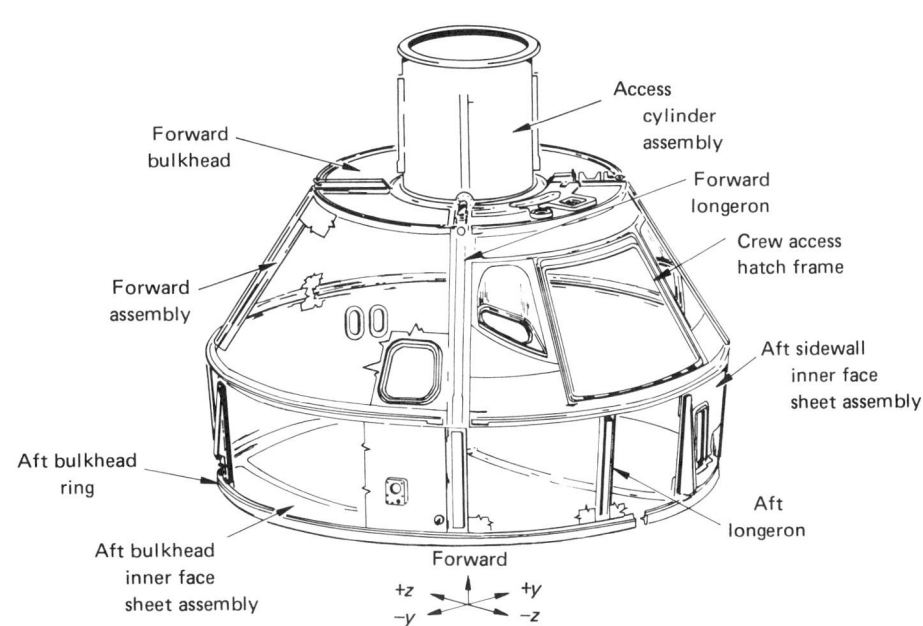

Forward bulkhead

Forward assembly

Aft bulkhead ring

Aft bulkhead inner face sheet assembly

Access cylinder assembly

Forward longeron

Crew access hatch frame

Aft sidewall inner face sheet assembly

Aft longeron

Forward

+z   +y

−y   −z

Diagrammatic views of the internal construction of the Apollo command module.

THE END OF THE SPACE RACE?

Both the Soviet and American space programs went through a period of appraisal and re-examination before they next sent a man into space. After the Apollo fire, manned flight was delayed for 21 months while NASA and North American Rockwell* completely reworked the command module. Unmanned flights were flown on 9 November 1967 (*Apollo 4*), 22 January 1968 (*Apollo 5*), and 4 April 1968 (*Apollo 6*) to check out the modified spacecraft. The Soviets carried out five unmanned launches prior to the joint *Soyuz 2* and *3* mission. On 27 October 1967, the U.S.S.R. sent *Cosmos 186* into a low circular orbit, and three days later it performed an automatic rendezvous and docking mission with *Cosmos 188.* Once *188,* launched for a direct, one-revolution rendezvous, came within 24 kilometers of *186,* the two spacecraft began an automated, preprogrammed closure and docking on the far side of the earth from the U.S.S.R. so that they would pass over Soviet territory in a docked configuration. The two spacecraft remained docked for 3.5 hours, after which they returned to earth, reentry commands having been given to each one day apart. A second automatic rendezvous and docking mission was conducted with *Cosmos 212* and *213,* launched on 14 and 15 April 1968. The five-day missions were successful, and the rigid docking was televised to ground control by onboard cameras. After an apparent final check-flight with *Cosmos 238* on 28 August, the Soviets launched *Soyuz 2,* which was to act as an unmanned target for Georgiy Timofeyevich Beregovoy, the pilot of *Soyuz 3,* who rode into orbit the following day, 26 October.[59]

Beregovoy's mission remains unclear. After making an automatically controlled rendezvous, the cosmonaut took control of his craft and guided it from a distance of 200 meters to within only a few meters of *Soyuz 2,* but he did not dock. While Western observers speculated over this non-event, the Soviets were preparing for a second flight in which rendezvous, docking, and crew exchange would take place.[60] Meanwhile, in the wake of the successful ten-day manned flight of *Apollo 7,* NASA was planning to launch the first circumlunar mission.

The December 1968 launch from Florida was a major step to realizing man's dream of traveling to the moon. *Apollo 8* demonstrated that the distance between the earth and the moon could be safely traversed. On Christmas Eve as they orbited the moon, Frank Borman, Jim Lovell, and William A. Anders shared their impressions of the stark lunar landscape, read a few

---

*In Mar. 1967, North American Aviation, Inc., and Rockwell-Standard Corporation merged to form the North American Rockwell Corporation.

94

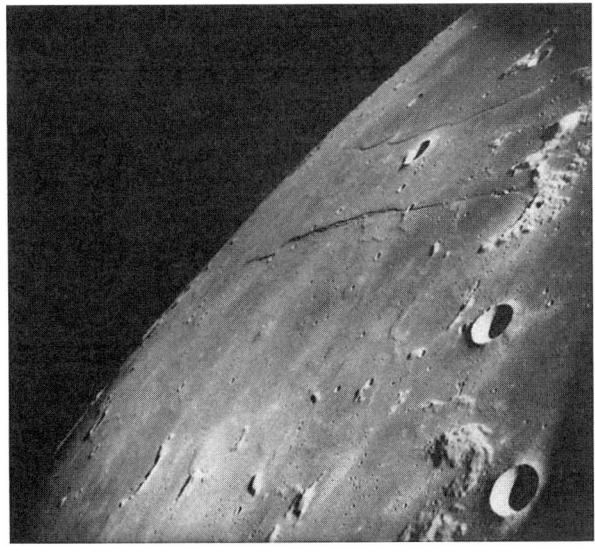

Stark lunar landscape described by Apollo 8 crew on Christmas Eve 1968.

verses from the first chapter of Genesis, and wished their earth-bound viewers a Merry Christmas. A *New York Times* article suggested that the space frontier was so vast that "there is no need here for wasteful rivalry deriving from earthbound nationalistic and political ambitions." But the *Washington Post* viewed the Christmas mission with a cynical eye; NASA was still racing to get to the moon before the Soviets preempted the feat. Columnist Joseph Kraft suggested a reappraisal of America's space goals now that the country was clearly ahead of the U.S.S.R. "There is no need for the United States to race Russia to every new milestone in space." He felt that the country needed a "program closely connected to explicit American requirements—a program of exploration for its own sake, not for the sake of beating the Russians."[61]

In Houston, *Apollo 8* was viewed as the pivotal flight in the Apollo Program. Christopher C. Kraft, Jr., Director of Flight Operations, later stated:

> It proved so many things that had a bearing on the progress of the program—
> things that might have been disproved. The navigation to and from the moon,
> the ability of the spacecraft systems to survive the deep space environment,
> all hinged on the Apollo 8 mission.

He also believed that the flight changed the competitive position of the United States and the Soviet Union in space. He had thought that "the Russians planned to fly a circumlunar mission, sending a manned spacecraft looping around and returning without orbiting the moon. That way they could say they sent the first man to the vicinity of the moon." Once *Apollo 8* made her voyage, "there was nothing left for them to do."[62]

But from the Soviet Union came another perspective. Boris Nikolaevich Petrov, Chairman of the Council for International Cooperation in Investigation and Utilization of Outer Space (Intercosmos) of the Soviet Academy of Sciences, called the *Apollo 8* flight an "outstanding achievement of American space sciences and technology" and praised the "courage of its three astronauts." Academician Petrov also indicated that the Soviet Union would continue to explore the moon, but with unmanned automatic spacecraft. "The major tasks still ahead in the study of the moon will . . . be carried out by automatic means, although that does not exclude the possiblity of manned flight."[63] Petrov's words would remain a puzzle. Had *Apollo 8* won the space race? Had the Soviets ever really been in the race to send a man to the moon? Surely Administrator Paine still had these questions in mind seven months later when he sought to renew NASA's search for a cooperative route to negotiations with the Soviets.

By 1969 Thomas Paine hoped that a change in Soviet-American space relations might be possible. Since the U.S. was clearly ahead in any race to the moon, an offer to cooperate would not jeopardize the lunar prize. And now the Soviet Union had more to gain from cooperation. By working with the nation that had led the way to the moon, the Soviets could create the image of technological parity. Paine perceived this period as an opportunity for new beginnings and began again the effort to discuss cooperation with Soviet space officials. Twelve years of bitter rivalry, during which each side had cooperated only in limited ways, could give way to closer relations if the Soviets were willing.

IV

Mission to Moscow

Between the spring of 1969 and the fall of 1970, the Paine-Keldysh correspondence had set the stage for serious discussions on developing compatible equipment and flight procedures. Tom Paine thought that cooperation in space was an important and timely idea and pushed for talks in furtherance of that goal—and he got them. Paine's success with the Soviet officials was vastly different from the experiences that had spanned the preceding twelve years. In this instance, the spirit of the past was definitely not the prologue.

Knowledge of the letters between the NASA Administrator and the Soviet Academician had been shared by a limited number of NASA people. As long as the communications were general and exploratory, action was concentrated in the offices of the Administrator and his Assistant for International Affairs. On 10 July 1970, however, President Nixon publicly confirmed his interest in pursuing discussions of space cooperation, stating that negotiations should be conducted at the technical agency level.[1] Thus, when talks with the Soviets appeared likely, NASA Headquarters geared up in preparation. Philip E. Culbertson's Advanced Manned Missions Planning Group in the Office of Manned Space Flight (OMSF) was one of the first to be drawn into the widening discussions, having been assigned to consider the development of compatible rendezvous and docking systems.

In mid-August, OMSF began to "work the problem,"* an exercise in defining the technical considerations that would be involved in any American-Soviet negotiations. Dale D. Myers, Associate Administrator for Manned Space Flight, sent a note on 19 August to Culbertson, who in turn assigned Eldon W. Hall and James Leroy Roberts of the Advanced Developments Office the primary responsibility for coordinating this effort among Headquarters and Center offices.[2]

---

*Working the problem, a commonly used phrase in NASA, has descriptive significance beyond the convenience of jargon; it means the analysis of systems and the manner in which they impinge or "interface" with one another. By laying out all possible factors on paper, the NASA managers and engineers can begin to see more clearly the nature of a given task. "Working the problem" is shorthand for the NASA approach to understanding technological relationships.

After a 12-day "quick look," Roberts submitted a draft report entitled "International Cooperation in Space," which presented his initial thoughts on developing joint systems. Roberts felt that the interest expressed "by the Soviets for discussion leading to the possibility of a common docking mechanism at space stations" came at an appropriate time since NASA was getting into detailed hardware discussions relating to the Space Shuttle (a reusable spacecraft) and Space Station concepts.* While Roberts and others believed that the Soviets might greatly benefit from an "open discussion of our system," they argued that "regardless of the Soviet intentions for the proposed discussions they should be pursued in depth."[3]

The Advanced Developments staff explored two possible types of missions employing compatible docking equipment—a rescue mission using either an Apollo or a Soyuz spacecraft to assist a disabled vehicle of the opposite type, or a mission to test out rendezvous and docking procedures. For several reasons, rescue possibilities appeared to be limited to an Apollo retrieving the crew of a crippled Soyuz. It would have been very difficult for the Soviets to accommodate all three Americans aboard their spacecraft unless they attempted an unmanned rendezvous with Apollo, and Soyuz was essentially an earth orbital craft, while Apollo was designed for lunar missions. Also, the opportunities during which Soyuz could provide assistance were limited since the two spacecraft normally flew in different orbital paths.[†] Roberts concluded that "while a mission of this type is not impossible it is highly improbable."

"With Apollo orbital and maneuvering capabilities we could provide assistance" to Soyuz, assuming an extravehicular transfer. Roberts went on to say that for NASA to seriously consider an actual rescue backup to a Soyuz mission, the Soviets would have to make their flight schedules and launch parameters available well in advance so the American agency could divert the necessary Apollo spacecraft and launch facilities in time for the Soviet missions. Roberts pointed out that such an equipment set-aside could also be used for an Apollo rescue, "thus negating consideration of a Soyuz mission as a back up for Apollo."[4] Space rescue was a far more complex and costly enterprise than it first appeared. Once the two countries shifted from

*Space Shuttle and Space Station were advanced programs in 1970. By the time of ASTP, Shuttle had advanced into the mockup stage. Space Station was terminated in 1972 because of cuts in NASA's budget.

†The problems of the *Apollo 13* flight in April 1970 were still fresh in the minds of NASA planners. At 56 hours into the mission, a service module oxygen tank had burst, forcing the cancellation of the lunar landing and emergency planning for the return trip. The spacecraft had to continue on, swing around the moon, and travel back to earth. A rescue capability limited to an earth orbit would have been of little assistance in this kind of emergency. Later, in the Skylab era, Soyuz might be capable of rendering aid in the event of trouble.

one-mission spacecraft to reusable craft such as the Space Shuttle, space rescue would become a more feasible and realistic topic for discussion.

While Roberts could see little justification for developing a compatible docking capability simply to provide a space rescue system, he did see some promise in applying a universal docking system to Skylab or Space Station. With the creation of standardized international hardware, it would be relatively easy for the Soviets to conduct joint missions with American space laboratories or vice versa. Roberts suggested further:

> It is essential for any fruitful discussion of common hardware to have a clear understanding of the Soviet system of rendezvous and docking. There is a possibility that our hardware may have to be modified to assist the Soviet spacecraft in rendezvous operations. The system can be made to work but an exchange of information by representatives as proposed is a necessary step in that direction.

Looking to the immediate future and the possibility of Soviet participation in Skylab, Roberts felt that it was "not likely that arrangements can be made and hardware requirements incorporated in time to meet the Skylab A mission." But he was of the opinion that "there should be sufficient time . . . to match the systems for later flights of Skylab and Space Station if there is a genuine interest in doing so."[5]

Implicit in Roberts' comments were several important "ifs." NASA could develop the necessary rendezvous and docking systems if the Soviets were genuinely interested in cooperation and if such participation could be integrated into NASA's schedule for manned missions. OMSF was not likely to recommend proposals that would seriously delay programs or adversely affect its budget. Clearly, those responsible for planning would have preferred to incorporate joint projects into future missions, thus giving them the opportunity to plan more leisurely and still not lose the opportunity to cooperate. Perhaps the Americans' biggest "if" concerning cooperation lay in the uncertain future of manned space flight in the post-Skylab era.

Culbertson responded to the Roberts memo with some suggested changes. He thought it might be a good idea to break the problem into three major areas—rendezvous, docking, and transfer. "In each case a brief description of the difference in technique and hardware (U.S.S.R. vs U.S.A.) could be given as available from open literature." Then it would be possible, he wrote, to describe solutions to these differences "in very brief fashion." Culbertson also cautioned against making the topics under discussion too complex. "I wouldn't use this memo as a mechanism for explaining the further opportunities for international cooperation. Let's keep it on one topic." He believed that one subject "should say something about early (Apollo) implications and follow on possibilities." Culbertson added one final caveat: "Let's also, in this memo, not question the U.S.S.R. motive. Leave that for other discussion."[6]

Following Culbertson's suggested format, Roberts sent memos to the Kennedy Space Center (KSC), the Manned Spacecraft Center (MSC), and the Skylab, Space Station, and Shuttle Offices at Headquarters. From Raymond J. Cerrato at KSC, he sought information concerning the technical feasibility of a standby rescue vehicle that could support Soviet space missions. Roberts especially wanted information about the problems associated with making a Saturn IB or Saturn V launch vehicle available for such an operation; for example, the lead time required for launch once the Soviets advised NASA of their intentions to conduct a manned flight.[7] Queries to MSC were much broader in scope. Jack C. Heberlig and Willard M. Taub were asked to provide answers to a number of questions in Culbertson's three categories. Specifically, they were requested to describe the known differences between American and Soviet hardware and techniques and to suggest possible steps toward eliminating those differences. The Office of Manned Space Flight that first week in September was doing its homework.[8]

One of the first responses to OMSF came from Skylab Program Director William C. Schneider. After taking "a fast look at the proposition of entertaining distinguished visitors in orbit," his Skylab office had concluded that "there doesn't seem to be anything that says it can't be done," but "there sure is a potful of things that would take a lot of joint planning. . . ." Schneider felt that an on-time launch, rendezvous, docking, and EVA transfer were all capabilities that had been proven within the Soviet and American programs. On the other hand, he did see areas in which considerable joint development would be necessary. We would need to interconnect the ground systems for tracking, mission control, and launch control, and develop a spacecraft-to-spacecraft voice communications link. After listing ten other topics that would have to be considered, Schneider said that his personnel would be glad to go into the subject of a joint mission at greater depth when needed.[9]

While there was limited enthusiasm for a joint flight in the Skylab Program Office, Paine on 4 September wrote a letter to Keldysh in which he proposed a Soyuz rendezvous with Skylab.[10] NASA was still awaiting Keldysh's response to the Administrator's earlier letter of 31 July, in which he had suggested joint talks on compatible docking systems. Meanwhile, Leroy Roberts was coordinating the collection of technical data, which no one was certain would ever be used.

On 10 September, Roberts circulated a new draft memorandum to Hall, Culbertson, and Charles W. "Chuck" Mathews,* which Roberts had prepared

*Mathews was Deputy Associate Administrator of OMSF and acting Space Station Task Force Director.

for Dale Myers' signature. Concentrating on the desirability of the Soviets providing information on docking mechanisms that might be used in future space stations, Roberts reported, "Soviet docking arrangements as we know them have been reviewed . . . and fruitful discussions at this time will be very helpful in defining design requirements for hardware still to be built for the space station."[11] Besides the work being conducted at MSC, North American Rockwell and McDonnell Douglas Aircraft Corporation were engaged in preliminary design studies of possible future docking mechanisms. Since these concepts were still in the drawing stage, it appeared to be an excellent time to obtain Soviet comments.

In addition to looking at future systems, Roberts appended to his memo the MSC materials comparing existing spacecraft. Will Taub,* one of the few NASA employees known to have followed closely the evolution of Soviet spacecraft, had prepared a series of sketches which compared the Soyuz and Apollo. These illustrations and MSC-prepared briefing charts permitted Headquarters personnel to develop a better understanding of the differences that existed between the American and Soviet approaches to space flight. These materials indicated that Soyuz was capable of either automatic or manual rendezvous and docking using radar and attitude control system responses from the target vehicle. The Soviet spacecraft could be flown unmanned or with crews of one, two, or three. Normal crew transfer from one Soyuz to another was an extravehicular maneuver, as demonstrated by the January 1969 flight of *Soyuz 4* and *5*.[12] Direct (internal) transfer would require modification of the docking end of the orbital module.

By comparison, Apollo rendezvous and docking maneuvers were conducted manually, not requiring target participation. While the Apollo command module usually was operated with a crew of three, that number could be reduced or the cabin structurally modified to accommodate five astronauts. Transfer between the command module and the lunar module was made through a passageway between the two craft, the probe and drogue assembly having been removed after docking. Although extravehicular transfer was possible, it had not been a feature of Apollo missions. Another significant difference between Soyuz and Apollo was the cabin pressures. The Soviets continued to use a pressure equivalent to one earth atmosphere, while the Americans still relied on their pure oxygen environment at a much lower pressure. In the opinion of MSC specialists, however,

---

*While many persons within NASA had followed the Soviet space program over the years, they had not concentrated sufficiently on technical details to develop an in-depth knowledge of the hardware. Taub had made an avocation of this subject and became especially useful in this early period.

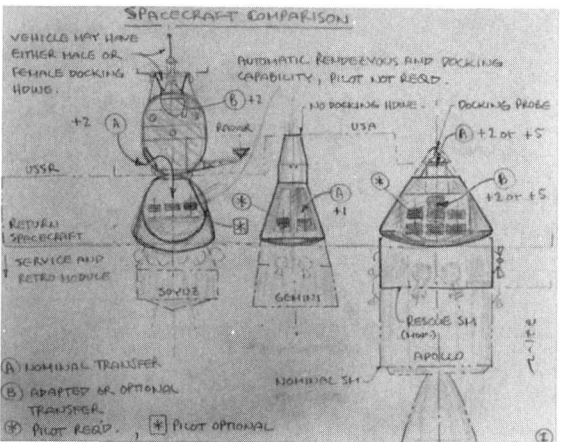
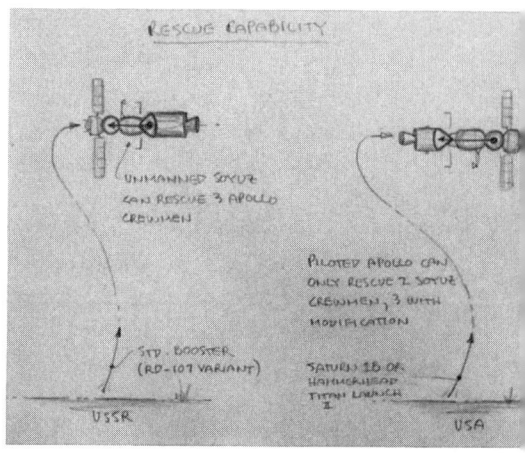
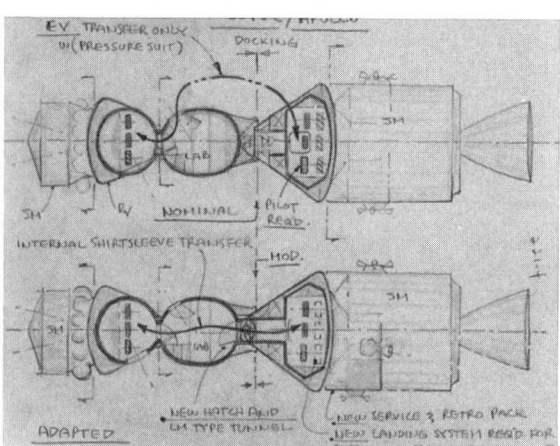
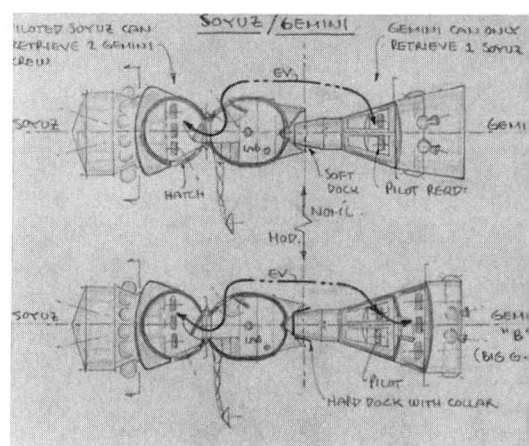

Four sketches by W. M. Taub outlining Soviet and American spacecraft characteristics and possible joint missions with existing spacecraft. Prepared in 1969 for G. M. Low.

none of the differences between the spacecraft posed a significant barrier to a joint mission.[13]

Chuck Mathews passed Roberts' material along to Dale Myers on 15 September. Since there still had been no response from Keldysh, Mathews commented, "I hear that this item has cooled a bit but I think it is still good to send this . . . along."[14] Myers signed the memo on the 17th and sent it to the Administrator's office.[15] OMSF and the Centers had investigated three possible types of cooperative missions—Apollo-Soyuz, Soyuz-Skylab, or future American-future Soviet spacecraft. Now the question remained as to the value of the exercise. When would the Soviets respond? What would they propose?

Acting Administrator George Low received a letter from Academician Keldysh on 23 September that brought an end to the suspense. Keldysh

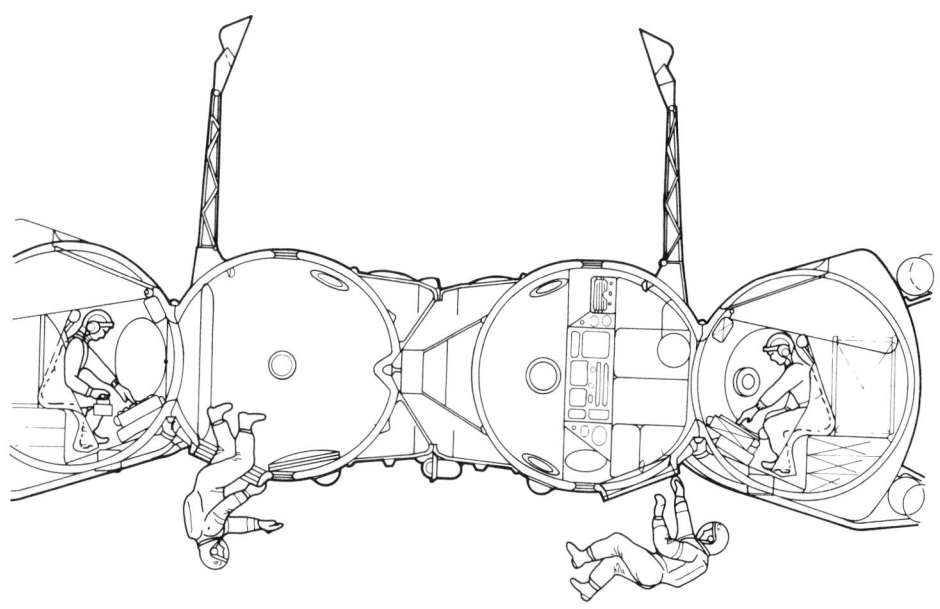

Artist's conception of Soyuz 4 *and* 5 *extravehicular transfer as prepared in 1969 by W. M. Taub at the Manned Spacecraft Center.*

suggested that either October or late November would be a suitable time for the first talks, and he proposed that they be held in Moscow.[16] Since President Nixon had given NASA the go-ahead to develop discussion with the Soviet Union, Low responded to Keldysh on the 25th, accepting the invitation and suggesting a meeting a month later.[17]

During the next several weeks, OMSF concentrated on preparing an agenda for the upcoming talks. On 23 September, Mathews, Hall, and Roberts met with Oscar E. Anderson, Jr., of the International Affairs Office to discuss the agenda and delegation for the Moscow trip. As the plans for the meeting went through several drafts, Low and Myers met to decide upon a suitable chief for the American group.[18] Since Low felt it was premature for the head of NASA to go to the Soviet Union, he selected Robert Gilruth, Director of MSC, because of his technical background and common-sense approach to complex negotiations.[19]

Low and Myers asked Gilruth to select the necessary technical specialists to complete the delegation. From MSC, Gilruth chose Caldwell Johnson and Glynn Lunney. Gilruth took only two men from Houston, because he felt that a small delegation would have a better chance for success. Since he wanted men with a breadth of knowledge, Johnson and Lunney were the obvious choices. Johnson, "a very, very talented mechanical designer," could discuss the mechanical and electrical questions associated with developing a compatible docking system. Lunney, "an expert flight controller," had the necessary background in orbital mechanics and mathematics to discuss the mission planning aspects of a joint flight. In an effort to include the Marshall Space Flight Center in the talks, Gilruth called Director Eberhard Rees at Huntsville, Alabama, and asked him to nominate one person who could talk about Skylab. Rees recommended George B. Hardy, Chief of Program Engineering and Integration for Skylab, who by virtue of his position had a broad understanding of the program. Arnold Frutkin, Assistant Administrator for International Affairs, represented Headquarters. William Krimer, an interpreter from the State Department, completed the six-man delegation.[20]

The news that they were going to Moscow came as a surprise to Johnson, Lunney, and Hardy. Lunney was presenting a speech on 7 October to the 1970 National Airport Conference in Oklahoma when he got the call telling him that he was going to the Soviet Union. "For me it was out of the clear blue sky. I did not know anything about [the proposed talks] until that time." These three specialists met with Gilruth on the 9th to discuss the nature of their presentations to the Soviets. They would seek to provide their counterpart specialists with enough information to give them a common basis for further discussions, but not so much as to overwhelm the Soviets or to encourage comments at home that they were giving away too much.[21] In Washington, Frutkin's staff was preparing a briefing to inform the press about the mission to Moscow.

The head of the International Office met the press at NASA Headquarters in mid-October and summarized the background to the talks. He explained that the emphasis on compatible docking systems just happened to be the specific American proposal to which the Soviets had responded affirmatively.[22]

> It is simply that the Soviets have chosen out of this long list of initiatives from the U.S. side this one case to explore in some depth at this time. It could have been something else. This one seems to be more meaningful to them.
> So just as I say we regard it as important, presumably also they regard it as important.

Frutkin took care to point out the very preliminary nature of the talks and to make certain that his questioners did not make too much of the space rescue capabilities inherent in the development of compatible docking systems. But reporters were especially interested in that aspect of the story because of announcements at the 21st IAF Congress in Konstanz, Germany, that the United States and Soviet Union had agreed to sponsor a space-rescue symposium.[23]   Frutkin cautioned that the IAF proposal was purely coincidental:

> When you see a release out of Konstanz that says the Soviet Union and the United States have agreed to a symposium of that sort, this is simply a shorthand way of saying that some individuals from the United States who are interested in space rescue on a professional basis are going to meet with some individuals from the Soviet Union who are interested in the same subject, and talk about this matter, just as they talk about a lot of other things. But there is no correspondence between their private professional discussions and our governmental official discussions, there is no relationship whatever.[24]

Despite Frutkin's statements about the nature of the discussions, the reporters still pressed him for a prediction on the earliest date that a joint mission might occur. The NASA representative responded that it was just too early to make such statements, but that Skylab was likely to be the first occasion. "We don't know how long—we don't know what the pace of our discussions is going to be." Reflecting on his experiences in negotiating with the Soviets, Frutkin said that such talks tended to progress slowly. He conceded that the question of timing was "very difficult to answer. . . ."[25] While the reporters went off to file their speculations about the future, Frutkin and his colleagues conducted a dress rehearsal of their presentations.[26]

Gilruth, Lunney, Johnson, and Hardy flew to Washington for the "dry-run" on 16 October. Johnson recalled that the Headquarters staff, especially George Low, seemed to be interested in the type of presentation that each man planned to make. Low appeared to be particularly curious about the extent to which each man could vary his approach and think on his feet. Since so little was known about what the Soviets wanted to discuss, it was very likely that each man would have to sense out his audience as he spoke. The key to success might lie with a flexibility of mind and ability to react quickly to whatever direction the discussions might take. During the two-hour meeting, the five men were also briefed by representatives from the State Department, the Department of Defense, and the intelligence service.[27]

STEPS FORWARD

As the American delegation left New York's Kennedy International on 23 October, each man wondered about the reception he would encounter in Moscow and reflected upon the warnings that had been given by representatives from the foreign service, defense, and intelligence communities. Gilruth later recalled that these professionals had been unanimous in their negative prognosis and had told the NASA specialists not to build up their hopes for early or easy agreements with the Soviets. It was the prevailing opinion that the Soviets would talk and talk, but in the end they would "break our hearts" by their refusal to cooperate.[28] Caldwell Johnson remembered the final leg of the journey, the flight from Copenhagen to Moscow, particularly clearly. The NASA group had the plane nearly to themselves, and there was a sense of solitude and uncertainty as they approached their destination.

Gilruth and his colleagues reached Moscow's Sheremetyevo airport late in the afternoon on 24 October 1970. The mid-afternoon sky had grown dark; visibility was limited by fog and a light mixture of snow and sleet. Looking back on his first steps onto Russian soil, Johnson commented, "I was awed by the situation and kind of uptight." Seeing a large number of military uniforms and figures in heavy overcoats, he became uneasy because as he later put it, "I grew up in that period of history where those things were impressed upon me and my generation . . . as bad things." However, his sense of concern quickly dissipated.[29]

Bob Gilruth sensed almost immediately that they were embarked upon a positive adventure. He had seen that same line of men waiting, but he had also noticed Cosmonaut Feoktistov, whom he had met the year before in Houston. Though his name momentarily eluded Gilruth, the smile did not. All the Soviets were smiling, and the motion picture cameras were at the ready. Instinctively, Gilruth felt that all would be well. He turned to the others and said, "It's going to be all right, because they are welcoming us in style."[30]

The airport greeting was warm and cordial. Immediately upon their entry into the airport terminal, the Americans were taken to a lounge where they exchanged introductions with Academician Boris Nikolayevich Petrov and his colleagues. After their luggage was gathered and cleared through customs, the Americans and their hosts were off to Moscow proper and their first introduction to the monumental Hotel Rossiya.[31] Tired by their long journey, the NASA contingent was ready to turn in for the night when Petrov asked Gilruth if 40 minutes would be enough time for them to freshen up before dinner. Following a pleasant meal at the Rossiya, the Soviets took their guests on a tour of Moscow, which included a ride up the Lenin Hills for a view of the city, Moscow State University, and the Moscow River. From there, they drove through central Moscow for a visit to the Space Monument on the Avenue of the Cosmonauts.

Looking back on that experience, Gilruth recalled that the night had an eerie quality. The lights, reflected from the low clouds and diffused by the mist sweeping around the monument, made this impressive structure all the more awesome. Although the Americans were bone-tired when they finally reached their hotel at 11 o'clock, they felt more sure of success. To Gilruth, it was like something out of a dream, a pleasant dream.

Early Sunday morning, the five Americans and their hosts went off to Zvezdny Gorodok (Star City), the cosmonaut training center, 40 minutes by car northeast of the capital. There they were greeted by Commandant General Andrei G. Kuznetsov, Major General Georgiy Timofeyevich Beregovoy, and Colonel Vladimir Aleksandrovich Shatalov. Following a brief greeting ceremony, the Americans were taken to the simulation facility where the cosmonauts practiced flight procedures. According to Lunney, the opportunity to examine the Soyuz simulators was one of the highlights of the visit.

The working part of the trip had begun. The simulators were arranged very much like those in Houston, with the training specialists seated at consoles where they could monitor replicas of the spacecraft control displays. Gilruth and Johnson were given a briefing on the simulators by Beregovoy, while Lunney and Hardy were accompanied by Shatalov. They were shown the general purpose trainer first. This simulator was situated in a vertical fashion, with the command/descent module positioned below the orbital module. Upon entering the latter, Lunney was impressed by the roomy feeling of the 2.2-meter by 2.65-meter interior. Although this cabin was the primary work and living station for the crew during a mission, Lunney was struck by the simplicity of the controls and instruments. As far as he could determine from his conversation through an interpreter with Shatalov, the crew seemed to be limited to controlling and monitoring the airlock, experiment, and biomedical functions carried out in the orbital module.

Passing through the airlock hatch that separated the orbital module from the command module during launch, docking, and EVA maneuvers, the three men entered the command module. "For three men this is a small volume," reported Lunney, "but it is only used during takeoff, landing and rendezvous, and periodically in orbit. . . ." He felt that the interior space was adequate since the cosmonauts flew in flight coveralls and wore suits only for EVA. Additional space was obtained by lowering the couches during flight; for landing, they were raised toward the control panel and supported by shock-absorbing attenuators during reentry.

Lunney and the others were particularly interested in the descriptions of the Soyuz control systems provided by Shatalov and Beregovoy. While occupied with the U.S. programs, the NASA representatives had followed the U.S.-U.S.S.R. competition only from accounts in American aerospace publications, but now they had the chance to hear those systems described

At left, illuminated and shrouded by mist, the space obelisk at the main entrance of the National Exhibition of Economic Achievements was one of the memorable sights of the October 1970 trip to Moscow. (Tass from Sovfoto). At right, replica of Vostok and launch vehicle displayed at the National Exhibition of Economic Achievements, Moscow.

At left, greeting at Star City. From left to right: V. A. Shatalov, G. T. Beregovoy, A. G. Kuznetsov, G. B. Hardy, W. N. Harbin (shaking hands with Shatalov), W. Krimer, R. R. Gilruth, A. W. Frutkin, and B. N. Petrov. At right, a group portrait against the pine forest backdrop at Star City. Left to right: W. N. Harbin, G. S. Lunney, V. A. Shatalov, a Soviet interpreter, A. G. Kuznetsov, B. N. Petrov, W. Krimer, A. W. Frutkin, G. B. Hardy, G. T. Beregovoy, R. R. Gilruth, C. C. Johnson, and K. P. Feoktistov (Soviet Academy of Sciences photos).

first-hand by men who had worked with and flown the Soyuz. Lying in the command module couches, the Americans could see and touch the controls, getting a better feel for the Soviet approach to manned space flight.

Lunney later thought about the briefing he received while lying in the commander's couch. Directly in front of him was the main control console. Starting at the upper left-hand corner of the instrument panel and proceeding clockwise, Shatalov explained the equipment. First, there was the rotating globe—the "space navigation indicator"—that gave the pilot his approximate position relative to the earth's surface. Adjacent to that instrument was a panel of lights that displayed the status of various space-craft systems. Next, Shatalov indicated a television screen through which the commander could observe the docking. Then there was a projection screen for displaying aspects of the flight program visually, while other data were presented on a digital data display. Above that latter unit was a chronometer to keep track of flight times.

Shatalov then pointed to an optical device located just above Lunney's right knee. This navigation sight, used in conjunction with the television display during rendezvous and docking, gave the commander a fixed view of the scene directly ahead of the spacecraft. Next to this apparatus were a series of gages, switches, and additional clocks. With these, the commander could keep track of cabin pressures, temperature, and power levels and could also monitor time-critical control commands. On either side of the main console were located "command signal" panels with rows of lights and push buttons that permitted the crew to execute specific commands to the spacecraft systems. Manual control of the Soyuz was accomplished through two hand controllers at the commander's side. Both Lunney and Johnson noted the large blank spaces on the walls of the command module covered by an off-white, felt-like padding. For Lunney, "the very strong impression was one of simplicity—no circuit breaker panels, no large number of switches, not many displays."[32]

After getting a general orientation to the various Soyuz systems, the Americans were given an opportunity to look at the docking simulators. This set of trainers consisted of two command module mockups—one for active and another for passive rendezvous. These two simulated spacecraft could be maneuvered into the docked position with other small models of Soyuz, viewed by the cosmonaut either on the television monitor or through the docking periscope. After watching these replicas of the regular flight systems, Lunney and his associates felt that they had a much better understanding of Soviet rendezvous techniques. There were still unanswered questions, but this introduction proved to be a great aid in the technical discussions that occupied the next two days.[33]

After the familiarization session with the spacecraft trainers, the men walked to the main building of the Cosmonaut Training Center, where they watched a motion picture about Yuri Gagarin's flight, following which they visited a manned space flight museum. Gilruth and the others saw the reconstruction of Gagarin's office, as well as all the memorabilia collected by the first man in space on his subsequent trips around the world. After the tour, the group retired to a mid-afternoon luncheon.

General Kuznetsov began the pre-meal formalities with a long, carefully prepared toast. He spoke directly to the NASA representatives and said in effect that he was relying upon them to exert their influence on the American government to ensure cooperation in space. He wanted the U.S. representatives to convey to their leaders the necessity for keeping space endeavors peaceful; he expressed his hope that space would not be turned into something evil. Lunney especially felt the personal nature of this message: "He was talking directly at and to us. He was saying to me that he was holding us responsible to see that space continued to be a peaceful place."[34] Shatalov followed with a toast comparing the histories of the United States and the Soviet Union, in which he stressed the similarities of the two countries and their aspirations. The vodka and the meal behind them, the Americans and Soviets walked about the grounds at Star City.

As if this were not enough to occupy a full day, the Americans were taken back to Moscow for a visit to the major space museum housed on the grounds of the Exhibition of Economic Achievements. Following a ten-minute stop at the Rossiya, their evening was capped by a trip to the Bolshoy Theater for a performance of Rimsky-Korsakov's opera *The Tsar's Bride.*

Monday, 26 October, was given over to discussions* of rendezvous experiences and techniques and to descriptions of spacecraft docking assemblies. Glynn Lunney gave the first presentation, describing the spacecraft hardware capabilities NASA considered essential for orbital rendezvous, communications, guidance, and propulsion systems. For an international rendezvous, he saw that a number of issues would have to be studied—compatible equipment to provide information on the range between spacecraft and their rate of closure; suitable docking lights, reflectors, and targets; and vehicle-to-vehicle voice communications. Lunney also summarized for the Soviets rendezvous techniques as they had evolved through Gemini and Apollo. While there were many adequate techniques, the specific approach to the problem would ultimately depend upon the degree of

---

*The Soviets present in addition to Petrov included K. P. Feoktistov, V. S. Syromyatnikov, V. V. Suslennikov, I. V. Lavrov, and N. Khabarin. Also joining the Americans was W. N. Harben, Science Attache, U.S. Embassy, Moscow.

automatic or manual control in a given spacecraft. Therefore, an accommodation would have to be reached whereby the basically automatic, radar-controlled rendezvous of Soyuz could be matched with the essentially manual approach of Apollo. These were by no means irreconcilable differences, but they would require much study. Lunney closed by telling the Soviets that NASA expected its future rendezvous techniques to be an outgrowth of the ones he had described.[35]

Next Feoktistov explained the Soviet methods of rendezvous, which were designed to work either manually or automatically from ground commands, though they definitely favored the latter approach. According to Feoktistov, the Soviets considered rendezvous in three distinct phases—delivery of the active spacecraft to the vicinity of the target spacecraft, automatic rendezvous maneuver to stationkeeping distances, and final approach to docking. Going into more detail, Feoktistov said that the first phase could be approached in two ways, direct ascent or rendezvous following placement of both ships in orbit. Direct ascent required precise timing, so that the second craft could catch the target within its first revolution. More satisfactory, they had found, was a rendezvous after the two vehicles were in basically similar orbits. The path of the active craft would be adjusted by engine burns generated by ground-based computers and transmitted by radio to the onboard guidance and propulsion systems. This maneuver would bring the two ships to a range at which a mutual automatic search would begin by the spacecraft tracking systems.

Phase two of the Soviet rendezvous process started when the radar antennas locked on and the guidance system oriented the ships in the proper attitude—nose-to-nose. The main engine of the active Soyuz would be fired automatically as directed by the guidance system to bring the two craft to within a range of 300 to 400 meters. During the third phase, the final approach to docking (*prichalivaniye,* literally mooring) would be completed by firing the 9-newton (2-pound) translational thrusters. While this final phase could be completed in either an automatic or manual mode, the Soviet specialists seemed to prefer the hands-off approach.[36]

When discussion turned to the docking systems used to lock spacecraft together following rendezvous, Caldwell Johnson presented a description of the systems NASA had used during Gemini and Apollo, to preface his outline of future docking concepts.* According to Johnson, the configura-

---

*Johnson defined the terms used in discussing docking equipment as follows: "The term docking as applied to spacecraft operations defines the mechanical, temporary joining together of two spacecraft, generally for the purpose of crew and cargo interchange. In that same context, docking systems refer to the collection of spacecraft equipment designed to perform the docking operation. Docking gear refers more specifically to the mechanisms that accomplish the mechanical joining."

tions of Gemini and the Agena target vehicle were nearly optimum for manually controlled docking. "The Gemini crewmen were in an excellent position visually to monitor the condition of the docking gear and to control their docking maneuvers." Furthermore, "both craft had full attitude and translation control capability, both as separate and connected vehicles." In Apollo, the command and service module geometry did not permit the crew to see the docking gear on either the command or the lunar module. This had not posed a serious problem, but Johnson noted the desirability of visual monitoring of the docking mechanism and process in future spacecraft.

After some further discussion of Gemini and Apollo docking experiences and a short film illustrating the final approach of the *Apollo 12* command and service modules to the lunar module, Johnson turned to a fuller description of future concepts for docking gear. He told the Soviets that previous systems had functioned satisfactorily enough, "but our experiences . . . have pointed out areas where we feel that the docking gear of future spacecraft can be significantly improved." He then went on to outline seven design features that he and his designers believed would greatly facilitate docking operations in future spacecraft. The first four criteria emerged from the experiences with Gemini and Apollo; the latter three came from studies of future systems. Johnson elaborated on each of these points in turn.

Safety, of course, was the preeminent consideration. The docking gear should be fail-safe, "at least to the extent that the gear suffer no damage during impact when the spacecraft are misaligned too greatly to allow capture," Johnson said, and there should arise no situation in which automatic disengagement would be prevented. A failure to complete docking should under no circumstances preclude another attempt at capture. Johnson and his colleagues also believed that the astronauts should be able to transfer from one spacecraft to another without donning a spacesuit. This "worthwhile convenience" of shirtsleeve transfer required that the coupled spacecraft contain compatible atmospheres. While the Apollo and Soyuz cabins had dissimilar environments, Johnson told the Soviets that NASA was planning to use sea level pressure in the future, thus eliminating any transfer problems from that source. Another transfer-related consideration centered on eliminating any docking gear that might block the passageway between spacecraft. The Apollo probe assembly, which had to be removed after the command and service modules were latched to the lunar module, had proved very inconvenient. "Every attempt should be made to select a docking gear that does not block the very passage it intends to effect."[37]

The docking gear of all spacecraft to date—American and Soviet—had employed some variation of the probe and drogue. The probe, or male-like configuration, on one spacecraft would enter the drogue, or female-like

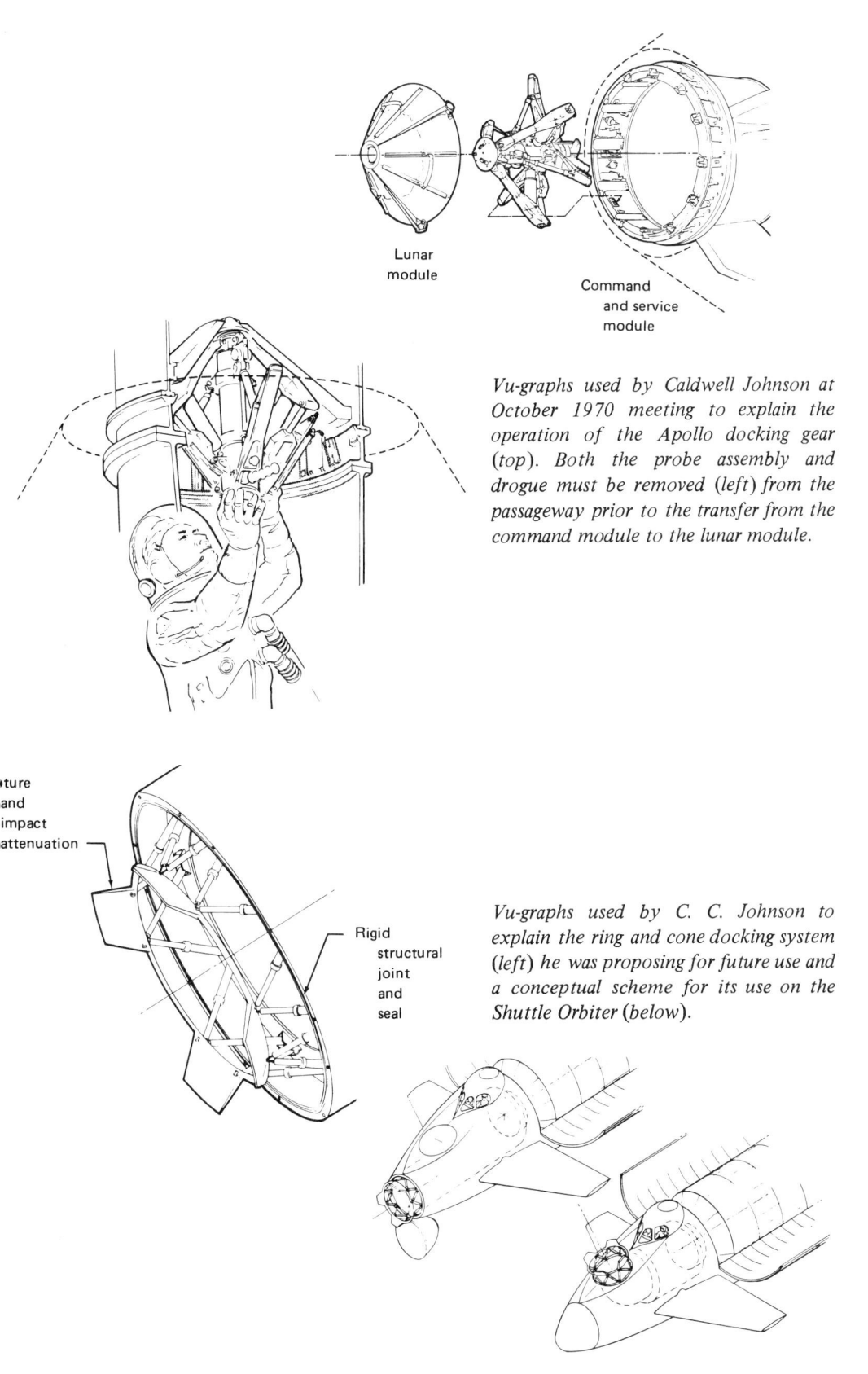

Lunar
module

Command
and service
module

Vu-graphs used by Caldwell Johnson at October 1970 meeting to explain the operation of the Apollo docking gear (top). Both the probe assembly and drogue must be removed (left) from the passageway prior to the transfer from the command module to the lunar module.

ture
and
impact
attenuation

Rigid
structural
joint
and
seal

Vu-graphs used by C. C. Johnson to explain the ring and cone docking system (left) he was proposing for future use and a conceptual scheme for its use on the Shuttle Orbiter (below).

configuration, on another spacecraft for docking. Johnson pointed out the shortcomings of such an approach:

> There was or is no manner in which two spacecraft with only "probe" gear can dock together, nor is there any manner in which two spacecraft with "drogue" gear can dock together. That constraint has not been inconvenient to our limited spaceflight activities to date; but, we think we should avoid that constraint in future docking gear. We think future docking gear should exist in a limited number of standard classes, and that any new gear of a given class should be able to dock properly with any other gear of the same class.[38]

Such an androgynous* docking gear should be designed so that either of the two spacecraft could dock and undock without the active support of the second vehicle.

As the final requirement in his list, Johnson saw the need for two structural modes for future docking gear, since subsequent spacecraft were expected to be much larger than the existing generation. Johnson doubted the desirability or practicality of using the docking mechanism to effect the structural joint between such craft. "We believe, rather, that the docking gear should be expected to provide only a relatively compliant structural joint; that the burden of rigid joining be assumed by the particular spacecraft's structural system." Caldwell then projected a Vu-graph of an androgynous docking system that combined all the design features he had mentioned.[39]

Described as a double ring and cone docking mechanism, this concept was one of Johnson's pet ideas. As was the case with many of his colleagues at MSC, Johnson had never really been satisfied with the Apollo probe and drogue arrangement. The history of that docking mechanism dated back to May 1962, when NASA and North American Aviation prepared a preliminary outline for space docking. As the investigation of docking gear progressed, seven different concepts were considered before the November 1963 decision to adopt the North American probe and drogue design.[40] Among the rejected ideas was the ring and cone concept designed in 1963 by Houston's Preliminary Design Section of the Advanced Spacecraft Technology Division. But rejection of this idea did not mean that it was dropped. Over the years, several men including Johnson continued to propose variations on this theme.

Johnson revived a variant of this docking gear in 1967 for the orbital workshop of the Apollo applications program, which became Skylab. He called his 1967 androgynous design a double interrupted ring and cone. The

---

Androgynous, a term taken from the life sciences, suggests the possession of characteristics of both sexes—sometimes called neuter.

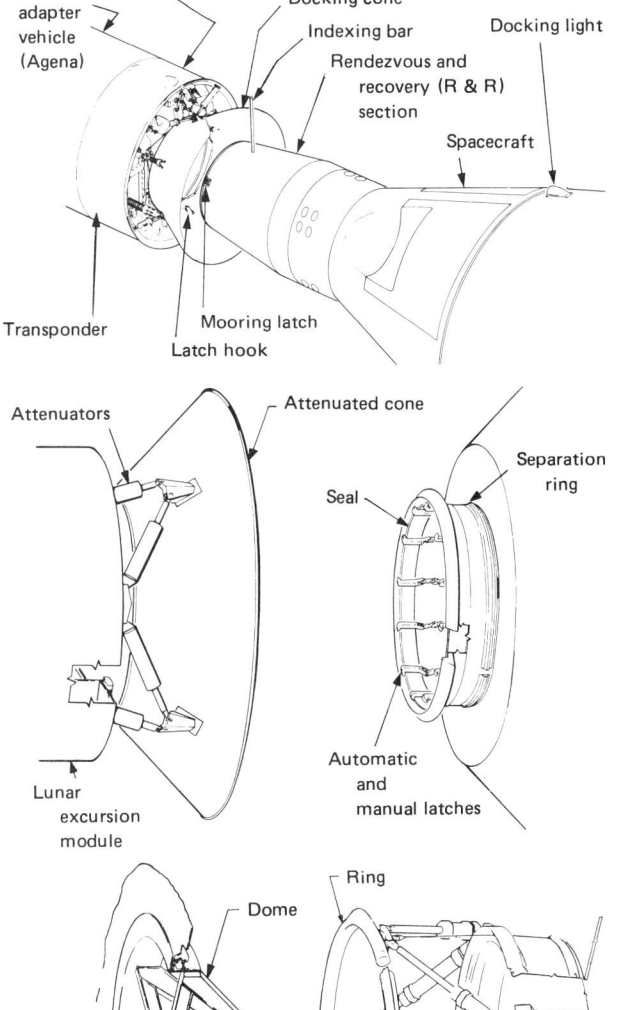

Docking
adapter
vehicle
(Agena)

Docking adaptor

Docking cone

Indexing bar

Rendezvous and
recovery (R & R)
section

Docking light

Spacecraft

Transponder

Mooring latch

Latch hook

Attenuators

Attenuated cone

Separation
ring

Seal

Automatic
and
manual latches

Lunar
excursion
module

Dome

Ring

Initial latch

Attenuator

Reel-in
cable

Lunar
excursion
module

Command
module

Evolution of Ring and Cone Docking System Concept

1961-1962. Solid cone as used in Project Gemini. Cone acted as guide for the active spacecraft during the docking.

Gemini concept as proposed in November 1963 for use in Project Apollo. When cone was sectioned to permit two to be put together, the sections became the guide fingers.

Cone and ring concept as proposed in November 1963 for use in Project Apollo. Attenuation system absorbed docking energies and permitted two systems to adapt to one another. Designed by J. Jones, W. Creasey, A. Bryant, and L. Ratcliff.

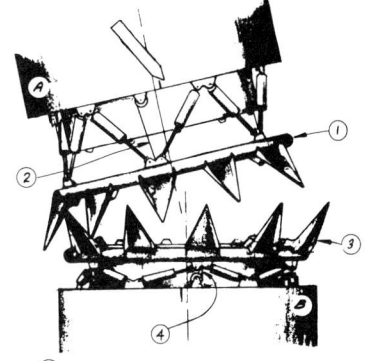

(1) GEAR FULLY EXTENDED
(2) NOMINAL TENSION ON REEL-IN CABLES
(3) GEAR IMMOBILIZED
(4) REEL-IN CABLE RELAXED

1967 double ring and cone docking system proposed by C. C. Johnson. After much analysis, Manned Spacecraft Center designers concluded that four fingers (or segments of a cone) would provide the best alignment for a universal docking system. J. Jones and T. Ross were leading individuals in the team which did the background work to this proposal.

cone was divided into 12 discrete fingers or guides so that the "cone" of one gear would match the "ring" of the mating gear and vice versa. "The fingers of one will exactly intermesh with the fingers of the mating gear." The proposed mechanism, which would not block the passageway between spacecraft, was androgynous, could accept a passive partner, and was fail-safe. Furthermore, Johnson had separated the capture latching mode from the structural latching mode.[41] His proposal was rejected again because the existing hardware was acceptable; the new concept did not possess sufficient superiority to merit such a change. Johnson had been working on a four-"finger" version of his earlier gear when he received word that he had been selected to visit Moscow. When Gilruth called his delegation together on 9 October, the designer had proposed to discuss his docking system with the Soviets, as being illustrative of one of the future approaches that NASA might take. Nobody knew how the Soviets would react to a discussion of hardware, but as it turned out, they were eager to talk about mechanical systems.[42]

While there were no specific Soviet comments regarding Johnson's presentation, they did give the Americans a detailed briefing on their Soyuz docking equipment. Vladimir Sergeyevich Syromyatnikov,* their 37-year-old mechanical design expert for docking systems, described the probe and drogue system currently used on Soyuz.[43] While similar in concept to the Apollo system, the Soviet "pin and cone" gear was not designed for internal transfer. Syromyatnikov told the Americans that the U.S.S.R. had adopted a docking mechanism without provision for internal transfer because it could be developed in less time. (This reinforced Johnson's opinion that there had been a "sense of urgency" associated with the development of Soyuz.) Returning to the main theme of this discussion, he reported that once capture was made, the Soviets employed an electric motor to retract the probe for final structural latching. In Apollo, the probe assembly was automatically retracted when the capture latches actuated the argon-gas-operated retraction mechanism. Lunney noted that the Soviet approach permitted repetitive docking and undocking, whereas Apollo was limited to two prime and two backup retractions. While the American system was sufficient for lunar missions, the heavier Soyuz docking equipment was more flexible.

The difference in approach to docking taken by the Soviets and NASA was also illustrated by the degree of precision required in the Soyuz docking

---

*While Syromyatnikov was unknown personally to the NASA representatives, his reputation as an aerospace engineer had been known to the NASA community since his appearance at the Fifth Aerospace Mechanisms Symposium at the Goddard Space Flight Center, 15-16 June 1970, where he spoke on aspects of the Soyuz docking system.

procedure. The Soviet docking equipment included electrical umbilical connectors contained in the face of the docking ring. These multiple prong and socket connectors required precise alignment, which the Soviets obtained by using 152-millimeter by 25-millimeter (6-inch by 1-inch) diameter guide pins. Once the head of the probe was engaged in the drogue, basic alignment having been accomplished by using docking targets, further alignment was completed by the guide pins of one craft entering sockets of the other craft. Like the American system, the Soyuz docking required matched pairs of spacecraft.

Syromyatnikov also talked briefly about a modified docking system that would permit internal transfers of crews and equipment. While this system had not yet flown, it appeared to be something that the Soviets planned to use in the relatively near future. Docking would be accomplished as in previous missions, but once the two ships were joined together the

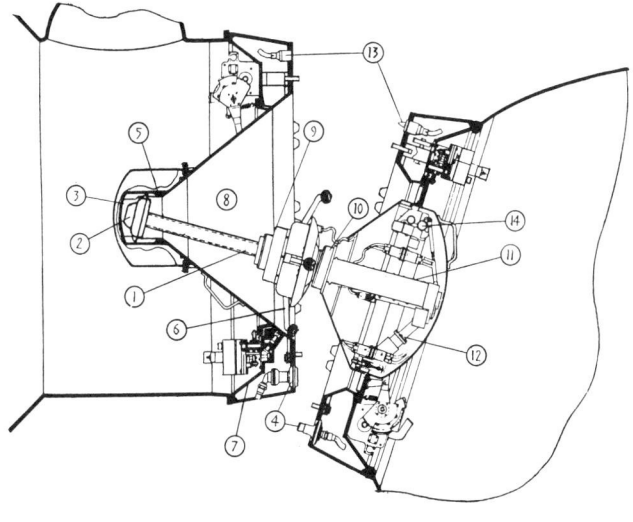

New Soyuz Docking Mechanism Described by V. S. Syromyatnikov at October 1970 Meeting in Moscow

Soyuz probe and drogue after initial capture: (1) probe; (2) probe head with capture latches; (3) drogue; (4) hydraulic umbilical connector; (5) stop; (6) alignment pins; (7) docking structural ring; (8) drogue cone; (9) electric drive for retraction of probe stem; (10) ball joint; (11) probe guide; (12) lateral shock absorber; (13) electrical umbilical; (14) electromechanical damper.

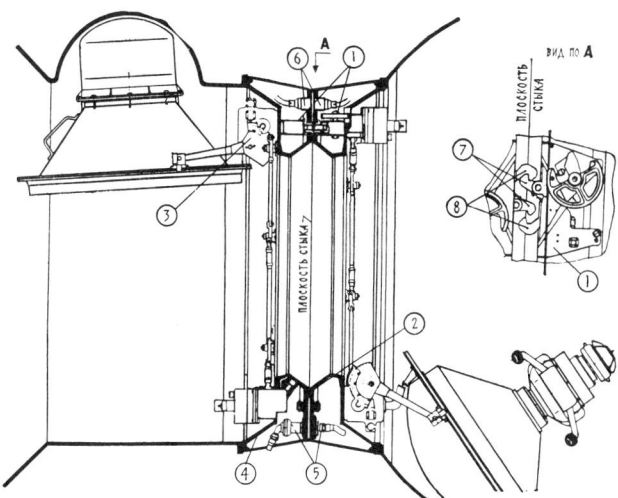

Soyuz docking assembly after latching has been completed and hatches have been swung open to permit transfer of crewmen: (1) peripheral latch; (2) docking interface seal; (3) hatch cover drive; (4) docking (structural) ring; (5) hydraulic umbilical; (6) electrical umbilical; (7) active hooks; (8) passive hooks.

probe and drogue assembly could be unlocked and swung out of the way. The passageway between the two vehicles would then be open for a shirt-sleeve transfer. This new mechanism was a real step forward from the first-generation docking system with its solid face, and the Soviets agreed to provide NASA with a fuller description.[44]

After a lunch break, the talks resumed with a discussion of Skylab by George Hardy. Skylab had grown from the desire to exploit more fully the launch vehicles and spacecraft that had been developed for the Apollo program. He told the Soviets that the Skylab hardware was designed in such a manner to permit it to be revisited, resupplied, and reused for extended earth-orbital missions. The flight lengths being projected for the three missions–28, 56, and 56 days–appeared to intrigue his audience, especially in light of the *Soyuz 9* flight the preceding June, which had lasted for a record 18 days. Hardy concluded with discussion on the rendezvous and docking operation associated with the program, showing the Soviets a model of the proposed spacecraft.[45]

Following a late afternoon adjournment, the Americans gave a party for their hosts at the home of the U.S. Embassy's Science Attache. During the socializing, Gilruth and Feoktistov shared stories and views on manned space flight. Since both men were among the "old timers" in their respective programs, they had a lot in common. Gilruth had wondered for years about who had been responsible for the development of the Soviet spacecraft, and he was particularly interested in Feoktistov's comments about having done the majority of the design work for Vostok, Voskhod, and Soyuz. But having risen to the position of Deputy Director of the Soviet manned space program, Feoktistov did not want to dwell on himself, so the conversation turned to other aspects of space flight–to Skylab, rotating space stations, and the ways one justifies manned space programs to scientists who prefer to use automatic probes. At evening's end, the Americans felt as if they had been to a reunion with old colleagues.[46]

Meeting again on Tuesday morning, the first hour was given over to further comments on Skylab by George Hardy and a description of Soyuz radio guidance equipment by V. V. Suslennikov. After that basic exchange of information, the men turned their attention to the topic of compatible systems to determine which aspects of that subject should be studied. At the end of this discussion, Feoktistov gave Gilruth a list of technical questions for which he felt the two sides should share answers. These questions indicated a basic concern in working toward compatible systems, and it seemed logical to all present that these problems should be divided into subject areas that teams of specialists could address. At Feoktistov's suggestion, three working groups were formed, following the precedent set by the 1962 Dryden-Blagonravov agreement. One working group would ensure the compatibility of overall methods and means for rendezvous and docking; another would concentrate on establishing compatibility between radio, optical, and other guidance and communications systems, and the final group would attend to compatibility questions related to docking assemblies and tunnels that might be created. The representatives then worked up a schedule of events that would guide their efforts during the next six months.

Assuming that there would be a joint meeting of the working groups in about six months, the two parties agreed to exchange further data. During November by correspondence they intended to trade technical materials on radio guidance and rendezvous systems, spacecraft atmospheres, and systems for voice communications. Later that winter, each side would send its counterpart a draft outline of those technical requirements that were considered essential to compatibility. This paper work would allow the two groups of engineers to get an idea of how the other worked. The spring 1971 meeting would then concentrate on further defining technical specifications for compatible systems, both sides having worked independently on preliminary designs. While all the men present were ready to get to work, no one expected their work to bear early fruit.

After lunch on Tuesday, Frutkin, Lunney, Feoktistov, and Ilya Vladimirovich Lavrov drafted an agreement incorporating the points discussed that morning. Feoktistov was ready once again with a draft. As Lunney later reported, they discussed the document for a relatively short time before coming to full agreement. Feoktistov's original proposal was "98% of what we signed the next day."[47] It became clear to the Americans that Feoktistov was a very efficient person and one of the prime movers behind the Soviet desire to develop complementary systems as soon as practical.

With their work out of the way, the Americans went on a tour of the lunar science laboratories and then had a brief discussion with M. V. Keldysh. The NASA representatives were impressed with both, developing an even deeper appreciation for the capabilities and accomplishments of Soviet space personnel. Gilruth especially understood Keldysh's comments about having to continually justify the space program to budgetary planners. After reviewing with his guests the progress that had been made, Keldysh invited them to dinner at the Prague restaurant. Following a pleasant evening of "shop talk" with Keldysh, Blagonravov, Petrov, and Feoktistov, the Americans returned to their hotel to rest up for their final day in Moscow.

On the morning of 28 October, the NASA and Soviet representatives assembled at the Presidium of the Soviet Academy to sign the "Summary of

Results." In contrast to the ornate physical surroundings, the ceremony was simple but impressive. There was no smugness in their sense of accomplishment, but a feeling that the time had come for the two nations to cooperate in space. In just three days, they had reached an agreement to work together; now they would have to make good their pact.

WORKING THE PROBLEM

When they returned to Houston, Gilruth, Lunney, Johnson, and Hardy sat down to discuss their accomplishments and the tasks ahead of them. The four men agreed that the discussion had been open and frank, and the problems they had anticipated had never materialized. Language differences had been their only barrier. Hardy felt that the Soviets "seemed very interested in achieving . . . and implementing some agreement to capabilities for compatible docking." He believed that they had done everything "that they knew how to do, to exchange information. . . . It was their suggestion . . . that we exchange additional information with more details." Hardy and the others, however, did not get a feeling for the Soviets' motives. Hardy continued, "I don't say this suspiciously, I just say it wondering. . . . it would seem to me that a rather significant policy decision on the part of NASA or maybe the Administration is in order." Now that the door was cracked, he saw the possibility of making an overture to engage in "a significant venture of some sort in the immediate future, or . . . to continue to discuss compatible docking in . . . the abstract."[48]

Caldwell Johnson was concerned about attempting to design systems in the abstract. He felt that considerable substance needed to be added to the discussions; for example, designing systems adaptable to current spacecraft rather than designing hardware for some unknown future vehicles. Gilruth suggested that the initiative lay with him and his three companions. "We're the ones who are going to have to determine whether or not it's feasible. And whether or not we want to do it." Policy decisions would follow after their recommendations. Realizing the significance of their position, the men agreed that they would have to "wring out" thoroughly any proposals and not "go off half cocked."

Speaking to the question of a real versus an abstract project, Lunney argued that a decision in favor of a more concrete effort was implicit in the schedule proposed by Feoktistov. Lunney believed that the Soviet Deputy Director realized the implications of the schedule. "I think he knew that we would have to go home and decide what applicability we were interested in." Hardy added that he remembered hearing Keldysh say that he had invited George Low to visit Moscow for wider ranging talks on cooperation in space science. While the NASA delegation had not commented on it at the time, Hardy felt that should Low accept the invitation and should the timing of his visit coincide with the January exchange of technical requirements, "then he could possibly bounce this thing around a little bit . . . to see if we're in fact on the right track or way out in left field." Gilruth concurred, and said further that it might be appropriate for Low to present a gift to the Gagarin Museum at the same time, since the U.S. was conspicuous for its failure to remember the first man in space.[49]

Pursuing this thought on the need for concrete discussions, Caldwell Johnson decided to set down on paper some ideas for possible missions. In a 3 November document, "Initial Efforts toward the Development of Compatible Rendezvous and Docking Hardware and Software for USSR and US Spacecraft," he presented several considerations to be studied by the personnel of the Spacecraft Design Office. He strongly felt that the designers should concentrate on developing hardware for a spacecraft currently being flown by the two nations, and he explained his rationale:

> Since the approved manned spaceflight programs of the US [are] comprised of Apollo and Skylab A, and possibly, exploitation of surplus Saturn and Apollo hardware; and, since the USSR manned space-flight program appears to be limited to earth-orbital missions utilizing a single or two docked Soyuz spacecraft, initial efforts toward the development of compatible rendezvous and docking hardware and software should emphasize those spacecraft and missions.

Johnson thought that this approach would not prevent consideration of rendezvous and docking between spacecraft still in the planning stages, but he felt that work on future systems should be limited to "development of generalized requirements and concepts rather than engineering solutions for hypothetical problems."[50]

The American designer believed that the initial efforts toward development of compatible systems should begin by studying the technical possibilities of two broad classes of Soviet-American flights—scheduled and non-scheduled earth-orbital missions. The scheduled flights provided three possibilities: Soyuz could dock with Skylab to demonstrate the feasibility of such an operation, conduct an experiment in cooperation with the American crew, or occupy Skylab after the NASA crew had departed. Or Apollo could dock with Soyuz or act as a propulsion stage to place the Soviet craft in a "different orbital situation." Finally, Soyuz could dock with Apollo to prove its ability as the active rendezvous partner. The non-scheduled possibilities were essentially rescues performed by one nation for the other.[51]

To implement these studies, Johnson drew up a list of tasks to be performed at MSC. These spacecraft docking studies called for further work on the double ring and cone docking gear, a clearer definition of the new internal transfer docking gear developed by the Soviets, an initial investigation of mounting the new Soyuz probe or drogue in the Apollo CSM, and a "first-cut" study of the technical feasibility of docking existing Soviet and American spacecraft. While Clarke Covington of the Advanced Earth-Orbital section of the Spacecraft Design Office supervised this investigation, René Berglund of the Advanced Missions Office collected materials to send to the Soviets in November. At Headquarters, George Low and Arnold Frutkin briefed the White House (Henry Kissinger) and the State Department (U. Alexis Johnson). Low confirmed the acceptability of the "Summary of Results" by letter to Keldysh and prepared a response to Keldysh's letter inviting the Acting Administrator to Moscow.[52]

As he had indicated to his visitors, Keldysh wanted Low to visit the Soviet capital to discuss the broader possibilities of cooperation in the space sciences. Low responded in late November, the day following the transmittal of the docking documents from MSC, saying that he would be very happy to travel to the U.S.S.R. for discussions with Keldysh and the Soviet Academy of Sciences. He had been "influenced by the technical discussions on rendezvous and docking which began so auspiciously in Moscow last month . . . it may be that we should give priority to a few selected items which could be defined and treated in a very concrete fashion analogous to the rendezvous-and-docking case." Low then went on to list four areas in which substantive cooperation could be undertaken—updating the mid-1960 agreements on developing better weather forecasting; broader sharing of scientific data (including the exchange of lunar samples); pooling knowledge of space biology and medicine; and jointly exploring the oceans by satellite.

In a letter sent to Washington on 4 December, Keldysh agreed to Low's agenda proposals and the mid-January meeting date his counterpart had suggested.[53]

With the initial docking studies underway, Low's pending visit to the Soviet Academy of Sciences in January 1971 would give him an opportunity to discuss further the topic of manned space flight with the Soviets. Indeed, the in-house studies at the Manned Spacecraft Center took on new significance, as Clarke Covington oversaw the preparation of a document that would outline the various docking methods for Apollo and Soyuz. By the end of December, NASA was preparing to suggest to the Soviets that a real test mission might be not only feasible but more desirable than drawn-out discussions about abstract, hypothetical missions at some unspecified time in the future.

V

Proposal for a Test Flight

In October 1970, Academician Keldysh responded to a September 1969 letter from Administrator Paine, agreeing that the "limited character" of Soviet-American cooperation in space science and applications could be broadened.[1] After the two sides had decided on a January 1971 meeting in Moscow to discuss this subject, NASA Acting Administrator Low set about choosing a delegation and determining the agency's position on those topics proposed for the agenda.[2] Acting on the advice of Arnold Frutkin, Low opted for a small delegation composed of individuals able to discuss a broad range of subjects rather than specialists.* Low and Frutkin thought it best to draft beforehand the agreements as they would like to see them signed, so that the Acting Administrator would always have in front of him the goals they wished to achieve. When he left Washington, he had a complete set of proposed agreements and a draft press release, as well.[3]

Before departing, Low was briefed by Under Secretary of State Alexis Johnson on the heightening diplomatic tension between the Soviet Union and the United States. The Soviets had just concluded a trial viewed in the U.S. as having anti-Semitic overtones, involving a group of accused airplane hijackers. Even as two of the Soviet Jews charged with the crime appealed their death sentences, the first ever levied for hijacking in the U.S.S.R., the Jewish Defense League had undertaken a campaign of bombing Soviet installations and intimidating Soviet personnel in New York and Washington. On 4 January, Soviet Ambassador Dobrynin delivered a note to the State Department accusing the American Government of "connivance" in these hostile acts and warned that the Soviet Government could not guarantee the safety of American officials and businessmen in Moscow.[4] Although Johnson told Low that he did not anticipate any difficulties for an official delegation, he did voice his concern about public statements that Low might

---

*Low was accompanied by Frutkin; John D. Naugle, Associate Administrator for Space Science and Applications; Arthur W. Johnson, Deputy Director, National Environmental Satellite Service; William Anders, Executive Secretary, National Aeronautics and Space Council; and Robert F. Packard, Director, Office of Space-Atmospheric and Marine Science Affairs, Department of State.

make at the end of the negotiations and cautioned him to check with the Embassy in Moscow before making a favorable release to the press if the diplomatic situation were to worsen.[5]

As preparations in Washington progressed for George Low's visit to the U.S.S.R., the manned spacecraft team in Houston was working on a set of alternative proposals for a flight using Apollo and Soyuz hardware. Shortly after returning from the Soviet capital in October, Bob Gilruth had suggested to Low that subsequent discussions with the Soviet Academy would be more productive if the two sides began talking about specific missions using existing spacecraft.[6] Gilruth and Caldwell Johnson had conducted an intensive feasibility study at the Manned Spacecraft Center (MSC) and presented their findings to Low on 5 January.

Based upon the rapid exchange of technical data and the tone of his recent correspondence with Keldysh, Low decided that it might be worthwhile to raise the possibility of a joint flight.[7] He was willing to increase the tempo of the compatibility talks with the Soviets, for both he and Frutkin believed that the whole approach of the U.S.S.R. toward cooperation had changed. Reflecting on the October 1970 meeting, Frutkin later said "that meeting was clearly different from anything we had ever had before with them." In the Dryden-Blagonravov era, meetings involved only the very senior personalities in the Academy of Sciences; "you didn't feel that you were dealing with people who got grease on their hands." October had been different. "The protocol was minimal, and business was clearly foremost," Frutkin added. He had been impressed by Suslennikov, Syromyatnikov, and especially Feoktistov, whom Frutkin had found "extremely able and very efficient . . . with no nonsense."[8] NASA's interest in obtaining more immediate results with the Soviets was boosted by this new working relationship.[9]

On 12 January 1971, a week before leaving for Moscow, Low and Frutkin flew to San Clemente, California, to discuss NASA's negotiating plans with the President's Foreign Policy Adviser, Henry Kissinger. Low briefly outlined the events leading to Keldysh's invitation and summarized his strategy for the meeting on space science and applications. In response to Low's request for the Administration's position on an actual test mission using Apollo and Soyuz spacecraft, Kissinger replied that as far as the White House was concerned Low had a completely free hand to negotiate in any area that was within NASA's overall responsibility. The President, Kissinger said, was in full support of these meetings and personally wanted Low to express to the Soviets his desire for cooperative efforts in space research and technology. Kissinger had only one request of the Acting Administrator; he would prefer that NASA personnel not contribute to the false notion that if

PROPOSAL FOR A TEST FLIGHT

they could reach technical agreements they could also solve political problems if given the opportunity. Kissinger felt that in the past some of the astronauts had tried to suggest that since it was easy to negotiate with the Soviets on space topics it should be equally simple in other areas. Such naivete on the part of highly publicized individuals only hampered the work of diplomats on both sides. In parting, Kissinger told Low: "As long as you stick to space, do anything you want to do. You are free to commit—in fact, I want you to tell your counterparts in Moscow that the President has sent you on this mission."[10]

Low and his party arrived in Moscow late Saturday afternoon, the 16th of January. Their reception at the airport was warm, and Keldysh was there to greet them. While the Americans waited for customs formalities to be completed, Low chatted with Keldysh and the Vice President of the Academy, Aleksandr Pavlovich Vinogradov, who had just returned to Moscow from Houston. They talked about the upcoming Apollo 14 mission, *Luna 16*—the topic of Vinogradov's presentation at MSC*—and *Lunokhod,* the unmanned moon rover that was still ranging widely over the lunar surface. There was no sign of any coolness or hostility, and once again it appeared that the desire to cooperate in space exploration outweighed any extraneous political events.[11]

Although Low asked for a reprieve from extensive sightseeing that night, he and his colleagues had a pleasant dinner with Keldysh, Blagonravov, and several other Soviets. Low and Keldysh talked about manned versus unmanned flights and the importance of space programs to the support of science and technology. Manned flights, they agreed, were essential "to lift the human spirit." They both felt that the United States and the Soviet Union must compete and cooperate in space—compete because they needed the contest to spur the nations on and cooperate because of the vastness of the universe and the number of problems that needed to be solved.[12]

*Vinogradov presented a paper, "Preliminary Data on Lunar Ground Brought to Earth by Automatic Probe 'Luna-16'," at the Second Lunar Science Conference sponsored by the Lunar Science Institute, held in Houston, 11-14 Jan. 1971.

Academician Aleksandr Pavlovich Vinogradov, left, examines a lunar rock collected on the Apollo 12 *mission. Assisting the visitor to the Manned Spacecraft Center is Dr. Michael B. Duke, center, curator in the Lunar and Earth Sciences Division. MSC Director Robert R. Gilruth looks on.*

At the Presidium of the Soviet Academy of Sciences, the Soviet and American negotiators face one another at the conference table in January 1971. Dr. Low and Academician Keldysh (below) headed the delegations and signed the agreements (Soviet Academy of Sciences photos).

After four days of detailed and physically exhausting negotiations,* Keldysh and Low initialed an agreement calling for fuller cooperation in five specific areas:

1. to improve the current exchange of data from meteorological satellites and consider alternative possibilities for coordinating systems;

*The Soviet delegation consisted of M. V. Keldysh; A. P. Vinogradov; B. N. Petrov; G. I. Petrov, Director, Institute for Space Research; I. P. Rumyantsev, V. S. Vereshchetin, I. V. Meshcheryakov, and A. I. Tsarev, Intercosmos; M. Ya. Marov, Institute of Applied Mathematics; Ye. K. Federov, Chief, and L. A. Aleksandrov, Deputy Chief, Main Administration Hydrometeorological Service; N. N. Gurovskiy, Chief, Directorate, Ministry of Health; O. G. Gazenko, Director, Institute of Medical-Biological Problems, Ministry of Health; Yu. A. Mozzhorin, Professor, Moscow Physics-Technical Institute; V. P. Minashin, Chief, Main Administration of Space Communication, and I. Ya. Petrov, Deputy Chief, Main Administration of Space Communication, Ministry of Communications; and K. G. Fedoseyev, Deputy Chairman of the USA Section, Ministry of Foreign Affairs.

2. to formulate cooperative provisions for a program of meteorological rocket soundings;

3. to study the possibility of conducting natural environment research by coordinated surface, air, and space measurements over international waters and specific ground sites;

4. to define and exchange information on the objectives of space, lunar and planetary exploration, to consider the possibility of coordinated lunar exploration, and to exchange lunar surface samples already obtained; and

5. to develop procedures whereby detailed space biology and space medicine data could be more regularly exchanged.[13]

Although the Soviets would have preferred to sign a more general set of statements, Low and the other Americans stressed the need for specific agreements. The U.S. delegation felt that the Soviets were surprisingly cooperative and open in their approach, aside from some routine haggling over wording. From the start, Keldysh had understood Low's concern for specificity and practicality in the agreements and had seen to it that a compromise was reached.[14] While the news media reported favorably on the proposal to exchange lunar samples, Low and Keldysh met privately to discuss an even bolder plan.[15]

A NEW PROPOSAL

Early on Wednesday, the 20th, while the negotiations were still in progress, Low and Frutkin met with Keldysh to talk about rendezvous and docking. Having been advised of the subject, Keldysh had asked Feoktistov to join them. Low said that NASA would like to propose the development of compatible systems for use with Apollo and Soyuz rather than with future spacecraft. He explained this idea in some detail, pointing out to Keldysh that the Americans did not yet want to make this a formal proposal but instead only wished to present it for the Soviets' consideration. Low remarked that Gilruth favored focusing on the development of equipment and systems for existing spacecraft to give the specialists in the two countries something much firmer with which to work.

Both Keldysh and Feoktistov were intrigued, and they said that although they were not free to commit their government to such a project they wanted to pursue this subject further and hear more details. Then they could advise their superiors and obtain a decision. Low agreed to send with the exchange of technical requirements scheduled for February a fuller description of the type of project he was proposing. Keldysh asked Low to refrain from mentioning this conversation publicly until there had been consultations internally. The two sides would subsequently make a public

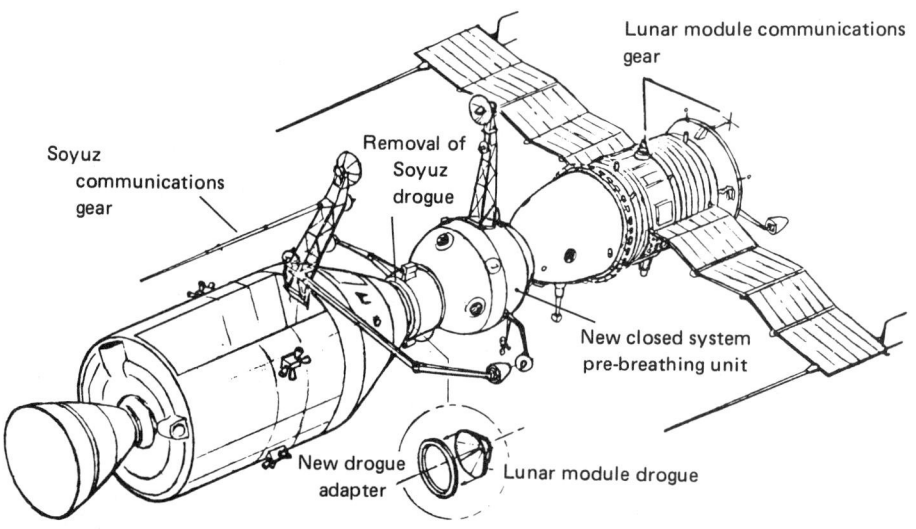

Mission profile

- Command and service module and Soyuz rendezvous and dock
- Soyuz pilot suits and pre-breathes
- Depressurize Soyuz orbital module to 5 psi air (assumes Soyuz capability to stop at 5 psi)
- Equalize pressure and open hatches
- Two command module crewmembers transfer in shirtsleeves to Soyuz with new pre-breathing unit (portable bottle backup)

- Plug in Soyuz intercommunication system; connect command module umbilical
- Close hatches; repressurize orbital module to 14.7 psi
- Visit in Soyuz orbital module; voice and TV over umbilical to command module to manned space flight network
- Return; pre-breathe on new unit
- Depressurize orbital module to 5 psi air
- Return to command module using pre-breathing unit

(a)

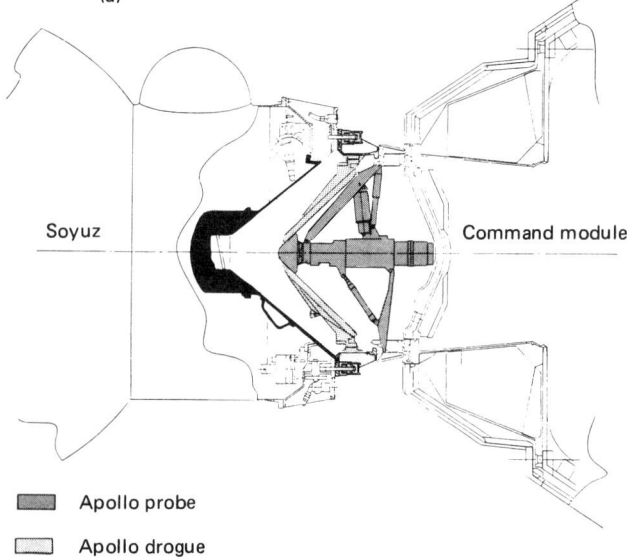

Conceptual drawings created by the design team headed by Clarke Covington these were done prior to the January 1971 Low-Keldysh meeting: (a) proposal for a minimum-modification approach to an Apollo-Soyuz docking mission; (b) Apollo drogue in Soyuz cone adapter, proposed to permit docking and transfer with minimum modification; (c) proposal for an airlock adapter that would facilitate transfer between spacecraft; (d) initial concept for an Apollo-Soyuz airlock adapter.

▨ Apollo probe

▨ Apollo drogue

▯ Apollo drogue adapter ring attached to Soyuz forming docking interface

▨ Soyuz drogue

(b)

announcement if this developed into a formal topic for negotiation. Low agreed to this arrangement.*[16]

Low based his discussions with Keldysh concerning a joint rendezvous and docking mission on the "USSR/US Docking Studies" prepared at MSC in late December 1970. In Houston, Clarke Covington had prepared materials on the two aspects of possible docking activities—the near and far term. For the former, he and his colleagues proposed feasibility studies of

*At this meeting in Moscow, Low had also presented to the cosmonauts a plaque designed by Gilruth to be placed in the Gagarin museum. As Low said to Gilruth in a 27 Jan. 1971 letter, "It was an emotional moment, and it was obvious that they were pleased at the recognition by us of their being first in space."

Mission profile

Command and service module launches first
Command and service module extracts air-lock adapter from spacecraft/lunar module adapter
Soyuz performs active rendezvous and docking
Pressurize airlock to 5 psi (P_{O_2} = 3.5 psi)
Two command module crewmembers transfer in shirtsleeves to airlock
Pressurize airlock to 14.7 psi air
Equalize pressures and open Soyuz hatch
Visit in Soyuz orbital module
Communications and TV from airlock
Return; pre-breathe on new unit
Command module crew returns to airlock; closes hatches
Depressurize airlock to 5 psi; breathe with mask and walkaround bottles

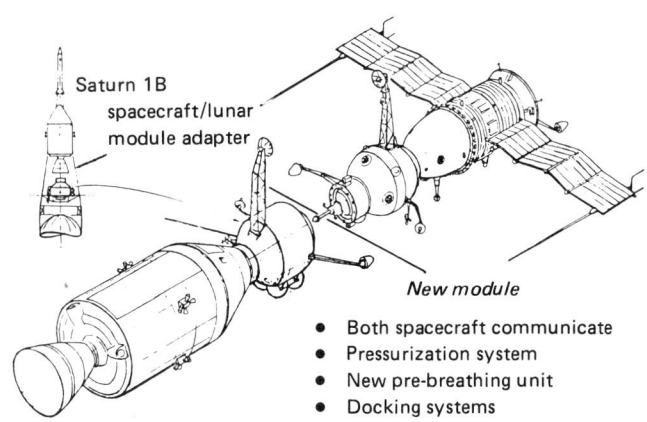

Saturn 1B
spacecraft/lunar
module adapter

New module
- Both spacecraft communicate
- Pressurization system
- New pre-breathing unit
- Docking systems

- Equalize pressure and open command module hatch
- Transfer to command module
- Close hatch

(c)

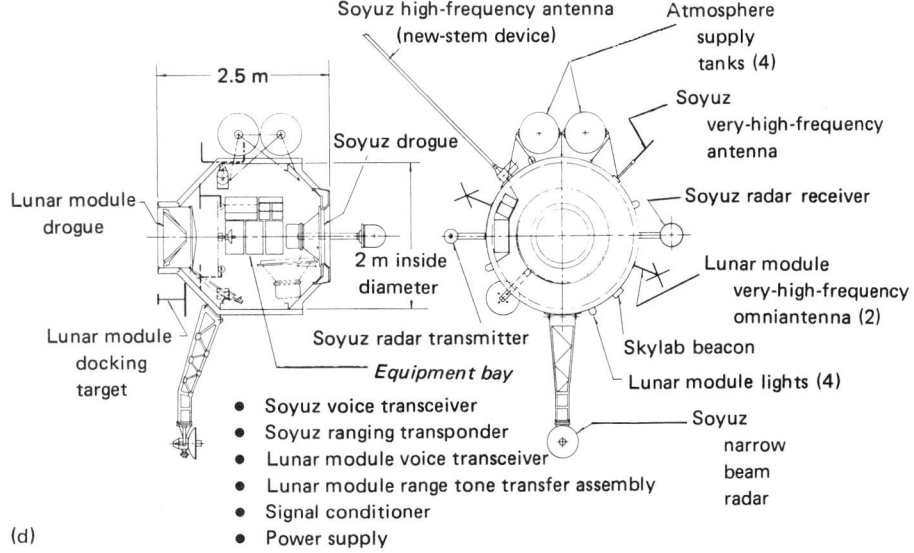

Soyuz high-frequency antenna (new-stem device)

Atmosphere supply tanks (4)

2.5 m

Soyuz drogue

Soyuz very-high-frequency antenna

Lunar module drogue

Soyuz radar receiver

2 m inside diameter

Lunar module very-high-frequency omniantenna (2)

Lunar module docking target

Soyuz radar transmitter

Equipment bay
- Soyuz voice transceiver
- Soyuz ranging transponder
- Lunar module voice transceiver
- Lunar module range tone transfer assembly
- Signal conditioner
- Power supply

Skylab beacon

Lunar module lights (4)

Soyuz narrow beam radar

(d)

specific docking missions and specific hardware systems that could be flown between 1972 and 1975. For the far term, the specialists suggested that the Joint Working Groups develop technical requirements and general concepts for the docking of future systems "as a continuing show of good faith."[17]

In effect, the MSC proposal inserted a new activity into the scheme of things as they had been agreed to earlier in Moscow. The primary focus of the October agreement had been work on compatible systems for future spacecraft. Now Caldwell Johnson, Covington, and their associates were pushing for a real mission using existing hardware. MSC specialists had listed several important guidelines. Such a joint mission should provide a public demonstration of a viable joint activity and as such should allow both countries to exhibit equal skill and effort. But above all, it should be an open, non-military enterprise that would continue NASA's philosophy of peaceful exploration in space.

To define the hardware needed for a rendezvous and docking mission, Will Taub had drawn a series of sketches showing variations on an Apollo-Soyuz mission. Covington used these in December 1970 when he briefed MSC management on five mission possibilities:

Concept 1   CSM and Soyuz dock without a crew transfer.
Concept 2   CSM and Soyuz dock with an extravehicular transfer.
Concept 3   CSM and Soyuz dock with an internal transfer, possibly without prebreathing; i.e., a "shirtsleeve transfer."
Concept 4   CSM and Soyuz dock to an adapter module that would permit shirtsleeve transfer.
Concept 5   CSM and Soyuz dock to a more elaborate "experiment module" that would permit extended scientific activities.[18]

Gilruth and his deputy, Chris Kraft, quickly decided that the fifth concept was too elaborate; they argued for keeping the system simple. They believed that in the absence of a political commitment from the Nixon Administration and because this was an unsolicited proposal, it would be best to suggest a "minimum meaningful" activity to the Soviets and then await their reaction. Thus, when Covington later briefed Headquarters before Low's trip, he dropped concepts 1 and 5 and replaced them with a new suggestion that called for both spacecraft, flying without structural modification, to rendezvous and stationkeep, but to make no attempt to dock. In addition, he described a possible rendezvous—with and without docking—of Soyuz with Skylab.[19]

Two important points had to be considered for any of the docking missions—the docking gear to be used and the impact of cabin atmosphere on crew transfers. For an Apollo-Soyuz linkup, the hardware proposals ranged from a simple adaptation of the existing gear to the creation of a special docking module with Apollo gear on one end and Soyuz gear on the other. The minimal changes to the docking equipment called for building an adapter that would permit the installation of a lunar-module-type drogue into the cone of the Soyuz. Then the Apollo could dock and latch its probe into this adapted Soyuz. This particular modification could be varied for use with either the solid face or the swing-away Soyuz docking mechanisms. A more elaborate alternative called for building an "airlock docking adapter," a mini-spacecraft that would be carried into orbit in the spacecraft/lunar module adapter (SLA) behind the command and service module (CSM).* Following the CSM's docking and removal maneuver with the airlock module, the Soyuz could dock with it, employing the standard Soyuz probe. While crew transfer in the simple system would be either internal or external depending upon the type of Soyuz docking interface, the airlock module concept assumed the use of the swing-away hatch on Soyuz.

Docking was only half the story; the differences between spacecraft environments had to be considered in any plans to transfer crews. Based on the rather limited information available about the Soyuz life support system,† NASA specialists assumed that crew transfer would likely occur at the normal operating pressures for both spacecraft, requiring the men moving from the higher to the lower pressure to pre-breathe. The cabin pressure of Apollo could not be raised above 414 millimeters of mercury (8 psi) because of structural limitations in the CSM, and the Soyuz cabin pressure could not be lowered much below that without significantly increasing the risk of fire, as the percentage of oxygen increased in the total volume of the remaining gases. While the obvious solution would have been compromise on cabin pressure at about 414 millimeters, this would have required substantial modifications, which at the time seemed to be contrary to the desire to make the fewest possible changes to the basic spacecraft. If the two spacecraft were flown with their standard atmospheres, oxygen would have to be pre-breathed prior to entering Apollo to prevent the bends. In an effort to provide for crewmember oxygen without adding additional oxygen to the Soyuz atmosphere, the MSC environmental control specialists fully expected to develop a new closed system portable life support mechanism to provide oxygen and recycle carbon dioxide for the Americans. Work had begun on such a unit in an effort to eliminate the problem of oxygen enrichment and any increased danger of fire during the pre-breathing

---

*The SLA, an 8.5-meter truncated cone between the service module and the launch vehicle instrument unit, enclosed the lunar module (LM) during launch and on its way to the moon.

†The information available to NASA included materials that had appeared in the American press over the years, those obtained during the October 1970 trip, and the report sent to Houston in the first technical exchange.

period. And as always, the risk of fire was the primary worry of the American environmental control systems designers.[20]

Developing an airlock module would have solved some of the problems involved in changing the pressure in either spacecraft. If the astronauts wanted to transfer to Soyuz, they would enter the airlock, close the hatch behind them, raise the pressure to 760 millimeters, and then enter the Soviet spacecraft. Going the other direction, they would enter the airlock after pre-breathing oxygen aboard Soyuz (or alternatively in the airlock itself) and when it was safe lower the pressure to 258 millimeters. Throughout this process, the pressure in each craft would remain virtually unchanged.

At this point, complexities of design seemed to abound. If the crewmen pre-breathed in the airlock module, then a full life support system would have to be included in that mini-spacecraft. If the pre-breathing occurred in Soyuz, a simpler life support system could be used in the airlock module, but the Americans would have to transfer in their suits, requiring provision for suit cooling circuits aboard Soyuz. Walter W. Guy of the Crew Systems Division urged the specialists to find a simpler way to conduct the transfers. Otherwise, life was going to be too complex for the crews.

In addition to making the transfer process somewhat easier and reducing further the possibility of oxygen enrichment to the Soviet craft, the airlock had several good features from a designer's point of view. In the first instance, all Soyuz docking aids could be secured on the exterior of the module, thus eliminating major changes to the CSM. Second, the interior surfaces of the airlock module would provide places for mounting various communication and power units that would otherwise have to be added to the CSM or to the Soyuz. But the airlock module was a new piece of hardware that would have to be designed, built, and tested. This was the major objection raised by both MSC and Headquarters.[21]

George Low, Wernher von Braun,* and others at Headquarters were interested in pursuing the simpler drogue-in-cone adaptation, and it was this type of system that Low had considered in January when he had talked to Keldysh in Moscow. So in February 1971, NASA transmitted two documents to the Academy of Sciences, the first fulfilling the 1970 agreement to exchange "technical requirements for rendezvous and docking." "Preliminary Rendezvous and Docking System Requirements for United States Spacecraft" was generated to provide the Soviets with only an "overview of NASA . . . requirements and systems," not specific solutions to compatibility issues. From this general paper, the planners hoped to move on to more detailed discussions.[22]

---

*Von Braun had been appointed Deputy Associate Administrator for Planning in Mar. 1970.

PROPOSAL FOR A TEST FLIGHT

The second paper—"A Concept for a Union of Soviet Socialist Republics/United States of America Rendezvous and Docking Mission"—was prepared by MSC personnel under the direction of René Berglund. Although drawn up in a relatively short time and based upon a still limited understanding of the Soyuz docking system, the document drafted by William K. Creasy and Thomas O. Ross, among others, was a rather detailed study of the docking interface for Apollo and Soyuz, presented as an illustration. Similar details were given for the necessary docking targets, communications equipment, and pre-breathing apparatus. These studies were designed to outline the way NASA would create compatibility and conduct a joint mission.[23]

In his letter of 17 February transmitting the two documents to Petrov, Gilruth explained why he sent the paper proposing a joint Apollo-Soyuz flight. Since this topic had been discussed by Low, Keldysh, Frutkin, and Feoktistov during January, the MSC staff had looked into the whole question of compatible systems. "In the process of our deliberation on this subject," Gilruth noted, "we have found the postulation of a specific docking mission and spacecraft configuration useful in understanding potential problem areas." He also told Petrov that analysis of such a typical mission concept—Apollo and Soyuz—should "be a beneficial way of assessing compatibility during the March/April Working Group Meetings."[24]

Gilruth then addressed the agenda for that spring gathering. "With regards to these detailed Working Group activities, I believe that a preliminary meeting . . . should be held to establish the types of spacecraft to be considered by the Working Groups." Not everyone need be present, Gilruth suggested, but he did "feel that the participation of the chairmen of our respective Working Groups would be most beneficial." For such discussions, Gilruth had appointed Glynn Lunney, Donald C. Cheatham, and Donald C. Wade to chair groups one through three respectively, and they would be joined by Arnold Frutkin, George Hardy, Caldwell Johnson, and René Berglund. "Should this suggestion meet with your approval," Gilruth continued, "I would like to invite you and your delegation . . . to the Manned Spacecraft Center, Houston, Texas, in March to conduct these discussions." Afterwards the full Working Groups could meet and begin their efforts.[25]

During the six months following the October meeting, MSC had begun to find some minor problems that would have to be worked out as they continued to expand the scope of their joint work—language and communications being two examples. Preliminary studies conducted at Houston, based on available Soviet data, opened as many questions as they answered. These new questions confirmed the necessity for additional information exchanges. There was also evidence that both sides would have to come to agreement on technical translation, so that each side could be assured that the other understood precisely what had been meant by specific words, phrases, and documents.[26]

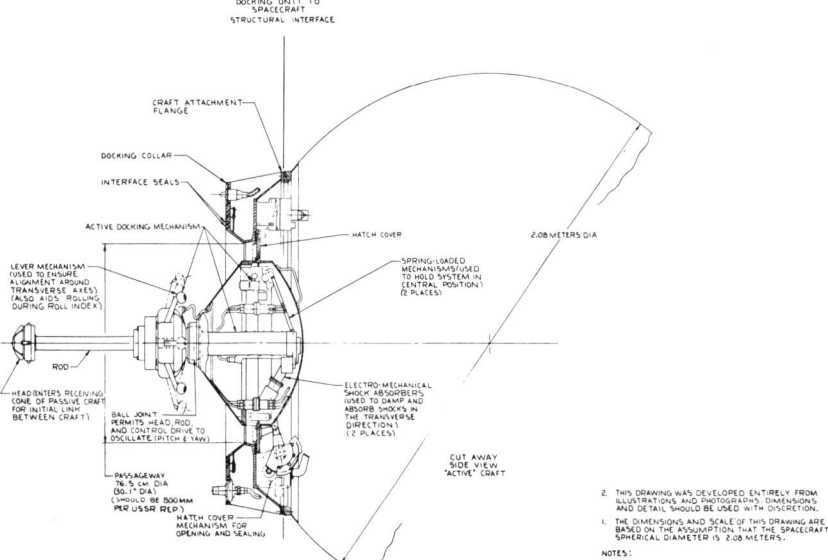

Two detailed views of the new Soyuz docking mechanism, prepared by W. K. Creasy and T. O. Ross at the Manned Spacecraft Center to give the NASA team a clearer understanding of how that system operated. One of the notes on the drawings reads, "The dimensions and scale of this drawing are based on the assumption that the spacecraft spherical diameter is 2.08 meters." As they later discovered, it was actually 2.2 meters. Otherwise their drawing was correct.

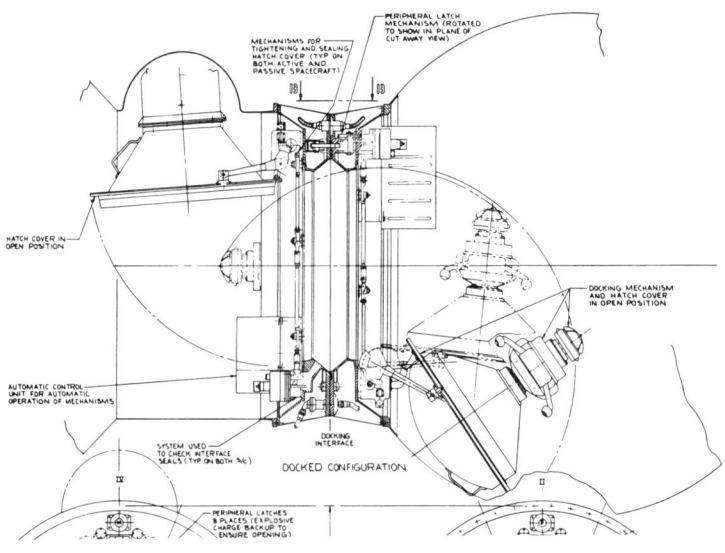

Then there was the time-gap problem in coordinating these communications. For instance, when Gilruth sent a draft of his letter to Petrov to Washington for approval on 12 February, Houston was still thinking about a meeting scheduled for March/April. But these plans were to be altered several times before the meetings finally took place. Academician Petrov sent his response on 15 March to Gilruth's letter dated 17 February. Petrov's response, along with seven documents that constituted the Soviet technical requirements for compatibility, was sent via diplomatic pouch from the American Embassy in Moscow to the State Department in Washington. That agency passed the material over to NASA Headquarters, where Frutkin's office received them on the 24th. The documents were next sent out for translation, and MSC finally received them at the end of the month. Gilruth got a preliminary briefing of their contents on 1 April and dispatched his reply on the 9th. For both the February letter from Gilruth and the March letter from Petrov, the turn-around time had been almost a month. Much faster communications would be essential to any joint enterprise.[27]

In his letter, Petrov approached the question of an actual test flight: "As far as your new proposal ... of an actual example of docking of the 'Soyuz' and 'Apollo' type spacecraft, it requires further study which our specialists are now engaged in." Noting that this was apparently "an intermediate solution" toward the development of compatible systems, Petrov felt that the two sides should stick to the schedule as agreed upon in the Moscow "Summary of Results." He did agree to the preliminary discussions suggested by Gilruth for planning the agenda more fully, and he proposed that they be held immediately before the Working Groups met. After asking Gilruth to select a date for the meetings, Petrov added, "From our point of view, the meeting of the Working Groups could be ... conducted in the middle of May."[28] Gilruth in turn suggested the period 17-21 May for their meeting and provided a summary of the agenda and the activities planned for the Soviets' stay in Houston.[29]

While his staff prepared for the Working Group meetings, Gilruth tried for an even earlier discussion with Petrov. The American Ambassador to Moscow, Jacob D. Beam, had reported to NASA via the State Department that one of Petrov's deputies had said that the installation of a compatible rendezvous and docking system on Soyuz and Apollo would be "difficult." Nevertheless, the deputy had indicated that this might be a proper topic for discussion during an upcoming visit by Petrov to the U.S. for an international symposium. Donald Morris, Frutkin's deputy, attempted to find a suitable time for Petrov and Gilruth to meet during this visit, but he was unsuccessful. Any consideration of the American proposal would have to wait until May.[30]

Hearing nothing to the contrary from the Soviets, MSC assumed that the May dates were acceptable and continued planning for the meetings. Late on the afternoon of 7 May, they received word that there was going to be a change. Leonard S. Nicholson, Berglund's assistant, remembered sitting in a briefing session in which Gilruth and Frutkin were being given a report on the preparations for the visit, when Frutkin received a call from his office. The U.S.S.R. delegation would not be coming.[31] The full text of the cable from Petrov, received the following morning in Houston, read: "To my regret I have to ask you to postpone meetings of our working groups [until] June due to engagements of our specialists. I shall let you know names of Soviet participants and desirable date of meeting in the near future."[32] The men gathered that Saturday morning were perplexed; Gilruth asked them to study the implications of slipping the meeting date to June and then report back to him by the following Friday.

Unknown to the Americans, the Soviets were planning another significant manned launch for early June. But the Americans also had a flight in the final stages of preparation. René Berglund reported to Gilruth on 14 May that "the unanimous conclusion of the working group chairmen and myself is that a meeting in June would be very inconvenient." Glynn Lunney was particularly concerned since Apollo 15 was scheduled to be launched on 26 July. As he was deeply involved with the mission as Chief of Flight Operations, any meeting within the last 30-45 days prior to launch would pose serious scheduling difficulties. Berglund told the Director that he and the chairmen were proposing that the meeting be delayed until early September, and they had drafted a letter to that effect. He continued, "There is some question as whether we should bother to reply at all until such time as Petrov proposes a date." Clearly there was some unhappiness, but Gilruth's calm and measured approach prevailed. NASA, he decided, should await the Soviets' next move.[33]

SPACE STATION I: PROMISES AND PROBLEMS

The Soviets, in fact, did have their hands full. They were preparing a second manned rendezvous with *Salyut I,* which they had placed in orbit on 19 April. Billed as the first "space station," Salyut was designed for long-term flights of approximately one month. As early as March 1971, the Soviets had begun to hint that they were preparing for a flight that would exceed the 18-day record mission of *Soyuz 9.* The unidentified "Chief Designer of Spaceships"* said that such a flight would be the prelude to creating a permanent space laboratory. The interview in *Sotsialisticheskaya Industriya* indicated that the Soyuz had "undergone necessary modifications

---

*Although the "Chief Designer" was tentatively identified by the *New York Times* as being M. K. Yangel, it was more likely that K. P. Feoktistov was speaking to the Soviet press.

to insure fulfillment of a long and extensive program" and suggested that the spacecraft, which were in "serial production," would remain the standard spaceship for some time.[34]

On 23 April 1971, the Soviets had placed *Soyuz 10* into orbit. Following an early morning launch, Soyuz began its rendezvous maneuvers and docked with *Salyut I* on the afternoon of the 24th. The final docking took place in two stages. The automatic systems brought the manned craft within 180 meters of the target vehicle, and then spacecraft commander Vladimir Alexandrovish Shatalov took over. After ninety minutes, he guided the Soyuz to a successful docking. The two vehicles remained joined for five hours and thirty minutes while a series of experiments were conducted with the flight systems of both Soyuz and Salyut. Much to the surprise of most observers, there was no attempt to transfer either Alexei Stanislavovich Yeliseyev or Nikolai Nikolayevich Rukavishnikov into the space station. After separation from Salyut, the crew of *Soyuz 10* conducted circular maneuvers around the station, taking photographs and transmitting live television pictures of it to the ground.[35]

Even as the three-man crew returned safely to earth, there was considerable speculation over the success of the mission. The Soviets had themselves given rise to the questions. After the mission, designer Feoktistov indicated that there had been some difficulties in the rendezvous and docking aspects of the flight. First, there had been a number of orbit changes during rendezvous. "In the course of this experiment 'Soyuz 10' changed its orbit three times and 'Salyut' station four times on commands from the earth." With respect to the docking, Feoktistov said:

> In servicing orbital stations . . . it will become necessary in the future to learn to dock a relatively small transport spaceship with a huge flying multipurpose laboratory. . . . The docking of this type is a more difficult task as compared with the docking of two "Soyuz" or "Cosmos" spaceships—craft of roughly the same mass.[36]

While second guessing continued over the "problems" encountered by *Soyuz 10,* the Soviets launched *Soyuz 11* on 6 June 1971. As the preparations advanced for the second rendezvous with *Salyut I,* Petrov cabled Gilruth on 24 May, proposing a 20 June arrival date in Houston for the Working Group members. Gilruth, wishing not to lose the momentum of the joint talks, accepted that date and requested information on the size and arrival time of the delegation.[37]

WELCOME TO HOUSTON

A 19-man delegation arrived at Houston's Intercontinental Airport at 8:30 on Sunday evening, the 20th, in very good spirits, basking in the reflected glory of their space station, which the *Soyuz 11* crew had manned on 7 June. A NASA party led by Gilruth met the Soviets and accompanied them on their hour-long ride south to the Kings Inn near MSC. Leroy Roberts, who was representing the Office of Manned Space Flight, recorded at the time that it was evident from the beginning that the Soviets had come with the intention of "getting down to business and getting as much accomplished as possible." They were particularly eager to get NASA's comments on the technical materials that they had transmitted to the U.S. in March.[38]

Monday morning was set aside for general introductory remarks by Gilruth and Petrov and for planning the week's activities. The MSC team had prepared a booklet in English and Russian that outlined the tentative schedule—both business and social—for the five-day visit. The Soviets requested only one minor change, to switch the summary presentations from Tuesday morning to that afternoon. That change would give them the morning to review the comments on their technical materials and to read the additional papers given them by the Americans.[39]

Monday afternoon was spent touring the center, with the press in tow taking pictures and watching the Soviets. Astronauts Fred W. Haise, Jr., Thomas K. Mattingly, and John W. Young assisted with the tour, which included the Visitor Orientation Center, the Mission Simulation and Training Facility, the Space Environment Simulation Laboratory, Mission Control, and the Flight Acceleration Facility. The Soviet visitors spent much of the afternoon at the Apollo simulator facility asking questions and taking turns performing simulated docking operations. They were also very interested in the display of Apollo docking hardware, and the Americans gave their guests an explanation of the equipment and its operation. Likewise, at Mission Control all of the Soviet questions concerning the staffing and operation of the center during missions were answered. A full day of activities was topped by a seafood dinner at Jimmie Walker's Restaurant on the Galveston Bay waterfront.

Since private consultations were scheduled for Tuesday morning, the U.S.S.R. delegation stayed at the Inn, studying the documentation NASA had prepared for its members. Meanwhile, the American Working Group members met with Gilruth to discuss for one final time the summary presentation that they were going to deliver to the entire Soviet delegation. Gilruth urged his chairmen to be flexible in their negotiating stance without yielding unnecessarily on essential points. The Americans and Soviets gathered in Room 966 of the Project Management Building shortly after lunch.[40]

Caldwell Johnson spoke for the American side and outlined the minimum requirements necessary for rendezvousing and docking U.S. space

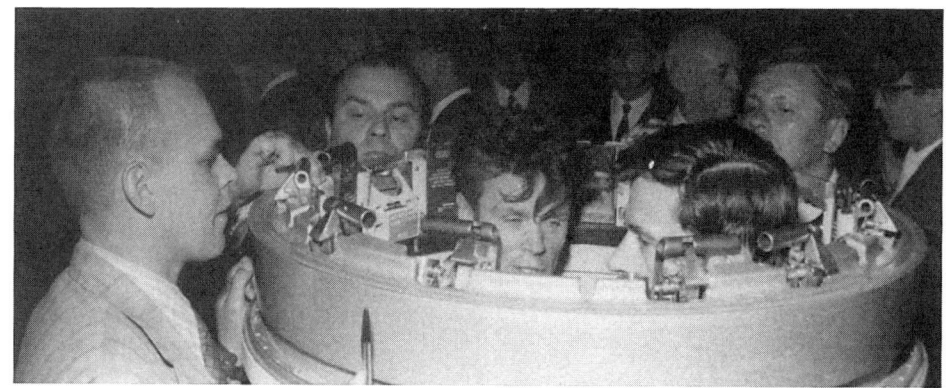

During the Manned Spacecraft Center, astronaut J. W. Young (right) and Soviet docking specialist V. S. Syromyatnikov discuss the inside of the Apollo docking tunnel (above), assisted by Yu. P. Khomenko, Soviet interpreter; at right, Don Wade (left) answers questions from V. S. Syromyatnikov and V. Zhivoglotov inspecting the Apollo docking probe; and at far right, Boris Petrov, head of the Soviet delegation, gets a close look inside the Apollo command module simulator, while Robert Gilruth, MSC Director, explains.

Soviets visit Houston, 21-25 June 1971

At left, from his ninth floor office, Robert Gilruth points out features of the Manned Spacecraft Center to B. N. Petrov (right). Partially obscured behind Petrov is Christopher C. Kraft, MSC Deputy Director; below left, the Americans (left) and the Soviets discuss agendas for the joint Working Group meetings; below, Working Group 2 takes time out for sightseeing in Houston.

vehicles—for example, the ability of one spacecraft to locate another in orbit, the status of control systems at the time of docking, and internal and external crew transfer. These requirements were, in other words, the ground rules for conducting manned rendezvous and docking, and the ever-present key element was crew safety. After talking about the design features of the American docking system and environmental control systems, Johnson listed a number of features that would have to be standardized before a joint mission could be undertaken.[41]

Petrov then spoke for the Soviets. He said that his specialists had read the documents given to them upon their arrival, and while they had some specific questions, he felt that both sides were in basic agreement on how to approach a joint mission. He then turned to his Working Group leaders, who discussed the topics related to their specialty. Valentin Nikolayevich Bobkov of Working Group 1 (responsible for rendezvous methods and overall compatibility) saw only two major topics that would require further discussion—the size of the hatch opening and the question of pre-breathing. Of these two, he expected that the hatch question was the one that would take some lengthy negotiation.[42]

The issue over the diameter of the transfer tunnel and hatch was indicative of the minor problems that could develop when the two sides failed to understand each other's thinking fully. NASA, in discussing future systems, had proposed the adoption of a 1.5-meter diameter for hatches and tunnels, which would permit easier transfer between spacecraft than had been experienced with the 0.8-meter tunnel used in Apollo. The Soviets, for reasons unknown to the MSC group, wanted to retain the 0.8-meter size. Only with the passage of time and many conversations would the question be resolved. For the June meeting, this would remain an unclear and unanswered problem.

Speaking for the Soviet guidance specialists assigned to the second Working Group, Viktor Pavlovich Legostayev said that there were virtually no differences in the two groups' approaches. Indeed, they expected to reach an early agreement in writing. The only difficulties came from the different terminology the two countries used, and they hoped to resolve that with relative ease. Legostayev suggested that this Working Group be divided into three subgroups to work on radio, optical, and target systems.

Syromyatnikov, the Soviet leader for Working Group 3, was equally optimistic in his predictions. Docking hardware terminology seemed to pose few problems, and Syromyatnikov and Caldwell Johnson seemed to agree on the approach to be taken in studying the technical considerations posed by mating two spacecraft.

These comments were followed by another statement by Academician Petrov, who said that considerable thought had been given to the

Apollo-Soyuz test flight that the United States had proposed. However, the Soviet Academy thought that the simple drogue-in-cone approach was not very productive. Clarke Covington remembered Petrov's reasoning–it was a dead end, with no application to future spacecraft. Petrov felt that such a flight would be considered a "space stunt"; instead, he suggested a test mission with a universal docking mechanism. The alternatives were Apollo docking with Salyut/Soyuz or Soyuz with Skylab/Apollo. Caldwell Johnson was surprised that the Soviets wanted to go ahead immediately and study the development of a universal docking mechanism. But he was pleased by the Soviets' apparent desire to attempt in the near future a joint mission that would create hardware with long-term utility.[43]

With opening remarks out of the way and agreement reached on Working Group agendas, the three teams assembled in their respective conference rooms to begin their deliberations. Leroy Roberts jotted down his impressions of the Tuesday afternoon sessions: ". . . got off to fast start–more agreements–good working relationship–no language problem–eagerness to find solutions. . . ."[44]

After the working sessions that day, the Soviets were given an opportunity to visit a suburban shopping mall, where they could make purchases and fill the requests of friends and family. In addition to space and Texas souvenirs, the Soviets made a number of specific purchases. High on their lists were children's clothes. To everyone's pleasant surprise, Penney's was having a sale, in which large quantities of children's garments were priced at two dollars each. Of the 19-member delegation, 15 bought something at this bargain table. A number of requests intrigued the Americans who accompanied the Soviets on their shopping trip. One man said that he had a little house in the country and wanted to change some things. Therefore, he needed something that would drill into concrete; the solution to his problem was a five-dollar star drill. One fellow purchased several pairs of tennis shoes with steel arch supports. And yet another shopper who needed a saw was quite pleased with his purchase of one with five interchangeable blades. Thus it went for several hours. Many things about the American consumer scene amazed, amused, or perplexed the Soviets. Free shopping bags were a surprise, as was being able to open packages to examine goods before paying for them. The use of credit cards by Americans disturbed the visitors, who reported that credit sales of major items–cars and appliances–were increasing in the U.S.S.R. but that credit purchasing often led to financial troubles. Finally, sales tax perplexed the Soviets and was never fully understood.

If the Soviets had a good time partaking of the Texas consumer economy, the Americans who waited on them and met them appeared to enjoy themselves equally. At one point during the evening, Academician

Petrov was shopping by himself in Woolworth's when three grade school children, a girl and two boys, approached him and asked him if he were from the Soviet Union. In English, Petrov replied that he was. The girl then gave an impromptu speech of welcome, saying that she was happy that they had come to the U.S. and hoped that their work would be successful. Petrov was moved to tears by this spontaneous greeting.[45]

The remainder of the June meeting followed this pattern of working during the day but sightseeing and socializing during the evening. The Working Groups began to concentrate on technical detail. Documentation of the technical agreements that the groups reached took a large part of the members' time, but it was very necessary to ensure the compatibility of hardware and systems. Not only did the two sides speak different languages, but also they had evolved different engineering styles and terminology. Once the negotiators reached agreement on a topic of discussion, a document had to be prepared in both English and Russian, verified as to meaning and technical content, and then signed by the engineers and interpreters. This could be a slow and tedious process, but it was an integral aspect of creating compatibility.

Working Group 1 members reached early agreement on the coordinate systems that govern a joint mission. A coordinate system is the mathematical method for exactly defining the position of a craft in space relative to a particular celestial body, and of the several possible alternatives, the Working Group chose an earth-centered system. The American representatives agreed to prepare a single document reflecting their negotiated understanding and mail a draft to the Soviets within two months. After review and assurance that the document was acceptable, it would be signed off by both sides and thereby become the standard reference document for the subject.

In turn, the Soviets were to prepare a single technical requirements document treating the combined subject of spacecraft atmospheres, hatches, and crew transfer. That paper, based upon exchanges in February and the deliberations in June, would be reviewed, exchanged, and verified after the fashion of the coordinate systems paper.[46] Other life support considerations that would have to be documented included cabin pressure limits, trace gas concentrations, carbon dioxide pressure limits, portable pre-breathing systems, drinking water quality, and color coding of equipment. In addition, after Working Group 1 members talked about communications between ground centers and locations for various types of equipment, they agreed that these topics also required further discussion. Next, they turned to consider real test missions.

After agreeing that they should jointly prepare models for different missions that might be flown, the delegates decided to base their planning on an experimental flight in which Apollo would dock with "a manned orbital scientific station of the Salyut-type." They also suggested that a subsequent experimental flight might be conducted with Soyuz and Skylab. Looking at such a test mission, the Working Group 1 signatories stated:

> In principle the technical feasibility to do this exists. For the purpose of a concrete study thereof, the parties have agreed to do some additional work on these problems with primary attention given to the following problems:
> (1) Location and design of docking assemblies
> (2) Atmospheric parameters
> (3) The need to provide airlocks
> (4) Location of equipment, apparatus, and components of the rendezvous and docking system.[47]

Working Groups 2 and 3 discussed, negotiated, and drew up agreements in their areas of responsibility in a similar fashion. Group 2 members concentrated on such subjects as requirements for light beacons, radio guidance and communications systems, and spacecraft attitude control systems. Working Group 3 reached agreement on the basic functions and design features of a universal docking system, as well as on the design approach to obtain necessary compatibility. They also agreed to discuss details about hatches, docking ring seals, and electrical connectors with Group 1. Likewise, they would hold a joint session with Group 2 to discuss questions associated with the conditions for initial physical contact between spacecraft. As in the case of the first Working Group, the other two divided among the Soviet and American teams the responsibilities for drafting and exchanging the necessary documentation.[48]

Gilruth and Petrov reported that the deliberations had been successful, and they stressed the possibility of a test mission. Such an experimental flight was technically feasible, and both parties agreed "that the technical and economic aspects of these possibilities should be additionally studied and discussed. . . ." To expedite their work in the months before the next joint meeting, which was tentatively scheduled for the end of November in Moscow, Glynn Lunney and Konstantin Davydovich Bushuyev were appointed Project Directors for their respective sides. They would act as focal points for all communications and technical exchanges. In a joint statement, Petrov and Gilruth reported that the meetings had been conducted in a businesslike atmosphere, and both men expressed their "gratification at the very rapid and substantive progress of their specialist working groups toward a comprehensive set of agreed requirements." While the reporters puzzled over the nature of the progress and fussed about not having an opportunity to grill Gilruth and Petrov, the American and Soviet delegations bid farewell.[49]

MEETING THE PRESS

Later, on Monday, 28 June, six Americans who had worked with the Soviets held a news conference in Houston. Until that time, there had been some speculation in the press concerning the nature and tenor of the joint deliberations. In fact, during the negotiations, the Soviet side had asked that the discussions as a general rule be kept confidential. Gilruth and Frutkin flatly declined to agree to such an embargo of information, and Frutkin stated that NASA could not and would not proceed with the talks on that basis. Only in the area of agreements pending official ratification, such as the June Summary of Results, did the agency reserve the right to remain silent. Once signed, however, those documents would become part of the public record.[50]

Characterizing the preceding week's activities as friendly, Gilruth commented to the press:

> It was a period of intense hard work covering very difficult technical areas. As you all know, rendezvous and docking is not simple for one country or one organization to conduct, and so I think everyone could imagine some of the complexity of trying to work out the arrangements between two different countries, particularly countries that speak such different languages.[51]

When Nick Chriss of the *Los Angeles Times* asked about the language barrier, Gilruth said that he would not be leveling if he did not admit that it was a formidable one. None of the American technical people spoke Russian and therefore had to rely upon interpreters. While they had been able to work around the language problem, it had been fortunate that several of the Soviets could speak and write English. Gilruth noted that there would be a need for additional simultaneous interpreters in the future, but English-Russian interpreters were not the most plentiful people in the world.

While careful not to give the impression that a joint mission was a sure thing, Gilruth answered a question raised by Paul Reiser of the Associated Press about timing. He said, "the mid-70s would be a reasonable time frame to think about. Certainly I don't believe it would be any sooner than that, and of course even that is contingent on the rate of progress we are able to make." Gilruth was quick to add that there had been no decision to conduct

The American team meets the press. Left to right: C. C. Johnson, G. S. Lunney, R. R. Gilruth, D. C. Cheatham, D. C. Wade, and R. A. Berglund.

a joint flight, only discussions on the merits of such an experimental mission. The desirability of an actual flight would be a topic for continued discussion.

To inquiries about the candidness of the discussions, the Americans reported that there was indeed a high degree of openness. Don Wade, chairman of Working Group 3, said that their candor had surprised him, but he added that this was relative to the pre-agreed list of discussion items, which they "stuck pretty close to." In those areas "they were very, very open with us," he said. Both publicly and privately, Wade's colleagues agreed with his appraisal.[52]

One of the recurring questions raised by the newspeople centered on the "give away" issue—giving away to the Russians hardware or technical know-how. Jim Maloney of the *Houston Post* asked about this first in the context of rendezvous and docking. He suggested that NASA was going to "donate" knowledge in this area, since the U.S. had much more experience than the Soviets. Peter Mosely of the Reuters news service asked if Gilruth would characterize the efforts so far as an exchange of technology. Gilruth answered that NASA may have had more rendezvous and docking experience but that the Soviets had had their share as well and clearly understood the flight mechanics involved. In response to the question of transferring technological knowledge, the MSC Director pointed out that the present talks were simply exchanges of views on how two nations might fly together. He did not anticipate any major changes in either nation's spacecraft as a result of the compatibility meetings. All that was really required, he said, was an agreement on the docking interface—the docking gear, and the like—and assurance that the interface requirements are adequate.

There were also questions about the wider implications of the negotiations. Jay C. Russell of KTRH radio of Houston asked Gilruth, aside from being able to sit down and work at making equipment that would fly together, "what does all this mean to the world?" Gilruth responded:

> Well, I think you'd have to decide that for yourselves. None of us here are politicians or politically inclined people. I think we all are impressed with the fact however, that we have been able to meet with the delegation from the Soviet Union in an area of great technical difficulty, work together, and with a friendly atmosphere come to a number of important general agreements and I think that it's always good when people can meet and work together in harmony.[53]

SOYUZ 11: TRIUMPH AND TRAGEDY

As the Soviets departed from Houston, *Soyuz 11* was completing its 20th day in orbit docked with *Salyut I.* This record breaking flight had been heralded by Keldysh as beginning a new era in space exploration. On 9 June, Blagonravov had declared in an article prepared for *Krasnaya Zvezda* that:

> In the opinion of Soviet scientists, such stations with replacement crews constitute mankind's main highway into space. They can become unique launching pads for flights to other planets. Large scientific laboratories will spring up for research into space technology and biology, geophysics and medicine, astronomy and astrophysics.... In time, such stations will be linked with earth not only by radio but by a regular space mail. By periodically putting small supplies of fuel aboard, it is possible to insure the station's long-term existence by switching on the engines and reestablishing the velocity lost as a result of braking in the upper layers of the atmosphere.[54]

The three-man crew of *Soyuz 11* (call signal "Yantar"), Georgi Timofeyevich Dobrovolskiy, Vladislav Nikolayevich Volkov, and Viktor Ivanovich Patsayev, had entered the space station on 7 June. The joined configuration of Soyuz and Salyut was 21.4 meters long with a total living space of 100 cubic meters, which gave the cosmonauts a place to conduct scientific experiments, relax, and sleep. For the next 23 days, each crewmember performed his scheduled experiments, which emphasized the study of human performance under, and reaction to, prolonged weightlessness. On the 29th, after completing their flight plan, the space dwellers transferred their scientific records, film, and log books to Soyuz in preparation for their return home.

At 9:28 in the evening, Dobrovolskiy undocked the ship and drifted free from the space station. After three additional orbits, the *Soyuz 11* crew notified ground control that they were beginning their descent. Mission

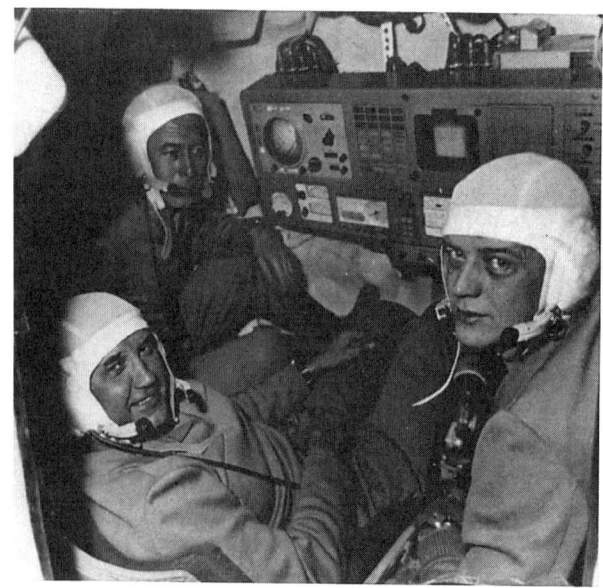

Crew of Soyuz 11 (*left to right*): *V. I. Patsayev, G. T. Dobrovolskiy, and V. N. Volkov train for their mission (Tass from Sovfoto).*

Control radioed: "Good bye, Yantar, till we see you soon on mother earth." Dobrovolskiy replied: "Thank you, be seeing you. I am starting orientation."[55] At 1:35 a.m. the retrorockets were fired automatically for a seven-minute burn, and the parachutes were deployed on schedule. Mission Control tried repeatedly to contact the crew at this time, but to no avail. When the recovery crews reached the descent vehicle and opened the access hatch, Dobrovolskiy, Volkov, and Patsayev were dead in their contoured couches.[56]

The accident was a stunning blow to both the Soviet Union and the international aerospace community. Once again, the experimental and risky nature of man's venture into space had been made clear. While the three bodies lay in state and a Special Commission investigated the cause of the multiple deaths, wide speculation spread in the West over the significance of the tragedy for the continuation of manned space flight.

One of the prevailing theories was that man might not be able to survive long periods of weightlessness. For several years, there had been a serious debate among scientists about the effects of prolonged weightlessness. During Project Gemini, there had been "signs" that the human heart grew lazy after an extended time in zero gravity. Then in July 1969, the monkey Bonny aboard the U.S. *Biosatellite 3* died of heart failure after recovery from a 9-day flight.

However, there were other theories regarding the Soviet disaster. George Low discounted the heart failure story, and Dr. Walton Jones, Deputy Director of Life Sciences in the Office of Manned Space Flight suggested that the men had died as the result of their cabin decompressing rapidly. The crew was found strapped in their seats with no apparent indication of any struggle. (The crew did not rely on space suits.) Dr. Jones said that this is how they would have appeared if a valve had leaked or the shell of the cabin had ruptured. In Houston, Dr. Charles Berry, flight surgeon to the astronauts, thought that the accident might have been caused by the release of a toxic substance. MSC Director Gilruth favored the decompression theory. Whatever the cause, both Soviet and American aerospace leaders realized the seriousness of the problem and its implications for manned flight in general and for the compatibility discussions in particular.[57]

As thousands of Muscovites filed by the funeral bier of the three cosmonauts on 1 July, Soviet President Nikolai V. Podgorny, Premier Kosygin, and Party General Secretary Leonid I. Brezhnev took turns standing watch as part of the honor guard. President Nixon on behalf of the United States told the Soviet leaders:

> The American people join in expressing to you and the Soviet people our deepest sympathy on the tragic deaths of the three Soviet cosmonauts. The

whole world followed the exploits of these courageous explorers of the unknown and shares the anguish of their tragedy. But the achievements of cosmonauts Dobrovolsky, Volkov and Patsayev remain. It will, I am sure, prove to have contributed greatly to the further achievements of the Soviet program for the exploration of space and thus to the widening of man's horizons.[58]

In addition, the President sent U.S. astronaut Thomas P. Stafford to Moscow as his official representative for the funeral ceremonies held in Red Square.[59]

Soviet space leaders were quick to reaffirm their plans to continue manned space flight. Writing for *Pravda* on 4 July, Petrov spoke of the conquest of space as a "difficult path," but he repeated Brezhnev's earlier statement—"Soviet science considers the creation of orbital stations with replacement crews to be man's highway to space." The scientist argued that man could play his most important exploratory role in the study of the earth and in astronomy from platforms positioned in "near-earth space." Furthermore, such earth orbital investigation is only valuable when it is conducted for extended periods on a regular schedule. Petrov said that "the seventies will be the epoch of development and broad application of long-term manned orbital stations with replacement crews, which will make it possible to switch from episodical experiments in space to a regular watch by scientists and specialists in space laboratories."

Summarizing the work conducted on board Salyut by the crew of *Soyuz 11,* Petrov restated the value of their contributions to science. In addition to the medical and biological experiments, they had carried out a number of studies related to weather and earth resources. According to the Soviet spokesman, the data returned in Soyuz would be used by students of agriculture, land reclamation, geodesy, and cartography, as well as by meteorologists to improve their forecasts. With words apparently aimed at domestic critics of the Soviet manned space program, Petrov reported:

> The experience of the cosmonauts' work has shown that the Salyut manned station is a space laboratory well adapted for experiments in orbital flight conditions. Such stations are opening broad prospects for the continuation and development of the research carried out by the first Salyut crew. . . . Ahead lie new flights into space and the creation of new inhabited orbital stations of the Salyut type. Undoubtedly, even larger and more complex manned multipurpose and specialized space stations will be built. But the significance of the work carried out by the first crew of the first manned orbital station . . . will never fade.[60]

The Special State Commission investigating the *Soyuz 11* deaths released a public statement on 12 July. After reporting that the flight had proceeded normally up to the beginning of reentry, the Commission stated:

> On the ship's descent trajectory, 30 minutes before landing, there occurred a rapid drop of pressure within the descent vehicle which led to the sudden deaths of the cosmonauts. . . . The drop in pressure resulted from a loss of the ship's sealing. An inspection of the descent vehicle . . . showed that there are no failures in its structure.[61]

The reasons for the "seal" failure were still under investigation, this terse statement continued.

While the official report apparently eliminated weightlessness and physical deconditioning as causes for the accident, the seal failure statement raised a new question. Americans preparing for Apollo 15 wondered if the Soyuz problem was of the type that might be experienced with an Apollo spacecraft. MSC Director Gilruth wrote Petrov shortly after the accident and asked him that question. Petrov reassured the Americans that "the drop in pressure resulted from a concrete failure of one of the elements of the descent vehicle system. Since it is a matter of specific and particular defect we are sure that it cannot be related to 'Apollo' spacecraft."[62] *Soyuz 11*'s misfortune did not affect NASA's plans for the launch of Apollo 15, but it did lead to some discussions outside the space agency on the safety of Soviet hardware.[63]

AFTER APOLLO: WHAT?

As the Soviets recovered from their tragedy and evaluated their manned space flight plans, NASA continued its preparations for Apollo 15. The agency's leadership was also looking with uncertainty to the future of its man-in-space efforts. Prior to his departure from NASA the previous autumn, Tom Paine had announced a reshuffling of the remaining Apollo missions. In a press conference on 2 September 1970, the Administrator had discussed these decisions, which reflected a husbanding of NASA's dwindling share of the national budget. The agency wanted to accomplish many goals, but these had to be attained with a limited number of dollars. As the 1969 Space Task Group studies had suggested, NASA would have to balance its present wants against future budgets. A shifting of current project monies would have to take place if NASA wanted not to jeopardize its plans for the future.[64]

Paine and his colleagues realized that during the 1980s there would be no manned missions to Mars, no other bold ventures equivalent to the lunar goal of the 1960s. Paine said that the principal decision facing the agency was "how best to carry out the Apollo and other existing programs to realize the maximum benefits . . . while preserving adequate resources for the future." NASA had decided to concentrate its manned efforts on three earth-orbit programs—Skylab in 1973 and Space Shuttle and Space Station in the 1980s. The earth-oriented unmanned program would include early development of the Earth Resources Technology Satellites and the Applications Technology Satellites. An unmanned planetary program would involve the Grand Tour flights to distant planets, the Viking Mars orbiters and landers, and the Pioneer missions to Venus, Mercury, and Jupiter. Add "a healthy aeronautical research program" to that list, and the demands on a shrinking budget were obvious.

One immediate way to conserve money was to reduce the number of Apollo moon landings. To pare $42.1 million from the fiscal year 1971 budget, two missions were canceled, and manpower levels at the manned space centers were scaled down accordingly. These decisions were taken not only reluctantly, but also against the advice of scientific agencies external to NASA. Apollo's remaining missions were redesignated 14 through 17, and the so-called "residual" hardware would be made available for Skylab, Space Station, and other programs that might follow the final lunar landing in 1972.

With astronauts Alan B. Shepard, Stuart A. Roosa, and Edgar D. Mitchell aboard, *Apollo 14* conducted a successful lunar exploratory trip in 1971. The 31 January-9 February mission was slightly marred by the failure of the probe assembly to operate smoothly as Commander Shepard tried to dock the CSM with the LM. Shepard and Mitchell spent their 9 hours and 24 minutes on the lunar surface exploring the terrain, but NASA hoped to increase considerably time spent on the moon during the next flight.[65]

Apollo 15 stressed longer EVA periods and use of the lunar roving vehicle. One hundred hours after a 26 July launch, David R. Scott and James B. Irwin separated their lunar module *Falcon* from the CSM *Endeavor* piloted by Alfred M. Wordon and headed for a touchdown in the mountainous Hadley-Appenine region near Salyut Crater. The results of their exploratory work were excellent, and Scott and Irwin set several records in the process. They had spent over 63 hours on the moon's surface, conducted a total of 18 hours and 35 minutes in extravehicular activity and traveled 28 kilometers in their moon buggy. But for all its success, *Apollo 15* did not bring much joy to the NASA people in Washington, Houston, and Huntsville, for only two more lunar missions remained.[66]

When James Chipman Fletcher was sworn in as the fourth Administrator of NASA on 27 April 1971, he became the head of an agency that was entering a new era. Fletcher, a physicist by professional training and a university president during the turbulent 1960s, had a personal background in the aerospace world and understood some of the problems that NASA

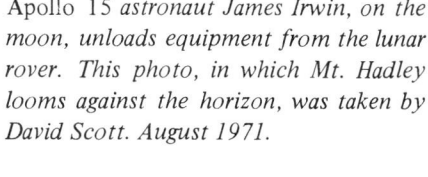
Apollo 15 *astronaut James Irwin, on the moon, unloads equipment from the lunar rover. This photo, in which Mt. Hadley looms against the horizon, was taken by David Scott. August 1971.*

would be facing in the years ahead. Reflecting the spirit of both the adventurer and the realist, he commented to the press after the announcement of his appointment that an important goal faced the agency—"to achieve . . . balance between manned and unmanned programs. It would be very exciting for man to go beyond the moon but that . . . is a little beyond the nation's budget right now." Such a statement might at first appear to have been somewhat flippant, but it could be taken as a manner of saying to the NASA team, "Do not despair; there is still important and exciting work to be done."[67]

Dr. Fletcher also announced very early that he supported closer cooperation with the Soviets. On 10 March, during a one-hour hearing before the Senate Committee on Aeronautical and Space Sciences, Fletcher told the Committee that the U.S. had made "some small steps" toward cooperation with the U.S.S.R.; now "we can make even larger steps." But the possibility of reducing the long hiatus between the Skylab missions in 1974 and the first Shuttle flights in the 1980s was another reason why Fletcher was interested in talks about a joint mission with the Soviets.[68]

At a pre-launch press briefing for Apollo 15, Dale Myers, Associate Administrator for Manned Space Flight, had spoken about the post-Skylab studies under way. He pointed out that there would be four Apollo CSM's left over, three from the canceled moon flights and one that had been set aside as a backup for Skylab. Studies conducted in Houston indicated that these spacecraft could be flown in earth-orbital missions for about $75 to $150 million each. One possible use for these CSMs would be to launch one a year, beginning in 1975, for earth resources surveying missions lasting from 16 to 30 days each. Of these four spacecraft, one could be set aside for a rendezvous and docking mission with the Soviets. Still another possibility would be orbiting a second Skylab, using the backup CSM for the flight

153

planned for 1973, but that would be very expensive and would require developing new mission goals for Skylab B.[69]

An interim manned program of some kind was highly desirable. In the first place, it would permit NASA to hold together its launch and flight control teams. Keeping these people together was as much a question of morale as it was money. The men working at Houston and Cape Canaveral were action-oriented; they needed the challenge of actual flights. And second, the crewmembers who trained for the last Apollo flights would still be eligible to fly in the Shuttle period, but they too might grow restless and disinterested if there were a four- or five-year break in flights. Availability of funds would determine the feasibility of an interim project for the space agency.

Myers said that there would probably not be money enough for both a full-scale Shuttle program and interim Apollo flights. If NASA decided to develop the Shuttle booster and orbital stages simultaneously, then there was little likelihood of any flights between the last Skylab visit and the first Shuttle launch. He pointed out, however, that a second approach might be taken. NASA could develop the Shuttle orbital spacecraft first, and while glide tests were being conducted with the early prototypes continue development on the reusable launch boosters. Under such a "phased approach," it might be possible to finance some other missions. But the key guideline was to undertake only those efforts that could be carried out without draining resources from the major effort—Shuttle.[70]

A STUDY TASK TEAM

As work progressed in Houston during the summer of 1971, two teams emerged. Most visible was the one under the direction of Glynn Lunney, comprised of the Working Groups that were organized to establish ground rules for working effectively with the Soviets. At the same time there existed a less formal organization, headed by René Berglund, charged with coordinating work within NASA and dealing with contractors after outside studies had been ordered. Membership in these two groups overlapped somewhat.

Shortly after the 21-25 June meeting with the Soviets, Berglund's team presented to Gilruth and his senior staff a paper outlining the basic hardware needed for a Soviet-American flight. Berglund proposed—for purposes of discussion—planning toward a mid- to late-1974 launch; everyone agreed that "with all that has to go on to make it work, this was an extremely tight schedule." Chris Kraft directed Berglund and Lunney to generate by September a realistic schedule and a cost figure for one CSM-Salyut flight. The earlier talk of four Apollo earth orbital missions was dropped.

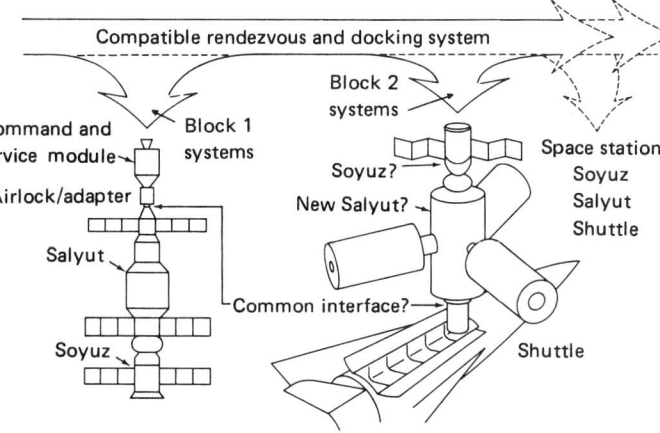

Caldwell Johnson's spacecraft designers were given an equally challenging assignment. Director Gilruth wanted the docking adapter design pushed ahead rapidly, with a working model prepared for the November meeting with the Soviets. Johnson quickly pointed out several tasks that would require further investigation. The first, called project engineering, was assigned to Clarke Covington, who had the overall responsibility for integrating the engineering done by the other designers and technical specialists. In addition, Covington was serving as a systems engineer to Lunney's Working Group 1 and to Berglund's Study Task Team. Covington became one of several engineers who found himself putting in 12- and 13-hour days.

Bill Creasy and his fellow mechanical engineers were working to a similar schedule on the design of a compatible docking system. Johnson, who believed that it would be most difficult to reach an agreement on the docking gear, wanted to proceed with a variation of that same ring and cone system he had illustrated for the Soviets the preceding October in Moscow. Since Petrov had rejected the simple adaptation of Apollo and Soyuz as a "space stunt" and since the Soviet space expert had proposed developing a universal docking mechanism, Johnson suggested that MSC draw up a "design specifically adequate to requirements of a particular CSM/Salyut mission, the design being representative only of the fundamental form and function of docking gear satisfying the requirements for compatible docking system for future spacecraft." Creasy was asked to conduct a preliminary design study to determine the nature, weight, and characteristic dimensions of the functional components of an androgynous docking mechanism. This study was to be of sufficient depth to allow a demonstration system to be built that would permit further engineering and development. While the preliminary design was to be adapted to a CSM-Salyut mission, it should be adaptable to future spacecraft as well.[71]

Responsibility for designing the airlock module was given to James C. Jones. Rejected earlier in favor of the simpler Apollo-Soyuz drogue-in-cone concept, this adapter had revived engineering interest, and preliminary designs were directed toward CSM-Salyut. These studies for the docking module (DM) were to be so detailed that the concepts could be engineered and developed by outside contractors. Jones was also assigned responsibility for the preliminary integration of the environmental control system into the DM and the first cut at designing a mounting for the airlock module inside the launch adapter.[72]

Building on these early design efforts, Berglund's team drew up a "Statement of Work," issued on 29 July 1971 to North American Rockwell, for a detailed study of all the elements required for an International Rendezvous and Docking Mission (IRDM). This four-month study was intended to expand upon the basic concepts and provide a fuller description of the hardware as it could be used in a rendezvous and docking mission and independent CSM earth survey. North American would consider which of the remaining CSM's (111, 115, 115A, or 119) would be best suited for modification and completion as the prime and backup spacecraft for a mission with the Soviets.[73]

In essence, the MSC Statement of Work and subsequent Document Change Requests told North American what NASA wanted; then the contractor was to carry out the engineering and development. For example, the agency documents proposed that the mission be 14 days long, with a joint docked phase of one to two days, after which Apollo would conduct earth survey experiments. The sequence of events during the mission was outlined:

> Saturn IB stage boost
> CSM separation, transposition and docking with extraction of the DM
> CSM transport of DM to a Salyut-type vehicle
> Rendezvous and docking of CSM-DM with Salyut-type vehicle (CSM active)
> Docked orbital operations (solar inertial attitude)
> Separation of CSM-DM from Salyut-type vehicle
> CSM maneuver to earth resources survey orbit condition
> Conduct earth resources survey activities
> EVA retrieval of experiment data
> CSM deorbit and entry[74]

With these guidelines, North American was to plot out the details of a joint flight and define all the hardware considerations involved in preparing a CSM and a DM for such a mission.

While the contractor personnel began their work, Gilruth created a formal Study Task Team at MSC to direct the IRDM study. René Berglund, appointed manager of this group, convened its first meeting on 4 August

1971 to discuss the general status of the IRDM work and the management philosophy to be used during North American's effort. At this meeting, the schedule for the next four months was mapped out so that the proper pace of activities could be ensured.[75] During August and September, work progressed on several fronts in preparation for the winter meeting with the Soviets.

DEFINING THE DOCKING MODULE

At the time the IRDM Statement of Work was issued, the docking module was still only partially defined. The initial ground rules for such a design were presented by Clarke Covington at an IRDM Study Team staff meeting on 16 August. First and foremost, the docking module should be built to accommodate (externally or internally) any additional equipment required by a joint mission so that the modifications to the basic CSM design could be kept to a minimum. There were a number of other fundamental considerations, too. Where possible, the DM should draw on the CSM for its power needs, the major exception being the DM life support system. The designers, James R. Jaax and Gerald P. Mills, agreed that normal crew transfers should not depend upon Salyut life support systems; the DM should have its own environmental control system. Covington and the Crew Systems Division engineers favored an independent system that provided the American crew with a sanctuary to withdraw to if there were difficulties during transfer.[76]

Since the Americans returning to the CSM would have to pre-breathe for four hours before reentering the command module, he "liked the idea of knowing that we were just minutes away from a U.S.-designed piece of equipment that you could jump back into. . . ." Even though the crew was a couple of hours away from stepping into the CSM, they would still be "in a piece of equipment that we understood . . . and which had been through our qualification program and our safety program."*[77] Further, the DM would have to be able to withstand the 760-mm-Hg pressure used in Salyut and accommodate the equipment required to communicate on the Soviet frequency. During August and September, several MSC teams worked further to define the preliminary design of the docking module, and Covington then took these materials and drew them together into a single document that could be passed on to the contractor.[78]

The "Docking Module Design Study" presented by Covington to MSC and North American representatives on 29 September was a full-scale outline

---

*The length of time required for pre-breathing was the subject of considerable discussion between the environmental control system engineers and the medical staff at MSC. The engineers wanted to reduce the time, but the doctors called for a conservative period of three to four hours.

of the design elements to be incorporated into the DM. In fact, the 110-page document really was quite specific on details, more so than might ordinarily have been expected. MSC was telling the contractor precisely what it wanted.[79] The DM was to serve five functions:

Primary Functions

- Serve as structural adapter between CM docking mechanism and new docking mechanism
- Serve as atmosphere adapter between CM and Salyut
- Provide habitable environment for crew while occupied
- House communications gear operating on the Soviet frequency

Additional Function

- Provide additional volume for 3-man CSM ERS [Earth Resources Survey] phase.[80]

This study also spelled out the basic dimensions for that new piece of hardware. The length of the DM from the CSM docking interface to the point at which the international docking mechanism would be attached was to be 2.54 meters,* with an additional 0.254 meter allowed for the new docking gear. The interior diameter was to be 1.42 meters with a hatch diameter at the CSM end of 0.84 meter, or the same as that used previously on the lunar module. At the Salyut end, MSC was proposing a 0.9-meter hatch.[81]

Hatch size was still a topic of considerable discussion in Houston. At the end of August, Glynn Lunney had written to Gilruth suggesting that a distinction be made between the hatch sizes used in an Apollo-Salyut mission and the diameters suggested for future systems. The planners looking forward to Shuttle and Space Station wanted a 1.5-meter hatch, but Lunney doubted that it was reasonable to impose that dimension on the designers preparing for Apollo-Salyut. He thought it was "fair to question whether this is the correct answer for the present and ... foreseeable future since the schedule for the large space station will remain unclear, but it must be at least 10 or 15 years away." Since the hatches on the docking module could be made to a smaller dimension without serious design impact on future efforts, NASA would propose a 0.9-meter hatch diameter to the Soviets at the next joint talks.[82]

While work on defining the docking module progressed at MSC, North American at Downey, California, moved ahead with their IRDM study. The Downey engineers had made an initial presentation in Houston on 24 August, containing materials that were being generated almost daily at MSC

---

*This is the space required to house two average size men in space suits, allowing for the inward swing of the hatches. The diameter of the hatch was defined by the minimum size that would accommodate a suited man and his portable life support system.

and North American. This process of evolving a document or presentation was called "iteration." Draft after draft was prepared, into which the latest findings or ideas were incorporated. Only after several iterations was a final report submitted.[83] The second status review made by the contractor on 29 September reflected the joint effort with MSC to that date.[84] Contractor

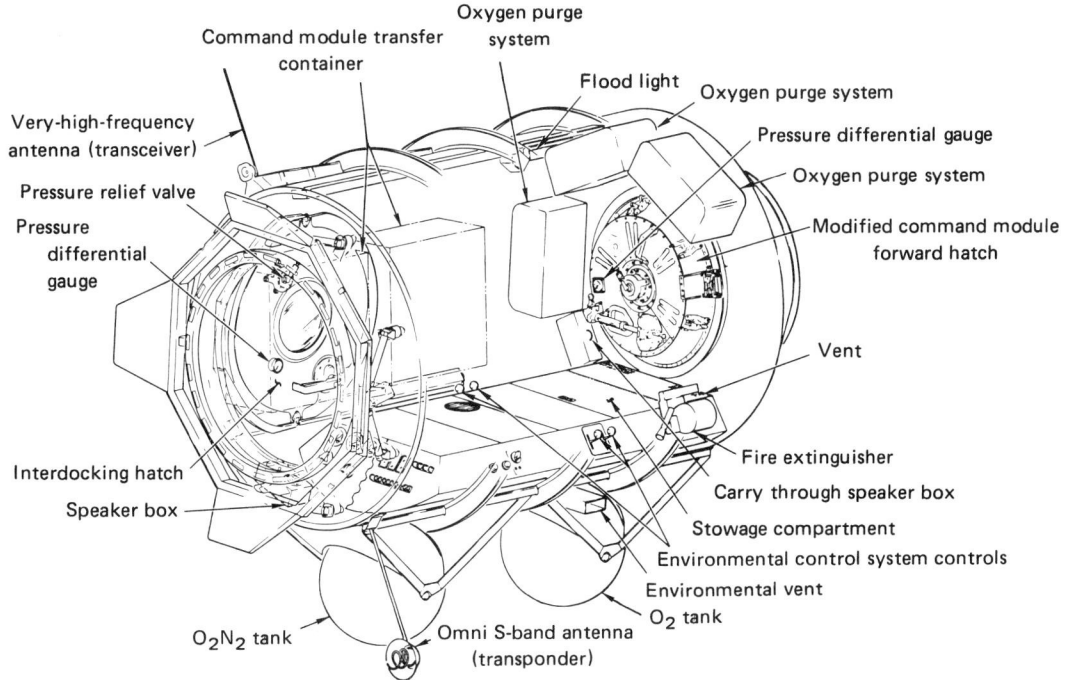

International docking module inboard profile

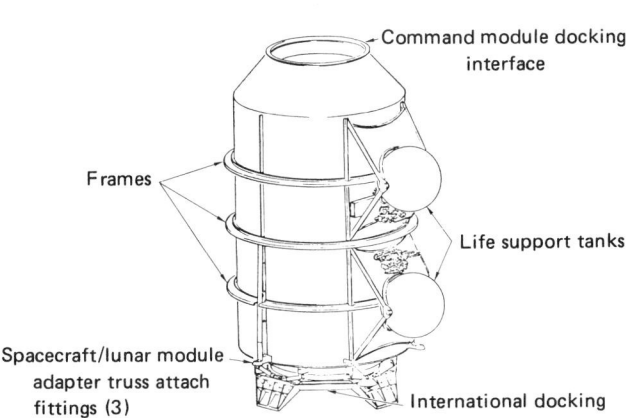

International docking module exterior

The November 1971 version of the docking module. Rockwell International assumed the use of four guides on docking gear and provided a porthole in the forward hatch for centerline television. While changes would continue to be made in this design, the basic ideas were taking shape.

personnel who came to the space center for this review also received a full presentation on the progress of MSC's Docking Module Study. They took this study home to California to use in preparing their final iteration of the IRDM study, due on 16 November.[85]

In five months, the combined NASA-contractor teams had drawn together a detailed outline of the multitude of considerations involved in the American half of a mission with the Soviets. Called "International Rendezvous and Docking Mission Final Briefing," this document began with a restatement of the basic objectives of such a flight. North American then reported that a "meaningful" dual mission could be performed. The modifications necessary on the CSM were reasonable, in terms of both expense and time. Looking at the docking module, the contractor reported the design to be straightforward, well within current engineering abilities, and the basic subsystems had been previously qualified in Apollo. Communications equipment that operated on the Soviet frequency was one new element that would have to be designed, manufactured, and qualified. While the DM could likely be ready to go in time for a 1974 launch, the Downey personnel responded that work on an international docking gear would have to be very carefully orchestrated to get it completed in time. The North American staff felt that a 1975 launch date would give them more flexibility and leeway but that they could have a spacecraft ready a year earlier if NASA so wished.[86]

The bulk of the final briefing was devoted to describing mission details, describing changes to the CSM, outlining the design and manufacture of the DM and its subsystems, and listing current Apollo hardware that could be used. Many highly technical orbital mechanics questions were addressed, not only to explain the launch time considerations for a joint docking but also to delineate such problems as the effect of lift-off schedules on the lighting available for the earth resources part of the proposed mission. The report looked into questions like the amount of reaction control system propellant that would be required, with equal attention being given to electrical power and other onboard consumables. The contractor also discussed possible areas for scientific experiments, describing some of the hardware that was available. Materials dealing with the docking module and its fabrication were equally detailed. A 249-page briefing, entitled "IRDM Programmatic Considerations Summary," and eight other sets of documents illustrated the technical feasibility of a rendezvous in earth orbit with the Soviets and testified to the ability of the NASA-industry team to work a problem in a short time.[87] René Berglund's Study Task Team had done its job, and Glynn Lunney's people were getting ready for a November departure to Moscow to talk turkey with the Soviets.

VI

Forging a Partnership

In planning for the third round of Soviet-American compatibility talks in the summer of 1971, Glynn Lunney wrote to Professor Bushuyev, expressing his condolences to the families and colleagues of the *Soyuz 11* cosmonauts. "This sad accident has further strengthened our emphasis on the solution of the common docking problems." Turning to the work being done in Houston, he commented, "As no doubt you are finding, there are many questions which arise as we have time to reflect upon and plan the work for our meetings later this year." One of these questions concerned the diameter of the Salyut port. Bill Creasy and his design colleagues had planned to propose a docking mechanism for the Soviets to study, but they needed to know what size gear would fit beneath the Salyut launch shroud, which provided the space station with aerodynamic streamlining. Lunney enclosed in his August letter a sketch that reflected the Manned Spacecraft Center's (MSC's) understanding of the dimensional liminations that would govern the mounting of such a docking system on Salyut, and he asked Professor Bushuyev to verify the sizes involved, which he did on 9 September.[1]

During September, Lunney again sent correspondence to Moscow regarding a proposed agenda for their joint meeting; NASA would prefer a two-part approach. "As we agreed in June," he wrote, "we have given priority consideration to a test mission between the Apollo spacecraft and the Salyut-type station," but our two countries must also continue "work on the technical requirements and solutions for long-term capability." To meet both needs, the NASA agenda separated the topics to be discussed into two categories—long range compatibility issues and a near term test mission. Lunney hoped that this format would clarify the distinctions between the immediate and longer range goals of the negotiations. He also pointedly played down the possibility of a joint mission with Skylab, by saying that it was much too early to talk about using such an untested, complex scientific space station.[2]

Bushuyev replied in October, agreeing that it appeared possible to look at both long range questions and an Apollo/Salyut mission "in parallel."[3] He

161

also sent lists prepared for each Working Group regarding documents that the Soviets believed could be put into final form at this meeting. A fourth list presented several general documents that they felt should be agreed upon ultimately. Finally, the Professor suggested the joint meetings be held from 29 November to 7 December in Moscow. Since this was well within the time for which NASA had targeted, Lunney accepted and advised the Soviets that the Americans would plan to arrive on the evening of Saturday, the 27th.[4]

PREPARATION IN HOUSTON

Technical Director Lunney had already prepared an outline to guide his associates as they prepared for their discussions with the Soviets. (See box on next page.) He thought that "this outline should facilitate a comparison between the near- and far-term activities; it should also form the basis for summarizing our results to NASA personnel, and for our discussions with the Soviets this fall."[5] On 6 and 7 October, a detailed technical review was held at MSC, followed two weeks later by a management review for Gilruth and Kraft. Lunney, the Working Group chairmen, Berglund, and Johnson then went to Headquarters on 10 November to brief George Low.[6]

The Headquarters briefing identified a number of items that would have to be cleared up in the joint negotiations. The long range questions naturally tended to be broader in scope than the Apollo/Salyut issues, which were concrete and specific. But some top level decisions were needed in both areas before the trip to Moscow. For example, the MSC representatives wanted to know Headquarters' position on scheduling a command and service module (CSM)/Salyut mission—1974 or 1975? Caldwell Johnson argued that a mid-1974 target date for launch would depend primarily upon the progress made in defining, developing, building, and testing a docking system. MSC also needed to know how and where Headquarters wished the docking system to be built. The Houston engineers assumed that the American half of the system would be built by a contractor, but would the United States try to build both halves, or would we negotiate a common interface specification and leave each side to fabricate its own part? The same questions arose concerning radio equipment. Would the U.S. lend the Soviets American receivers, or would we give them the technical specifications for the hardware and let them build the radios? Though this issue embraced the sticky subject of technology transfer, NASA knew that the radio and frequency used in Apollo would never be incorporated into Shuttle. So if they should be asked to build receivers for the Apollo frequency, the Soviets would be building a piece of hardware that was obsolescent, but which might contain technological concepts that were not.

Apollo-Salyut Test Mission Planning Activities—Manned Space Center

A. Objectives
B. Schedules
C. Mission model
   1. Summary (include questions for the Soviets)
   2. Overall mission
      a. Assumptions (include questions for the Soviets)
      b. Ground rules
      c. Profiles (altitude, phasing, etc.)
      d. Timelines
      e. Consumables
   3. Docked mission
      a. Assumptions (include questions for the Soviets)
      b. Ground rules
      c. Timelines (procedures, equipment transfer, cooperative experiments, restrictions, etc.)
D. Technical subjects*—requirements and solutions
   1. Atmospheres, life support
      a. Composition and characteristics of atmosphere
      b. Crew transfer (nominal and others)
   2. Constraints
      a. During docking
      b. While docked
   3. Coordinate systems
   4. Guidance systems
      a. Optical
      b. Radio
      c. Lights
      d. Docking targets
   5. Control systems
      a. During docking
      b. While docked
   6. Communications
      a. Air to air
      b. Air to ground
   7. Docking mechanism
      a. Functions
      b. Capabilities
      c. Parameters (geometry, kinematic envelope, etc.)
E. Other subjects
   1. Training
      a. Crew
      b. Mission team
   2. Mission control
   3. Mission rules, contingency procedures, etc.

*Each of these subject areas should be organized with separate paper on at least the following topics: (1) subjects and issues for discussion with U.S.S.R.; (2) recommended position and/or numerical values relative to discussion subjects; (3) expected implication of "recommended position" to the U.S. program; (4) technical analysis, trades, other options; (5) recommended methods to implement solutions (e.g., exchanged hardware); (6) qualification, testing; (7) launch checkout requirements, and (8) questions for the Soviets.

Because MSC specialists wanted to develop an equal-partner relationship with the Soviets, they preferred to develop common specifications for that basic interface, the docking mechanism. Uniformity was absolutely necessary in the docking mechanism; elsewhere, compatibility was all that was required. The actual detailed execution of the design could vary as long as the functioning of the system met agreed specifications. When it came to equipment like the Apollo radio, MSC preferred to loan the Soviets the hardware, if possible, and save everyone time and needless work. Further away from the interface, understanding some systems would likely have to be based upon mutual assurances. MSC pointed out to Headquarters, "Conduct of such a mission warrants a measure of trust and the need to accept less than-100% knowledge and understanding of each other's equipment." But the critical areas in which full disclosure was necessary would be an issue to be resolved within each of the Working Groups.[7]

Looking at the long term, the planners saw here some important considerations that would also demand comment from Headquarters. On the American side, long-range requirements were being drafted with Shuttle-era spacecraft in mind (Shuttle Orbiter, modular stations, and space stations), but MSC was still unable to state specific needs for a number of Shuttle subsystems—communications, guidance, and tracking. The Houston representatives told the Washington staff that "In a number of technical areas, we should not agree on requirements (step A) [with the U.S.S.R.] until our long-term programs are better defined." But the Soviets would seem prepared to finalize their technical requirements after their next meeting and move on to step B, the preliminary design of hardware. Caldwell Johnson had felt all along that there was a problem of semantics in using the word "future." NASA tended to reserve this adjective for concepts that were still rather nebulously defined, while the Soviet engineers used future to describe any spacecraft that had not yet been flown.[8]

Even though the Americans were in the dark about the Soviets' plans for the 1980s and unclear about details of their own next generation of hardware, Lunney and the others were sure that an actual test mission would have specific benefits. First, the Soviets and the Americans would learn to work together. Second, jointly designing a docking mechanism would be an opportunity to work out the specific issues involved in bringing two different engineering approaches together in a compatible piece of hardware. And third, an Apollo/Salyut mission would provide NASA time to define more fully its requirements for Shuttle-era subsystems. Clearly, the Manned Spacecraft Center favored a test mission not only for its educational value but also because it would permit NASA to "work the problem" of creating compatible systems with the Soviets in their discussions of future systems without prematurely foreclosing flexibility in Shuttle design.[9]

164

ROUND THREE—MOSCOW

Glynn Lunney, Bob Gilruth, and their 18 companions arrived in Moscow on Saturday evening, 27 November 1971, where their welcome by members of the Soviet delegation was given considerable attention by the Soviet news media. The Americans were processed quickly through immigration and customs formalities, arriving at the Hotel Rossiya about 90 minutes after landing. On the way to the hotel, Lunney and Bushuyev, accompanied by an interpreter, had a pleasant chat about their past work and plans for the coming week, a discussion which was continued later that evening at a Soviet-hosted dinner at the Rossiya. Sunday was essentially a free day, and most of the Americans went on a special bus tour of Moscow, which included the People's Exhibition of Economic Achievements. Lunney noted that the space display at the exhibition grounds had some new exhibits—two full-scale Soyuz in docked configuration, the *Luna 16* lander, which had visited the moon and returned with a small sample of lunar soil, and a replica of the moon rover *Lunokhod.*[10]

On Monday morning, the NASA delegation went to the Institute of Automatics and Telemechanics, a 30-minute bus ride from the hotel. The Institute, sponsored by the Soviet Academy of Sciences and devoted to the study of automatic control systems (cybernetics), was also home base for Academician Petrov. The NASA group gathered with the Soviets for a plenary session in a large lecture room. After introductory remarks and some discussion of the week's agenda, Lunney gave the Soviets two papers. One summarized the present status of the American long-term technical requirements and the other details of a possible Apollo/Salyut mission.[11] When the Soviets reciprocated at the end of the morning session, the two groups spent the remainder of the day translating and studying. The Soviets were reluctant to begin any detailed discussions until they had an opportunity to more fully understand this new material. While one of the American interpreters read aloud in Russian to the Soviets from the NASA papers, a quickly transcribed version of shorthand notes taken from a verbal translation of the Soviet materials was prepared for the Americans.

In addition to these basic documents, Lunney and his colleagues argued for and obtained a chance to present for the entire Soviet group highlights of the U.S. mission model and docking mechanism studies. The quick summary gave everyone, including Academician Petrov and his executive staff, a basic understanding of the NASA ideas for a joint mission. With this background, the three Working Groups could go their separate ways, but they would be negotiating within a more clearly understood framework.

Lunney reported that in Working Group 1, which he chaired, the Soviet side had "very capable experts on the subjects of life support and mission planning." At various times during the week, the men were able to divide into smaller subgroups to discuss specific topics. With the aid of the interpreters, "a good deal of understanding was reached, and several enclosures on specific subjects were prepared for inclusion in the minutes." Looking at the experiences of the other groups, Lunney commented, "Working Group #2 also used the splintering technique because of the multitude of systems that were covered. . . . Working Group #3 on the docking mechanism tended to work more as a group . . . because of the nature of their [topic]." He believed that by following the Low precedent— preparing ahead of time documents similar to those agreements that were desired—the NASA representatives in Working Groups 1 and 2 "were able to lead most of the discussions and focus on the parts of the problem that we felt [were] significant."[12]

The remainder of the week (29 November-3 December) was spent in Working Group sessions, with the specialists devoting most of Friday and part of Saturday documenting their results. Those who could get free that weekend were taken to Star City, where they toured the cosmonaut training facilities. Lunney saw K. P. Feoktistov there, and the designer-cosmonaut gave his American friend an in-depth briefing on the Soyuz control systems. After a stand-up buffet luncheon given by the Commanding General of Star City, the Working Group members returned to Moscow. Sunday was spent sightseeing, with a trip to the Zagorsk monastery, about 81 kilometers from Moscow.

On Monday, 6 December, the delegations met a final time at the Institute to verify and sign the minutes of their meetings, with the executive staffs reading and authenticating their minutes and those of the Working Groups. The Summary of Results, which included the minutes of each Working Group as attachments, was signed at the House of Scientists that evening.* Lunney subsequently commented that this whole procedure was "a fairly tedious process, [but] sufficient time must be programmed for this not-very-productive necessity."[13]

ISSUES AND ANSWERS

The negotiations conducted in Moscow had indicated by their variety and scope the growing complexity of the joint effort. At the executive level, the question of more rapid and frequent communications between the Soviet and American Technical Directors had been raised. In the November briefing

*The names of the signatories to these and subsequent joint minutes are presented in appendix C.

to Headquarters before the trip MSC had pointed out that preparations for a test mission could never be conducted by the slow process of exchanging letters through diplomatic mail: "If such a test mission is to be developed, we need to establish a method for more timely communications with the Soviets."[14] In Moscow, Chris Kraft had raised this matter, urging NASA and the Soviet Academy to establish weekly or biweekly telephone conference calls between the Technical Directors, with these discussions being confirmed by telex. At first, Petrov had balked, saying that it would be too expensive. He argued besides that they would have to bring in their telephone people before they could have direct telephone conversations with the Americans. For this reason, he could not discuss the topic. Kraft insisted that telephone conferences and telex exchanges had been required in the American manned space program since Project Mercury. Gilruth added, "It was essential to permit an easy flow of information and to establish a system for reassurance that progress was being made." He told the Soviets that they "had to agree to this point or the mission would be impossible." Kraft and Gilruth hammered away on this subject for some time, and finally Gilruth told the Soviets that should they be unwilling to agree to the telephone conversations the NASA delegation might as well pack up and go back to Houston. After some hesitation, the Soviets decided to try the telephone/ telex approach, and this agreement had been included in the Summary of Results.[15]

During the executive group meetings, having declared that "a test mission appears technically feasible and desirable," the two sides did determine that it would be necessary to make an early decision about the practicality of scheduling a flight for 1975. With the American side proposing that the launch occur in the spring or summer of the year, the parties had included in the Results an agreement that each side would send the other by 1 April 1972 "a statement of its position on the prospects for the actual conduct of the test mission in 1975" and their concepts of such a mission. To pace the implementation of these decisions, the executive staff had drawn up a preliminary list of milestones, or major events, for the planning, design, and implementation phases of preparing for a test flight. (See box on next page.) This schedule was patterned after a standard NASA format, and the original draft would be subjected to further discussion prior to the April deadline. As the list grew, the need for closer communication became even more apparent. It also became clear that all letters, telexes, and telephone conversations should be coordinated by the Technical Directors. Many hands would work on the joint project, but they would have to be carefully orchestrated to assure success.[16]

Various questions and issues had been raised in each of the Working

Proposed Preliminary Schedule

	1972	1973	1974	1975

Concept agreements: ▼ Exchange project concepts

Mission

Equipment   ▼ December 1971 meeting should provide the basic understandings and the exchanges necessary to define mission adequately for approval in spring 1972 and to permit signing certain documents at next meeting in May or June 1972   **LAUNCH PERIOD**

Schedule

Program orientation   ▼ Official go-ahead

U.S. vehicle flown:

Apollo command and service module (CSM) — Modifications / Test

Apollo docking module — Design / Manufacture / Test / Launch checkout

Interface equipment:

Docking system — ▼ ▼ Design and development / ▼ Qualify / ▼ Checkout

Tracking system — ▼ ▼ Coordinate requirements CSM–ID&D / Qualify / Compatibility test / Checkout

Communications — ▼ ▼ Coordinate requirements CSM–ID&D / Qualify / Compatibility test / Checkout

Lights — ▼ Coordinate requirements Design and development ▼ Manufacturing and test

Docking contact criteria — ▼ Coordinate requirements

Docking target — ▼ Coordinate requirements Fabrication and test

Control system — ▼ Coordinate requirements ▼ Simulation test

Crew transfer equipment and conditions — ▼ Coordinate requirements ▼ Fabrication and test

Mission documentation:

Preliminary documents for mutual signing:

Organization plan   ▼

Mission objectives and requirements   ▼ — Revisions as required

Launch window plan   ▼

Trajectory plan   ▼

Preliminary documents for mutual signing:

Crew activities plan   ▼

Mission operations plan   ▼ — Revisions are required

Contingency plans   ▼

Detailed procedures   ▼

▼ Design reviews and agreements to be scheduled as required

CSM–ID&D   Command and service module–international design and development

Groups, and for the most part they had been resolved as the talks progressed. Group 1 completed the general documentation of its agreements on life support systems, coordinate systems, constraints on spacecraft configuration, and communications links between ground control centers. With respect to the proposed Apollo/Salyut test mission, the two sides spelled out the objectives of such a flight and listed the project documents that would have to be prepared for the mission. The chairmen of the Group agreed to a mutual exchange of data on launch windows within two months, on program information for the test mission by April, and on communications channels for the respective control centers within three months.The Americans also planned to provide a draft of an interface organizational plan for the project.[17]

Working Group 2 had also come to a number of significant decisions. They developed a list of guidance and control systems and other onboard equipment in the Soviet and American spacecraft that would have to be made compatible. The preparation of documentation covering the subjects of docking lights, docking targets, and contact conditions between spacecraft, as well as the technical data on control systems and radio tracking, progressed satisfactorily. This group planned to reorganize the documentation into two volumes covering general requirements for the future and specific demands on the systems proposed for Apollo/Salyut. For the test mission, the two sides would need to develop communications and tracking systems to an agreed set of technical requirements. An Apollo-type VHF ranging system would be installed in the Salyut as a backup system, and the Soviets had said they would study the issue of building their part of the onboard communications system versus using equipment provided by NASA.

Group 2 had also delved into the control and guidance problems relating to docking. For example, by considering the relative velocity of the two spacecraft, the docking system engineers established numerical values for the force with which the two vehicles might dock. The two sides also concurred on docking targets. One would be mounted in the center of the Salyut docking hatch, providing the Apollo CSM pilot with a dynamic visual reference for alignment. A second target, of the passive type used in the Apollo program, would be placed on Salyut where it could be seen from the command module through the crewman optical alignment sight. In addition, each side had been assigned work on control stabilization requirements for the two spacecraft and had been asked to look further into the design, development, evaluation, and installation of the docking target concepts.[18] Meanwhile, Group 3 had concentrated on the problems of creating a universal docking mechanism.

AN INTERNATIONAL DOCKING SYSTEM

Before the winter meeting, both Caldwell Johnson and Vladimir Syromyatnikov had been thinking about what they would like to incorporate into a compatible docking system. Johnson had been urging that the group accept the double ring and cone docking concept he had described to the Soviets on his first trip to Moscow. In June, he had had an opportunity to chat with Soviet docking specialist Valentin Nikolayevich Bobkov during a free-wheeling round table conversation among the engineers. In addition to Johnson and Bobkov, George Hardy, Robert "Ed" Smylie, Edgar "Ed" Lineberry, Leroy Roberts, Ilya Vladimirovich Lavrov, and Igor Petrovich Shmyglevskiy took part in the shop talk. While Shmyglevskiy was the only one present who spoke both English and Russian, Bobkov could read and write English, so the eight men drew sketches, translated words verbally and on the drawing pad, and made hand gestures to understand one another.

During this conference, Bobkov had indicated that the Soviets also favored some version of the double ring and cone. Bobkov illustrated through rough sketches that the overall diameter of the docking system could not exceed 1.3 meters, because any larger system would require a change in the launch shroud. When Johnson raised the question of altering the shroud, the Soviets stressed the major impact that such a modification would have. In addition to having to design a new shroud, they would have to test out the launch aerodynamics of the altered hardware. The Americans had hoped to argue for a larger tunnel, but such a change appeared to be too great for their counterparts. The Soviets in turn understood the American's thoughts on an airlock module.[19]

After the June meetings in Houston, Johnson had put Bill Creasy and his mechanical designers to work on the preliminary design of a docking mechanism. By the time the NASA delegation left for Moscow, Creasy's crew had designed and built a 1-meter double ring and cone docking system that had four guide fingers and attenuators on both rings, so either half could be active or passive during docking. The Structures and Mechanics Laboratory at MSC made 16-millimeter movies demonstrating this system in action, which Johnson took to Moscow in November, along with a booklet describing the system and a model of the capture latches. He had gone prepared to sell his idea.[20]

Once he was in the U.S.S.R., Johnson discovered, however, that his job was that of an engineer, not a salesman. Since October 1970, Syromyatnikov had been working on a variation of NASA's ring and cone concept. Instead of the four guide fingers in the American proposal, Syromyatnikov suggested three, and in lieu of hydraulic shock-absorbers, he proposed electro-

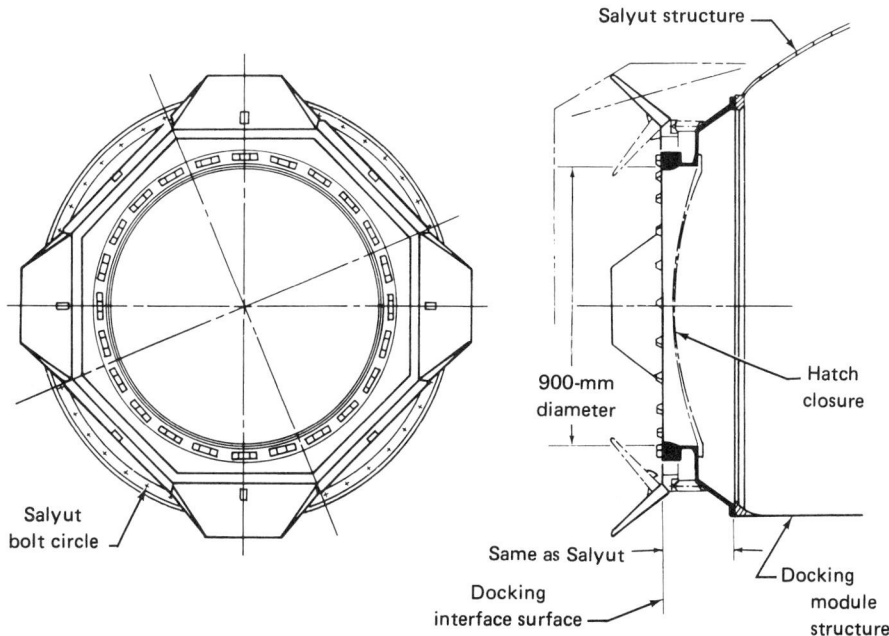

NASA proposal for a four-guide docking mechanism is shown to the Soviets during the November 1971 meeting in Moscow. Note the 900-millimeter-diameter tunnel.

mechanical attenuators. In essence, the Soviets had accepted the idea of using a set of intermeshing fingers to guide the two halves of the docking gear from the point of initial contact to capture. The concept of using shock absorbing attenuators on the active spacecraft's capture ring to buffer the impact of two spacecraft coming together was also acceptable. Both groups of engineers planned to retract the active half of the docking gear using an electrically powered winch to reel in a cable. Once retracted, structural or body latches would be engaged to lock the two ships together. Three basic issues had to be resolved—the number of guides, the type of attenuators, and the type of structural latches—before the design of a universal system could proceed.[21]

Johnson, Creasy, and the other engineers in the Spacecraft Design Division had wanted to use four guides because they believed that it provided the best geometry when using hydraulic attenuators. As Bill Creasy subsequently explained it, the most probable failure situation using hydraulic attenuators would be a leak that would cause one shock absorber to collapse on impact. A study of various combinations had led the MSC specialists to conclude that four guides and eight shock absorbers was the

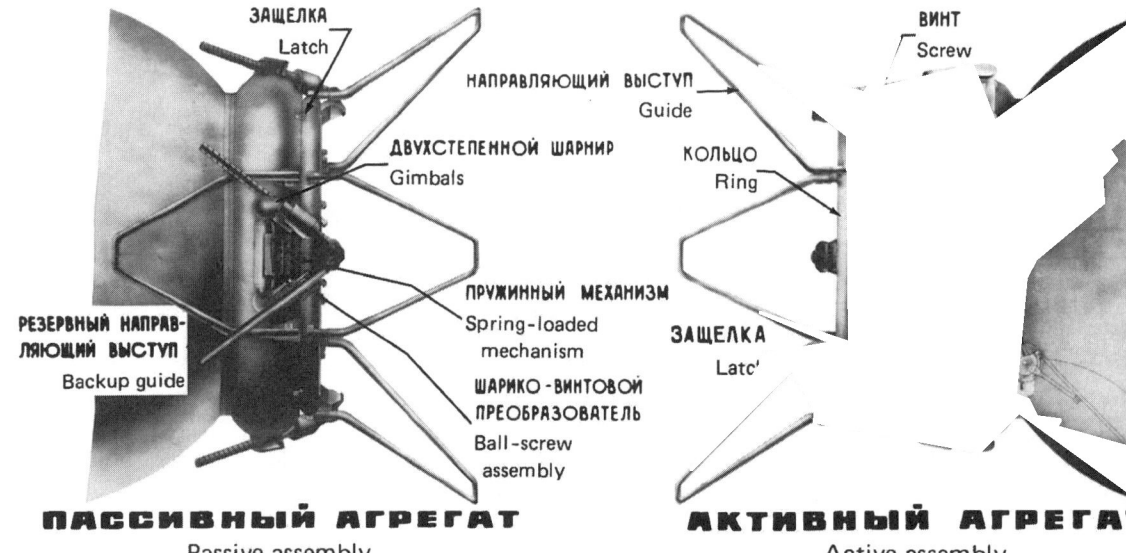

ЗАЩЕЛКА
Latch

НАПРАВЛЯЮЩИЙ ВЫСТУП
Guide

ДВУХСТЕПЕННОЙ ШАРНИР
Gimbals

РЕЗЕРВНЫЙ НАПРАВ-
ЛЯЮЩИЙ ВЫСТУП
Backup guide

ПРУЖИННЫЙ МЕХАНИЗМ
Spring-loaded
mechanism

ШАРИКО - ВИНТОВОЙ
ПРЕОБРАЗОВАТЕЛЬ
Ball-screw
assembly

ВИНТ
Screw

КОЛЬЦО
Ring

ЗАЩЕЛКА
Latc'

ПАССИВНЫЙ АГРЕГАТ
Passive assembly

АКТИВНЫЙ АГРЕГА'
Active assembly

Docking system proposed by Soviet designer Syromyatnikov during the November 1971 meeting in Moscow.

optimum design. Creasy pointed out too that the most likely trouble with an electromechanical system would be a freeze-up or binding of one of the pairs of attenuators. Thus, the Soviets had sought to minimize the number of pairs in their system for the same reason that the Americans had preferred a larger number—to limit the probability of something going wrong.*[22]

As Johnson talked this out with Syromyatnikov, it became clear that they both wanted to stay with systems that would give them maximum confidence in the design. But they agreed that a compromise could be reached. Johnson reported on his discussions with Syromyatnikov:

> Since there was no conflict in principle, nor was there envisioned to be a conflict in subsequent engineering detail between interfacing features of the proposed US and proposed Soviet docking mechanisms, and since the US had no significant engineering or hardware equity in its proposed design, and since the USSR had considerable equity in its proposed design, the Soviet design was selected as a baseline for the next phase of study.[23]

By the end of the November-December meeting, the two Group 3 teams had signed a set of minutes outlining the basic concept for a universal

*At an earlier meeting, V. Zhivoglotov had told R. D. White that the Soviets were opposed to a system using eight attenuators because the electrical device they had planned to employ to dissipate the docking energies could not be used with eight attenuators but it could be used with the six shock absorbers.

androgynous docking system. The formal statement read, "The design concept includes a ring equipped with guides and capture latches that were located on movable rods which serve as attenuators and retracting actuators, and a docking ring on which are located peripheral mating capture latches with a docking seal." Basic information on shapes and dimensions of the guides were also included in the minutes. They were to be solid and not rodlike, as first proposed by the Soviets, and three in number. As long as the requirement for absorbing docking forces was met, each side was free to execute the actual attenuator design as it best saw fit. The Soviets planned to use an electromechanical approach designed for the Soyuz docking probe, and the Americans proposed to stick with hydraulic shock absorbers similar to those used on the Apollo probe. This proposal also called for developing docking gear that could be used in either an active or passive mode; when one ship's system was active, the other would be passive.

Looking into the detailed design of the mechanism, the two sides had further agreed that the capture latches would follow the design developed at MSC and the structural latches and ring would follow the Soviet pattern. These paired sets of hooks had been successfully used on both Soyuz and Salyut. In addition, Group 3 concurred on details regarding the alignment pins, spring thrusters (to assist in the separation of the spacecraft at undocking), and electrical connector locations. To evaluate the docking system concept and to ensure the establishment of compatibility at an early point in the development, the men planned to build a two-fifths-scale test model, the exact details of which would be decided at the next joint meeting.[24]

Upon his return to Houston, Caldwell Johnson prepared a memorandum to document some of the informal understandings reached in Moscow. He indicated that this reflected "upon the manner in which the two countries will conduct and coordinate the next phase of the engineering studies of those systems. . . . The understandings . . . were reached more often than not outside of formal meetings, and so are not likely otherwise to be reported." For example, in the area of hatch diameter, he noted that "it became apparent from the beginning . . . that a hatch diameter greater than about 800 mm could not be incorporated into the Salyut spacecraft without great difficulty," but MSC had "long since reconciled itself" to a test hatch diameter of less than 1 meter. Johnson went on to comment that "the capture ring assembly had variously been called ring and cone, double ring and cone, and ring and fingers. It was agreed henceforth to call the capture ring 'ring' and the fingers 'guides.'" Thus it went—negotiation, understanding, compromise, and accommodation.[25]

The docking talks in Moscow had convinced Johnson that frequent face to face communications were necessary. He believed that an acceleration of the design process was also in order if they were to settle on a single design by June 1972. It would not be possible to develop a design through correspondence or meetings every six, three, or even two months. He saw two possible alternatives:

> Accelerate the iterative process by very frequent and informal, face-to-face negotiation between the key designers—each having authority to make technical decisions on the spot; or, assign design responsibility for each interfacing element of the mechanism to one country or the other. Of the two alternatives, only the first is practicable, since apportionment of responsibility would likely take just as long as the design process.[26]

While tentative arrangements had been discussed for telephone and Teletype exchanges, Johnson thought that limited progress had already been made when it came to "the exchange of [a] vast amount of technical detail data such as drawings, diagrams, performance analyses, etc., that are necessary for each side to understand the nature of the interfacing system." The concerns expressed in Johnson's memo reflected the thoughts that other members of the American delegation had as they returned home. A joint mission with the Soviets was clearly feasible from a technical standpoint, but the key to such a complex project would be creating the proper management format. That task would fall on Glynn Lunney's shoulders.

After a day of shopping for gifts in Moscow, the NASA delegation had left the Soviet capital on 7 December via England, where they briefly visited the Royal Aircraft Establishment at Farnborough. This pleasant diversion, according to Lunney, took "many of us back to the NACA days." In Houston, the Working Group members soon found that they had their work cut out for them. On 16 December, Lunney distributed a memo outlining tasks to be done and clarifying who was responsible for each. Lunney anticipated convening special joint sessions for the Group 2 members involved in radio tracking discussions and for the Group 3 members concerned with the docking mechanism. "In considering our experience so far in these discussion," Lunney commented, "we have found it absolutely essential to have well-prepared documents for each meeting in order to efficiently conduct and steer the discussions and resultant agreement documents." He believed most of the documents that had to be ready in the near future were relatively "straightforward" and easy to prepare, but they must schedule the work carefully and pursue it in a businesslike manner.[27] (See box on facing page.) While Lunney and his colleagues began to work their specific tasks, the Office of Manned Space Flight (OMSF) staff in Washington was looking more deeply into the costs of an international flight.

Schedule for Next Six Months *

Mid-February (two months)
    Long-term
        Exchange papers on:
                • Connection of liquid cooled garments LOG's to other ship
                • Quantitative characteristics and possible solutions for pre-breathing
    Apollo/Salyut
        Exchange papers on launch window constraints and solutions to daylight Apollo launch
        U.S. provide agendas for "Special Subject Meetings" in March
        Finalize Working Group 2 documents agreed to in December meeting
    March 1 meeting
        Special meetings in Houston on:
                • Docking mechanism
                • Radio tracking system
Mid-March (three months)
    Apollo/Salyut
            • U.S.S.R. comments on U.S. proposal for communication between control centers
            • U.S. proposal of organization plan
            • Exchange outlines for project documentation
            • U.S.S.R. provide Salyut life support system data
Mid-April (four months)
    Apollo/Salyut
            • Each country exchange "Statement of Position"
            • Each country exchange "Project Technical Proposal Document"
May to June (five to six months)
    Next full-scale meeting

*Informal planning schedule distributed by G. S. Lunney, Dec. 1970, after trip to Moscow.

ESTIMATING THE COSTS OF A MISSION

The possibility of flying a joint mission with the Soviets in mid-1975 posed some interesting problems for Dale Myers' staff in OMSF. When they began to look at this problem in the fall of 1971, it became apparent that they would have to make some quick decisions about this yet-to-be-authorized project if they were to meet the proposed launch date. MSC would need to start the development work on the docking module and the docking system in early 1972. And modification of the CSM should start immediately. Limits on time and money were not the only problems. OMSF had been advised by North American Rockwell that beginning in October 1971 the labor force that had been building the command and service modules (CSMs) would be reduced. A decision on which CSMs to set aside for an international rendezvous and docking mission (IRDM) had to be made quickly. Gilruth had requested that 115 and 115A be completed because they were of the most recent series of CSMs, with a scientific instrument module (SIM) bay into which earth resources survey instruments could be placed. The older 111 CSM was closer to being ready for launch, but it did not have a SIM bay. CSM 119, the Skylab backup and rescue spacecraft, could not be allocated to IRDM until the final Skylab visit, then scheduled for 1974. When the money, time, and labor issues were balanced against the wishes of the mission planners, some hard choices had to be made.[28]

Dale Myers had written to Bob Gilruth four days before the delegation's departure to Moscow to ask him to look over a list of "Suggested Guidelines for a Minimum Cost International Docking Module." This list, prepared by William C. Schneider, Director of the Skylab Program, reflected OMSF's concern for keeping the IRDM equipment simple and cost effective. Schneider, drawing from his experiences with Skylab, suggested that the module be kept as small as practical and that it be designed with a high safety factor. He thought it best to follow the Gemini design principle of placing many systems, particularly wiring, on the outside of the docking module, thus lowering flammability concerns. At the end of his recital of 20 items, he said:

> The fundamental, you can see, is *keep it simple.* Of course, that's how Skylab started in 1966. I have no solution to maintain that posture other than a generalized observation that an active Headquarters staff is invaluable in detecting and controlling policy variations. . . .
>
> I strongly urge that the Skylab system of PRR, PDR, CDR[*] be adhered to and that short cuts be resisted despite the immediate lure of maintaining schedule. Each time we've rushed, cancelled, or hurried by one of these milestones, I've come to regret it later on.

Schneider had additional thoughts when it came to keeping costs to a minimum. He proposed that Marshall Space Flight Center in Huntsville, Alabama, develop and build the docking module; according to Schneider they had a proven capability (Saturn launch vehicles, Apollo telescope mount, multiple docking adapter for Skylab), existing facilities, and the

---

*Preliminary Requirements Review, Preliminary Design Review, and Critical Design Review were elements of the NASA spacecraft development cycle, which had evolved since the early days of Apollo.

proper labor mix. These elements would permit Marshall to do the job more cheaply than MSC and a contractor. Furthermore, he believed that with Shuttle Orbiter and Skylab drawing heavily on Houston's personnel, the docking module development "probably would not receive much attention or would divert talent from the other tasks." Schneider could see only one area in which MSFC might have some difficulties—working with the Flight Operations Directorate at MSC. To solve that problem, he recommended that Clifford E. Charlesworth, Eugene F. Kranz, or Glynn Lunney be transferred to Marshall as "Module Manager to insure a clean interface."[30]

Myers sent Schneider's list of 20 guidelines to Gilruth, with the request that the MSC program plan include these points, but Schneider's other thoughts about building the docking module at Marshall were not included.[31] Gilruth responded that his team basically agreed with Schneider's guidelines but countered that these points had already developed somewhat differently. He enclosed the fourth revision of the "International Rendezvous and Docking Mission Guidelines and Constraints Document" for OMSF's perusal. Gilruth told Myers that MSC "would be glad to discuss the guidelines and the method of implementation in detail with you and your personnel at the appropriate time."[32] Implicit in his remarks was the idea that the IRDM was a Houston project. It involved Apollo spacecraft, and MSC knew how to get the job done. Only Frutkin, the interpreters, and several secretaries from Washington had joined the Houston delegation that went to Moscow in November. As the joint effort progressed, Marshall would be noticeably absent during the negotiations. The Americans might fly with Salyut, but it was not likely that the Soviets would rendezvous with Marshall's Skylab. At the November-December meeting, the Soviets and Americans ruled out a union with the first Skylab; if such a mission was ever undertaken, it would be with "a Skylab or another type [of station] to put into orbit after 1975."[33]

Continuing his dialogue with Gilruth, Myers sent his comments on the International Rendezvous and Docking Program Plan to MSC on 14 December 1971. Myers agreed that this document could serve as the basis for further discussions with the Advanced Missions Program Office at Headquarters, and he advised Gilruth that Phil Culbertson's staff would "work with you and your people in finalizing such a plan." OMSF and Advanced Planning had some specific items that they wanted Houston to look at again. MSC had proposed that North American Rockwell undertake developing the docking module on a sole source procurement plan. Myer's staff questioned the justification for not soliciting other contractors in open competition, and they wanted Gilruth to think about competitive selection. Likewise, OMSF preferred that the prebreathing requirement during transfer be eliminated, if possible, and that the planing schedules be further refined.[34] Gilruth's staff worked on these problems throughout December and into February 1972.

MSC's studies of the costs of an International Rendezvous and Docking Mission and the best way to contract for its equipment produced an avalanche of paper. Data indicated that such a mission, using CSM 115 and 115A, would cost in excess of $267 million and could run nearly as high as $280 million for three docking modules (one test, one backup, and one flight), seven docking mechanisms (two flight, four test, and one spare), and experiment packages. These investigations convinced the Center management that experience would produce economy in this case, so North American Rockwell should develop and fabricate the docking module and docking mechanism. As the builder of the CSM, Rockwell would be able to work with the command module/docking module interface with minimum difficulty. In addition, they had the Apollo manufacturing equipment and the necessary labor skills, if the job were begun before the company started laying off their experienced employees. However, the ultimate decisions about how much money NASA could afford to allocate to the mission and who the contractor would be had to come from Headquarters.[35]

Dale Myers met with the top management* on 24 February to discuss the cost of the proposed docking mission, and they reached three key decisions. First, the planning effort was to be oriented toward a program that would include a demonstration flight, but the total program effort was not to exceed $250 million. Based upon the data already generated, this ceiling precluded the use of either CSM 115 or 115A. Second, Houston would have to base its planning on the use of CSM 111 as the likely flight test vehicle and CSM 119 as a potential backup vehicle (assuming that it was not flown during Skylab). The budget included the necessary modifications for CSM 119 to make it flight ready, but it did not cover the expense of an actual mission based on 119. The final decision made on 24 February concerned experiments. Since the 111 and 119 service modules did not have scientific instrument bays, the experiments would have to be much simpler than the earth resources survey originally proposed. Of the $250 million total, $10 million were allocated for developing experiments that could be housed in the command and docking modules. No more work on CSM 115 and 115A was contemplated.[36]

Managing the development of the IRDM hardware was the task of the Manned Spacecraft Center and its new Director, Christopher C. Kraft. Effective 14 January 1972, Robert Gilruth had assumed the position of

*Those present were Administrator J. C. Fletcher, G. M. Low, W. H. Shapley, and A. W. Frutkin.

Director of Key Personnel Development for NASA, and Deputy Director Kraft had moved into the number one position. Like his predecessor, Kraft was an old-timer in the American space program, joining NACA in 1945 and becoming one of the original members of the Project Mercury team. Before becoming Gilruth's deputy in 1969, he had been Director of Flight Operations in Houston. The tasks facing his center in 1972 included preparing for Skylab, developing the multipurpose Space Shuttle,* and proceeding with Apollo/Salyut—whose teams were already preparing for the next round of discussions with the Soviets as Kraft settled into his new office.[37]

NEGOTIATION BY TELEPHONE

On 27 January 1972, Glynn Lunney wrote to Professor Bushuyev, proposing a list of questions to be discussed in their first telephone conference.[38] The basic purposes of the conversation were to clarify arrangements for the March meeting of Group 3 in Houston, to clear up some technical questions associated with the design of the docking mechanism, and to discuss arrangements for the Group 2 meeting tentatively scheduled for June in Moscow. During January and February, the number of letters between Houston and Moscow had increased, indicating the growing complexity of the joint effort. The specialists needed faster answers; at the request of the Soviets MSC initiated a telephone call to Bushuyev on the morning of 2 March. Because the overseas circuits were busy, nearly 40 minutes passed before the NASA party reached the Academy of Sciences. For 75 minutes, the two sides struggled with the initially awkward process of talking through an interpreter over a not-too-perfect overseas telephone connection. On the Soviet side, the process was complicated by the fact that they were using two telephone handsets, but in Houston a conference arrangement circumvented the necessity of passing the phone from one person to the other.[39]

Glynn Lunney was in Washington at the time of the call so Caldwell Johnson spoke for MSC.† Bushuyev and Johnson discussed several questions associated with fitting the docking mechanism under the launch shroud of Salyut. The Soviets agreed to ease up on their height requirement for the docking mechanism, a change that would be discussed at the Houston talks

---

*The Space Shuttle had received Presidential approval on 5 Jan. 1972.

†Present in Houston for the telecon were C. C. Johnson, D. C. Wade, R. D. White, J. C. Jones, J. C. Waite, W. K. Creasy, R. Reid, E. N. Harrin, and W. Karakulko. On the Soviet side, K. D. Bushuyev, V. P. Legostayev, V. S. Syromyatnikov, I. V. Lavrov, V. N. Bobkov, and B. P. Artemov were near the phones. Harrin, Karakulko, and Artemov acted as interpreters.

scheduled for later in march. No date was selected for the Moscow visit of Working Group 2, but the American side restated its desire to hold the meeting after the 16 April launch of Apollo 16. After several minutes of speaking with Syromyatnikov about other docking mechanism questions and with Legostayev about Group 2 matters, Johnson bid Professor Bushuyev *do svidaniya* [good-bye]. The Professor in turn wished his best to Johnson and asked him to convey greetings to Lunney.[40] (See box on facing page.)

The first telecon was helpful but difficult. The Americans sent a transcript of the tape recorded conversation to Moscow, and the Soviets sent their version to Houston. Thereafter, exchanging minutes became another way to assure clear understanding of such communications. Nevertheless, the Soviets, and particularly the Professor, were not satisfied with the telephone as a medium for discussing technical matters. As a result, Lunney on his return to MSC wrote to his counterpart:

> It is my strong personal belief that continued exchanges like the tele-conference and probably more frequent meetings are essential to the success of the project. The difficulties and dangers of this mission will be reduced in direct proportion to the increase in knowledge and understanding between us and our colleagues.[41]

Stressing this point further, Lunney suggested that the Group 2 meeting in Moscow be preceded by a similar telecon. The Americans were determined to establish fast and reliable communications with the Soviets. The work in the immediate weeks ahead would stress the necessity of spelling out specifications for the docking system.

DESIGNING THE INTERFACE

During the spring months of 1972, the personnel of the MSC Engineering and Development Directorate pursued the design of an international docking system. Working concurrently with the North American Rockwell team and the Soviet Group 3 members, the Houston engineers were attempting to ensure the speedy development of hardware. Starting a contractor to work on a project before design was firm was not unusual. In the Apollo program MSC had followed the same approach in the design and development of the command and service modules, the lunar module, and their various subsystems. The iterative process of design helped to ensure the timely delivery of hardware and the maintenance of tight schedules.

The 27 March-3 April visit of the Soviet Working Group 3 members to Houston was essential to the NASA plan of having North American Rockwell start the detailed engineering of an Advanced Missions docking system. The four-man delegation led by Syromyatnikov quickly got down to the task

Excerpt from Transcript of Telecon Between
Manned Spacecraft Center, Houston, Texas, U.S., and Intercosmos,
Moscow, U.S.S.R. from Approximately 9:30 to 10:45 CST

American Translator:	Good evening!
Soviet Party:	Hello!
American Translator:	This is the Manned Spacecraft Center speaking. May we speak to Professor Bushuyev?
Soviet Party:	Hello!
American Translator:	Hello, can you hear me?
Soviet Translator:	I hear you well.
American Translator:	Good! This is the MSC NASA USA, may we speak to Professor Bushuyev?
Soviet Party:	Prof. Bushuyev to the telephone? I will ask him.
American Translator:	Oh! That is you.
Soviet Party:	Yes.
American Translator:	Mr. Caldwell Johnson will now speak to you.
Mr. C. Johnson:	Sdravstvuite! [Hello!] (In English) Greetings from the MSC, Houston, Texas, to our Soviet colleagues in Moscow.
Soviet Party:	Hello!
American Translator:	Mr. Caldwell Johnson will now speak to you in English, and after, if you like, I will translate it to you in Russian. Will that be convenient for you?
Soviet Party:	Hello!
American Translator:	Hello!
Professor Bushuyev:	This is Professor Bushuyev speaking. I would like to propose the following. Do you hear me well?
American Translator:	Yes, can you speak a little louder?
Professor Bushuyev:	I would like to make the following proposal.
American Translator:	Please.
Professor Bushuyev:	We have worked on the questions initiated by Dr. Lunney in the letter of 27 Jan. and propose to (or offer to) lay out the answer to these questions with the aid of our translator.
American Translator:	Good.
Professor Bushuyev:	If Dr. Lunney agrees, I will transfer the phone to Dr. Lunney's friend, a co-worker in our delegation Mr. Artemov.
American Translator:	It will be . . .
Professor Bushuyev:	What?
American Translator:	Who? Artemov?
Professor Bushuyev:	Mr. (tape garbled) of our delegation who participated in the meeting which took place in Moscow at the end of Nov.
American Translator:	Good! One second.
Professor Bushuyev:	I will transfer the telephone to Mr. (tape garbled).
American Translator:	Excuse us please. May we break in?
Professor Bushuyev:	Yes.
American Translator:	Mr. Caldwell Johnson will now speak, whom you know; he had been in Moscow. He would like to say something first. All right?
Professor Bushuyev:	All right. (Some Soviet speech, but unintelligible.)
Mr. Johnson:	Prof. Bushuyev . . .

of joining in defining the dimensions and specifications of the docking system. This information was spelled out in the minutes of their meeting and in four sheets of engineering drawings.* Bill Creasy and several of his colleagues worked with Yevgeniy Gennadiyevich Bobrov at the drafting table to lay out these first Soviet-American engineering drawings. Larry Ratcliff drew the capture ring and guides on drafting paper, and Robert McElya supplied the details of the structural interface ring, while Bobrov prepared a similar drawing for the structural latches. T. O. Ross then took these drawings and conducted a dimensional analysis to be sure that all items were compatible.

On 3 April, the two sides completed their drawings and wrote their minutes. These drawings were a blending of the way in which the Americans and the Soviets usually presented data on paper. Creasy said, "Their drawing procedure is different from ours and sometimes we joke and say that . . . [these Group 3 drawings] must violate the drawing conventions of at least the U.S. and Russia and probably several other countries." But each side could understand and work from the information as recorded, and that was the important point. Looking back on this effort, Creasy commented that, despite five subsequent updatings of the April drawings, the basic work only required some minor refinements and adding the tolerance dimensions.[42]

Agreement on technical specifications for the docking system cleared the way for NASA to begin discussions with Rockwell about building the docking module and the docking system and modifying the CSM. As MSC engineers worked with the potential contractor in drafting a statement of work for Apollo/Salyut test mission hardware, the procurement staff in Houston drew up their contracting plans.

At Headquarters, the agency's senior staff was looking into various political aspects of conducting a joint mission, and two issues were paramount in these discussions. First, congressional authorization and appropriations would have to be obtained before NASA could begin to modify or build the necessary hardware. Second, a bilateral agreement between the United States and the Soviet Union would have to precede the request for funds. George Low was given the task of determining how to resolve this issue.

APRIL IN MOSCOW

Following the November-December 1971 meeting in Moscow, NASA Headquarters has recommended to the White House that a formal agreement on an Apollo/Salyut mission be included on the agenda for the May Summit meeting between President Nixon and Premier Kosygin. After several

---

*Following the formalization of ASTP in June 1972, these drawings became part of Interacting Equipment Document 50 004, "Apollo Soyuz Physical Interface Requirements."

discussions with the White House during the ensuing months, Henry Kissinger asked NASA to make a firm recommendation by 15 April concerning the feasibility of conducting such a flight. In Glynn Lunney's opinion, the Soviets would have to agree to three basic management documents before NASA could make a positive recommendation to the President. Draft versions of Lunney's documents—a project technical proposal, an organizational plan, and a project schedule—were ready for transmittal to Moscow by the end of March. Fletcher and Low decided that Low, Lunney, and Frutkin should visit Moscow during the week of 2 April to discuss these documents and reach a common position on the most important points. Low remembered that they especially wanted "to determine whether the Soviets really understood what we were talking about."[43]

Fletcher and Low also decided not to publicize this trip;* insofar as MSC was concerned, Lunney was visiting Washington, and Low was supposedly on leave "to take care of family business." To further assure that no one would know their destination, Low's secretary went to a commercial travel agent to get his tickets instead of buying them through the NASA travel office. Low felt that the semi-clandestine nature of the trip lent some excitement to his normally closely regulated life. On this occasion only Fletcher, Mrs. Low, and Low's secretary would know where he was.[44] That is, they thought that was the situation until the Sunday morning newspapers appeared on 2 April.

John Noble Wilford, on page one of the *New York Times,* reported an interview with Academician Petrov, in which Petrov mentioned an upcoming meeting with NASA officials. He pointed out that the negotiations thus far had "considered only the technical aspects of solving these problems" of a joint mission and that neither government had yet approved the flight. When did he expect such approval? Petrov said, "This would depend much on the meetings that will take place next week and probably on the joint meetings of all the working groups of engineers afterwards." When Wilford asked if the necessary arrangements could be made in time for the Summit discussions, Petrov replied, "I would not like to guess on that. I know that on a government level there are a lot of very important problems to discuss, and whether [a joint mission] is one of them depends on the leaders, not us."[45] Fletcher and Low held their breath and waited to see if anyone would follow up the story. No one did.

Low, Frutkin, and Lunney departed Washington on Easter evening and arrived in Paris early Monday. After a short layover, they continued aboard an Aeroflot jet to Moscow, where they were met by Petrov, Bushuyev, and

*At the request of the White House, this trip was not publicized because NASA planned to discuss a possible agenda item for the forthcoming Summit meeting.

Vereshchetin. During their ride into the city, Petrov told Low that Keldysh had been hospitalized. Vladimir Alexandrovich Kotelnikov, acting in the capacity of Academy President, would be negotiating on behalf of the Soviets. Because of a schedule conflict, he would not be able to meet with the Americans until Tuesday afternoon. In their free time, the three Americans visited the American Embassy, where they were invited to lunch with Ambassador Jacob D. Beam and his guests later that week. Jack Tech, the science attaché at the Embassy, later asked Low if he knew who the guests for Thursday's luncheon would be, and Low replied that Ambassador Kaiser and his son would be joining them. Tech responded by asking if he knew who Ambassador Kaiser's son was. When Low confessed that he did not, Tech dropped the bombshell—Robert Kaiser was the *Washington Post*'s Moscow correspondent. Low went immediately back to Ambassador Beam and said that in light of the desire of the White House and the State Department to keep their visit quiet, he questioned the wisdom of dining with the press. The Ambassador assured Low that the luncheon would be a social affair and that there would be no need to discuss his mission. Furthermore, Beam said that he would take personal responsibility if there were any leaks. Although he was extremely skeptical about this whole idea, Low saw no way to avoid the invitation.[46]

For about two hours on Tuesday afternoon, the American trio met with Kotelnikov, Petrov, Bushuyev, Vereschetin, and I. P. Rumyantsev. After a typical Moscow lunch at the Club of the Scientists, they continued their discussions with Petrov until 7:00 that evening. The two groups reconvened the next morning and continued their negotiations until early afternoon. When the Americans adjourned, they ate a quick lunch at the American Embassy snack bar while they rewrote their version of the Summary of Results. The afternoon session with the Soviets lasted only 2 hours, and based upon the revised American draft and the basic understanding reached that morning, the two sides were able to conclude the substance of the talks. Frutkin and Vereshchetin completed the final editing of the agreement Thursday morning.

Low, Frutkin, and Lunney attended their obligatory noon meal on Thursday, which proved to be uneventful, while waiting for the English version of the Summary to be typed at the Embassy. The three returned to the Presidium of the Soviet Academy of Sciences (where all the discussions had been held) for the signing of the documents. The usual ceremony, in which both sides signed two English and two Russian copies, took place in Kotelnikov's office. The Acting President told Low that according to legend Napoleon had slept in this room during his last night in Moscow 160 years earlier. There was a farewell dinner on the night of the 6th, and Low and his colleagues departed for home the next morning.

The Americans' basic purpose for these meetings had been to obtain assurance from the Soviets that there could be agreement on the organizational structure to conduct a joint mission and that the mission could be carried out according to a specified timetable. Low in his opening remarks on Tuesday had told the Soviets that NASA was sure that a joint mission was technically feasible, but the agency was not sure that in managerial terms it was possible. Thus, Low's goal for the Moscow meeting was to gain this assurance. Before the two sides pursued this point further, Kotelnikov said that he had an important statement that he would like to make.

Kotelnikov told the NASA people that in re-evaluating the proposed test mission the Soviets had come to the conclusion that it would not be technically and economically feasible to fly the mission using Salyut. Salyut had only one docking port and the addition of a second port would be very difficult technically and very costly in both time and money. Therefore, the Soviets proposed to conduct the test flight using Soyuz, which could accept all the modifications necessary for such a mission. They were quite forceful in stating that there would be no changes in any of the agreements made thus far.

Surprise was perhaps the mildest word for the Americans' reaction. Nevertheless, Low quickly responded and told Kotelnikov that barring any technical difficulties, the switch from Salyut to Soyuz would be acceptable.[47] He turned to Lunney and asked him if he saw any technical reason for opposing such a change, and Lunney could think of none. Operationally, this would present a simpler mission since it would involve only two coordinated launches—Apollo and Soyuz—and not three—Apollo, Salyut, and Soyuz. Low and Frutkin tried to think through any "political" implications and found none. It would still be possible to exchange crews, which would be the major public impact of the mission, and such a mission would give the Americans an added advantage—not calling attention to the fact that the Soviets already had a space station flying and NASA did not.

After Low agreed to this change, he took the opportunity to raise an issue that was of concern to NASA—the lack of Soviet responsiveness to the proposals concerning regular, direct voice communications between the two sides.[48] Low mentioned that he was interested in more than just the basic issue of communication; he said that if this unresponsiveness was indicative of their attitude for the future, it would be very difficult to conduct a joint mission. Kotelnikov quickly understood why Low placed such importance on this issue and said it would be settled immediately. After considerable debate and discussion, the NASA position on regular communication between Lunney and Bushuyev prevailed.

On Tuesday afternoon, the discussion turned to the "Apollo/Salyut Test Mission Consideration," which was essentially a summary of the organization plan. The Americans had hoped to agree on this plan in detail. As Lunney was presenting the document, the discussion fell apart and became quite confusing, with an inordinate amount of time being spent quibbling over the exact wording of each sentence. "We quickly saw," Low reported, that we would be in Moscow for weeks rather than days were we to proceed this way." Low called for a short recess, so the Americans could discuss their strategy.

Before the Americans went off to themselves, they showed the Soviets a draft version of the Summary of Results that they hoped would be the basis for their mutual agreement. Low told the Soviets that it was essential to reach an accommodation and full understanding of the "12 principles governing mission conduct" that were a part of the "Apollo/Salyut Test Mission Consideration" document, which Low now suggested might be included in the Summary. The Soviets said they would look at these materials while the Americans held their private discussion.[49]

After the recess, the Americans and Soviets resumed the negotiations, reviewing the 12 principles and the Summary of Results until Wednesday. The negotiations were long and difficult, and sometimes when it appeared that agreement had been reached in English on a specific point, the material when read back in English after being translated into Russian sounded like the text of a completely different agreement. Low continually had to emphasize the necessity of having complete concurrence on the substance of the text. At one point in the negotiations, he told the Soviets that unless he could come away from this meeting with a firm agreement on the basic principles of organization, documentation, and scheduling, he would be in no position to recommend the test mission to President Nixon. He stated further that he would even go so far as to make a negative report. On the other hand, he expressed a willingness to stay in Moscow until they were able to hammer out the necessary words.

On Wednesday when the three Americans returned from lunch with a freshly typed copy of the Summary of Results, Yu. V. Zonov translated the English draft and then called a recess so that the Soviets could discuss the document in private. The Soviets seemed amazed that anyone could have completely recast an entire document in such a short time. When they came back, the Soviets told the visitors that the revised paper, with some minor editorial exceptions, was completely acceptable to them. The alterations were performed by Vereshchetin and Frutkin.[50]

The Summary of Results that emerged from these efforts was the keystone in the negotiations for a joint test mission.[51] Without the basic understandings that were forged at that time, the subsequent work would have been difficult, to say the least. In all, seventeen points (see box on facing page) illustrated the level of trust and understanding that would have to

17 Points of Agreement Negotiation in Moscow, 4-6 Apr. 1972

A. For the preparatory (pre-launch) period—
1. Regular and direct contact will be provided through communication links and visits as required.
2. A complete project schedule will be developed and commitments will be made on both sides to meet this schedule in order to avoid costly delays to either party.
3. Arrangements will be made for necessary contact and understanding between specialists engaged in developing and conducting the project.
4. A comprehensive test, qualification, and simulation program will be developed.
5. A sufficient level of familiarization and training, where applicable, with the other country's vehicle and/or normal training equipment must be defined and provided for safety-of-flight assurance. The necessary training exercises will be conducted in each country for the other country's flight crew and ground operations personnel.
6. The parties recognize in particular that they must jointly make a concerted effort to arrive at a full agreement on the engineering aspects of the mission during the meeting of the work groups in July 1972.
7. Two years prior to the flight, responsible persons who will directly participate in the flight operations should be included in the working groups in order to assure a proper level of mutual understanding and continuity of personnel into the real-time operation.

B. For the mission operation—
1. Control of the flight of the Apollo type spacecraft will be accomplished by the American Control Center and that of the Soyuz by the Soviet Control Center, with sufficient communications channels between centers for proper coordination.
2. In the course of control, decisions concerning questions affecting joint elements of the flight program, including countdown coordination, will be made after consultation with the control center of the other country.
3. Joint elements of the flight will be conducted according to coordinated and approved mission documentation, including contingency plans.
4. In the conduct of the flight, pre-planned exchanges of technical information and status will be performed on a scheduled basis.
5. The host country control center or host country spacecraft commander will have primary responsibility for deciding the appropriate pre-planned contingency course of action for a given situation in the host vehicle. Each country will prepare detailed rules for various equipment failures requiring any of the pre-planned contingency courses of action.
6. In situations requiring immediate response, or when out of contact with ground personnel, decision will be taken by the commander of the host ship according to the pre-planned, contingency courses of action.
7. Any television downlink will be immediately transmitted to the other country's control center. The capability to listen to the voice communications between the vehicles and the ground will be available to the other country's control center on a pre-planned basis and, upon joint consent, as further required or deemed desirable.
8. Both sides will continue to consider techniques for providing additional information and background to the other country's control center personnel to assist in mutual understanding (including the placement of representatives in each other's control centers).
9. As a minimum, flight crews should be trained in the other country's language well enough to understand it and act in response as appropriate to establish voice communications regarding normal and contingency courses of action.
10. A public information plan will be developed which takes into account the obligations and practices of both sides.

be established before a joint mission could be carried out. Of these points, the most difficult to negotiate were ones relating to crew training and the public release of information about the flight. After much dialogue, it was decided that the candidate crews would have to be identified one to two years before the flight so that they would have adequate time to train on the other nation's hardware. On the point of releasing information about a joint mission, the Soviets agreed that everything during a normal flight should be released immediately. In case of a major disaster, they would be willing to release information just as they had done in the case of *Soyuz 11*. Their main concern seemed to lie with the minor abnormalities during a flight that might be blown out of proportion or misunderstood. In his turn, Low had stressed an absolute need for NASA to continue its policy of disclosing all information available at the control center and tracking stations. At the conclusion of the discussions, the two sides agreed that they would develop a public information plan that would take into account the "obligations and practices" of both sides.

Looking back on that experience in Moscow, Low was optimistic when he returned to Washington. He had reached the conclusion that the two sides were ready now to undertake a test mission. As for hardware matters, they had reached an understanding on all issues that had been identified so far and did not foresee any new problems that they would be unable to handle. On the management side, the Soviets and the Americans had decided on such matters as regular and direct contact through frequent telephone and telex exchanges, the requirements for and control of formal documentation, joint reviews of design and hardware of various stages of development, the requirement for joint tests of interconnecting systems, early participation by flight operations specialists, and the like. Based upon all these agreements, it was George Low's recommendation that the United States government execute an agreement for a test mission.[52]

TESTING THE AGREEMENT

Working Group 2 was scheduled to have a joint meeting in Moscow early in May, and MSC believed this session would provide an opportunity to test the recent agreements reached in Moscow. On 10 April, three days after his return to Houston, Lunney sent a telegram to Bushuyev. MSC would call Moscow on "Friday, April 14, 1972, at 7:00 AM Houston time, 4:00 PM Moscow time" to discuss the agenda items outlined in this telegram.[53] The first attempt to establish a telephone connection with the Professor was unsuccessful, because the Americans tried to tie Lunney, who was at Kennedy Space Center (KSC), into the line for a three-way conversation. On the second try, only the Houston people made the connection with Moscow. After initial greetings to Bushuyev, the remainder of the conference call was conducted by the respective Working Group 2 chairmen, Legostayev and Cheatham. They agreed that the Americans should visit Moscow on 15-20 May. Cheatham proposed that the May agenda include discussions of communications and television links between spacecraft, an exchange of data on the Apollo and Soyuz control systems characteristics, and further study of the docking target system. After a discussion of the radio frequencies to be used between spacecraft and between ground control centers, the two men answered each other's general questions. While the telephone connection was still less than satisfactory, this second telecon was more successful than the first and helped both sides prepare for the upcoming meeting, the starting date of which was later advanced to the 10th.[54]

Group 2's spring meeting was an important one in which a full agenda was addressed and progress made on a number of key issues. While discussion continued on the external lights, docking targets, coordinate systems, and other topics related to docking, the main subject was spacecraft-to-spacecraft radio communications and distance ranging. At previous meetings there had been considerable discussion about radio frequencies: Would each side exchange radio equipment for its frequency or give the necessary data to the other group so they could build the equipment? The Soviets had advised NASA at the November-December 1971 meeting that they would continue to use the 121.75-megahertz (MHz) frequency for their voice communications. The Americans in turn advised the Soviets that they had yet to determine which frequency they would use but would do so by March 1972.[55]

While the obvious choice would have been to continue using the Apollo voice frequency, the Department of Defense was eager to have NASA abandon its use of frequencies in the 225- to 400-megahertz bands. The Apollo voice frequency had been loaned to NASA in 1958 by the military for Project Mercury, and they had been pressing the space agency since then to give it up. A 1968 agreement between NASA and the Department of Defense called for NASA to withdraw from all military frequencies by 1975. In an effort to save from $500 000 to $700 000 for new radio equipment, MSC had worked with NASA Headquarters during early 1972 to obtain Department of Defense approval for use of the 259.7 and 298.6-megahertz frequencies for a joint Soviet-American test project. This agreement had been tentatively reached just before the American delegation left for Moscow.[56]

A second issue that remained to be resolved both internally and with the Soviets centered on the "build versus exchange" question. At first glance, it seemed that it would be simpler for each country to give its radio equipment to the other for installation into their respective spacecraft. On the American side, this exchange appeared to be complicated by the fact that the Apollo VHF transceiver also embodied another assembly that provided a backup distance ranging capability between the CSM and the lunar module. This little unit, called the Range Tone Transfer Assembly, had been added after the original design of the transceiver in 1962, and it was rather sophisticated in terms of its solid-state circuitry. There was some concern at NASA and in the Defense Department that providing this hardware to the Soviets for a joint mission might also constitute a giveaway of valuable technological information. This problem of possible technology transfer had not yet been resolved by the time of the May meeting of Working Group 2. The Americans asked the Soviets to postpone a decision on radio transceivers, and they agreed to do so.

This "exchange-build" issue serves to illustrate how difficult the negotiations could be. Just defining the nature, scope, and implications of the many technical considerations involved in compatibility was a complex, time-consuming task, recalled R. H. Dietz of Group 2. And this process became even more complicated when neither side had a clear understanding of its own goals for a particular topic. In preparing for the negotiations, the Americans had drafted three Interacting Equipment Documents in two versions—the first could be used if a decision was made to exchange equipment, and the second was ready if they decided to build. While this applied to only three of the twenty-six documents that NASA took to Moscow, everyone would prefer to avoid double efforts.[57]

Not all the communications raised at the May meeting proved as thorny as the "exchange-build" issue. Considerable progress was made on the topics of cable links between spacecraft for voice communications after docking and communications systems for future missions. Donald Cheatham, the American chairman of Working Group 2, felt that the meeting was basically successful. The two sides had sufficient time to work out the points of agreement, and as a consequence they got all of the primary issues clearly defined and resolved. He felt that this session was a good indication on the Working Group level that there would be no irreconcilable differences in working out the technical aspects of a joint mission. The way seemed clear for the government-to-government agreement at the May Summit in Moscow.[58]

THE NIXON-KOSYGIN SUMMIT

Upon his return to Washington from Moscow in April, George Low had informed Henry Kissinger that from NASA's point of view a joint mission in

1975 was a realistic goal and that no additional meetings between the Soviet Academy of Sciences and NASA would be required before placing the topic on the agenda of the May Summit meeting. Low felt that the agreement between the two governments could be relatively short and straightforward. While Low was communicating with the White House, NASA's public position on the topic was one of silence.

Low recalled that from mid-April to mid-May reporters had exhibited keen interest in the possibility of a joint docking mission proposal being part of the Nixon-Kosygin talks.[59] In the many interviews between NASA officials and the press, "there was never any hint about the 4-6 April meeting, nor was there ever any hint that during the meeting the Soyuz spacecraft was substituted for the Salyut," Low said. He believed that the agency had been able to keep discussions about its work leading up to the Summit to a minimum "only because a very small number of NASA people had been involved in the activities. . . ." While their participation in the business of summitry had been successful, Fletcher and Low were puzzled over how slowly work on the Summit-level space agreement was going at the State Department.[60]

It was not until the week before the Summit meeting that the State Department and the White House began to coordinate with the Soviets the draft language of the document of space. On 20 May, the Soviets responded to the American proposed text with a much lengthier document which, among other things, included the text of the Low-Keldysh agreement of 21 January 1971 and the agreement hammered out in April 1972. When the Soviet response was received in Washington, Secretary of State Rogers and Kissinger were immediately contacted aboard their airplane over the Atlantic en route to Salzburg, Austria. Kissinger asked the State Department staff to contact Low and have him help them work out a suitable alternative to the Soviet proposal without significantly revising the text.

Low went over to the State Department at about 2:30 Saturday afternoon and worked with the staff there until the middle of the night. In the process of that lengthy session, they were able to revise the preamble of the agreement, while retaining the sense and meaning of the Soviet draft. In only one area, communications satellites, did they make any major change. This had not been part of the Low-Keldysh agreement because Low had told the Soviets that this was an area of commercial enterprise in the United States. Since NASA was not empowered to negotiate for these private companies, Low had this section eliminated from the draft sent to Kissinger that night.

After Low's Saturday session with the State Department, NASA had no additional information about the status of the space agreement, except for "persistent signals" that it was scheduled to be signed on Wednesday, the 24th. On Tuesday, Low left Washington for San Diego where he was scheduled to give a speech—"NASA Looks Ahead in the 70s." During the course of the evening after dinner, Low received a trans-Atlantic telephone call patched through the State Department operations center that involved himself, Frutkin, and a State staff member. There were still some questions about the wording of the text, and the three men worked out a final version just in time for Low to make his way back into the ballroom where he was being introduced as the evening's main speaker.[62]

President Nixon and Premier Kosygin signed the "Agreement Concerning Cooperation in the Exploration and Use of Outer Space for Peaceful Purposes" at 6:00 p.m. Moscow time on the 24th. Later that afternoon (Washington time), Vice President Agnew introduced NASA Administrator James C. Fletcher to the press at a briefing held in the Executive Office Building. The reporters were given the text of the space agreement while Fletcher made the following statement: "We . . . are extremely pleased that President Nixon's meeting with officials of the Soviet Union in Moscow has brought to fruition the most meaningful cooperation in space yet achieved by our two nations." He noted that they had been discussing the possibilities for such cooperation for some time and that this agreement molded these technical discussions into a "definitized program." Of the various planned enterprises, "the most dramatic . . . will involve the rendezvous and docking of a U.S. spacecraft with a Russian Soyuz spacecraft in 1975."[62]

The space agreement was only one of a host of important issues discussed at the Summit. The Soviet and American leaders agreed on ways of working together to protect the natural environment, to advance health, to cooperate in science and technology, to prevent incidents at sea, and to expand trade between the two nations. President Nixon spoke over radio and television to the people of the Soviet Union on the evening of 28 May. He noted that one of his principal aims as President had been "to establish a

President Richard Nixon and Premier Aleksey Kosygin sign a five-year agreement between the United States and the Soviet Union on cooperation in the fields of science and technology, 24 May 1972.

better relationship between the Soviet Union and the United States." Our two great nations, which have never faced one another on the battle field, "shall sometimes be competitors, but . . . need never be enemies."[63] Nixon felt that it was "most important" that the two countries had "taken an historic first step in the limitation of nuclear strategic arms."[64] This agreement was signed on the 26th of May, the product of the Strategic Arms Limitations Talks (SALT) begun in 1969. However, it lacked one important element that did exist in the Apollo-Soyuz test flight agreement–the Apollo-Soyuz accord was tied to a specific timetable. The engineers of the U.S. and the U.S.S.R. would have to work hard and without discord if they were to meet it. The concrete goal of flying together by a given date promised to guarantee success, whereas the general agreement to limit strategic arms carried no such inherent assurances. The task ahead of Glynn Lunney and Professor Bushuyev was a challenging one–the forging of a partnership.

VII
Creating a Test Project

The 1972 Nixon-Kosygin accord on space was the watershed in bilateral discussions between NASA and the Soviet Academy of Sciences. Both the Low-Keldysh agreement on space science and applications and the test mission now had the official imprimatur of the two respective governments. Prior to the Summit, work on the mission had been managed by NASA's advanced planners, but now that the decision had been made to fly, the mission planners, the flight operations staff, and the engineering and development personnel—a large and to a great extent new team of individuals—took the prime roles.

On 13 June 1972, Dale Myers sent a memorandum to the Center Directors at Cape Kennedy, Huntsville, and Houston, outlining the organizational policy decisions that had been made in preparation for the July plenary meeting with the Soviets. The Apollo-Soyuz Test Project* was scheduled for mid-1975. With a Saturn IB, the Americans would launch command and service module (CSM) 111, reserving CSM 119 as the backup vehicle, if it were not flown during Skylab.[1] Myers advised the centers that effective 11 June the management of the joint project had been transferred from the Office of Advanced Missions to the Apollo Program Office. Philip Culbertson and his staff were directed to assist Rocco A. Petrone and the Apollo team. In Houston, work on modifying the CSM would be handled by the Apollo Spacecraft Program Office, under the direction of Owen G. Morris. Glynn Lunney, who had been assigned as Special Assistant to Morris the preceding March, was given primary responsibility for overseeing ASTP. Preparation of the Saturn IB launch vehicle would be carried out by the Saturn Program Office at the Marshall Space Flight Center (MSFC), and all launch related activities would be the responsibility of Kennedy Space Center (KSC).[2]

During May, René Berglund had suggested to Chris Kraft that Glynn Lunney be given the responsibility for managing the ASTP contract with

---

*Although used unofficially after the May Summit, Apollo-Soyuz Test Project did not become the official designation for the joint Soviet-American flight until 30 June 1972.

North American Rockwell. Berglund thought that the task of negotiating with the Soviets and the spacecraft contractor should be in the hands of a single individual and organization.[3] Lunney seemed to be the logical choice on the American side for directing the development of the mission and the spacecraft hardware, as well as managing the U.S. negotiations with the Soviets. At age 36, Lunney had worked with first NACA and then NASA for 17 years, coming to Houston in 1962. At MSC, Lunney had been a flight director for the *Gemini IX* through *Gemini XII* missions and filled a similar role in ten of the Apollo flights, being the lead flight director for the unmanned AS-201 flight, *Apollo 4,* which was the first launch of the Saturn V, and the manned voyages of *Apollo 7* and *10.* As Chief of the Flight Directors Office, Lunney gained national recognition as the leader of the team that worked out the return trip plans for the *Apollo 13* crew following the inflight explosion that incapacitated the electrical and oxygen systems of their service module. His performance during those trying days and during the early negotiations with the Soviets had indicated to the Manned Spacecraft Center's (MSC's) Director that Lunney was the individual to manage ASTP for NASA.[4]

Glynn Lunney and his colleagues worked hard to structure a basic organizational plan for the next joint meeting. They hoped that the working procedure developed in Houston at that summer session would serve as a model for the many meetings that would follow. Included in Lunney's plans was a proposal for a more detailed schedule of activities for the next three years and a scheme for documenting in English and Russian all technical agreements. The American plan for documentation suggested two series of reports that would be approved jointly and signed by the appropriate Working Group members and the Project Technical Directors. ASTP Documents would codify the basic understandings for conducting the mission, while Interacting Equipment Documents would record specific technical data required to ensure compatibility, lay out test plans, and present hardware specifications and drawings in standardized format.

One presentation, the "Proposed Operating Plan for US/USSR Meeting on the Apollo/Soyuz Test Project, Houston, Texas, July, 1972," was typical of the work done in Houston. Besides this document, which outlined the scope of each working session, tentative agendas and milestones for the various discussions were also presented.[5] Other pre-meeting documents considered such logistical matters as transportation between the Soviets' motel and the meeting sites, plans for refreshments and meals, public affairs arrangements for photographs of the groups at work, as well as assignments for interpreters, translators, and Russian language typists. Similar attention was given to preparing after-work activities for their Soviet guests.[6]

CREATING A TEST PROJECT

Efforts put into arranging the summer meeting were indicative of the NASA way of working. As the frequency of the joint negotiations increased and as the size of the NASA team expanded, so did the amount of paperwork and the number of briefings and reviews. The pyramid, which reached its apex in Lunney's office, expanded downward at MSC to include engineers and specialists in nearly every division. Fifty-eight key individuals were invited to a 2-hour ASTP briefing on 13 June. Starting with Director Chris Kraft, his deputy Sigurd Sjoberg, and their technical assistant George Abbey, the list of invitees included nearly all those directorate and division chiefs whose organizations would participate in or support the joint mission. Donald K. Slayton, Director of Flight Crew Operations, and his deputy Tom Stafford, attended the briefing, as did Alan Shepard, Chief of the Astronaut Office. Also present were the chiefs of the Flight Crew Integration, Crew Training Simulation, and Crew Procedures Divisions. Flight surgeons and members of the medical research team were on hand. From the Engineering and Development Directorate came Max Faget, accompanied by his division chiefs. The Flight Operations Directorate was represented by flight controllers, computer analysts, landing and recovery specialists, mission planners, and the flight support team. In addition to these individuals, Apollo Program Office, Skylab Program Office, and Science and Applications Directorate representatives were there.[7]

This June briefing was a method of getting the word out; each division chief would in turn inform his subordinates of the tasks that lay ahead. As those tasks were apportioned, the number of memoranda and reports would increase dramatically as the various teams kept their colleagues informed of their progress. Distribution lists were compiled and periodically revised, and reams of paper were fed into photocopying machines. All the paper that was circulated had a purpose--to get the job done and see that all the work expended was as efficient as possible among such a large group of people. MSC was now doing what it had been established to do—plan, develop, and fly a manned mission in space. This briefing was just one step in ensuring "a more widespread understanding of [the] project," which Glynn Lunney believed to be "very important to [the] timely, successful execution" of ASTP.[8]

While preparations for the joint meeting progressed, the Apollo Spacecraft Program Office completed contractual arrangements with North American Rockwell.[9] On 30 June, the Procurement Office mailed a letter contract* to North American containing a statement of work. Basic tasks

*This letter contract was a ninety-day commitment on the part of NASA, issued to start the described engineering and manufacturing. A negotiated and definitive contract was issued on 6 Oct. 1972. A fuller account of the contracting activities is presented in source note No. 9.

included the necessary modifications to CSM 111, essentially the same type of spacecraft flown on *Apollo 12, 13,* and *14,* so that it would meet the requirements of ASTP. North American also agreed under this contract to develop and fabricate the docking module (DM), docking system, and the support structure for the DM in the spacecraft lunar module adapter. In conjunction with the engineering, fabrication, and assembly, the prime contractor was further assigned major portions of the ground testing for the CSM, DM, and docking mechanism and a host of other activities that were necessary to prepare the spacecraft and its systems for the flight and to check it out after the mission. William B. Bergen, President of the Aerospace Group, accepted the contract on behalf of North American Rockwell on 6 July, the day the Soviets arrived.[10]

JULY IN HOUSTON

Between 6 and 18 July, the days were busy ones for the ASTP Working Groups. Each group had a full agenda that it wished to see completed, but the work went slowly—at some points tediously—because of the language barrier. M. Pete Frank, chairman of Working Group 1, for example, recorded in his notes:

> Don't seem to be able to complete anything quickly.
> Spent a lot of time on correcting *or* modifying unimportant trivia.
> Translations cannot be trusted....
> Russian language takes about twice as long to say as does English.
> Much time was spent trying to understand jargon....[11]

R.H. Dietz, commenting on the communications equipment talks, stated, "It was found that diagrams, pictures, etc., were very useful in our discussions. In fact, the written word would cause many hang-ups, but when a picture or diagram was used, immediate understanding usually resulted. This applies to both sides."[12] As the two teams became more deeply involved in the nitty gritty details of laying out a flight, specificity of meaning and complete understanding of the other side's approach were essential. And they were difficult.

Some Americans found this laborious point-by-point negotiating frustrating, having brought to the meetings expectations about how much work needed to be accomplished and at what pace. But the Soviets introduced a different perspective to the same sessions. Glynn Lunney later reflected on this problem:

> we had the experience at a couple meetings in the beginning where we would
> be planning to cover certain ground, and from our point of view it seemed

198

that they were not quite as prepared or ready ... as we were. Again in retrospect, we have consistently had the same class of problems when we deal with new contractors, in the sense that when we get on board with a contractor we all have a certain way of doing business.... And after a while we come to take it for granted and expect everyone else to know what it is.... To some extent we had that kind of problem in ASTP in the beginning.[13]

While the Soviets would have to master the NASA method of working a problem, the Americans also had to prepare themselves for the surprises that their counterparts could spring. Several individuals at MSC were still trying to adapt to the switch from Salyut to Soyuz when they learned almost by accident that the Soviets planned to have two spacecraft ready for launch to ensure mission success.

There had been considerable discussion during the first day's negotiations in Working Group 1 about who should launch first—Apollo or Soyuz. Clarke Covington recalled the events leading up to the discovery of the two Soyuz spacecraft and launch vehicles. "We had drawn up a proposed mission plan that showed Apollo going first ... primarily because Apollo had extended mission time when compared to Soyuz; 11, 12, 13, 14 days. They had this man-day limitation which limited them to about 4 days plus a day margin." Having taken this difference in capability into account, the MSC mission planners figured that Apollo should launch first; then it could wait for the Soviets if their first launch attempt were unsuccessful. According to Covington:

They came in with a plan where Soyuz was launched first. We discussed this for several hours, not really knowing or understanding their reasons. They went through a lot of flight mechanics type [explanations]. All of our Working Group 1 people were ... trying to understand why they wanted to go first. It was really ... illogical. We were convinced, this was privately, that they had a fear of failure.[14]

At the end of the day, Covington received a copy of the Soviet "Project Technical Proposal" in Russian. Believing that he ought to know what was in that document before starting the next day's discussions, he went looking for a translator. Covington talked Ross Lavroff, one of the interpreters working with MSC, into orally translating it. They went to Covington's office at about eight in the evening, and Lavroff began the all-night task. After several hours, Covington heard Lavroff read something to the effect that the Soviets proposed to launch Soyuz, which would be followed by the launch of Apollo. Should all possible alternative launch attempts prove unsuccessful, the first Soyuz would deorbit, and the second one would be prepared for launch. Since there had been no mention of a second Soyuz during that first day of talks, Covington asked Lavroff to read that part again. Surely, he had mistranslated; there must be another way of interpreting that passage. Lavroff read it again. If the first Soyuz has to come down, then the second will be launched. Covington was astonished.

Early the next morning, Covington cornered Lunney before he began his "tag-up" staff meeting, which preceded each day's negotiations, to tell him that the Soviets had not one but two Soyuz craft and launch vehicles allocated for ASTP. Lunney, also amazed, said that he would indeed ask the Soviets about this interesting piece of news. At the morning session, the Soviets responded. Yes, they would have a second Soyuz ready for launch. American Working Group 1 members just "sat around dumbfounded. . . ." Covington said, "that totally cratered our argument as to the need for Apollo to go first. If they are willing to launch two spacecraft then there wasn't any reason why they should not go first." But why had they not revealed the details in the beginning? No one at MSC ever knew why; they just accepted it as one of the Soviet idiosyncrasies around which they would have to work.[15]

Despite the surprises and the frustrations, a considerable number of points were settled at the July meeting. Lunney and Bushuyev signed the three basic documents that they had agreed to prepare in April—the "Project Technical Proposal," describing the mission plan and the hardware; the "Organization Plan," stating how the project would be controlled and managed both prior to and during the flight; and the "Project Schedules," a time table for Working Group activities. In addition, the two Technical Directors decided to expand the number of Working Groups from three to five. Working Group 4, chaired by R. H. Dietz and B. V. Nikitin, would concentrate on the communications and tracking questions that originally had been part of Group 2's purview, while Working Group 5, under the direction of R. E. Smylie and I. V. Lavrov, would handle the environmental control and crew systems topics originally assigned to Group 1.[16]

Lunney and Bushuyev, together with Kraft and Petrov, met with the press on 17 July to talk about results. Lunney described the mission profile as it had evolved to date:

We've agreed that we would launch the Soyuz spacecraft first from the Soviet Union. There would then be three launch opportunities for the Apollo spacecraft, the first one occurring about 7½ hours after the launch of the Soyuz. We envision a rendezvous sequence by Apollo spacecraft, the command service module, which would take approximately 1 day. At the end of that 1 day, we would arrive at stationkeeping conditions, and perform the final docking, and we would perform that actively with the Apollo spacecraft as the active docking spacecraft. After this docking we have planned for a period of approximately 2 days, about 48 hours, during which time an agreed schedule of exchange of crews between the two ships would occur. We have also allowed time for the performance of . . . joint experiments in any scientific activities that we will be able to fly on either or both of our spacecraft. After this period of approximately 48 hours, the vehicles would undock and then would carry out any objectives that either country would have on its own for this particular mission. We have not finalized for the Apollo how long we would stay up because it depends on precisely what experiments we would carry, but as that period drew to a close, we would plan on a relatively normal de-orbit sequence with the landing in the Pacific Ocean.[17]

This mission plan represented the detailed agreements that had been reached by the Working Groups.

Group 1 members involved with the mission profile had agreed that the launch would take place in July 1975. While 1 July was "tentatively selected as an arbitrary launch date for purposes of compatible trajectory plans," dates between July and October would be considered. The Soviets agreed to provide the Soyuz orbit parameters to NASA within six hours after their launch so the Americans could update the Apollo trajectory. Information that might affect the rendezvous would have to be provided with maximum accuracy and immediacy. A major issue involved the question of attitude control during the mission's docked phase, but this problem remained unresolved for a time. Both Apollo and Soyuz had limited amounts of fuel for their attitude control systems.

As to crew and flight controller training, the specialists decided that mandatory joint crew training would include familiarization with both spacecraft and preparation of a "Crew Activities Plan" and a "Detailed Operational Procedures" document. At the next meeting, they proposed to exchange plans and schedules and also decided to develop training procedures for the flight controllers that would embody joint simulations between the Soviet and American control centers.[18]

Members of Working Group 1, who would soon splinter off to become Group 5, wrestled with questions concerning life support systems and crew transfer. Noticeably absent from the Soviet delegation that summer was Ilya Vladimirovich Lavrov, who had been Ed Smylie's counterpart and the sole Soviet negotiator in the earlier discussions on environmental control systems (ECS). Yuri Serafimovich Dolgopolov acted in Lavrov's place. And to the Americans' dismay, Dolgopolov, who specialized in water, food, and waste management aspects of the life support system, did not have a fundamental knowledge of the Soyuz systems that provided oxygen and eliminated carbon dioxide and other waste products from the spacecraft atmosphere.[19]

During the ECS discussions, Yuri Stepanovich Denisov assisted Dolgopolov, but the Americans were still perplexed. A few days before the meeting began, they had received the first schematic diagram of the Soyuz ECS. It indicated the presence of oxygen and nitrogen in pressure bottles in place of

Picture time in Houston for members of Working Group 1 (left), July 1972: seated (left to right), T. P. Stafford, L. A. Gorshkov, N. K. Latter, and M. P. Frank. Standing (left to right), J. H. Temple, O. G. Sytin, E. G. Lineberry, Yu. S. Denisov, and R. J. Ward. Working Group 5 (below) was photographed just before the Saturday lunch break: seated (left to right), T. Holmes, R. E. Smylie, and Yu. S. Dolgopolov. Standing (left to right), J. D. Hollan, J. R. Jaax, W. W. Guy, and W. R. Hawkins.

the chemical bed oxygen regeneration system used in previous Soyuz spacecraft.[20] When Ed Smylie and Walt Guy asked for an explanation, the two Soviets were at a loss to explain whether it represented the system currently in use or whether it was a proposal for the future. Without a definitive answer, the Americans asked for more details to be provided at the next meeting. From these discussions, Smylie and his Manned Spacecraft Center (MSC) colleagues were beginning to get an understanding of the nature of the problem experienced during the *Soyuz 11* reentry. That fatal leak, still shrouded in mystery, could not have been corrected on board. Once pressure was lost, there appeared to be no way to recover because the cosmonauts carried no bottled oxygen. As a consequence, the Americans asked the Soviets to consider the inclusion of an emergency repressurization system for the ASTP flight, if such a system was not already embodied in the Soyuz ECS design.

Walt Guy recalled that in July 1972 the Americans still believed that the Soviets had bottled nitrogen aboard Soyuz for leakage makeup. Only much later did they learn that there were no gas stores on board at all. When this became clear, the Americans understood why the Soviets insisted upon a virtually leakproof spacecraft. According to Guy, "They had implored us from the beginning to tighten our leakage [rates]," which were "10 times greater than theirs." Where the Soviets tolerated virtually zero leakage, it was not unusual for the American spacecraft to lose up to a "tenth of a pound per hour, [and] that is not really very many pounds in a normal mission." In Apollo, the Americans compensated for this loss by carrying extra oxygen.[21]

Even though Dolgopolov had not prepared any documentation prior to the joint July sessions, he, Guy, and Jim Jaax drafted four Interacting Equipment Documents that described spacecraft environment. In addition, the Americans gave the Soviets data on space suits and prebreathing units that they expected to use. In large measure, the U.S. desire to understand the Soyuz environmental control system arose from the need to make some immediate decisions about the design of the docking module.

Concurrently with the Soviet-American meetings, the first Preliminary Design Review (PDR) with the engineers from North American Rockwell was held at MSC on the 13th. At the PDR, Guy and Jaax had to make some changes in the docking module ECS schematics to reflect the changes negotiated the night before with Dolgopolov. These modifications were a result of the Soviets' concern about the safety of some of the valves that led to overboard gas vents. Where the Americans relied upon the basic design of the valves to ensure safety and carried extra gas in case of valve failure, the Soviets wanted a redundant valve in the overboard system as an extra precaution.[22]

While the life support people were exploring these topics, the combined Group 2 and 4 team were negotiating an agenda full of items that were an extension of their work done during May in Moscow. In addition to completing interacting equipment documentation covering external lights, docking targets, control system use, and associated rendezvous and docking questions, they solved the thorny problem of "exchanging versus building" communications and tracking equipment. NASA agreed to provide the Soviets with the Apollo lunar module VHF transceivers and range tone transfer assemblies. R. H. Dietz told his counterparts that the Americans would manufacture the Soviet frequency VHF/FM radio communications equipment for the command module. Other topics included cable communications and electrical connectors between the two spacecraft and the type of television cameras that would be transferred from one craft to another.[23]

Group 3 concentrated on spelling out more fully specifications for the docking system. Some refinements were made in the guides and other parts of the mechanism; as with the other groups, a schedule for the upcoming months was written, indicating documents to be prepared and tests to be conducted. After the team had a thorough look at the American two-fifths-scale docking system, which helped the designers discuss the operation of the

Working Group 2, July 1972: seated (left to right), I. K. Kupriyanov, D. C. Cheatham, V. P. Legostayev, and H. E. Smith; standing (left to right), Dmitri Zareschnak, R. L. Berry, C. E. Manry, O. I. Babkov, R. Reid, V. A. Podelyakin, K. J. Cox, I. P. Shmyglevskiy, and S. E. Snipes. In another part of Working Group 2 activities, V. P. Legostayev (left) receives an explanation from D. C. Cheatham on the operation of the proposed centerline television docking guidance system. Standing behind Cheatham is Ralph Sawyer, Chief of the Telemetry and Communications Division at the Manned Spacecraft Center.

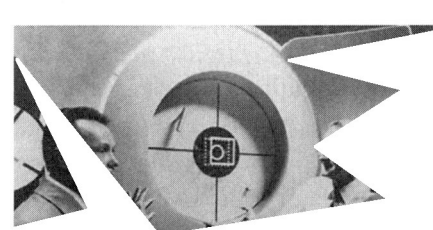

At left, Working Group 3 members po with the American two-fifths-scale moa of the ASTP docking system. Left right: A. Sementovsky, E. N. Harrin, V. Bagno, D. C. Wade, V. S. Syromyatnika Ye. G. Bobrov, W. K. Creasy, and L. Williams. Kneeling by the docking har ware: K. A. Bloom (left) and R. D. Whi In the lower photo, C. C. Johnson (fo ground) is explaining an idea about t compatible docking mechanism to Sov and American members of Worki Group 3. Around the table from the l are L. G. Williams, D. C. Wade, W. Creasy, K. A. Bloom, E. N. Harrin (p tially obscured), V. I. Bagno, B. S. Chiz kov, Ye. G. Bobrov, V. S. Syromyatnik and R. D. White.

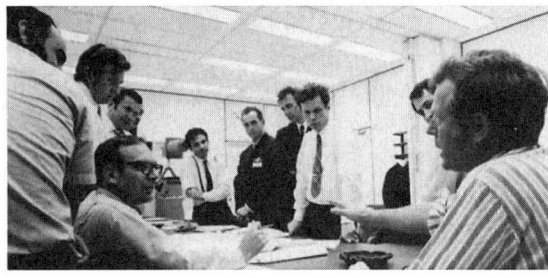

Working Group 4 (below): seated (left to right), M. W. Hamilton, E. E. Lattier, O. I. Babkov, and O. G. Ispolatov. Standing (left to right), P. W. Shores, V. A. Raspletin, R. H. Dietz, A. S. Morgulev, A. D. Travis, and Dmitri Arensburger. At right, Ye. G. Bobrov, Soviet docking specialist, questions Bob White (left) and Ezi Harrin about the operation of the Apollo docking probe. L. G. Williams and W. K. Creasy stand in the background.

mechanism and decide on refinements, they scheduled joint model tests for December. Then the engineers would be able to see just how the interfacing elements of one country's system mated with those of the other. The Soviets said they would draft the "Test Plan for Scale Models of Apollo/Soyuz Docking System" (IED 50 003), while the Americans drew up the dimensions of the model and the test fixtures.

The last day of joint activities in Houston, 18 July, was given over to final preparations of the documents.* With the departure of the Soviets, the MSC personnel directed their full attention to preparing for the next sessions, scheduled for October. The afternoon of the 18th and most of the 19th were spent in debriefings.[24] Working Group chairmen summarized the events of the preceding two weeks for their colleagues and Director Kraft. Looking at the overall rate of progress, Lunney thought that the Soviets were gearing up more slowly for ASTP than was NASA, but this was due in large part to the difference in building cycles for the two types of spacecraft. NASA was faced with a declining work force at the North American Rockwell factory in California, since CSM 111 and the modification kit for CSM 119 would be the last of Apollo. Lunney reported that we "wanted to get all the work done as soon as we could." The Soviets, on the other hand, were building their Soyuz on a serial production basis with no signs of diminishing their activity. They could take a more leisurely approach to finalizing design elements. NASA's expeditious building of hardware was entirely consistent with the agency's *modus operandi,* according to Lunney.

> In general our approach is a go-go thing. For good reasons—that is not to say that it should be hasty—but generally we've found that it doesn't do you any good to dilly-dally around and leave problems hanging around for a long time. They just multiply, and you end up with more snakes than you can kill in one week, if you don't start stomping them as soon as they show up. So we just plunged into this thing with our typical gung-ho approach.[25]

During the next 70 days, MSC was kept busy. As engineers and designers spent late hours at their drafting tables, development tests were conducted and more design reviews were held.[26] Lunney chaired ASTP staff meetings at 1:00 p.m. every Friday to summarize the week's activities and assign "action" items for the next week.[27] (See box on next page.) As the liaison between MSC and NASA Headquarters, Lunney also kept Apollo Program Director Petrone informed of the progress being made on ASTP. Headquarters in turn kept the White House and the Congress advised.[28] This cycle completed, it was time for the next round of talks.

*At this meeting MSC with the support of Boeing began a "History Data File" for each meeting, which preserved all jointly signed documents and materials exchanged, as well as other materials of historical and managerial importance.

Chronology of Manned Spacecraft Center Activities—July-Oct. 1972

Date	Location	Meeting/test
6-18 July	Houston	Joint U.S./U.S.S.R. meetings.
18-19 July	Houston	MSC management debriefings.
22-28 July	Houston	Thermal vacuum test of ASTP docking gaskets at JSC's Space Environment Effects Laboratory (SEEL).
25-26 July	Houston	Working Group 4 and Apollo Spacecraft Office staff discuss work for October joint meeting.
27 July	Houston	Preliminary Design Review (PDR) no. 2.
28 July	Houston	ASTP staff meeting.
2 Aug.	Houston	Crew Systems Division meeting with North American Rockwell (NAR) representatives to discuss life support system breadboard test.
2 Aug.	Houston	Meeting to discuss service module reaction control system (RCS) design and external thermal constraints.
4 Aug.	Houston	ASTP staff meeting.
8 Aug.	Houston	Service Module RCS design and thermal criteria meeting.
11 Aug.	Houston	ASTP staff meeting.
18 Aug.	Houston	ASTP staff meeting.
21-24 Aug.	Downey (NAR)	PDR no. 3.
24 Aug.	Houston	Telecon with the Soviets.
25 Aug.	Houston	ASTP staff meeting.
28 Aug.-30 Sept.	Houston	Thermal vacuum tests of ASTP docking gaskets at JSC's SEEL facility.
1 Sept.	Houston	ASTP staff meeting
5 Sept.	Houston	October agenda telexed to the U.S.S.R.
6 Sept.	Downey	Walk through review of plywood mockup of docking module; representatives were present from NAR and the Astronaut Office.
9-15 Sept.	Houston	Space Environment Test Division and Engineering and Development Directorate hold meeting on testing to be conducted on docking module.
13 Sept.	Houston	ASTP staff meeting.
13 Sept.	Houston	Telecon between G. S. Lunney and G. Jeffs, NAR, to review activities.
13 Sept.	Houston	ASTP management review for R. A. Petrone, with representatives present from Headquarters, KSC, and MSFC.
20-21 Sept.	Downey	Design Review of docking system layouts and assembly drawings.
22 Sept.	Houston	ASTP staff meeting.
27 Sept.	Houston	Technical director and Working Group chairmen meeting to discuss work plans for October meeting in Moscow.
28 Sept.	Houston	Telecon with the Soviets.
29 Sept.	Houston	ASTP staff meeting.
5-6 Oct.	Downey	ASTP docking module crew station mockup review.
9-19 Oct.	Moscow	Joint U.S./U.S.S.R. meetings.

OCTOBER IN MOSCOW

A 27-person NASA delegation arrived in Moscow shortly before 4:00 p.m. on Saturday, 7 October 1972. After a day of sightseeing, the three Working Groups (1, 3, and 5) began 11 days of work on the agenda items drafted during a September exchange by telex and telephone.[29] H. E. "Ed" Smith represented Group 2 in some of the negotiations, while George Jeffs, manager of the CSM program at Rockwell, and Clarke Covington acted as advisers to Lunney and the three chairmen. By then a sort of routine had emerged—work the technical problems at home and prepare for the next meetings; attend the joint sessions and define the next phase of activity. The Americans were ready.

Subjects discussed by Working Group 1 this time fell into five categories. First came trajectory considerations: specialists outlined the paths of the two spacecraft from launch through rendezvous and docking. In addition to investigating alternative launch opportunities so that the mission could still be conducted if there was an abort during initial launch attempts, the two sides examined which Apollo revolution—the 14th or the 29th—would be selected for docking.

When the Soviets discovered that they had a ceiling of approximately 225 kilometers due to the weight of Soyuz as configured for ASTP, they asked the Americans to consider lowering the docking orbit to 222 kilometers from 232.[30] But they made their request in the guise of a technical problem in orbital flight mechanics. After considerable confusion and much dialogue, Covington asked his Soviet counterpart, "Tell me, is the main reason you want to fly at a lower altitude because you have got a weight problem and your launch vehicle can't get you any higher than that?" Yes, came the Soviet reply. Looking back, Covington said:

> It was no real problem to us, but we just couldn't understand why they wanted to do it. . . . It was no big deal for us but it was a big deal for them. They seemed to embarrass easily about the capability of their spacecraft, which they had no need to do. Their spacecraft was designed for a different thing that Apollo was. The Apollo spacecraft is way over designed for this mission, it was built to go to the moon and back. We just had an inherent capability greater than theirs.[31]

But the Americans were realizing that early Soviet boasts of leadership in space still echoed in the background. When it came to making changes, the Soviets would always prefer a technical rationale to directly admitting limitations or asking for assistance.

But these difficulties did not hinder the work of Group 1. Under the chairmenship of M. P. Frank and V. A. Timchenko, they went on to discuss problems related to finding common formulas to determine the earth's atmosphere and gravitational field so as to achieve compatible trajectory calculations. They needed to be certain that when they placed their spacecraft in orbit the mathematical computations would permit the Apollo crew to complete rendezvous. Besides the trajectory issues, progress was made on four other topics—mission requirements, contingency plans for abnormal situations, onboard flight and activities plans, and crew training and mission operations.[32]

Frank observed that the Soviet delegation, which included 15 people, represented systems planning, flight control, flight crew activities, and mission documentation. In the past, NASA had been concerned over some of the Soviet specialists' lack of expertise, but seemingly the correct personnel were now involved. Pete Frank also indicated that while the two sides approached the organization of flight-related data in completely different ways, they basically agreed on the technical information involved. After discussing crew training with Astronaut Tom Stafford and Cosmonauts A. G. Nikolayev and A. S. Yeliseyev, Working Group 1 scheduled the initial training session for mid-1973 in Houston, with a second meeting planned for autumn in the U.S.S.R. The Soviets were considering selecting two prime crews and two backup crews to train with the U.S. astronauts. The second set of Soviet crews would be trained as standby in case a second Soyuz must be launched.[33]

Ed Smith, who had replaced Cheatham as American chairman of Working Group 2, discussed a number of guidance and navigation issues with V. P. Legostayev, V. A. Podelyakin, and I. P. Shmyglevskiy. One of their major decisions related to the proposed development of a "centerline television" system for docking, mounted in the front of the docking module, transmitting to the Apollo crew a television image of a docking target on the Soyuz half of the docking mechanism. This system was being evaluated to determine if it would give the Apollo crew a better docking approach than the externally mounted passive target that had been used throughout the lunar module dockings. But for technical and financial reasons, the Americans wanted to drop the new idea and stay with the external target. The Soviets agreed. Canceling the television system also meant eliminating a glass viewing port in the DM hatch—a desirable change since it simplified the design, removing a minor element, which if damaged, could have posed a threat to the crew.[34]

Smith and Legostayev proposed to convene meetings of Groups 2 and 4 in Houston at the end of November, even though the Soviets had indicated that it might be difficult to complete all their preparations by that date. They had not anticipated the depth at which the Americans were pursuing these discussions but consented to make every effort to complete their work in time. On the other hand, the Americans in Working Group 3 had been quite pleased with the preparations the Soviets had made for their October session; they had fully understood the expectations and needs of their NASA counterparts.

PRELIMINARY SYSTEMS REVIEW (STAGE I)

Under the direction of V. S. Syromyatnikov, the Soviet Group 3 team had readied their documentation in both English and Russian and had prepared their two-fifths-scale model of the docking system for the joint meeting. Some of the Americans observed that while the U.S.S.R. mechanism was more complex mechanically than the American one, it was suitable for the mission and "sophisticated" in its execution. The two sides reviewed and signed the two-fifths-model test plan and scheduled the test for December in Moscow. Another important Group 3 milestone was the completion of the first part of the docking system Preliminary Systems Review (PSR).[35]

The PSR was planned to be a "formal configuration review . . . initiated near the end of the conceptual phase, but prior to the start of detail design" work on the docking mechanism. As part of their presentation to the Preliminary Systems Review Board (the Board being the Technical Directors), Don Wade and Vladimir Syromyatnikov included all the test data, specifications, and drawings for the docking system, as well as a design evaluation for the mechanism. After hearing their report, Lunney and Bushuyev felt three problem areas needed further study. First, the requirement for a spring thruster designed to help separate the two spacecraft had caught their attention, since the failure of this thruster to compress properly could prevent completion of docking. Second, Lunney and Bushuyev emphasized the importance of an indicator that would verify that the structural latches were properly in place. The American system provided information on the functioning of each latch but did not indicate that the interface seals were compressed, while the Soviet system gave data on compression of the seals but none for the latches. To assure the structural integrity of the transfer tunnel, it was important to know that all eight latches were closed.[36] The third problem area also dealt with those structural latches. Was it possible that they could be inadvertently released? Bushuyev and Lunney called for a thorough re-evaluation of all these issues and advised Group 3 to present their specific recommendations to them in December and January.[37]

Don Wade, the American chairman, was not completely happy with the

Two views of the Soviet two-fifths-scale model of their version of the ASTP docking system, used in the October 1972 Preliminary Systems Review in Moscow (Soviet Academy of Sciences photos).

manner in which the PSR had been executed. In his opinion, toward the end of the review session, matters became "pretty fouled up." Nevertheless, the teams did achieve their goals at the PSR, and Wade believed that the second phase of the review scheduled for December would be accomplished in a more orderly fashion. But the development of international space hardware was not an easy task. The Americans and Soviets were still learning to work together, and while Lunney had commented that working with the Russian specialists was quite similar to working with a new contractor there was one significant difference. In the Apollo-Soyuz Test Project, the two teams were equal partners. Neither side could tell the other how to do its work. Instead, engineering agreements had to be negotiated, undoubtedly a new experience for both countries. Two men were particularly aware of this difference; Lunney and Bushuyev kept their teams in line as the process of learning to work together continued.[38]

COMPATIBLE ATMOSPHERES

Ed Smylie and his NASA colleagues working on environmental control problems with the Soviets were quite satisfied with their progress. Ilya Vladimirovich Lavrov was present at the Moscow sessions, and it was apparent to the Americans that he had sparked considerable activity on his team's part since the July meeting. With few preliminaries, the October talks tackled the question of spacecraft pressures. The Apollo command module would continue to operate at its standard pressure, 258 mm Hg. But at the Soviets' suggestion, the pressure of the Soyuz would be lowered during the docked portion of the flight to eliminate lengthy pre-breathing periods. While the "most rational decision to make would have been to utilize the same atmosphere with regard to total pressure and partial pressures of oxygen," the Apollo could not have an operational pressure higher than 325 millimeters, and the Soyuz for reasons of fire safety could not have an oxygen content higher than 40 percent. In addition to greatly simplifying the onboard hardware and equipment with which the crews would work, this important Soviet design concession also meant that the transfers between the two ships would be easier, reducing time and procedures.[39]

Group 5 also discussed the Soyuz life support system and modifications that would be required to lower pressure during the joint phase of the mission. Lavrov indicated that the Soviets had put aside all considerations of using a pressurized oxygen system and instead would modify the existing potassium superoxide oxygen generating system. In addition, they said that they would develop an emergency pressurization system as NASA had suggested. At this meeting, the Americans finally understood that this capability had not been present on earlier Soyuz flights and that in fact the Soviets had never carried any pressurized gases on their spacecraft.[40]

The specialists also discussed such contingency situations as extravehicular transfer and the return of mixed crews. After reviewing the life support systems of both spacecraft, they came to the conclusion that there was no real need to consider external transfer on this mission, thus eliminating the need for special equipment related to space suits. Return of a crewmember from one nation in the spacecraft of the other was considered a possibility in an emergency situation, but it would have to be accomplished without the use of any supplementary equipment. The Soviets requested more time to study this contingency and agreed to report back in March 1973. Lavrov also indicated that they had definite plans for vacuum chamber tests (similar to the American tests of an early breadboard mockup of the DM) of an orbital module/docking module combination. The Americans said they would provide the Soviet team with drawings so they could build a DM boilerplate model for these tests.[41]

The meeting in Moscow had gone well. From his vantage point on 25 October, Lunney said at a combined debriefing session and ASTP staff meeting in Houston that he had been pleased with their progress. He noted that the Soviets had been well prepared, providing adequate translators, interpreters, and general logistical support for the gathering. Reporting on future activities, he said that Groups 2 and 4 would meet in November, Group 3 would hold tests in December, and the next plenary sessions were scheduled for March 1973 in Houston.[42]

After getting back to Texas, Lunney and his ASTP team prepared for two major reviews of hardware design with North American Rockwell. On 8 November, the Critical Design Review (CDR) for CSM 111 and the Preliminary Design Review for the docking module and docking system were convened at the contractor's Downey factory. Basically, the CDR was held to conduct a detailed check on the CSM engineering specifications and drawings prior to their release to the production engineers who would then oversee the manufacturing and assembly. The PDR for the docking module and the docking system was another in the series of basic checks on the detailed design tasks related to those pieces of flight hardware. For these reviews, there were 24 teams, each of which was co-captained by a NASA and a contractor employee. Following two days of study, Lunney sat as chairman of a Design Review Board, which met at 8:30 a.m. on the 10th.*[43] While these reviews were in process, Rocco Petrone briefed the Office of Manned Space Flight (OMSF) Management Council on the major accomplishments of the October trip to Moscow and Ed Smith and R. H. Dietz prepared their Working Groups for the sessions slated to begin on 24 November.[44]

THANKSGIVING IN HOUSTON

As originally planned, the Soviet delegation would arrive in Houston on 22 November so they could join in an American Thanksgiving dinner before starting to work on Friday, the 24th. Because of a 19-hour mechanical delay in Moscow, however, the visitors did not arrive at Houston Intercontinental Airport until 3:00 Thanksgiving afternoon. Ed Smith and his Group 2 colleagues listened to the Oklahoma-Nebraska football game on a portable radio while they waited. When the party did finally deplane, they were all rather wilted and exhausted. Ed Smith rode to the Kings Inn in Clear Lake City with the Group 2 specialists, and R. H. Dietz and A. Don Travis accompanied Group 4. All plans for a festive meal were canceled, and the Soviets were given the remainder of the day to recover from their trip. Looking back on that occasion, Don Travis recalled that he ate a "hell of a lot of turkey" over the next few days.[45]

Work began at 9:00 a.m. on the 24th. Viktor Pavlovich Legostayev was in charge of the nine-man Soviet delegation, and he told Smith, Working Group 2 chairman, and Dietz, Working Group 4 chairman, that he would

---

*The members of the board were as follows: from MSC, G. S. Lunney, T. P. Stafford, D. R. Scott, R. A. Colonna, F. Miller, R. P. Burt, E. W. Sievers, and A. Dennett; and from Rockwell, G. Merrick, E. P. Smith, and C. Helms.

initially give the majority of his time to communications issues because Group 4 had the greatest number of agenda items to be completed in the time alloted for their stay. Once the communications and tracking discussions were satisfactorily under way, Legostayev turned the Soviet part of the talks over to Boris Viktorovich Nikitin and assisted Group 2.

Ed Smith found his three Group 2 counterparts to be first-rate engineers and good people with whom to work. Smith, who had the personal sense of precision and perfection that guidance and navigation demanded, was full of praise for Legostayev:

> As had been the case in the past, his knowledge of the analytical problems associated with guidance and control is excellent. His knowledge of English is good and improving. He discusses technical problems in English easily, and only in the case of large meetings does he use Russian. He is at ease and very congenial in either large or small groups and appears to be an excellent leader and organizer. He is competent in problems of hardware integration but appears to prefer analytical work.[46]

From their discussions, Smith got the feeling that Legostayev was the designer of the Soyuz automatic rendezvous and docking control system and that he might have been the chief designer for all the attitude control systems used on that spacecraft. The Americans found Legostayev willing to listen to both sides of an issue before he made his own position known. Like most of his Soviet colleagues, he did not have the authority to make decisions on the spot during a meeting, but his recommendations appeared to carry considerable weight with his superiors.[47]

In his post-meeting report to Lunney, Smith indicated that he had been equally impressed with Shmyglevskiy and Podelyakin, who, in addition to being experts in their respective fields of guidance and control and docking targets, also had a good command of English. Nearly 95 percent of Group 2's negotiations were conducted in English, a factor that speeded their work considerably. Elsewhere, the language barrier was more of a problem, and many American engineers began to learn Russian. Despite their studies, difficulties with preparing joint documents would continue to be a primary concern.

During their November meeting, Group 2 looked at three basic topics—control systems, rendezvous analysis and tracking requirements, and docking targets. Smith and his NASA colleagues exchanged functional descriptions of the Apollo and Soyuz control laws used in the flight systems with Legostayev and Shmyglevskiy. Group 3 had asked for this information so that they could complete the computer program for simulating the docking of the two spacecraft. During their discussion of control systems, Group 2 outlined and agreed on the procedure that would be used for controlling the ships when they were docked. This meeting gave the chairmen a better understanding of the conditions that flight crews could anticipate prior to docking and the manner in which the two ships would act after they were joined together as one orbiting mass.[48]

American specialists brought their latest revision of rendezvous trajectory and tracking requirements. A normal flight path dictated the need to begin VHF radio tracking at 236 kilometers, but the Americans wanted to extend the tracking range to 266 kilometers to account for trajectory dispersions that might occur if the launch of either spacecraft was delayed to one of the alternate opportunities. For optical tracking, the Soviets and Americans planned to exchange samples of their different exterior coatings so that the reflectivity of the ships' surfaces could be determined. Optical tracking with the Apollo sextant appeared to pose no problems for the normal trajectory, but some of the flight paths dictated by alternate launch times might cause some difficulties if the Soyuz were lost in the brilliance of the sunlit earth. These issues were placed on the agenda for further study.

Podelyakin described for the Americans the docking target installation that they were planning to build for Soyuz. MSC personnel in turn presented the North American Rockwell proposal for ensuring proper alignment of the docking target and the Apollo alignment sight. They also considered various methods for aligning the two craft if the Soyuz target failed to deploy properly. This subject was also placed on the agenda for March, when the group would hold a Preliminary Systems Review of the docking alignment system.[49]

Simultaneously with the Group 2 effort, the members of the Working Group 4 communications team worked literally days, nights, Saturdays, and one Sunday to complete all the items on their list. Part of the difficulty in negotiating arose from the Soviets' fixed 3 December departure date. Dietz said in a report that the selection of an arbitrary date for completing the work without taking into account the anticipated workload placed an unnecessary strain on the support people (translators, typists, and drivers), as well as on the delegation itself. Glynn Lunney subsequently wrote to Professor Bushuyev, stressing the need for adequate time. In addition to setting aside one or two days at the end of a meeting for cleaning up the documents and signing them, Lunney suggested that documents scheduled to be signed should be made available in draft form at least one month before the session. If documents were understood beforehand, the meeting time could be put to better use. Documents introduced for the first should be presented in both languages to prevent a similar waste of time.[50]

Despite the tight schedule, Group 4 accomplished all its major goals. After reviewing the Soviet antenna data, the Americans concurred with their counterparts' wish to build the Soyuz antennas for both the 121.75-mega-ahertz (VHF/FM) and the 296.8-megahertz (VHF/AM) systems. Agreements were also reached concerning signal characteristics for the radio communications and ranging systems, compatibility test plans for those systems, and installation of the Apollo VHF/AM equipment aboard Soyuz. The specialists also completed a definition of the cable communications system that would be used between the two craft, and they finished preliminary talks about the communications links that were necessary between Houston and Moscow mission control centers. From the American vantage point, the Group 4 activities were extremely productive. The depth of system definition was sufficient to permit the detailed design of the communication gear. Lunney, in his post-meeting letter to Bushuyev, said that he was very pleased with the progress made and asked the Professor to thank the Soviet specialists for their hard work and dedication.[51]

PRELIMINARY SYSTEMS REVIEW (STAGE 2)

Working Group 3 tests of the two-fifths-scale model and the second part of the Preliminary Systems Review for the docking system was the last joint activity scheduled for 1972. The Americans arrived in Moscow on the 6th of December and worked through the 15th. MSC specialists were becoming seasoned travelers. During the October meeting in Moscow, most of the Americans had been infected by an intestinal parasite, which severely debilitated some of them after their return home. As a consequence, the team that went to the U.S.S.R. in December took along some pans in which they could boil their drinking water, hot plates, and some American style food—crackers, peanut butter, and canned dinners—to which their stomachs were more accustomed. With these supplementary rations and other items obtained from the American Embassy commissary, Don Wade reported that his team returned in much better health. Group 3's work went better than it had in October, too.[52]

Wade said that the Soviets were well prepared for the meeting. They had a team ready to begin the tests of the two-fifth's-scale docking systems. And following Lunney's suggestion, the Soviets provided the Americans with English and Russian versions of the materials they planned to disucss. There were some minor problems along the way, but all in all the trip was very successful. The PSR went especially smoothly, with the changes in the engineering drawings being studied and accepted by Syromyatnikov and Wade, who in turn recommended their acceptance by the Technical Directors at their March 1973 meeting.[53]

Testing the scale models at the Institute of Space Research in Moscow went equally well. In his report to Chris Kraft, Lunney pointed out that these tests had indicated compatibility of the two systems in both the active and passive modes and that Working Group 3 had anticipated the minor problems experienced during the exercise and had already accounted for them in the revised drawings. Summarizing, Lunney said:

> The meeting is considered to be a very successful one and, while we haven't placed a great deal of importance on the results of the model tests, I believe the accomplishment of this first major hardware related milestone on schedule is in itself significant. We have identified no major problems and are proceeding on schedule.[54]

Critics and supporters alike were surprised at the basically cooperative attitude of the Soviets and the progress being made toward the joint flight. There had been some frustrating moments for Glynn Lunney and his associates, and there would be times of tension and disagreement in the future, but how things had changed since the days of October 1957 or April 1961. Given the background of competition that had produced Sputnik, Gagarin's orbital mission, and the American resolve to beat the Soviets to the moon, the cooperative aspects of ASTP boggled some minds. One pair of critical writers found it all hard to believe:

> Subsequent meetings were reported as going equally well. This applied both to specific "Working Groups," which met with increasing frequency either in the USSR or the US, and to large "plenary" meetings of full delegations from both sides concerned with the project. Within the compass of the ASTP project itself, these meetings were marked by exchanges of technical data and information, as well as a degree of personal contacts among the specialists involved almost without precedent in US-Soviet relations including those of the war-time alliance.[55]

The Soviets and Americans accomplished many things, but they still faced a multitude of tasks.

YEARS OF INTENSE ACTIVITY

At the end of 1972, 26 months since NASA's first visit to Moscow to discuss cooperation and six months since the Summit officially created ASTP, Lunney could reflect upon the project's accomplishments with a positive frame of mind. A mission had been defined. Hardware design and development were well along. And Working Group activities during the thirty months that remained until launch would follow a pattern established during 1970-1972 and the schedule negotiated by Bushuyev and Lunney. More than anyone else, Lunney was responsible for maintaining the pace of the joint effort. From his office on the seventh floor of the Program Management Building at MSC, he had to exercise considerable diplomatic and managerial skill to keep his NASA, contractor, and Soviet teammates moving along to the July 1975 deadline. After the *Apollo 17* flight, Lunney was given a more direct line of authority for reaching that goal.

The sixth and final lunar landing, successfully completed by the Apollo crew on 19 December 1972, closed out another chapter in NASA manned space flight operations. With the return of *17's* Eugene A. Cernan, Ronald E. Evans, and Harrison H. Schmitt, OMSF reorganized in preparation for Skylab and ASTP. Dale Myers announced in January 1973 that Rocco Petrone would be leaving the Apollo Program Office to become the Director of the Marshall Space Flight Center. Petrone was replaced by Chester M. Lee, who moved up from Apollo Mission Director. At Houston, Lunney succeeded Owen Morris as Manager of the Apollo Spacecraft Program Office, which in addition to ASTP had the responsibility for managing the command and service module aspects of Skylab, scheduled to be visited for the first time on 25 May 1973. Chet Lee and Glynn Lunney now directed the team that would carry the Apollo half of ASTP to completion.[56]

Lee and Lunney worked well together. While Lunney was concerned mainly with the technical aspects of ASTP, Lee had to worry about technical, political, economic, and public relations considerations. A 1941 Naval Academy graduate with 24 years of service, Lee spent the latter part of his naval career working on the Polaris ballistic missile weapon system. Captain Lee, as he was called by this NASA colleagues, joined the space agency in 1965 as Chief of Plans in OMSF's Mission Operations Directorate. Lee and Lunney shared more than the same managerial problems—both men liked good cigars and had a reliable sense of humor. But the two men shared another more important trait—an honest, straightforward manner of dealing with other people. This characteristic was a very valuable one for NASA when Captain Lee talked to members of the press and Congress.

On 2 October 1973, Chet Lee gave a typically candid briefing to members of the Manned Space Flight Subcommittee of the House of Representatives Committee on Science and Astronautics. George Low, Gene Cernan, and Lee had traveled to Capitol Hill that morning to provide the congressmen in closed session with detailed background on ASTP and to relieve one particularly nagging concern. Chairman Olin Teague and Representative Don Fuqua had corresponded with NASA about the scientific experiments planned for ASTP. As Fuqua stated their worry, "Our concern has been in the event of any reason it were not possible to conduct a joint mission with the Soviets NASA should be prepared to justify the mission on its merits."[57] Clearly, confidence in the Soviets' ability—politically and technically—to perform the joint mission was not universal. The Manned Space Flight Subcommittee wanted some assurance that the scientific program planned for the Apollo part of the flight would help justify the $250 million total cost.

The timing of Chet Lee's presentation was significant. On the day before, NASA had celebrated its 15th anniversary, and Lunney had arrived in Moscow with a 47-member delegation for a meeting that would culminate in a Mid-Term Review of ASTP; and on the day of the briefing, tests of the full-scale Soviet and American docking systems began in Houston. Lee and Cernan were scheduled to leave for Moscow on 3 October, and Low would follow them in about ten days' time. Although the congressmen were primarily interested in the experiments program, Lee gave them a complete status review so they would have a better context within which to judge ASTP and the scientific experiments.

He began with a report on the new hardware designed for the mission. The joint design work on the docking system was complete, as was the design effort on the docking module. Modifications to the CSM, which Lee pointed out was left over from the Apollo program, had been made with the exception of those that would be required by the experiments hardware and the modified high gain antenna needed for communicating with the Applications Technology Satellite (ATS) for improved television, radio, and scientific telemetry transmissions to the ground. Lee indicated that ATS-F was very important to the success of the scientific experiments. Apollo had been able to broadcast picture, voice, and data from the moon on an almost uninterrupted basis. Skylab was able to communicate from its 438.2-kilometer orbit for an average of 28 minutes per 93-minute revolution. But ASTP at an altitude of 225 kilometers would have ground station coverage for only about 15 minutes per 88-minute revolution. This limited ability to transmit to receiving stations would severely hamper the amount of data that could be gained from some of the experiments. With ATS-F, which was scheduled for launch in 1974 (at which time it would be called ATS-6), the communications coverage would be extended to about 49 minutes per orbit.

Reporting on the status of other hardware elements, Lee told the congressmen that the first of five docking systems had been completed by Rockwell International* for use in the development tests. While the joint dynamic tests were scheduled for mid-November, the first round of docking seal tests had been completed and the results reviewed in Moscow at the end of June. Though some minor design changes were being made as a result, confidence in the seal used in the docking system had increased consider-

*North American Rockwell Corporation had been renamed in Feb. 1973.

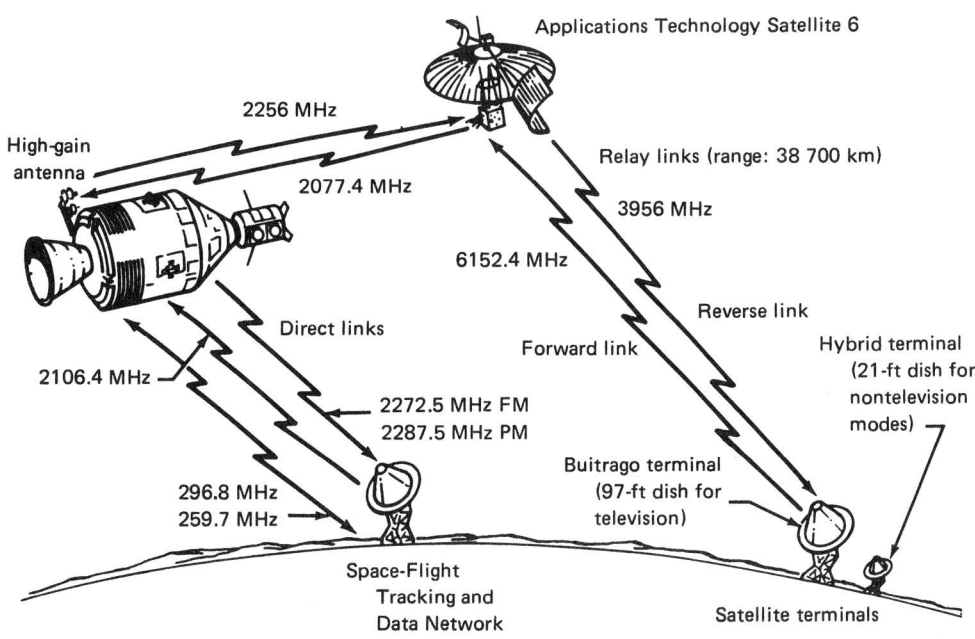

Apollo communications links.

ably. Fabrication of the docking module was also on schedule. He pointed out that this was largely because of the decision to build the life support system and electrical control equipment into a panel that could be constructed separately and then installed into the spacecraft. Lee could give a very favorable hardware status report.

Lee was equally optimistic when he talked about operational planning. The "Joint Crew Activities Plan," which presented the details of the crew actions during the flight, had reached the point where for a first launch opportunity it could be used that very day. The experiments would have to be worked into it, but basically the activities plan was ready to go, Lee said. An early completion date for the Crew Activities Plan had been set because "we recognized that with the language difficulties and numerous joint activities planned we needed an early start. . . ."[58]

Representative Bill Gunter questioned Lee's optimism. Lee responded by saying, "we are on schedule and . . . we are satisfied with [our] progress, but we do have some qualms." When asked how one could be on schedule and still be experiencing delays, George Low explained:

> The hardware is on schedule. The paper work is flowing a little more slowly than we like to see. This has not yet hurt us; the project [director's] concern is that as we get closer to the launch, there won't be this kind of luxury of time. We have to work things out now. The paper, too, will flow faster.[59]

The Soviets had been slow in providing some essential documents the two sides had agreed to prepare and exchange. They had never refused to provide information; they were just slow. For example, at Moscow in June at the very last minute, the Soviets presented a Working Group 4 report that was to have been delivered 13 months earlier, expecting the Americans to sign it.* They would not.[60] To Low, Lee, and Lunney, it appeared to be partly a problem generated by Professor Bushuyev's lack of freedom to make decisions on the spot. Whether in Moscow or Houston, the Professor had to refer to his superiors before he could provide many kinds of information. Lee had reported to Fletcher and Low:

> Professor Bushuyev frankly admits that because of the Soviet internal system he does have a problem in meeting commitments on documentation and providing replies to specific questions and requests for amplifying information, but that he does not have this problem to the same degree with hardware.[61]

At other times, the Soviets just did not provide in their documents the detail necessary to satisfy NASA. When the specialists from Houston explained why they needed specific points of information, the Soviets provided the additional data, but seldom did they give all the information the first time. Many Americans were frustrated by this tooth-pulling contest.

The *Soyuz 11* was a good example of this problem. To get a better understanding of the failure that led to the tragedy, Glynn Lunney had asked Bushuyev about the technical details of the accident several times, and still he had not received a clear explanation. He had pressed the point in Houston during the March 1973 talks, and Dave Scott had raised the issue again for Lunney at the June meetings in Moscow.[62] When Lee and Lunney raised the topic a third time in Houston during July, the Professor told them that he had already explained in March the nature of the failure and the corrective actions taken to assure that it would not be repeated. Lunney firmly explained to Bushuyev that more details were required to satisfy safety and reliability requirements for the joint mission and to assure both supporters and critics of ASTP that the American crew would not be in danger when Apollo docked with Soyuz.

Chet Lee had reported that "from his information it was difficult to reconstruct the failure and [the Soviet explanation] provided little on the corrective action." Therefore, Lunney requested a fuller and more comprehensible explanation. Bushuyev was very hesitant to promise this, and according to Lee he "appeared to stall by stating the Soviets should then get copies of the Apollo failure reports." After Lunney and Lee showed

*The results of this meeting are summarized in appendix D.

Bushuyev a copy of a message from Keldysh acknowledging receipt of the *Apollo 13* accident report, the Professor promised to work on this request. Significantly, he would not agree to put this matter into the formal minutes of the meeting, but he did assent to its being included in a letter Lunney planned to write to him.[63]

Captain Lee, with Lunney's support, had recommended to the Administrator that a Mid-Term Review might be useful for working out some of these problems:

> Glynn Lunney and I have discussed this at some length. We agree that perhaps a meeting between Mr. Myers and Academician Petrov or Dr. Low and Academician Keldysh under the category of a "Review of the Status and Report on ASTP" might be most helpful in avoiding future problems and delays in the Working Groups' progress, particularly as we move into the more specific plans for the mission.[64]

Lee was convinced of the genuine desire on the Soviets' part to make the mission a success. He was also impressed by the rapport that had developed between the Americans and their Soviet colleagues and "in particular, the frankness, confidence and personal working relationships between" Lunney and Bushuyev. Still, he believed that NASA should continue

> to carefully, but frankly, pursue answers, information and agreements on issues that may be touchy but are related to the mission. In this manner, we will not only provide greater confidence of ASTP success, but we can also gradually eliminate some of the time consuming barriers to smooth and expeditious working relationships with the Soviets in space cooperative efforts.[65]

In his testimony before the Manned Space Flight Subcommittee, Low said that NASA's desire to build a solid basis for present and future cooperation was "one of the reasons for my going over there in two weeks for this Mid Term Review." He also stressed to his audience on 2 October that while Lee and Lunney were probably getting less cooperation than they would have liked, "from the management point of view we are getting far more than we expected to." Despite the delays, the Soviets had met every obligation they had agreed to in April 1972.[66] Still, the concern over the Soviets' possibly defaulting or failing to fly was the reason Low, Lee, and Cernan were giving their briefing to Representative Teague and his associates. Chet Lee turned to a discussion of the proposed package of ASTP experiments.

Lee's presentation and the committee members' comments that followed it dealt less with the actual experiments themselves than with the merits of spending $250 million to fly $10 million worth of experiments in the event of the Soviets' failure to rendezvous with Apollo. Once Lee had stated that there had been 146 responses to the request for experiment proposals and that a large number of excellent candidate topics had been selected for further evaluation, the conversation turned to possible means of adding to Apollo's scientific payload. Captain Lee saw three possible ways of increasing the scientific examinations of a unilateral mission—load the backup docking module with additional experiments, create a scientific instrument module bay in the prime CSM, or revisit Skylab, which would have been in unmanned orbit for nearly a year and a half. In addition to the unfavorable impact on the launch timetable, all of these alternative plans would have been expensive and probably caused the project to run over its $250 million budget. Each alternative would involve extra engineering and careful balancing of payload weight and launch vehicle capacity.

George Low looked at the entire project from a political perspective. NASA had sought authorization to conduct a rendezvous and docking mission with a Soviet spacecraft for the purpose of developing a common system for working together in space. At the same time, NASA had pointed out that whatever flies in space should get maximum return for the investment. That is why the agency set aside $10 million for scientific studies. Low continued:

> We have discussed it with the Congress since then on the basis that . . . for any less we could not do a decent experiment package. . . . That is how the $10 million were arrived at. You asked the question, what would we do if the Russians for some reason would not fly with us, political, technical or otherwise, and would the mission in itself with the $10 million worth of experiments . . . be worth flying without that rendezvous. I think that answer would depend very much as to when this would happen. Were it to happen now when we have spent a substantial sum of money, which is still a small fraction of the $250 million, we might well decide and discuss with the committee the possibility of cancelling it altogether. Because I am not sure whether it is worth remaining funds to be expended to go up there in 1975 for the $10 million worth of experiments alone without the rendezvous and docking.[67]

On the other hand, if the spacecraft were on the launch pad and ready to go and for some reason the Soviet portion of the mission were canceled, then NASA would likely want to go ahead with the flight but only after consulting with and obtaining the approval of the Congress and the executive branch.

Representative Teague wanted to know if the American public should be advised ahead of time that NASA had alternative plans for the mission. Representative John W. Wydler saw some dangers in such a course of action. "What would our national reaction be . . . if the Soviet Union were to announce their alternative plans for the project if it doesn't come off?" He thought the American public would assume that the Soviets did not expect the U.S. to fly. "I think that would be something that could be very easily misunderstood from the point of view of the other side if you started to plan what you are going to do if this mission doesn't happen."[68] In George Low's position, the most logical course to follow was to develop contingency plans but to assume that the Soviets did indeed plan to fly in 1975. None of the alternatives seemed as desirable as the basic idea of a joint mission. Essentially, NASA had faith that the Soviets would meet their commitment. It was a gamble, but the risk seemed to be a reasonable one.

Apollo and Soyuz at Mid-Term

The Mid-Term Review was another NASA tool that the Americans inserted into the joint project. As the name indicated, this examination at the mid-way point gave management an opportunity to ask questions of the technical teams and to evaluate their progress. George Low wanted an ASTP Mid-Term Review because Glynn Lunney had expressed his concern several times during 1973 over the Soviets' inability to meet deadlines in some areas. Lunney had already discussed this with Professor Bushuyev in their formal meetings and in private communications; early in August, Lunney drafted a letter to his counterpart in which he noted that despite the excellent progress of the work at the July meetings he must "amplify . . . [his] concerns regarding some of [their] discussions." While both he and Bushuyev agreed that they were continuing to meet their major milestones, Lunney said:

> Despite these very significant accomplishments, I am still concerned about the delay we are experiencing in obtaining pertinent technical and program related data from your side. As I discussed [earlier], we have experienced a delay in exchange of material of up to 9 months. This has occurred even though we have signed minutes committing ourselves to specific dates for these exchanges. Our experience indicates that the need for rapid exchange of information and reports greatly increases as the time for flight approaches. For example, as we approach the launch date, the preparation and negotiation of documents such as the Safety Assessment Reports will have to be completed in a very short time rather than the 6 to 9 months currently required.[1]

Lunney went on to address problems being encountered by Working Group 4. "I think we both agree that the work of this group has not been satisfactory, and this has been due to a lack of timely preparation, primarily, on the USSR side." But the American Technical Director "was pleased to hear from [Bushuyev] that [he was] considering steps to solve this problem."[2]

Looking ahead, Lunney also felt that more data would be needed on the *Soyuz 11* hardware failure. A detailed written report describing the problem and the corrective steps subsequently taken was in order. The

Apollo and Soyuz spacecraft as configured for ASTP

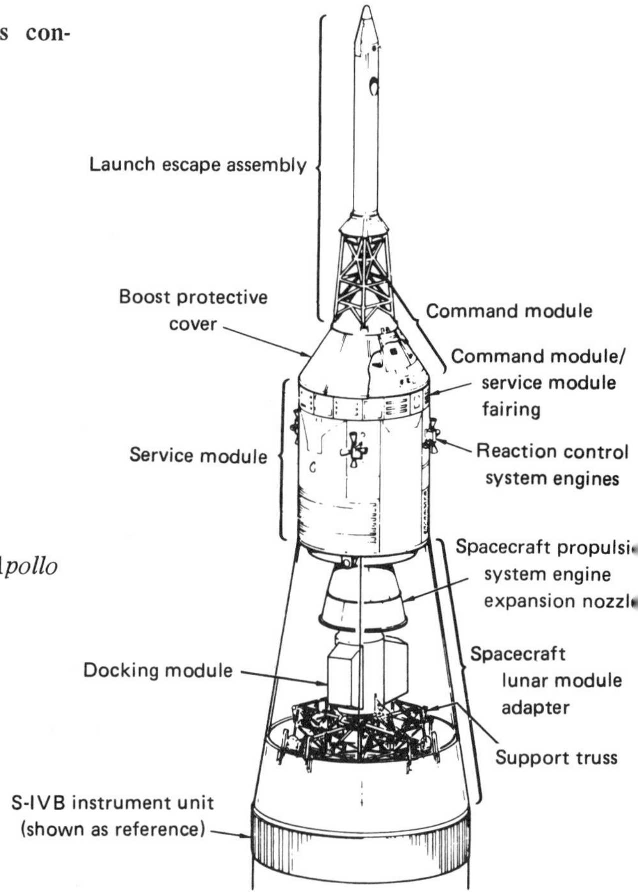

Launch escape assembly

Boost protective
cover

Command module

Command module/
service module
fairing

Reaction control
system engines

Service module

Launch configuration of the Apollo spacecraft.

Spacecraft propulsi
system engine
expansion nozzl

Spacecraft
lunar module
adapter

Docking module

Support truss

S-IVB instrument unit
(shown as reference)

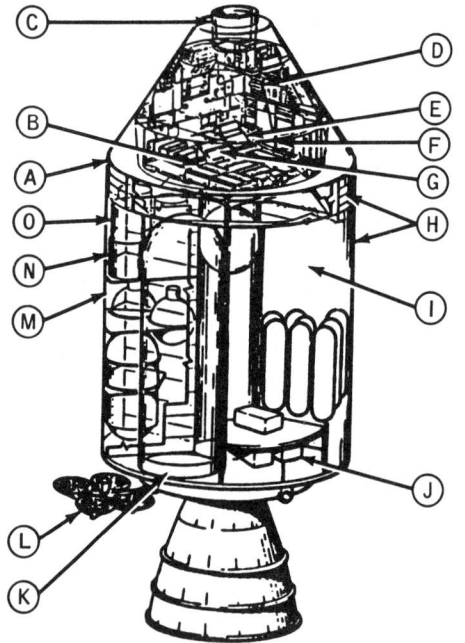

C
B
A
O
N
M

D
E
F
G
H
I
J
L
K

A   Provisions for experiments added.
B   Stowage provisions modified.
C   Docking module umbilicals replaced Skylab tun
     umbilical.
D   Displays and controls added for compatible dock
     system, docking module, experiments, and ATS
     communications.
E   Modified unified S-band equipment and premodu
     tion processor added for ATS-6 communicatio
F   Extravehicular activity station deleted.
G   Videotape recorder added.
H   Receiver and antenna added for Doppler tracking
     periment.
I   Descent battery pack deleted.
J   Experiments and remotely controlled covers add
     (service module bay 1).
K   Power amplifiers added for ATS-6 communicatio
L   High-gain antenna added for ATS-6 communicatio
M   Water storage tank deleted.
N   Insulation added adjacent to reaction control syst
     thrusters.
O   Three fuel cells installed instead of two.

Major Apollo spacecraft changes from Skylab configuration.

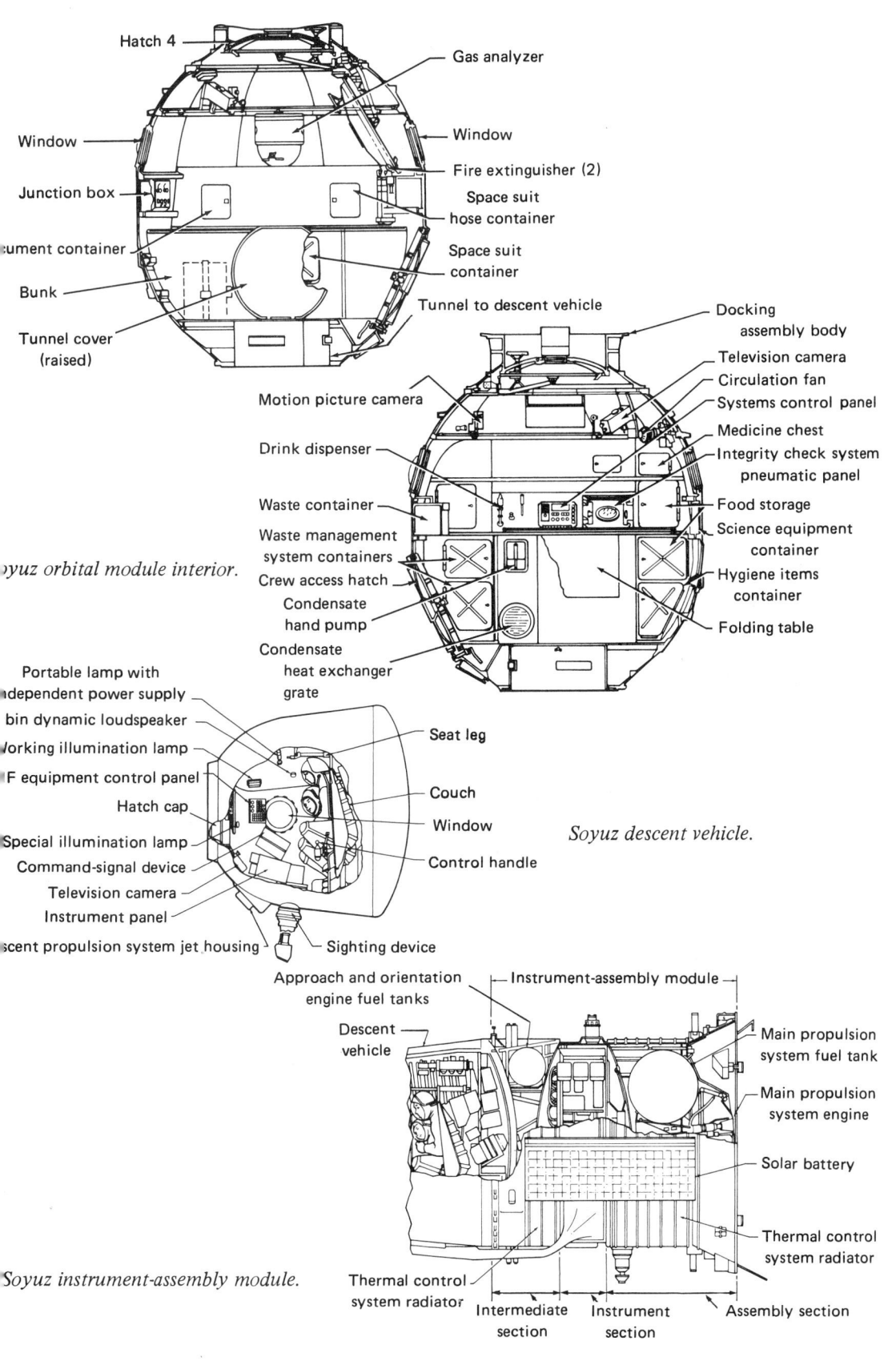

Hatch 4

Gas analyzer

Window

Window

Fire extinguisher (2)

Junction box

Space suit
hose container

...ument container

Space suit
container

Bunk

Tunnel to descent vehicle

Tunnel cover
(raised)

Docking
assembly body

Television camera

Circulation fan

Motion picture camera

Systems control panel

Medicine chest

Drink dispenser

Integrity check system
pneumatic panel

Waste container

Food storage

Waste management
system containers

Science equipment
container

Crew access hatch

Hygiene items
container

Condensate
hand pump

Folding table

Condensate
heat exchanger
grate

...oyuz orbital module interior.

Portable lamp with
...ndependent power supply

Seat leg

...bin dynamic loudspeaker

...Working illumination lamp

Couch

...F equipment control panel

Window

Hatch cap

...Special illumination lamp

Control handle

Command-signal device

Television camera

Soyuz descent vehicle.

Instrument panel

...scent propulsion system jet housing

Sighting device

Approach and orientation
engine fuel tanks

Instrument-assembly module

Descent
vehicle

Main propulsion
system fuel tank

Main propulsion
system engine

Solar battery

Thermal control
system radiator

Soyuz instrument-assembly module.

Thermal control
system radiator

Intermediate
section

Instrument
section

Assembly section

Americans believed that they "should institute a policy of exchanging such information as part of [their] process of developing mutual confidence in the success and safety of the joint flight." Furthermore, he argued that this mutual understanding should be extended to "Salyut and Skylab programs as they relate to the overall confidence in the Apollo and Soyuz spacecraft." Lunney raised this issue because Bushuyev's assessment of *Salyut 2*, launched on 3 April 1973—reportedly a normal mission—did not agree with reports from American tracking stations that indicated that the space station ·had broken up into many pieces. Lunney hoped "that during our meeting in October we will be able to further discuss and understand this problem." He added that he had been informed that Low had proposed that he and Academician Keldysh conduct a review of the entire program during the Mid-Term Review.[3]

With only 21 months remaining until the scheduled launch, Low thought that October 1973 would be an auspicious time to scrutinize the flight preparations. On 14 August, Low wrote to Keldysh: "it seems to me that it would be wise for you and me to meet at an early date to assess the progress of the ASTP project in mid-course. In particular, I believe we should try to give special consideration to those areas which could most likely present difficulties in the months ahead." He wanted "to discuss in detail four subjects"—Soviet hardware failures (*Soyuz 11* and *Salyut 2*), joint participation in test and flight preparation activities, project milestones, and the preparation of documentation. Closing his letter, Low asked if it would be possible to visit some of the Soviet space facilities during his visit. "I would appreciate your suggesting an itinerary, but I would, of course, be interested in visiting installations of the sort you visited at NASA last October."[4]

Keldysh's favorable and warm response, dated 30 August, arrived in Washington in early September. He said that he had "attentively read [Low's] letter" and agreed that "in such a complex and responsible task, from a technical and organizational point of view . . . questions could arise, which would require additional consideration." He asked that the issues raised by Low be "studied with full attention" before they met in October for their review.[5]

About a week before his scheduled departure for the Soviet Union, Low received a telephone call from Chet Lee, who was already in Moscow. He reported that Keldysh was ill and would be unable to participate in the review.* Lee added that the Soviets still wanted to have the meeting and

*Earlier in 1973, the Houston-based heart surgeon Michael E. DeBakey had flown to the Soviet Union to operate on Keldysh. After a short recuperation, Keldysh had plunged back into work, so that by the fall of the year he was worn down and exhausted. His physicians ordered him to rest and to refrain from participation in taxing activities.

were counting on Low's visit, having arranged tours of several space facilities for him. This trans-Atlantic conversation was followed by an official telegram from Keldysh in which he indicated that he had delegated the Soviet chairmanship for the Mid-Term Review to Boris N. Petrov. When the text of the cable was delivered to the American Embassy in Moscow for transmittal to Washington, V. S. Vereshchetin had told Jack L. Tech, the Science Attache, that Petrov rather than the higher-ranking Kotelnikov would substitute because of Petrov's close familiarity with the day-to-day management of the project and not because of any downgrading of the review. After consulting Arnold Frutkin, Low decided that they should still travel to Moscow.[6]

MID-TERM REVIEW

Low and Frutkin arrived in the U.S.S.R. on the evening of 14 October. Early the next morning, Low met with Lunney, who told him that the work had gone extremely well these past two weeks and that much had been accomplished. The Americans left the Rossiya Hotel and went to the Presidium of the Soviet Academy of Sciences where the Mid-Term Review would be held. That day the teams were kept especially busy with reports to Low and Petrov. After Lunney and Bushuyev told the chairmen that all technical aspects of the program were on schedule, a spokesman from each Working Group presented a detailed schedule of activities and statement of progress in terms of those schedules. Notebooks of Vu-graphs had been prepared in both languages so that all present* could follow the proceedings. R. H. Dietz recalled later that Low appeared eager to determine the exact status of each group's work. His questions were searching and detailed. Low did not want any problems to appear unexpectedly, and he was taking a strong personal interest to demonstrate to all involved that NASA's top management expected ASTP to succeed.[7]

The Technical Directors also reported on a number of important decisions that had been reached during October. Lunney and Bushuyev had agreed to reciprocal participation of specialists as observers during the life support system tests in Moscow and Houston, to joint docking seal tests, and to the participation of American specialists in the pre-flight checkout of the VHF/AM equipment at the Soviet launch site. These and other understandings reached made Low and Lunney more confident. Still, they pursued

---

*Americans participating in the review included Low, Frutkin, Lee, Lunney, M. P. Frank, R. H. Dietz, R. E. Smylie, T. P. Stafford, and E. A. Cernan. Soviets in attendance included Petrov, Bushuyev, Vereshchetin, Abduyevski, A. S. Yeliseyev, I. P. Rumyantsev, A. A. Leonov, V. A. Timchenko, V. P. Legostayev, V. S. Syromyatnikov, B. V. Nikitin, Ye. N. Galin, I. V. Lavrov, and Yu. V. Zonov.

the four discussion topics that had prompted Low's request for the review in the first place.

Project documentation was discussed during the main meetings, during executive sessions, and in private between Low and Petrov. The Soviets had made considerable progress in catching up in all areas of documentation, but Lunney was still concerned that as time grew shorter there would be less time to prepare new documents. Bushuyev believed that the solution to the difficulty was better forecasting of documentation needs. Low and Lunney agreed but added that this was "not the complete solution because we [could] not possibly foresee all problem areas." Petrov then indicated that he understood Low's point and promised to keep an eye on the situation personally.[8]

Low also received the information the Americans had sought about *Soyuz 11.* During the course of the technical sessions preceding the review, Professor Bushuyev had made a detailed presentation about the failure—post-flight investigation, experimental reenactment of the failure, and steps taken to make certain that it could not recur. According to those present, the release of this information was a personal triumph for Bushuyev and his team since they apparently had to convince many people in the U.S.S.R. that the Americans needed to know all the details. The highly favorable opinion the Americans held of Bushuyev as a tough-minded negotiator and strong-willed manager was reinforced by his report.

The fatal cabin depressurization occurred when a "breathing ventilation valve"* located in the interface ring between the orbital module and the descent module opened inadvertently during the downward path of the descent vehicle, Bushuyev said. At approximately 723 seconds after retrofire, the 12 Soyuz pyro-cartridges fired simultaneously instead of sequentially to separate the two modules. The force of the discharge caused the internal mechanism of the pressure equalization valve to release a seal that was usually discarded pyrotechnically much later to adjust the cabin pressure automatically. When the valve opened at a height of 168 kilometers, the gradual but steady loss of pressure was fatal to the crew within about 30 seconds. By 935 seconds after retrofire, the cabin pressure had dropped to zero and remained there until 1640 seconds when the pressure began to increase as the ship entered the upper reaches of the atmosphere.

The extent of tissue damage to the bodies of the cosmonauts caused by the boiling of the blood during the 700 seconds they were exposed to the vacuum could have been misinterpreted initially as being the result of a more catastrophic and instantaneous decompression. Only through analysis of the

---

*This valve combined the functions of the Apollo pressure equalization valve and the landing ventilation valve.

telemetry records of the attitude control system thruster firings that had been made to counteract the force of the escaping gases and through the pyrotechnic powder traces found in the throat of the pressure equalization valve, were Soviet specialists able to determine that the valve had malfunctioned and had been the sole cause of the deaths.[9]

Further information presented by the Soviets on the valve and seal failure cleared up the "mystery" of *Soyuz 11*. Several factors had led to the confusion that surrounded this topic. First, early reports from the Soviets had indicated that the problem was one associated with the spacecraft's *germetizatsyia,* which could be translated to mean either the failing of a seal or the loss of air tightness. Thus the Americans were unable to grasp exactly what had happened. Second, the U.S. team thought they had understood I. V. Lavrov's private remarks to Ed Smylie in December 1971 to mean that the problem lay with the pressure equalization valve, but other Soviet reports had indicated that the trouble started in the seals that guaranteed the hermeticity of the hatch between the orbital module and the descent vehicle. That latter explanation had been given to American reporters by cosmonaut Shatalov as late as June 1973 when they visited Star City.[10] Bushuyev's explanation ended the speculation, especially since Houston's environmental control experts could analytically verify the information given them as entirely consistent with the telemetry data reported by the Soviets.

American specialists could also tell Lunney that, as they had thought all along, the problem was not one that could pose a real threat to the safety of the crews during the docked phase of ASTP.[11] Nevertheless, this presentation on *Soyuz 11* and the fact that the Professor had been able to release the exact details, even though it did not immediately affect the safety of the American crew, was an important step forward in forging a partnership. Both sides had to establish faith in the other's hardware and believe that it was safe. The Soviets had opened up and talked about an extremely painful subject. It had taken two years for them to do so, but the resulting level of candor, coming as it did at this crucial Mid-Term Review, indicated that both sides were reaching the level of trust necessary to build a genuine space partnership.

Bushuyev also told the Americans that once the problem was recognized and verified experimentally, the Soviet designers had modified their hardware. They had tested the altered system in two Cosmos flights–*Cosmos 496,* flown 26 June-2 July 1972, and *Cosmos 573,* flown 15-17 June 1973. The results of these flights had given them confidence in their solution to the problem, and on 29 September 1973, Lt. Col. Vasily Grigoryevich Lazarev, a test pilot and physician, and Oleg Grigoryevich Makarov, a civilian spacecraft engineer, completed a two-day test flight aboard *Soyuz 12.* Soviet reports indicated that the cosmonauts had worn space suits during launch and reentry, and beginning with this flight, two-man crews would become standard for Soyuz so there would be room to store suits.[12] During October, the Soviets reaffirmed their plans to fly two or three manned Soyuz flights in 1974, and they suggested that these ASTP-related missions would fly in the configuration planned for the joint exercise.[13]

During their executive session, Low told Petrov that he greatly appreciated their report on *Soyuz 11* and asked him about those additional failures that had been reported by the Western press during the summer of 1973. Petrov told Low that *Salyut 2* was an updated version of the Soviet space station and because of the changes in the design there had been no plans to send men to occupy it. He said further that the 3 April-28 May flight had been designed to test the automatic control system; there was no need to have a crew board the station. While this might have seemed strange to the Americans, the Soviets seemed to rely more heavily on test flights, as opposed to NASA's use of earth-based simulations. On the subject of *Cosmos 557,* which had been launched on 11 May, Petrov stated that this flight was not related to the manned space flight program.

A TOUR OF SOVIET SPACE FACILITIES

Following the review, Low was taken to see several space-related facilities. On the morning of 16 October, after making a brief courtesy call on Academician Kotelnikov, Low went to the Institute of Geochemistry and Analytical Chemistry. Director A. P. Vinogradov, ill with a bad cold, instructed his Deputy Director to give Low a tour of the Institute where lunar samples from *Luna 16* and *20* and from *Apollo 11* through *17* were housed and studied. Next, Low visited the Institute of Space Research and met its new director, Professor R. S. Sagdeyev, with whom he discussed the four spacecraft the Soviets had launched to Mars in July and August. Low's activities of the 16th were finally completed when he called on Academician V. A. Kirillin, the Deputy Chairman of the Soviet Council of Ministers and the Chairman of the State Committee for Science and Technology. After Low gave a brief report on the progress of ASTP and other joint projects, Kirillin asked the NASA Deputy Administrator for his views on the practical benefits derived from the exploration of space. They spoke of communications, weather, and earth resources, as well as the potential long-range results of some of their scientific efforts in space. In reemphasizing his point that the future of space must be practical, Kirillin said that one important aspect of earth resources would certainly be the study of geology from space.[14]

On the 17th, Low and part of the NASA delegation visited Star City. He recorded in his trip report, "I saw more of Star City this time than I had during my previous visit. Of major significance is the amount of new construction underway.... A new training building is being put up especially for ASTP...." That four-story building was to include classrooms, lecture halls, and display rooms for spacecraft subsystems. In addition, the Soviets were building a new hostel and dispensary for the United States team, as well as other buildings to house simulators and a new and very large centrifuge. When the NASA contingent visited the Soyuz simulator, Alexi Arkipovich Leonov, the Soviet ASTP crew commander, briefed them on the changes that had been made following *Soyuz 11.* Besides removing the third couch, engineers had installed pressure suit connections and new pressure relief and shut-off valves. Valeriy Nikolayevich Kubasov, the second member of the Soviet ASTP crew, gave the visitors a brief description of the Soyuz space suit, which was modeled by a technician. This relatively lightweight garment was the same type they planned to use in the joint flight. Leonov pointed out that it took about 5 minutes to don the pressure suit, and Low noted that since it was only worn for about 2 hours at a time there were no provisions for waste removal. The last thing Low did at the Cosmonaut Training Center was to tour the Salyut mockup, with the assistance of K. P. Feoktistov.[15]

On the morning of the 18th, the Americans were taken to the Soviet Mission Control Center at Kaliningrad, by car about 45 minutes northeast of Moscow. The center was situated within a large complex of buildings, and the Americans were told that the facility had been used in the past for unmanned flights but that *Soyuz 12* had been directed from here. The Soviets planned to direct future Soyuz missions, including ASTP, from Kaliningrad. Low and his colleagues were met by Dr. Abduyevski, Deputy Director of the Control Center, and cosmonaut Yeliseyev, who had been selected as Soyuz Flight Director for ASTP.

Yeliseyev conducted a briefing, using wall charts in Russian and English to explain how the control center functioned. He also described the flow of information within the center and the organization of the flight controllers within the mission operation control room. When the cosmonaut led the U.S. team onto the balcony overlooking the control room, they saw a facility that was strikingly similar to Houston control. "As we entered," Low reported, a video "playback of the *Soyuz 12* countdown was in progress. Across the top of the front wall were a number of clocks showing Moscow time, elapsed time, station acquisition time, and station loss of signal time. The top of the center screen was a world map with a lighted dot indicating the spacecraft location." On a screen to the right was a television picture of the spacecraft and booster at the launch site. From a typewriter keyboard at the back of the room, a technician typed a message that appeared on the bottom half of the right screen—"Welcome American colleagues."

Low was favorably impressed. He further described the facility:

> On the floor were four rows of consoles. The very back row, which is out of sight from the balcony, is for the people who set up the communications and data flow within the Control Center. Also the project director (Bushuyev) will sit in this back row. The flight director is on the next row from the back and is the focal point for all activity in the Control Center. To his left and right, and in the two rows of consoles in front of him, are the various support functions, which are pretty much the same as the functions within our own control center, except that there is no launch vehicle console. Each console has a number of television screens, and the flight controller at that console can call up all sorts of displays.... The communication system allows the flight director to talk to any or all of the other consoles as well as to the back rooms.[16]

The Americans learned that the control center takes over after the spacecraft has separated from the launch vehicle in orbit. Until that time, the flight is under the full control of the launch center. During a question and answer session, the Soviets responded fully to all of the technical queries raised by their visitors. George Low's trip to Moscow had been both useful and informative. Petrov had told the NASA representatives that Star City, the Kaliningrad Control Center, and the Baykonur launch complex would be open to American specialists as necessary. Low was especially pleased to hear this since Tom Stafford, the American ASTP crew commander, had expressed a strong desire to see the actual Soyuz flight hardware during the pre-flight checkout. But still to be decided when the U.S. team departed were the details concerning access for American newspaper and television reporters to those same facilities.

PUBLIC INFORMATION PLANS

October in Moscow was notable for more than just the Mid-Term Review. It was also the first time that NASA Headquarters public affairs personnel attempted to negotiate with the Soviets. Before this, John P. Donnelly and his deputies, Alfred P. Alibrando and Robert J. Shafer, had been participating in the public affairs planning process from a distance. But once they began to take a more active role, expressing their desire for face-to-face discussions with their counterparts, they discovered that their requests—even their very presence—were regarded as an intrusion by Glynn Lunney and the others in Houston who were managing the negotiations. It appeared to Donnelly that the Johnson Spacecraft Center (JSC) was reluctant to share that responsibility with him because the technical teams feared that the introduction of new faces would tend to slow the negotiations. But Donnelly was eager to participate because he was concerned that the technical personnel working for Lunney, in their efforts to meet the launch deadline, might make agreements with the Soviets that could undermine NASA's public affairs policy of full disclosure. Looking back, Low explained the different motivations underlying the negotiation objectives of the Public Affairs and Program Office personnel:

> The project people had essentially one basic goal and that was to make the project succeed. Anything which would make attaining that goal more difficult would and should be opposed by the project people. Thus, a negotiating position which might "upset" their Soviet colleagues would be something that the project people would want to avoid if at all possible. The Public Affairs people on the other hand saw a tremendous opportunity for the United States to show "detente" in its best light. They also saw the need to maintain NASA's open position with the world press and the credibility which NASA has achieved in dealing with the news media. To attain these Public Affairs' goals might entail taking very hard negotiating positions— harder than the technical people would like to have seen on a non-technical issue.[17]

These essentially opposing positions led Low to annunciate two principles in a number of meetings that he held with Donnelly, Shafer, Frutkin, Lee, and Lunney. First, Donnelly could not do anything that would cause the overall negotiations to come to a halt or to fall apart. As Low reported, "this meant that Donnelly would have to check with me before getting himself into a position where hard lines would be drawn—lines that would lead to a major confrontation." In Low's view, "the Public Affairs people did a remarkable job in avoiding such confrontations." Second, the public exposure of the project—especially television—was a major objective of ASTP, accorded as high a priority as everything else in the project except flight safety. This meant that the project people subsequently had to alter flight plans and the like to accommodate in-flight television as required by Public Affairs. Low pointed out that "This was a change from the way we had operated in previous programs, a change which I believed to be necessary for this special project."[18]

Release of information about the joint mission was an area in which NASA personnel had anticipated possible difficulties from the earliest stages. One line in the Low-Kotelnikov agreement of April 1972 had addressed the issue of public information: "A public information plan will be developed which takes into account the obligations and practices of both sides." That phrase combines both genius and difficulty. It gave both sides what they wanted—control over mission-related news—but it did not explain how those two sets of obligations and practices would be reconciled. George Low and his colleagues in the American space agency firmly believed that they could not enter into any agreements that would lead to the alteration of NASA's policy of immediate and full public disclosure. This "real-time" release of audio, video, and other news materials had provided momentary embarrassments in the early days of the program (the failure of MA-1 shortly after lift-off in July 1960 or the sinking of Gus Grissom's spacecraft after his suborbital flight in July 1961), but live television had also covered the most dramatic moments of the space age as well (man's first steps onto the lunar surface or the repairs the first Skylab team made to their damaged laboratory).

Traditionally, the Soviets had released information about their missions only after the fact. And they had not engaged in extensive use of television, preferring instead to tell the space story through newspapers and motion pictures. Therefore, NASA and the Soviet Academy had to reconcile two issues—real-time versus after-the-fact news coverage and reliance upon different media forms. As Headquarters and JSC public affairs representatives were to discover, their requirements for live television broadcasts from Apollo and Soyuz were to be often in conflict with the Soviet desire to make motion pictures of the same events. Skillful negotiations were required to satisfy the obligations and practices of both sides.

Before they could discuss such matters with the Soviets, the Americans needed to agree among themselves. As the Assistant Administrator for Public Affairs, Donnelly had been interested in the public information aspects of the joint mission since the early days of the talks, but his team at Headquarters had only begun to work actively on the topic at the time of the May 1972 Nixon-Kosygin Summit. On 19 May, Bob Shafer had written a memo to Donnelly in which he outlined actions that would have to be taken once the international rendezvous and docking mission was officially announced in Moscow. "First of all, I think we've got to come to terms with the White House and State on the overall public affairs/public information responsibility for the mission. We'll have to take the initiative on that as soon as possible. . . . we need a meeting with Lunney and those of his superiors and subordinates who are actively working with the USSR team." Objectives of such a meeting were:

> to establish our responsibility for public affairs planning and implementation concerned with IRDM [International Rendezvous and Docking Mission]; to define the interface between Public Affairs and project management; to get a better understanding of their working relationship with members of the USSR team so that we can pattern ours accordingly wherever appropriate; and to identify what we believe to be sensitive areas we must accommodate in one manner or another as we proceed so that we do not unwittingly disrupt the progress of the cooperative effort.[19]

Once the role of the Washington Public Affairs Office was finally clarified, Shafer would then recommend that a meeting take place with "appropriate representatives" in the U.S.S.R. to address the development of a public affairs plan. He continued, "All of this, it seems to me, is urgent," and he thought that the Headquarters public affairs staff should be ready to talk with the Soviets "by the end of September at the very latest."[20] But planning public affairs activities for *Apollo 17* and Skylab took much of Donnelly's and Shafer's time, so they were unable to touch base with the people in Houston for nearly a year, and it was not until October 1973 that they had an opportunity to meet with the Soviets.

Meanwhile in Houston, Lunney and the JSC public affairs staff had already taken the initiative in developing procedures for the release of newsworthy information generated during the joint meetings. When the idea of a Public Affairs Plan was first raised, Lunney recommended that the proposed plan be broken into two parts—pre-mission joint activities and actual in-flight joint activities. He reasoned that a single document would be too much to negotiate at one time. Furthermore, it was still too early to clearly define all of the flight-related public affairs activities. By starting with the pre-mission issues, the two teams could learn more about each other's obligations and practices and give the flight planners an opportunity to more fully map out crew activities that would affect the second part of the plan.[21]

By January 1973, John E. Riley of the JSC Public Affairs Office had developed a draft of the first half of the Public Information Plan. This early version stated: "NASA proposes that the ASTP public information activities be governed by two documents." The first was planned to "deal with pre-flight activities, including actions of Joint Working Groups; hardware development and manufacture;[*] training of flight crews, engineers, flight controllers and other personnel involved in the mission; simulators and tests; and control center preparations requiring joint activities." The other document's purpose was to cover flight and post-flight activities.[22] After several versions of Part I of the plan had been drafted, JSC's Public Affairs Officer, John W. King, forwarded the document to Chet Lee in Washington.[23]

Lee circulated the proposal at Headquarters, seeking comments particularly from the Public Affairs and International Affairs Offices. By mid-February, Donnelly and Frutkin approved a revised draft of Jack King's information guideline, and they sent a copy to the American Embassy in Moscow for comment.[24] Lee responded to this "most recent draft of the

---

*Hardware development and manufacture was dropped from subsequent JSC drafts.

proposed . . . information plan" by saying that "it contains new provisions I believe are not conducive to continued smooth relations with our Soviet counterparts."[25] Four major changes bothered him. One of these related to the issue of status reports. The earlier drafts had provided that status reports of "joint working group meetings and joint activities . . . be issued *by the host country* and the *contents approved mutually* prior to release." In the Donnelly-Frutkin approved version, the document read: "*each country may issue* status reports and . . . the *substance of reports* will be *provided in advance* to the *head of the other side's delegation.*" Lee argued that this was "a clear deviation from our methods of operation with the Soviets in our joint meetings to date." He believed that the existing system should be continued since it had "functioned smoothly and to our knowledge has not put any undue constraint on information released to the press."[26] In a joint memo to Lee, Donnelly and Frutkin responded that they did not want the "ASTP Information Plan [to] make US media residents in the USSR dependent on the Soviets for news of ASTP activities." In their opinion, the original proposal did so since the host country could determine the content and frequency of status reports, "based upon its obligations and usual practices." As an alternative to their proposal, Donnelly and Frutkin suggested the following, which Lee found acceptable:

> During meetings of Joint Working Groups and during joint activities of flight crews and other mission personnel, the US and USSR heads of delegations may issue joint status reports to the news media. Joint status reports are expected to be the usual procedure, but if either side wishes to issue status reports to news media on its own side, in accord with its normal obligations and usual practices, it may do so after notice of the substance of the release to the head of the other side's delegation present.[27]

After similar horse trading at Headquarters on the other three points, which all dealt with different aspects of the same question—equal treatment of American and Soviet press representatives in the Soviet Union—JSC was permitted to give NASA's draft of "ASTP Public Information Plan Part I" to the Soviets during the March 1973 meetings. Later during the July sessions in Houston, the Soviets said that they had no basic objections to the draft text but that they did want to modify some of the language. They would submit their comments by the end of August. In July, there had also been some discussion on the joint production of a post-flight motion picture that would summarize the project. That film, the signing of Part I, and discussions of the content and schedule to be followed in negotiating Part II were placed on the agenda for October.[28]

Although Glynn Lunney had planned to send only Jack King to Moscow, John Donnelly asked that he too be permitted to participate.

Donnelly specifically wanted to go because there had been no response from the Soviets since the July meeting regarding the public affairs topics. And he insisted that Bob Shafer accompany him so that they could discuss television issues with the Soviets. Both men were worried that if left to the technical people, ASTP might occur in the dark, and they wanted the broadest possible television coverage for this mission. After considerable discussion, during which Donnelly and Shafer took their case to George Low, the two men departed for Moscow with King.[29]

Upon their arrival, they were met at the airport by Nikolai Vasilyevich Khabarin from the Council for International Cooperation in Space Exploration and Use (Intercosmos). During their ride into the city, Khabarin grilled Donnelly, apparently so he could determine who this new American was, how much authority he had, and where he fit into the NASA hierarchy. Shafer recalled later that the question and answer session was getting nowhere until Arnold Frutkin's name came into the discussion. Khabarin asked Donnelly how his position compared to Frutkin's. Donnelly told him that they were at the same level, both being Assistant Administrators. Khabarin responded that Frutkin reported directly to George Low, and Donnelly came back with, "So do I." This discussion, which went on to include questions concerning the relative sizes of the staffs working for the two men and so on, gave the Soviets some understanding for how these new faces fit into the NASA scheme of things.

Donnelly compared his first meeting that October with his Soviet counterpart, Igor Pavlovich Rumyantsev, to the sparring two boxers do the first time they meet in the ring. "We were feeling each other out. Clearly we didn't trust them, and they didn't trust us."[30] Questions of trust were to surface several times during this meeting. Shafer recalled that Rumyantsev came into the room where they had all gathered and made a formal statement about how good it was for them to be together and to be working towards this joint flight. But he wanted to know why NASA had called for this meeting and what exactly they wished to discuss. Donnelly explained that they were there to complete work on Part I of the Public Affairs Plan, to discuss the joint movie, and to begin work on outlining Part II. The Americans spent the rest of the day explaining to the Soviets what they meant by public affairs and what NASA hoped to accomplish in negotiating both halves of the plan. Rumyantsev, an Intercosmos staff member, was a professional negotiator in international matters, but he was not an expert on public affairs. It took a while for him and the other Soviets to fully comprehend what Donnelly and Shafer meant by full and open disclosure of information to the press. It also took time before they were convinced that neither NASA nor the American government in any manner managed the press. The obligations and practices of the United States and the Soviet Union were quite different and not easy to reconcile.

Donnelly's negotiating stance with Rumyantsev was by his own admission hard-nosed. And as a result, the process was a slow one. After several days of talks, Donnelly discussed their progress on Part I with George Low on the evening of 15 October. Low then decided to meet directly with Petrov, who as Chairman of Intercosmos was Rumyantsev's boss. On the 17th, Frutkin and Low met in their hotel room and tried to clarify a plan for their discussions with the Soviets. That next morning saw Frutkin and Low come to agreement over the Public Information Plan with Petrov, Vereshchetin, and Rumyantsev.* While Low and Petrov did not sign the resulting document, preferring to wait two weeks for formal ratification, Donnelly and Rumyantsev affixed their signatures to "ASTP Public Information Plan Part I," ASTP 20 050, as an indication of good faith, as did Lunney and Bushuyev. Final ratification of this much-debated plan came in November when Kotelnikov notified Low that the Soviet side accepted the modified language drafted in Moscow.[31]

Donnelly and Shafter learned from their trip to Moscow that negotiation was more art than science. Two other sticky topics discussed during October illustrated that point. The Soviets dearly wanted a jointly produced motion picture describing ASTP. A jointly produced movie would be another visible indication of cooperation, and equally important, the two countries would share the cost of producing the film. Furthermore, the idea of a movie was particularly attractive to the Soviets since they could show it in State theaters as a major feature attraction, but NASA did not expect U.S. movie houses to desire such a production, and it seemed equally unlikely that the television networks would buy the lengthy documentary. Such a film fit one system, but it did not meet the obligations and practices of the other. Lunney advised Bushuyev back in September 1973 that the Americans did not favor this project, but Donnelly and Shafer had to tell them again that NASA would not enter into such an enterprise. Being the bearers of such bad news did not enhance their rapport with their newly found colleagues, nor did their insistence on a second issue—equal treatment for American newsmen covering ASTP in the U.S.S.R.[32]

Early in their talks on the 10th of October, Rumyantsev had told the Americans that it would not be possible to invite every American correspondent who resided in Moscow to all ASTP press conferences. When Donnelly asked why, Rumyantsev said that the room where such gatherings

*Also present were Donnelly, Lee, Lunney, Bushuyev, A. I. Tsarev, and V. I. Kozorev.

were held was too small to accommodate them all. Donnelly said that the Soviets would simply have to find a larger room, but his counterpart replied that it was impossible to alter the location of the briefings—all press conferences were held in that room! He indicated that the Soviet solution to the space problem was to limit the size of the press delegation. Donnelly was told that in previous technical negotiations, the Americans—over the objections of Eli Flamm, the Press Attache at the American Embassy—had agreed to limits on the size of press contingents, as long as equal numbers from both sides were permitted to attend. Donnelly argued against such restrictions, saying that they were only valid when genuine physical restraints existed, such as those at the training facilities at Star City. But he was against arbitrarily imposed limits, holding them unreasonable and contrary to the spirit of the Information Plan they were trying to establish. According to Donnelly, this was nothing less than "censorship by selectivity." The men suspended their negotiations for the afternoon at a loss for agreement.[33]

On the following morning, Rumyantsev approached the Americans. As Shafer recollected, the Soviet negotiator proclaimed, "Mister Donnelly, there is an answer! It is called a pool!" In making their proposed alteration, the Soviets had a completely different understanding of that concept than did the Americans. "Their suggestion was that we dictate the pool—that we go to the U.S. correspondents and say, 'Form a pool and take it here.'" Donnelly told Rumyantsev that press pools in the West did not work that way. The news representatives selected the members of a pool delegation when they had been informed that a particular activity would allow only a small group to attend. NASA could not and would not determine pool membership. After a discussion that lasted nearly the whole day, a breakthrough occurred when Donnelly inquired if the source of their problem lay with the size of the Soviet delegation and not with the size of the American contingent. Rumyantsev replied that Donnelly was beginning to understand.[34] To take into account the Soviets' desire to limit the number of Soviet correspondents that might be invited to a news briefing, the following language was drafted into Part I of the plan: "For each joint-activities event, each side may designate the number of accredited press from its side to be invited, taking into account its own customs and traditions."[35]

Donnelly also persuaded Rumyantsev to accept another principle—"with the exception of situations in which physical limitations make it impossible, all accredited U.S. correspondents would be invited to pre-mission news events."[36] "In situations where physical or technical limitations require, the host country may propose that the news media establish pooled coverage."[37] Despite this agreement, Shafer later wrote Donnelly that the question of full representation for American media personnel in Moscow had been the "principal issue which divided the two sides during our negotiations of Part I ... [and] seems likely to reappear from time to time."[38]

The Americans' concern about equal treatment for the American press was well founded. American correspondents, with their noses for news and penchants for investigative reporting, did not always have the best of relations with the Soviet government. These newsmen were seldom happy with the handouts they received from government news agencies, and the Soviets rewarded only those reporters whose stories were positive. Americans often criticized the Soviet practice of late night phone calls to select reporters concerning news events that would occur the following day. In the case of ASTP, they wanted free access to news events, and they expected NASA to protect their interests. This posed several problems. NASA could try to guarantee them full access to joint events, but the agency could not assist them in their desire to cover unilateral Soviet activities. The Information Plan stated: "Decisions related to news media access to independent activities are the unilateral responsibility of each country in accordance with its established traditions and practices."[39] Nor could NASA shield the resident media representatives in Moscow from non-resident correspondents who managed to get special visas that allowed them to interview cosmonauts or members of the Soviet Academy. Donnelly argued that the resident press would have to fight those battles through their home offices; after all, competition was one of the aspects of a free press.[40] Though they could not protect American correspondents from each other, NASA public affairs people could ensure that they had equal access to information.

Equal access to ASTP news events only came with much hard work. The talks held in October 1973 were just the beginning. Part II of the Public Information Plan (especially the discussion of real time television) was to involve far more complex and lengthy negotiations. A final agreement on the mission-related news coverage would not be completed until three months before the launch, and drafting a plan was only the first step. As Jack Riley discovered in November 1973, a formal plan did not exist for the Soviets until it was officially ratified. A couple of days before the end of the astronauts' first visit to the Soviet Union in November 1973, Valentin Ivanovich Kozorev, Scientific Secretary of Intercosmos, approached Riley to tell him that the Soviets would like to use one of the photographs that they had received from NASA during the June-July 1973 cosmonaut visit to JSC to illustrate an article they planned to publish. Kozorev had been instructed to obtain Riley's permission to use that photograph. Riley reported on this conversation:

> I responded that they were free to use any of the photos provided by NASA and that there were no restrictions on their use for news purposes. I said that we planned similar use of the photos we received from them.
>
> Kozorev thanked me profusely and then said that he regretted that he could not be as generous as I had been. He said that we would be given five or six photos before we left and that we would require permission from them on a picture by picture basis.[41]

Kozorev referred to the photographs taken at the control center in Kaliningrad following the Mid-Term Review. He said that "Dr. Lunney" had asked the Professor for permission to release those illustrations to the American press but Bushuyev still had not secured authorization. When Riley mentioned that the release of such items should be covered by the photography exchange section of Part I of the Information Plan, he "was told for the first time flat out . . . that the Soviets did not consider the plan to be in effect yet." Kozorev indicated that the Soviets had sent the plan to Washington for Low's signature but it had not been returned yet. Until they had a signed copy in their hands, the plan was not operative.

Kozorev apparently believed that his position would cause Riley to reconsider his "generosity," because he again asked about releasing the photographs they had received from Houston. Riley told Donnelly in a memo, "I got the impression that he was somewhat ill at ease with his position and would have felt justified if I had changed my mind and insisted that they too would have to get permission for each individual photo." Instead Riley told Kozorev that they both knew that the plan had been approved and were only waiting for formal notification. "I intend to operate under the spirit of the plan even though formal signed documents were not yet available, and I repeated that they were free to use photos obtained from NASA."[42]

Kozorev and Riley also had a second discussion dealing with the participation in ASTP news conferences of correspondents from countries other than the United States and the Soviet Union. Several days before the 29 November briefing marking the end of the astronaut familiarization tour, Kozorev asked Riley whether NASA objected to newsmen from other countries attending. Riley told him that it was NASA's policy to welcome any accredited reporter, irrespective of nationality. Again Kozorev thanked the American public affairs representative and added that he would tell the several foreign correspondents that they could participate. Early on the morning after this press conference, however, Riley received a telephone call from a West German reporter who asked if there would be an opportunity for him to talk with the astronauts before they departed. Riley later informed Donnelly:

> I responded that we were leaving that day and that a news conference had been held the previous day. He said that he knew about that conference but when he asked to attend, he was told by Soviet authorities that NASA had requested that only American and Soviet correspondents be permitted to cover the conference and, therefore, they could not permit him to attend.[43]

Riley passed the reporter's complaint on to Donnelly with the information that several East European reporters had been present during the news session with the crews. Riley went further to note that Eli Flamm at the Embassy could not understand the exclusion of this particular individual since he normally had an excellent relationship with the Soviets. By early 1974, Donnelly and the others working on the NASA public affairs team had learned that they had a difficult task ahead of them.

A REPORT TO CONGRESS

Despite these difficulties with public affairs, George Low was still genuinely optimistic about the prospects for a successful flight. And even on topics such as public affairs, there was hope since NASA had not given up any of its traditional openness and since the Soviets seemed willing to negotiate in good faith. So upon his return from the Soviet Union, Low touched base with Chairman Olin Teague of the House Committee on Science and Astronautics. In a letter, he told Teague about his various chats with the Soviet space leaders, summarized the results of the joint talks, and described his visits to the space facilities. Among the significant results produced by the Working Group sessions, Low noted:

> It was agreed to conduct five joint scientific experiments on the mission involving biological interaction, microbial exchange, a multipurpose furnace, artificial solar eclipse and ultraviolet absorption.[*]
> It was agreed that there would be reciprocal participation of US and USSR specialists in preflight fit checks at the launch site of compatible hardware such as TV cameras, speaker box, etc. in the flight Soyuz and Apollo spacecraft.
> With regard to the Apollo VHF/AM communication equipment, it was agreed that the US specialists will participate in the checkout of the equipment after delivery to the USSR and also during the preflight checkout of this equipment in the flight Soyuz at Baikonur, the Soviet launch site.
> In addition to these agreements, improvement was noted in the prepara-

*Detailed descriptions of ASTP experiments are presented in appendix E.

tion of plans and documents, particularly in the Communications Working Group. All documentation is essentially now on schedule.[44]

Low advised Teague that these agreements would "materially contribute to a successful mission and ... [were] a good indication of the Soviets' commitment to making this mission a success." Turning to the Mid-Term Review, the Deputy Administrator reported that "The Project Technical Directors ... and the Working Group Chairman made detailed presentations" to Academician Petrov and me "and responded to all questions." As a consequence of this exercise they had "concluded that the progress made and the quality of the joint work to date [gave them] high confidence that the scheduled launch date [would] be met."[45]

Privately Low was equally confident of success, particularly considering the international scene at the time. On 6 October, the fourth major war between the Arab states and Israel had erupted when troops from Egypt, Syria, Iraq, and several other countries attacked. The Yom Kippur War had raged throughout the stay of the Americans in Moscow, with the U.S.S.R. and the U.S. airlifting arms to the opposing sides. And on 17 October, the Organization of Arab Petroleum Exporting Countries had announced a coordinated program of oil production and export cuts to those nations that supported Israel.[46] Jim Jaax of Working Group 5 recalled that he and his colleagues only learned about the war when one of the interpreters read of the conflict in a Soviet newspaper during a bus ride from the Rossiya Hotel to the Institute of Space Research. In their isolation, they had had no other indications that a war was being fought in the Middle East.[47]

Potentially as disruptive to the Soviet-American space efforts had been the U.S. National Academy of Sciences' protests to the Soviet Academy concerning the "heightened campaign of condemnation" being waged against dissenting Academician Andrei Dmitriyevich Sakharov. The President of the American Academy had cabled Keldysh in September 1973 regarding the matter and had subsequently published the text of his message in *Science* on 21 September.[48] Low, commenting on these problems, said, "Although we were in Moscow during an international crisis and during the exchange of letters between the U.S. and Soviet Academies on the Sakharov affair, neither of these subjects came up at any time during our visit."[49] Low noted that one *New York Times* article concluded: "the warm treatment of Mr. Low and a team of American specialists, working with their Soviet counterparts to prepare for the Apollo-Soyuz mission, was read as a deliberate gesture by Moscow to emphasize its interest in Soviet-American cooperation and detente despite the frictions of the Middle East conflict."[50]

At the end of 1973, a successful flight in July 1975 seemed probable. The Soviet and American teams had made considerable technical progress and, despite the tight schedules and heavy work loads, were confident. ASTP appeared to be politically possible as well, since major international crises had not intruded into the world of the Working Groups. The year 1973 had also seen the two crews begin to work out the details of joint training. The day of rendezvous was approaching.

IX

Preparing for the Mission

With a 30 January 1973 announcement, the U.S. was first to make public their ASTP crew assignments. Brigadier General Thomas P. Stafford, a veteran of three flights and Deputy Director of Flight Crew Operations since 1971, would lead the prime crew. The Command Module Pilot, Vance D. Brand, had been backup Command Module Pilot for *Apollo 15,* and at the time of his appointment to the ASTP crew was backup commander for the second and third manned Skylab missions. Donald K. "Deke" Slayton would fill the position of Docking Module Pilot. Since a heartbeat irregularity had deprived him of a flight on Project Mercury, Slayton as Director of Flight Crew Operations had played a key role in the management of crew selection and training at NASA. In March 1972, following a comprehensive series of medical examinations, Slayton was restored to full flight status. At 48, Deke was six years older than his crew mates and the oldest man yet to be selected for a space trip.[1]

Stafford's crew was backed by Alan L. Bean, Ronald E. Evans, and Jack R. Lousma. Bean, the fourth man to walk on the moon, had been in the space program since October 1963. This exacting, hard working naval officer was scheduled to command the second Skylab crew, which was preparing for a July 1973 launch. Evans, a Navy captain, had been Command Module Pilot for *Apollo 17,* and Lousma, a Marine Corps major, was preparing to accompany Bean on the flight to Skylab.[2]

Richard H. Truly, Robert F. Overmyer, Robert L. Crippen, and Karol J. Bobko would assist the flight crews in their training. These four support crewmen had transferred to NASA in 1969 following the cancellation of the Manned Orbiting Laboratory, a Department of Defense program. During the preparations for ASTP, they would stand-in for the prime crews in a number a time-consuming but critical activities, such as mission planning and lengthy manned tests of the flight hardware. During the flight, Truly, Crippen, and Bobko would act as spacecraft communicators from the mission control center in Houston. Overmyer, who was to work extensively with the Soviets in mission planning and crew training, would be one of the technical advisers at the mission control center in Kaliningrad during the flight. Together these ten men would work as a team for the American half of the joint flight.[3]

On 1 February, Glynn Lunney introduced the American ASTP astronauts to the press. "The naming of the crew . . . is always an exciting time for us in the manned space business and I think especially in this project . . . [since] it indicates the progress that has been made on the planning for this activity," Lunney said. He turned the microphone over to Stafford, who indicated that it was great to have been named to a crew and that he was looking forward to getting away "from some of the paper work for a while and get[ting] back to simulation and training." For those critics who saw ASTP as simply an easy orbital flight, Stafford had a few words of caution.

> The mission . . . is probably going to be one of the [most] difficult the manned space flight team has ever undertaken because it involves a different country, a different language, different operating techniques, and it's just . . . slow and painstaking . . . to work out all these [details].[4]

Stafford saw ASTP as a great challenge and a means of opening doors to a better future. Brand, who was fully occupied with training for Skylab, told the reporters that he agreed with Stafford's evaluation of the mission. He hoped that his Skylab training in the command module simulators would help him in preparations for ASTP. Once the last two flights to the space station were completed, he would turn his full attention to the joint mission, concentrating especially on learning Russian.

Since his restoration to flying status, Slayton had been working for a place on the ASTP crew. During the summer of 1972, Slayton, Bobko, and Crippen had been studying the Russian language. Bobko and Crippen spent their spare time on the language during a 56-day Skylab Medical Experiment Altitude Test, and Slayton had thought that some knowledge of the language might improve his chances for selection, as well. In his remarks to the press, Slayton began by thanking all those who had over the years tried to get him certified once again for flying and especially Dr. William K. Douglas and Robert R. Gilruth, who had worked to keep him flying 12 years earlier. "If I had no other reason to fly this mission," Slayton added, "I'd want to vindicate their good judgment." He also thanked Dr. Charles A. Berry of NASA and Dr. Hal Mankin of the Mayo Clinic for their efforts that led to his being available for this crew.

> And third, of course, and not least, . . . on behalf of all the crew I'd like to thank Chris Kraft for putting us on the flight. I think Chris had a tougher decision in getting the crew [for] this flight than I ever had picking flight crews, because we've got 39 guys . . . who would have like to flown it.[5]

Reflecting on the twelve years that he had sat behind a desk and watched other men fly, Slayton said that all in all he had been "pretty fortunate" in working for NASA. He had missed out on a lot of the adventure of space flight but he had also missed the tragedy—the snow goose that had wrecked Theodore Freeman's T-38 jet trainer, C. C. Williams' "bum aircraft," and the fire that had gutted Apollo 204. He told his audience that he had stayed with the space program because he was there to fly. He had expected to be returned to flight status all along; it had just taken longer than he had anticipated. For the "last 20 or 30 years I've been paid to fly, which is the thing I love most." Now, Deke Slayton was looking forward to his first space flight as a "mature rookie"; he hoped "to fly a couple of more after this one."[6]

Soviet crew announcements for the 1975 flight came on 24 May to coincide with the opening of the 1973 Paris Air Show. Alexei Arkhipovich Leonov and Valeriy Nikolayevich Kubasov were chosen as the prime crew. Leonov was a veteran of the *Voskhod II* flight, during which he had performed the first extravehicular excursion. Kubasov had been the backup technical scientist for *Soyuz 5* and flight engineer on *Soyuz 6*. He would fill that role again in the ASTP mission, while Leonov would command their craft.

Prime crew members for the second Soyuz were Anatoliy Vasilyevich Filipchenko and Nikolay Nikolayevich Rukavishnikov. Filipchenko, who had become a cosmonaut in 1963, was the backup command pilot for *Soyuz 4* and command pilot on *Soyuz 7*. Rukavishnikov joined the cosmonaut team in 1967 and became the test engineer for *Soyuz 10*. Backup crewmen were Vladimir Aleksandrovich Dzhanibekov, Boris Dmitriyevich Andreyev, Yuri Viktorovich Romanenko, and Aleksandr Sergeyevich Ivanchenko—all rookies who had joined the cosmonaut corps in 1970. This public announcement of crew assignments was a first for the Soviets, who in the past had never identified cosmonauts until they had actually flown.[7]

THE 1973 PARIS AIR SHOW

According to *Aviation Week and Space Technology,* a full-scale representation of the Apollo command and service module (CSM) and docking module joined with a Soyuz spacecraft formed the "focal point for the 30th Paris Air Show" held on 25 May to 3 June 1973.[8] The Soviets and Americans, represented by Igor Gregoryevich Pochitalin and Charles A. Biggs, had decided at the December 1972 Moscow negotiations to prepare an exhibit that would show the aerospace world the progress made toward the joint flight.[9] Although plans for the display came too late for the show management to provide space in one of the permanent pavilions, they agreed to set aside some land usually reserved for parking on which the two teams could erect a temporary fabric dome suspended from a geodesic frame. Inside the 930 square meters, NASA planned to display a refurbished CSM that had been used in vibration tests, and the Soviets would assemble a Soyuz from leftover test hardware.[10]

Soviet and American workers took five days to put together the dome and mate the spacecraft. American co-director Biggs quipped to an *Aviation Week* writer that he hoped the ASTP crews would get the job done a lot faster when they met in orbit. Pochitalin estimated that 2000 to 3000 people a day would visit the exhibit to see the mockups and the captioned photographs explaining the project in English, French, and Russian.[11] To help publicize the presentation, cosmonauts Leonov, Kubasov, Filipchenko, and Yeliseyev met with the *Apollo 17* crew—Cernan, Evans, and Schmitt— and their ASTP crewmate Tom Stafford. Brand and Slayton were unable to leave Houston because they were involved in the first manned visit to Skylab. Stafford and Leonov managed to fill in for the missing astronauts, posing for photographers and answering questions from the media and visiting dignitaries.

The ASTP exhibit in France was well received by the public until the very last hours it was open. The joint pavilion, the first exhibit to be seen upon entering the show grounds, stood in stark contrast to the military and commercial rivalry involved in other displays. In an editorial, *Aviation Week's* Robert Hotz indicated that the dramatic Apollo-Soyuz "docking display" was a symbolic expression of growing cooperation in space. A happy and friendly event, the Paris presentation attracted 400 000 visitors, surpassing the unexpected attendance thirteen times. The camaraderie displayed by the cosmonauts and astronauts was a welcome sign; they had not always been so close.[12]

Workmen rest alongside the Apollo and Soyuz mockups, as the joint ASTP exhibit is being set up for the opening of the 1973 Paris Air Show.

RIVALS AND FRIENDS

In the dozen years that had followed Yuri Gagarin's flight, the astronauts and cosmonauts had met a number of times. But these first meetings had been shadowed by the cold war. John Glenn and Gherman Titov had been the first rival spacemen to meet and exchange views, at the May 1962 COSPAR gathering in Washington. After the two men and their wives toured the capital and made a social call on President Kennedy at the White House, the space travelers held a news conference. Titov was circumspect in answering questions about his Vostok craft and would discuss space cooperation only in the context of disarmament.[13]

Three years passed before the next meeting. In June 1965, a very cool handshake was exchanged by three Americans—Vice President Hubert H. Humphrey, astronauts James A. McDivitt, and Edward H. White—and Yuri Gagarin. This encounter at a Paris Air Show luncheon took place after a formal meeting between these men had failed to materialize.[14] In September of the same year, Gordon Cooper and Pete Conrad had a much warmer conversation with Leonov and Belyayev at an international meeting in Athens. As they exchanged lapel pins, the men agreed that they would have to meet again and compare notes about space flight.[15]

As the years passed, the cosmonauts and astronauts began to socialize more freely. At the 1967 Paris Air Show, Mike Collins and Dave Scott drank a vodka toast with cosmonauts Belyayev and Feoktistov, and Scott called for "greater cooperation between the United States and the Soviet Union." To which, Belyayev replied, "Yes, in space."[16] In 1969, McDivitt, Scott, and Schweickart gave a tour of the *Apollo 8* command module to cosmonauts Shatalov and Yeliseyev, who in turn treated the Americans to vodka and caviar served aboard a Yak-40 airliner being displayed at the Soviets' Paris Air Show pavilion.[17]

A month later, in July 1969, *Apollo 8* commander Frank Borman and his family were given an extensive sightseeing trip in the Soviet Union. Titov, Feoktistov, and Beregovoy escorted the Bormans around Star City and other space facilities. During their visit, Borman renewed the subject of cooperation, mentioning the possibility of joint missions in very general terms.[18] American astronauts hosted a reciprocal goodwill trip for Beregovoy and Feoktistov at the end of October. During their two weeks of crisscrossing the United States, they visited an American Institute of Aeronautics and Astronautics conference at Anaheim, California, had a brief chat with President Nixon, and were guests of honor at a dinner in Houston thrown by 30 astronauts.[19]

By the time NASA and the Soviet Academy of Sciences began talking about a rendezvous and docking mission, meetings between astronauts and cosmonauts were almost commonplace. During 1970-1971, a half dozen meetings took place, in addition to Stafford's trip to Moscow for the funeral of the *Soyuz 11* crew and his subsequent visit to Star City in October. The atmosphere was considerably more friendly than in the Mercury-Vostok days. For example, when V. A. Shatalov was in Houston in 1972, he visited with Dave Scott and his family in their home, where they compared notes on such topics as child rearing and education. And when Scott led a delegation to Moscow in June 1973 as Lunney's Technical Assistant, Shatalov and seven other cosmonauts gave the astronaut and two members of his team a complete tour of the facilities at Star City, including an opportunity to examine the Salyut trainer.[20] Though from different social, economic, and political worlds, the astronauts and cosmonauts had much in common, both as professionals and human beings.

SPACECRAFT FAMILIARIZATION

At the March 1973 Working Group meetings, Shatalov, in charge of cosmonaut training, and Bob Overmyer worked to pull together a "Crew and Ground Personnel Training Plan," ASTP 40 700, which defined the study and practice sessions that would be held for crews, flight controllers, and other control center staff. They agreed that there would be three training sessions in the U.S.S.R. and three in the United States. Instead of trying to second-guess the curricula for the second and third meetings, they planned to let the host country advise its guest team a month or so in advance of the training agenda. Any updated material should be added to the training plan by a document change notice. The length of the meetings would be kept flexible in an effort to provide an adequate stay for the orientations but not waste time. Cosmonauts would visit Houston in July, and the astronauts hoped to travel to Star City in the fall.[21]

In anticipation of the familiarization visits, the Soviets lowered another barrier in their space program during June 1973—they invited a delegation of American aerospace writers to visit Zvezdny Gorodok (Star City). Donald Winston and Robert Hotz of *Aviation Week* and John Shaw and Jerry Hannafin of *Time* were among those who toured the facilities. The correspondents were impressed by the vitality of the Soviet space program. Shaw reported: "Unlike the Johnson Space Center[*] in Houston, where major retrenchments are underway, Star City is rapidly expanding—a sure sign of the Soviet Union's continued dedication to the exploration of space."[22] *Aviation Week*

*The Manned Spacecraft Center had been renamed in honor of Lyndon B. Johnson in a congressional act signed by President Nixon on 17 Feb. 1973.

agreed: "The building activity underscores Russia's determination to retain capability in manned space missions despite a series of set backs that has forced an unscheduled two-year hiatus in manned orbital flights by that country."[23] From their conversations with General Shatalov, the reporters learned that the cosmonauts were making special preparations for work with their American counterparts.

Ten cosmonauts—the ASTP crews, Yeliseyev, and Shatalov—and four Soviet training specialists arrived in Houston with the rest of the Working Groups on 8 July 1973. At the request of the Soviets, a large block of time was set aside for them to listen to taped recordings of actual Apollo air-to-ground conversations. While getting a better idea of what they would be hearing during the mission, they also reviewed the "Glossary of Conversational Expressions between Cosmonauts and Astronauts during ASTP," which was a step toward standardizing the mission language. This work was followed by a series of video taped presentations on the command and docking modules that had been prepared by Rockwell International and narrated by Alex Sementovsky, one of Rockwell's Russian-speaking engineers.

Each of the video lectures was followed by a question and answer period. By presenting the basic material in Russian the first time, considerable training time was saved. These tapes, covering the design and operation of the Apollo spacecraft systems, were supplemented by handouts with the same material and illustrations, and both were taken home so the cosmonauts could spend as much time with the topics as they felt was necessary. After participating in a discussion with other Working Group 1 members on the "Joint Crew Activities Plan," ASTP 40 301, each cosmonaut was given a ride in the command module simulator, so he could get a better understanding of how some of the command module systems worked and observe the simulator's capabilities. Following that exercise, the Soviets had an opportunity to examine the docking module mockup and study its systems.[24]

On 14 July, Lunney and Bushuyev accompanied the cosmonauts to Rockwell International's factory at Downey. Once there, the Professor and his comrades were able to observe work being done on CSM 111, examine a high fidelity mockup of the docking module, and study the effects of reentry on several command modules stored at Downey. Their factory tour ended with a demonstration of the Apollo docking and entry simulators. The Soviets returned to Houston for another week of activities before departing for Moscow on 21 July.[25]

Reporters speculated that the Soviets had left when they did to avoid having to accept Glynn Lunney's invitation to watch the 28 July launch of

The Soviets are shown the Apollo 17 *heatshield, which had been removed from the command and service module in the rear. From left to right, A. S. Ivanchenkov, A. S. Yeliseyev, N. N. Rukavishnikov, V. N. Kubasov, K. D. Bushuyev, T. P. Stafford, A. Tatistcheff, and C. W. Helms.*

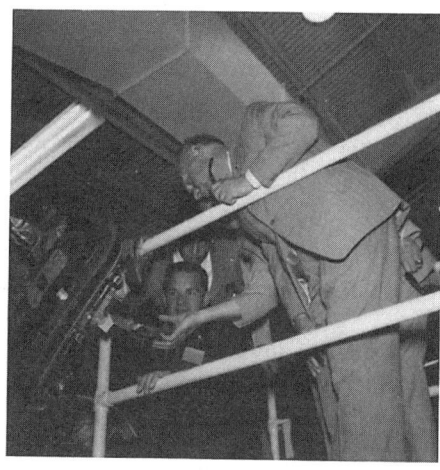

Tom Stafford, behind Professor Bushuyev, explains the functioning of the hatch quick opening mechanism to the Professor and Cosmonauts Kubasov and Ivanchenkov.

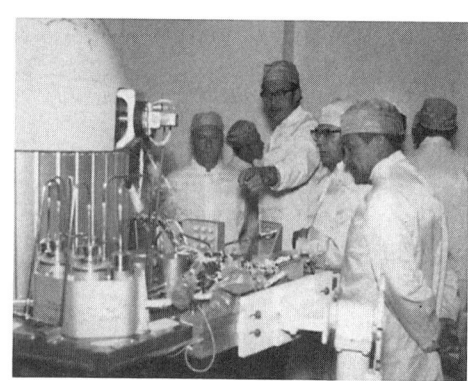

George Merrick, Vice President, Space Division, Rockwell International, explains the cryogenic equipment to be installed in the service module. His audience consists of, from left to right, Bushuyev, Sementovsky, and Filipchenko.

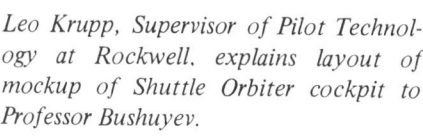

Leo Krupp, Supervisor of Pilot Technology at Rockwell, explains layout of mockup of Shuttle Orbiter cockpit to Professor Bushuyev.

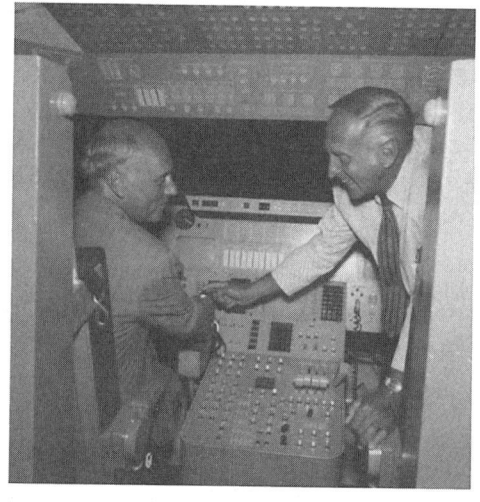

the second Skylab crew. Bushuyev, when questioned at the press conference closing the two-week stay, said that they had accomplished all of their objectives and that remaining for a third week would have presented "difficulties for some of [our] side because some of the participants in our delegation have duties at home which cannot be postponed."[26] Whether the Soviet spokesman was making excuses or whether some members of his team were going home to prepare for the launch of *Soyuz 12* is unclear. Nevertheless, the termination of the meetings was completely in line with the agreement not to waste time. The cosmonauts had completed their work, and they and the astronauts would begin readying themselves for their Moscow session.

DO YOU SPEAK RUSSIAN?

During the winter of 1972 the Manned Spacecraft Center (MSC) created the civil service position of Russian Language Officer, and Nicholas Timacheff filled that slot on 2 January 1973. His duties were varied and included supervising the interpreters who worked during the joint missions, the contractor who provided translators who in turn worked on the documentation and the Russian language training for the American flight crews. In addition, Timacheff and his assistant, Donalyn Epstein, filled in as interpreters at meetings and telecons, reviewed movie scripts, and oversaw the compilation of a commonly accepted English-Russian/Russian-English glossary for ASTP.

Language training was a major challenge for the crews, despite Tom Stafford's comments to the press during the July meeting that Russian was *nyet problém*.[27] During the fall of 1972, language training had been discussed in Washington and Houston, and all parties agreed that a formal program of instruction was needed to supplement the personal studies in which some of the astronauts were engaged. There appeared to be three possible approaches—enroll the astronauts in a formal course such as those offered by the State Department Foreign Service Institute or the Department of Defense language schools; contract with a university to provide instruction; or bring instructors to the space center to work with the crewmembers. The language schools required a full year of residential study, and the crew obviously could not leave their other activities for that length of time. Johnson Spacecraft Center (JSC) management also preferred to keep the work within government circles in an effort to keep costs down; that ruled out universities. Early in 1973, Lunney, Timacheff, and the others finally agreed to try having instructors from the Foreign Service Institute work at JSC with Slayton and Stafford for short stretches to see if this approach would be satisfactory. Vance Brand and the backup crewmen would begin language studies once their commitments to Skylab were completed.[28]

By the time the Soviets arrived for the July training sessions, Slayton had received nearly 140 hours of Russian instruction, and Stafford 115. Between August and the November trip to Star City, Slayton raised his total to 245, and Stafford to 225. A year earlier, Deke had noted in a memo that he hoped "all will consider adequate" 300 hours of language training.[29] But having nearly reached that point, Slayton and Stafford realized that many, many more hours of studying Russian would have to precede the flight.

STUDYING IN STAR CITY

The American ASTP crews visited the U.S.S.R. in mid-November 1973 for eleven days of spacecraft familiarization. In addition to the Soviet crews and training specialists, the U.S. prime and backup crews, two support crewmen (Overmyer and Bobko), Gene Cernan,* Nick Timacheff, and John E. Riley, a Public Affairs Officer from JSC, were present at the Yu. A. Gagarin Cosmonaut Training Center in Star City. Following the pattern set by the JSC crew training staff in July, the Soviets presented nine video-taped lectures to the visiting astronauts. Starting with a description of the Soyuz flight from launch to rendezvous, undocking to landing, the television tapes covered a number of significant aspects of the Soviet craft. This set of lectures and their subsequent study of the mockups and trainers gave the astronauts a better feel for the Soyuz flight control systems and onboard displays and the environmental controls for oxygen generation, temperature levels, and food, water, and waste management. Details of radio and television communications equipment closed the presentations.[30]

During the course of their stay, the astronauts had ample opportunity to become acquainted with the Soyuz general purpose and docking simulators and the Soyuz and Salyut mockups. They listened to recordings of air-to-ground conversations from an earlier Soyuz mission and discussed an even longer list of common terms that would be used during the flight. Finally, they went over with their hosts the "Joint Crew Activities Plan" and the "On Board Joint Operations Instructions." The Soviets gave them copies of the video tapes, hardbound copies of the scripts, which were illustrated with line drawings, and tape recordings of the air-to-ground communications.[31]

---

*Cernan had replaced Dave Scott as Lunney's special assistant after Scott left in Aug. 1973 to become Director of NASA's Flight Research Center, Edwards, Calif.

PREPARING FOR THE MISSION

Rest stop during trip from Star City to Moscow provides crewmen with a chance for a snowball fight, November 1973.

In addition to the classroom instruction, the astronauts participated in cosmonaut physical training activities and social events. During non-working hours, the two space teams jogged, swam, and shared steam baths. At one point during a rest stop on their journey between Star City and Moscow, they engaged in a snowball fight, a rare treat for the men from semitropical Houston. Their crowded agenda also included a trip to the ballet, where the astronauts were literally showered with bouquets of roses by young ballerinas. But the snow covered fields of Russia were quickly followed by the mild winter of Gulf Coast Texas. And with the return to Houston came more detailed preparations and training for the flight.

TESTING HARDWARE

While the crews were beginning to study spacecraft systems related to the mission, progress continued in readying the hardware they would ultimately fly. From 16 September to 24 December, Working Group 3 conducted tests with the developmental version of the docking system. This first piece of full scale equipment had been built from the engineering drawings as they had been perfected to that point, and while it was still far from being a flight-ready item, this version of the docking system was subjected to careful analytical and operational scrutiny on the computer-driven Dynamic Docking System Simulator (DDTS). Housed in the JSC Structures and Mechanics Laboratory, the DDTS was capable of duplicating the most severe impacts and thermal conditions that could be anticipated when the docking systems were brought together in space. The test program consisted of 236 test runs, which subjected the American and Soviet gear to temperature ranges of −50° to 70° centigrade and to the active and passive modes.

A team of eight Soviet specialists, led by V. S. Syromyatnikov and Ye. G. Bobrov, worked with the regulars of Group 3, plus the test personnel in George E. Griffith's Structural Test Branch, computer people from the Spacecraft Systems Laboratory, and contractor employees who operated the DDTS. The extremely complex facility had been completed just before the Soviet team arrived, and some initial problems were encountered with the simulator, taking time away from the scheduled testing. According to Bob White, who had overall responsibility for the tests, "this caused team members from both countries to dedicate many extra hours at night and on weekends for make-up testing. Everyone worked admirably without complaining and a strong sense of mutual respect became discernible." Although the schedule was very demanding, one weekend was set aside for a private tour of the State Capitol and the Governor's office in Austin, as well as the engineering college and the Lyndon B. Johnson Library at the University of Texas.

During this fourteen-week evaluation process, a number of minor changes were incorporated into the design. But since there were no failures or major problems with either the U.S. or U.S.S.R. docking system, manufacture of the flight hardware could proceed on schedule. While the flight and backup systems would be subjected to a much more rigorous quality assurance program while being manufactured, the dynamic docking tests of the prototype system had been an essential step in defining the characteristics of that production equipment. With all the extra work completed, the Soviets departed on Christmas day, and many of the Americans, accompanied by their families, traveled to the Houston Intercontinental Airport to wish their friends a safe journey.[32]

During mid-January joint meetings held in Houston, the Soviets participated in tests of the docking module environmental control systems (ECS). As in the case of the development of the docking system, the breadboard version of the docking module ECS was designed to check out

Interior view of environmental control system breadboard. Test specialist Tom Wilks prepares for test of the system under simulated space conditions.

the operational and functional characteristics of the prototype equipment prior to fabrication of flight hardware. Although the test hardware was different in appearance from the final docking module and although the Apollo and Soyuz transfer tunnels were simulated by two pressure vessels, the ECS test hardware functioned like the real thing. The ECS breadboard was placed in the Life Systems Laboratory vacuum chamber, a horizontal cylinder 2.44 meters in diameter and 5.8 meters long. The chamber, divided by a bulkhead into two compartments, consisted of a manlock passageway and a test chamber to house the test article. Like most test facilities at JSC, the vacuum chamber was equipped with a closed circuit television system, which permitted remote viewing of the manlock compartment, the bread-board ECS, and the test chamber interior, and also equipped with a multi-channel intercommunications system, which linked all test personnel.

The joint tests, which ran from 16-23 January, were divided into two major categories—manned simulated mission tests and unmanned functional performance tests. During the manned tests conducted on 16 January, the performance of the system was demonstrated by simulating three transfers, during which the docking module environment reflected extreme situations that were not likely to occur during flight. By testing extreme cases, the suitability of the systems was scrutinized and the acceptability of manned operation under low or high oxygen pressures was determined. Later, unmanned functional performance testing was conducted to establish the leakage rates for the test chamber and to verify the major failure protection systems included in the docking module ECS. Results from these exercises indicated that the environmental control system met all the design specifications and that the transfer procedures were adequate in both normal and emergency situations. In addition, all of the safety equipment, such as the overpressure valve, performed successfully.[33]

Next came familiarization training for the American ASTP crews with this hardware. After the completion of the last Skylab visit on 8 February, all of the crewmembers were given briefings on the docking module systems. On 25 February, they participated in a two-hour walk-through of the ECS

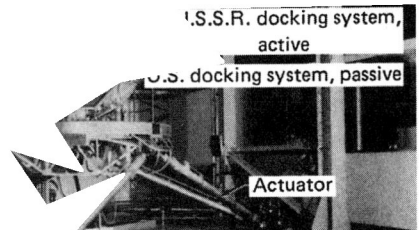

Docking systems installed on a simulator.

breadboard and test setup. The following day, after a four-hour Skylab debriefing in which all ten ASTP crewmen were involved, Brand and Evans took the first turn in the vacuum chamber to learn the ECS equipment and to practice transfer. Stafford and Slayton went through the same four-hour experience the next day, as did Bean and Lousma on 5 March. The crews were also increasing the number of training hours spent in the command module procedures simulator and command module simulator. And if that were not enough, they had met on the 4th with their new Russian instructors and had begun a new series of intensive lessons.[34]

Slayton and Stafford had not resumed their language studies after their last return from Star City, and the other crewmembers needed to begin learning Russian. During the November 1973 training sessions in the Soviet Union, the U.S. astronauts had discovered that the cosmonauts had made significant progress in their English studies. When Stafford asked Leonov how they had made such advances, he told the Americans that each member of the Soviet prime crew had his own individual instructors. They were studying language six to eight hours a day. Stafford cabled Washington through the American Embassy in Moscow and requested that the State Department's Foreign Service Institute provide the astronauts with two full-time Russian instructors starting early January. Stafford later told Chris Kraft that the American crew was going to look bad if its members were unable to communicate satisfactorily with their Soviet counterparts. They must get some full-time language training.[35]

Given the need for additional instruction and the desire to keep abreast of the progress being made by the Soviets, Nick Timacheff was authorized to locate professional teachers who could work with the astronauts. Timacheff screened a number of applicants during the post-Christmas convention of Slavic language professionals in Chicago. Four teachers were selected for their knowledge of contemporary vernacular Russian as opposed to the language as spoken by diplomats. Anatole A. Forostenko, Vasil Kiostun, James D. Flannery, and Nina N. Horner would learn as they taught, since they would have to teach their students the Russian equivalents of NASA's aerospace jargon. While Nina Horner concentrated on lengthy hours of classroom instruction, Forostenko, Kostun, and Flannery accompanied the astronauts on many trips and worked out with them in the gym in an effort to keep them thinking Russian—even when they were playing handball or lifting weights. Starting on 4 March, the prime and backup crewmembers received 3 or more hours of language instruction daily, five days a week. In their spare time, if they were not flying T-38s to keep their reactions sharp, they had cassette tape recorders by their sides to keep their ears sharp.

While the crewmembers studied, Working Group 5 specialists led by Walt Guy went to Moscow to observe testing of the modified Soyuz life support system. American technicians visited a Red Air Force base about 6 kilometers from Star City where the Soviets had their vacuum test chambers. The main test chamber used for the Soyuz tests was composed of a horizontal manlock and a vertical cylinder that was sufficiently large to hold a stacked descent vehicle, orbital module, and docking module simulator. Guy noted the similarities and differences between this test facility and the one in Houston, a major difference being the lack of communications headsets among the test personnel. He commended, "The close proximity of the test crew to each other and the exceptionally quiet test environment . . . made the public address system quite acceptable."

Among the systems evaluated during the tests were those for lowering and raising the Soyuz cabin pressure to determine the effect that transferring men from one spacecraft to another had on the gas composition under normal and abnormal conditions. Guy, Group 5's American chairman, reported later that he came away from the tests with no doubts that the Soviet ECS would work satisfactorily. He had been somewhat concerned about the basic uncontrollability of the chemical bed oyxgen system, but after prolonged simulated transfers into the docking module mockup followed by flushing all the docking module gases into Soyuz, the Soviet ECS proved capable of removing the carbon dioxide and other effluents from the atmosphere. At the Americans' request, the trial runs involving four men were longer than the transfers planned for the actual mission; therefore, this was an excellent evaluation of its capabilities. After working with the Soviets in their laboratory, the U.S. team grew confident that there would be no problems with the U.S.S.R. equipment. This was exactly why Glynn Lunney had wanted his men to participate in such activities.[36]

DEVELOPING FLIGHT PROCEDURES

At the next meeting of all the Working Groups in April 1974, V. F. Bykovskiy joined Leonov and Kubasov as their instructor in developing flight procedures that would be practiced at the sessions scheduled for mid-summer in Star City and September in Houston. Their work began on the afternoon of 15 April, when the prime crews and Bob Overmyer spent an hour discussing the training that remained before the flight. Then after meeting briefly with the press on the morning of the 16th, they got down to work. Astronauts Bean and Evans worked with the Soviets in the command module simulator until lunch, and Overmeyer and the American prime crew spent that afternoon with them, evaluating docking procedures. The remaining eight work days were just as busy. Wednesday was spent with survival training and a review of the Russian-English glossary. On Thursday, the Soviets and Vance Brand received a briefing on the 16-millimeter movie camera, after which they worked out in the gym. They went over the flight plan with the other astronauts and training personnel that afternoon. The entire team met for briefings on the docking module and trainer on Friday.

Nearly all the second week was devoted to learning to operate the docking module equipment during transfers, using the high fidelity mockup to verify the in-orbit transfer activities. They worked through the hundreds of necessary procedural instructions, like those required to establish the integrity of the seals and latches connecting the module to the spacecraft. Besides the 12 hours spent in the docking module trainer, the cosmonauts worked some more on the flight plan, practiced with the communications equipment, received an Apollo television camera briefing, and spent 3 hours in the command module simulator with Brand and Bean, working on the final stage of Apollo's rendezvous and approach to Soyuz.[37] Stafford, speaking to the media on 26 April, said that the training session had been a "rewarding experience." In the case of crew transfers, considerable progress had been made. "For example," Stafford indicated, "the second day we tried it, we did it in about one-third the time that we did the first day." The reporters, it seemed, were also interested in the language question.

Bruce E. Hicks of United Press International asked Tom Stafford how the Americans were handling Russian. The general responded in Russian, and Leonov translated for him, replying that Stafford said that he understands the Soviet crew. Leonov added, "We've no problem in language." He went on to say that during the transfer training they had worked out the communication format that they planned to use during flight. "Our work is considerably better when the American crew speaks Russian and our crew speaks English," the Soviet commander said. "This forces us to maintain a considerable amount of discipline, to be attentive to each other, and to speak . . . much more clearly."[38]

Each of the subsequent meetings between the astronauts and cosmonauts stressed language exercises. At the 24 June to 11 July training session in Star City, every crewman received 10 hours of communications practice, which generally involved speaking over an intercom to a counterpart in a different office. Reading from a script, they could simulate the conversations that would pass between Apollo and Soyuz. This experience not only gave them an opportunity to improve their pronunciation but also introduced Leonov and Kubasov to Russian spoken with a Weatherford, Oklahoma, accent. "*Soyuz, ehto Apollon. Stuikovka na pyat minut. . . .*" This is what Stafford might say as his spacecraft closed the gap between the two ships. Leonov would reply, "Apollo, this is Soyuz. I understand; docking is in five minutes."

But there was time for some horseplay during that summer session in Russia. Ron Evans had brought half a suitcase of fireworks with him in

Testing and crew training constituted heavy workloads in ASTP during 1974. At left, Soviet and NASA engineers stand in front of a test chamber housing a Soyuz mockup during the March 1974 tests of the life support system at Star City. From left to right: V. V. Novikov, R. L. Grafe, D. F. Hughes, W. E. Elliss, R. E. Mayo, E. N. Harrin, W. W. Guy, the Soviet facility engineer, two test crewmen, and two military assistants to General V. N. Kholodkov (Soviet Academy of Sciences photo). The following month in Houston, crewmen Leonov and Slayton are in the docking module for checkout and familiarization training.

In July 1974 the crews met in Star City for more training. At left, an overall view of spacecraft simulators at Star City, with Soyuz in the foreground and Salyut beyond. Below, N. N. Rukavishnikov (rear) explains Soyuz communications equipment to Deke Slayton.

In September 1974 in Houston, Soviet and American crewmen practice in the docking module mockup, rehearsing their conversation during a transfer operation.

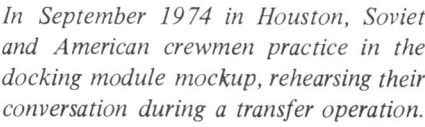

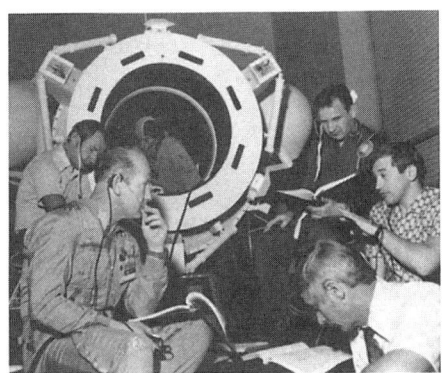

anticipation of the Fourth of July. After attending the American Ambassador's cocktail party, the Soviet and American crews returned to the astronauts' hotel at Star City. About dusk, Evans began the show, setting off a string of firecrackers. Stafford recalled that "it sounded like a machine gun and all the lights were going on in the building across the lake. Pretty soon you could see all these flashing lights. . . ." Once the police arrived, they formed a huddle. At this point, Stafford cried, "Hey, let's give them a bottle rocket!" With that, they fired a small rocket from a mineral water bottle, and it arced over the heads of the policemen. Finally one of the officers approached the Americans. Stafford called out in Russian, *"Dobriy vecher. Kak vi pozhivaete.*[*] . . . It is the day of our revolution," he explained. The official nodded that he understood the astronaut's explanation and retired with a look of amazement on his face.[39]

At the September exercises, the Americans had created a more elaborate simulation of inter-spacecraft communications. Working from a script prepared by the American crews and their language instructors, Leonov and Kubasov sat on one side of a glass-partitioned laboratory in the Flight Crew Training Facility at JSC and the Americans on the other side. This additional practice with flight conversations, coupled with further language training for both crews and more experience with hardware mockups, improved their ability to communicate with one another. As with all other aspects of preparing for the flight, learning Russian and English was an essential expenditure of long hours and hard concentration for both sides.[40]

QUESTIONS ABOUT SOYUZ

Concurrent with their training exercises in 1974, the astronauts became involved in a renewed controversy over the flightworthiness of the Soviet spacecraft. The first phase of this debate opened with the publication of two articles in *Aviation Week* that argued that Soyuz was a very marginal design when compared to Apollo, especially in the area of guidance and control. These stories, which came from conversations with some NASA astronauts, made several strong judgments about the quality of Soviet hardware, stating: "In some areas, Soyuz capability is below that available in the Mercury spacecraft flown by American astronauts almost 13 years ago."[41] At the time these items were written, the NASA team still did not fully understand the operation of the Soviet spacecraft control systems. In fact, when the astronauts returned to Star City in June, they were given at their request a

*Good evening. How are you?

briefing on these systems by V. P. Legostayev to clear up several points of confusion resulting from data presented to them the preceding November.

Legostayev and his colleagues had created control systems for Soyuz that were sufficient for the earth orbital missions that the craft was expected to fulfill. From the start, Soyuz was basically a vehicle designed to be controlled automatically from the ground. As various unmanned Cosmos flights had indicated, Soyuz-type ships could be directed to rendezvous and dock from the ground. When manning Soyuz, the cosmonauts acted more as systems monitors than as pilots. This approach to manned space flight and the limited activities demanded in earth orbit meant that the Soviet designers did not need to develop the more complex guidance and navigation equipment that had been required for Apollo's trips to the moon. They relied instead on sun sensors and earth horizon observation, plus limited use of gyroscopes for navigation and guidance.

In this area, Soyuz and Apollo represented completely different approaches to a problem. For American astronauts who were used to having their hands on the controls and flying by the seats of their pants, Soyuz was not the kind of ship with which they would feel comfortable. Being a passenger was not their cup of tea; thus, it was not unreasonable for them to make negative comparisons of Soyuz to Apollo. But such value judgments were at best subjective. The Soviet approach was not worse than the American way of flying—it was simply different. When *Aviation Week* took the facts of the difference in design and coupled them with astronaut opinions, it sounded as if the Soviets had an inferior spacecraft. Not surprisingly, the Soviets were offended by these comparisons, and Glynn Lunney cautioned his people to take care how they evaluated Soviet hardware when talking to the press. He advised them to stick to the facts and to beware of editorializing. He suggested that only a "damn fool" would mix fact with opinion.[42]

Phase two of the argument over the reliability of Soyuz started with the apparent failure of *Soyuz 15* to complete its mission. Successful flights of *Soyuz 12* (September 1973), *Soyuz 13* (December 1973), and *Soyuz 14* (launched for a 14-day mission during the June-July 1974 visit of the astronauts to Star City) had helped to reassure many of those public figures who were still worried about the *Soyuz 11* tragedy. When *Soyuz 15* failed to dock with *Salyut 3* and returned to earth after just two days and made a night landing, cries arose from Capitol Hill and in the news media, questioning once again the wisdom of the joint flight.[43] Most vocal among the congressional critics was Senator William Proxmire, who wrote to Administrator Fletcher asking for a complete safety review of Soyuz prior to the ASTP mission. "In particular," he recommended "that the National

Aeronautics and Space Administration go slow and proceed with all due caution." And he felt that "present plans for a joint space mission should be seriously re-examined in light of the continuing difficulty in the Soviet program." Proxmire said that he did not want the space agency to take chances with the lives "of our astronauts for the sake of some untangible diplomatic benefits of detente."[44]

While the Senator's concern was understandable, he was ill informed if he believed that anyone within NASA was about to gamble with the safety of either the American *or* Soviet crew. The whole purpose of exercises such as the safety assessment reports was to identify problem areas and establish that such potential trouble spots would not affect the execution of the mission. Lunney, in a regularly scheduled telephone conversation, had discussed the *Soyuz 15* mission with Bushuyev on 27 August, the day after it was launched, and the Professor had told him that it was in no way related to ASTP, contrary to some media speculations. Bushuyev said it was a test of automatic docking systems.[45]

At their 26 August-20 September 1974 meetings in Moscow, Lunney and Bushuyev talked at considerable length about *Soyuz 15* and the *Cosmos 638* and *672* flights. The latter, flown in April and August, were unmanned tests of Soyuz as modified for ASTP, and they were unqualified successes. As recorded in the joint minutes, the objective of *Soyuz 15* was:

> the testing of a system of automatic approach and docking. This system is not used in the Apollo-Soyuz program. All Soyuz 15 systems that are analogous to those used in the ASTP flight worked in a satisfactory manner. During the final phase of approach not all of the monitored parameters for approach and docking were within the prescribed range. The crew, therefore, in accordance with previous instructions, switched off the automatic approach and docking system. Following completion of the planned flight program, the preplanned night landing was achieved for the purpose of verifying the feasibility of . . . a night landing.[46]

In addition to the discussions between the Technical Directors, the American crews were briefed by the cosmonauts.

On 11 September 1974, Tom Stafford raised the *Soyuz 15* issue in a Houston press conference. He said that he had been given the full story by General Shatalov at the beginning of their current training session. Since there was still some concern about the flight, however, Stafford gave the floor to Shatalov who had agreed to answer queries from the press. Shatalov told the reporters that *Soyuz 15* had been a test of a system to permit automatic docking with Salyut, since one of the long range goals of the Soviet space program was the use of unmanned resupply craft that could dock with the space station and automatically transfer fuel and supplies. Cosmonauts G. V. Sarafanov and L. V. Demin had flown the mission to observe the functioning of the new system, and the Soviet spokesman reminded the press that this was the traditional role of the test pilot. Further, he pointed out that while NASA might rely more heavily on ground-based simulations, the Soviet space engineers had traditionally flight tested their spacecraft. In the case of *Soyuz 15,* when it became apparent that the systems were not working properly, the flight was terminated. Again, these experiences were not unusual in the business of flight testing hardware.[47] Phase three of the Soyuz reliability debate came with the successful flight of *Soyuz 16.* Manned by the number two ASTP prime crew, Filipchenko and Rukavishnikov, this flight was a full dress rehearsal of the Soviet half of the joint mission. From the afternoon lift-off at 12:40 Moscow time on 2 December to the morning landing at 11:04 on the 8th, the flight of Soyuz was nearly perfect, and the results of the test of life support, docking, antenna deployment, and ground control systems were excellent. Shortly after launch, the Soviets had notified the Johnson Space Center, so the Spaceflight Tracking Data Network could begin tracking the spaceship.[48]

Lunney's team in Houston had known that the Soviets were planning a manned flight for the end of 1974. In fact, the Soviets had been prepared to give the Americans advance notice of the launch.

> It was agreed that during the upcoming manned Soyuz flight which is a precursor test flight for the ASTP mission, the American side will perform Soyuz spacecraft tracking with their own ground tracking stations and the two sides will subsequently compare tracking data. The American side will be informed about the launch date and planned orbital parameters 5 days prior to launch. State vectors of the spacecraft will also be provided after insertion into orbit.[49]

Subsequently, the Soviets added the restriction that this information would be given to NASA only if the agency agreed to withhold it from the press until the flight had actually begun. After lengthy discussions, which involved George Low, Glynn Lunney, Chet Lee, Arnold Frutkin, and John Donnelly, it was concluded that NASA's tracking *Soyuz 16* could be considered a joint activity.[50] To withhold details from the public concerning such an exercise would not be consistent with the agency's traditional practice of providing information. On 11 October, Lunney telexed Bushuyev:

> We appreciate Soviet desire to make own announcement of launch notice and launch. However, because of our own involvement in this activity, we would find ourselves in a difficult position if we could not report this information to our press. Therefore, we prefer to receive no information in this case until you have released it or we can release it. When we learn of the launch under these conditions we will initiate tracking activities. . . .[51]

At 6:35 Houston time on the morning of 2 December, V. A. Timchenko called JSC. The security guard who took the early morning call said that Mr. Lunney was not yet in his office. At the Soviets' request, the guard notified the U.S. Technical Director that Moscow would be contacting him by telephone at 8:15. Less than two hours later, Timchenko and Lunney were talking about the mission manned by Filipchenko and Rukavishnikov. Lunney in turn advised the tracking team, giving them the data provided by Timchenko. These mathematical statements of the spacecraft's location and velocity at a given time would permit the tracking stations to follow its path, an exercise that was essential for the rendezvous part of the joint mission.[52]

Life Support System Operation Timeline: Checkout of ASTP Modifications to Spacecraft During Soyuz 16 Flight*

Ground elapsed time (hr:min)	Operation
	Pre-launch preparations
−3:30	Descent vehicle gas analyzer, orbital module gas analyzer, and pressure integrity check unit activation.
−2:30	Crew in pressure suits ingresses vehicle: SC connects his PG to OM fan assembly, activates OM panel and PG fan assembly, and begins pre-launch OM examination. FE ingresses DV, connects his PG to DV fan assembly, activates CSD and PG fan assembly, and begins pre-launch DV examination. Following examination, SC deactivates PG fan assembly and OM panel, disconnects PG from fan, transfers to DV, connects PG to the DV fan assembly and activates it. DV RA activation.
−1:30	Begin status and operations check of DV systems. Close hatch 5. Close OM ingress hatch. Pressurize OM with 125 mm Hg of oxygen. OM pressure integrity by launch team.
−0:40	Begin PG pressure integrity check.
−0:30	End PG pressure integrity check.
−0:05	Lower PG visors.
−0:00	Launch: December 2, 1974, 12:40 Moscow time.
	Orbital flight
0:14	Raise PG visors.
0:16	Switch PICU to pressure leak monitor mode.
0:30	Remove PG gloves.
0:49	End pressure integrity monitoring of modules. Activate GMSS automatic controls. Close "TANK" valves.
1:30	Activate OM RA. Equalize DV-OM pressure and open hatch 5. DV RA "OFF." Transfer to OM. Set OM PVV to "CLOSED" position. DV-OM pressure vent test. Remove PGs and begin drying.
5:00	End PG drying.
6:43 to 8:28	DV-OM pressure vent to 540 mm Hg.
10:40 to 18:50	Sleep.
28:37 to 28:53	Corrective pressure vent from 540 to 510 mm Hg.
34:30 to 42:20	Sleep.
44:30	Open bypass valve (initiation TCS LML coolant flow through Apollo radio station transceiver mounting assembly).
48:00	Close bypass valve.
51:00	OM RA CO_2 absorber on.
51:10	Don PGs.
51:30	Switch OM RA to minimum flow mode from OM panel. Transfer to DV and close hatch 5.
52:10	Open hatch 5.

PREPARING FOR THE MISSION

Bushuyev gave Lunney brief reports on this ASTP precursor flight during telecons on 3 and 8 December. Subsequently during the winter meetings in Houston, he provided full details of the *Soyuz 16* mission to the American members of Working Group 4. On 31 January through his interpreter, Yu. S. Zonov, the Professor told Lunney that the flight had been a complete dress rehearsal for the Soviet portion of ASTP. The Soyuz spacecraft was identical to the one that would be flown in July, and the Soviets had designed the December flight plan to check out key parts of the ASTP plan. Of particular interest to the Americans were the reports provided by Bushuyev on the functional tests of the modified life support system. (See box below.[53])

Life Support System Operation Timeline: Checkout of ASTP Modifications to Spacecraft During Soyuz 16 Flight*—Concluded

Ground elapsed time (hr:min)	Operation
52:50	Switch OM RA to automatic control mode and activate CO_2 absorber.
53:25	Remove PGs, begin drying.
56:40	Stow PGs.
58:20 to 66:00	Sleep.
82:00 to 89:40	Sleep.
102:21	DV OM pressurization to 830 mm Hg.
104:00	Disconnect removable condensation collector, transfer it to OM, and connect DV collector.
105:50 to 113:25	Sleep.
119:20	Transfer to DV and close hatch 5. Jettison APDS mock-up ring.
120:10	Open hatch 5.
123:30	DV-OM test pressure vent from 805 to 760 mm Hg.
130:00	Switch TCS ERL external line coolant temperature setting from 7°C to 5°C.
130:00 to 137:00	Sleep.
137:00	Switch gas temperature setting at heat exchanger-condenser output from 20°C to 15°C.
	Descent preparations and descent
137:10	Set OM PVV handle to "ELECT CONTROL" position.
137:40	Don PGs.
138:10	Transfer to DV; close hatch 5. Connect PG to GMMS; activate PG fan. Activate DV RA. OM pressure vent by 125 mm Hg. Monitor hatch 5 pressure integrity.
139:20	Monitor PG pressure integrity.
141:41	OM pressure vent.
141:47	Lower visors.
141:53	DV-OM separation.
142:24	Landing: December 8, 1974, 11:04 Moscow time.

*List of abbreviations:

APDS	Androgynous-peripheral docking system	OM	Orbital module
CSD	Command signal device	PG	Pressure garment (space suit)
DV	Descent vehicle	PICU	Pressure integrity check unit
ERL	External radiator loop	PVV	Pressure vent valve
FE	Flight engineer	RA	Regenerated assembly (oxygen generator)
GMSS	Gas mixture supply system	SC	Soyuz commander
LML	Living module loop	TCS	Thermal control system

Bushuyev also called to Lunney's attention the fact that *Soyuz 16* had been placed into an initial orbit different from the ASTP rendezvous orbit so that the Soviets could test the spacecraft's maneuverability. The Professor went on to provide the Americans with details regarding the docking system evaluation:

> As I said before, for the imitation of the operation of docking assembly of the Apollo spacecraft, we made a special technological ring which corresponded to the docking ring of the American assembly. During the flight, we tested the following items: the opening and the closing of the [capture] latches. The retraction of the ring with the guides. The alignment of the ring. The opening of the [structural] latches. The closing of the latches. The undocking. The reserve opening of the active hooks. . . . During the process of opening the hooks and undocking, the movement of the hooks was done not to the end but to the position of intermediate. This was done specially so we could do the final separation of the ring with the help of the pyrotechnics [i.e., to test the emergency release system] .[54]

All the tests of the docking were carried out successfully with no problems.

Bushuyev was very confident that Soyuz was ready for the joint mission. After a nearly perfect flight by *Soyuz 16,* he had good reason to be optimistic. In fact, he commented that both Filipchenko and Rukavishnikov, veterans of earlier Soyuz flights, had indicated that all the changes incorporated into the spacecraft had made it a more flexible ship to fly.[55] Filipchenko and Rukavishnikov spoke with the press on 13 December when the Soviets conducted a post-flight news conference, a check out of their public affairs procedures for ASTP. The two crewmen, plus Petrov, Beregovoy, Flight Director Shatalov, and Bushuyev, met with several hundred correspondents. Bob White, American Working Group 3 chairman in Moscow for the pre-flight tests of the docking systems, also attended. He noted that this was the first time the press had been able to directly ask questions of a Soviet crew after a mission and the first time most of them had been permitted to visit Star City where the press conference was held.[56] The Soviets' optimism over *Soyuz 16* was soon shared by Lunney, his Working Group chairmen, and the crews. Soyuz was ready; the Soviet reports, joint test data, and safety assessment reports proved it. This evaluation was presented to the U.S. Aerospace Safety Advisory Panel (ASAP).

THE WATCHDOGS CONCUR

Created in the wake of the Apollo 204 fire, the Aerospace Safety Advisory Panel acted as an independent body reporting to the Administrator of NASA on the flight readiness of every manned mission from the standpoint of safety.[57] As such, the panel looked over the shoulders of NASA and contractor personnel while they prepared for flights to make certain that all possible safety precautions were taken. Since early in 1973, this body had been conducting reviews of all ASTP-related activities that might affect the safety of the mission.

The panel concluded that the Apollo spacecraft (CSM 111 and 119), the launch vehicle, and ground support equipment appeared to be ready for the mission. They noted that modifications necessitated by the joint mission had been completed and subjected to detailed safety assessments and hardware qualification tests. Panel members were of the opinion that appropriate attention had been given to the effects of equipment aging during storage, a matter of some concern for both the CSMs and the launch vehicle, SA-210.

Turning to the new hardware, the panel was equally satisfied. For the docking module, ASAP commented, the designers had applied safety margins significantly greater than those used in prior manned vehicles. The 15.8-millimeter aluminum plate from which the docking module was constructed possessed inherent strength considerably greater than that required by any loads likely to be encountered during the mission. In a similar fashion, the high pressure gas vessels used in the docking module environmental control system had been designed with a safety factor of four. The reliability of the docking module and its subsystems had been proven by mathematical analysis and qualification testing that provided "a basis for confidence in the flight systems meeting mission requirements."[58]

Of equal interest to the panel was the docking system, because it constituted the direct interface with Soyuz. In view of ASAP's concern, Charles D. Harrington, a member of the panel, observed the Moscow portion of the compatibility testing in mid-November 1974. Commenting on this experience, the panel reported:

> This ... provided further insight into the Soyuz hardware, joint working relations between technical and management personnel, and the joint testing program. The Panel examined the test program and its results to assure that the qualification testing was adequate and that no residual safety problems for the flight personnel could be identified. Of the many key system components, the docking system seals, locking latches, and alignment pins and sockets were of particular interest. Development tests and qualification tests have been conducted on these items to assure proper operation within the joint phases of the mission. All known problems have been resolved.[59]

Turning to the sensitive topic of Soyuz flight readiness, the panel indicated that its members had discussed at length with the Working Group chairmen the adequacy of Soviet management in the areas of design, testing, fabrication, and check-out. The chairman said that they "had found no management situations that would compromise NASA's ability to provide for crew safety during the joint phase of the mission." Since the panel did not have firsthand data concerning Soyuz, they had to rely upon the judgment of those who had been working with the Soviets. Considering that the Soyuz design had a long test and flight history, the panel concluded that the spacecraft was suitable for the joint mission. They did not see any circumstances that might endanger the crews, noting that almost all of the Soyuz systems were designed to operate automatically or semi-automatically with a minimized role for the cosmonauts. These elements and the testing program for the new onboard systems gave the panel reasonable confidence in the Soviet spacecraft. After looking at all aspects of the mission, ASAP stated, "confidence in crew safety for the joint phases is essentially equal to that for prior manned earth orbital flights."[60]

Presentation of the Aerospace Safety Advisory Panel's findings was made in Washington on 5 February 1975. In addition to Administrator Fletcher and other senior officials of the space agency, staff members from the Senate Committee on Aeronautical and Space Sciences and the House Committee on Science and Astronautics were present during their report. On the following day, four members of the panel, two staff members, and a consultant traveled to Houston to talk with the two Technical Directors about specific aspects of ASTP that still concerned them. Prior to meeting with the Soviets, panel members Howard K. Nason, Charles Harrington, Herbert E. Grier, and Lieutenant General Warren D. Johnston met with Glynn Lunney. Chairman Nason, president of the Monsanto Research Corporation, told Lunney that the panel would like to ask the Soviets some specific questions in an effort to clarify a few points. General Johnston, Director of the Defense Nuclear Agency, in particular had a specific query that he wanted Lunney to have translated into Russian so it could be presented to Bushuyev.[61]

Lunney, sensitive to the anxiety that the appearance of a hitherto unfamiliar group asking probing questions might cause among the Soviets, suggested that there might be a better approach. He volunteered to ask the Professor to give his views on each of the areas of concern, thereby obtaining the information without appearing like an inquisition. Lunney added that the ASAP members might want to "put the shoe on the other foot" when they worried about the reliability of Soyuz. He said that sometimes American problems had to be resolved in a manner that might appear to an outsider to be unorthodox and unacceptable. He cited as an example a "crew alert" light that had indicated a problem during a checkout of the ASTP spacecraft at the Cape. NASA's solution had been the reasonable one; they had disconnected the warning light and isolated the wiring leading to it when it was determined that indeed nothing was wrong. This was an acceptable procedure that the ASAP members could understand, but would it be fully comprehended by observers from another country? He asked them to reflect on how they might react if they were a Soviet safety board and they had found that the Americans planned to fly a spacecraft with a cabin atmosphere of 100 percent oxygen when a possible short could cause a fire. Johnson and the others indicated that they understood. Since they really only wanted to reassure themselves on a few points, they would let Lunney ask the questions.

Upon his return to the joint meeting site, Lunney asked the Professor and Alex Tatistcheff, Lunney's interpreter, to join him in his temporary office to discuss the impending meeting with the Safety Advisory Panel. Tatistcheff, in an effort to allay any concern on the Professor's part, was careful to point out that although in Russian there was only one word for both safety and security (*bezopasnost'*), in English these were two different words. The panel was simply a committee of technical experts selected by NASA's Administrator to provide an independent evaluation of the safety precautions for all manned flights. It was not a body involved with any of the American intelligence or security organizations. Once this linguistic distinction was made clear, the Professor said that he was willing to speak with the panel members but that he would prefer not to be placed in the position where he might be required to present a lengthy defense of Soyuz. Lunney assured him that the Safety Panel would not expect him to engage in such an exercise, because there was adequate information available in the various ASTP documents. After the mission, Bushuyev quoted Lunney as having said, "You see, neither of us has any doubts about this, but members of the commission [ASAP] hear only my voice. For them, your opinion, your arguments will be very authoritative." Bushuyev added, "I agree."[62] With the ground rules for the meeting established, Lunney brought the two groups together.

The early minutes of the gathering were very formal, and the Soviets were slightly defensive in their reactions. Lunney introduced the members of the Soviet delegation to the panel, and Nason introduced in turn his group and gave a brief explanation of the background and purpose of the panel. Responsible to the Administrator, they were just one more element of the overall agency effort to reduce accidents. In the case of manned flights, their goal was to be as certain as possible that every step had been taken to eliminate all flight hazards. In the case of ASTP, Nason pointed out that they were interested in the dangers posed by fire, toxic fumes, and an undocking of the spacecraft caused by a failure of the latches or inadvertent detonation of the pyrotechnics. A related area of interest was the ability of the crews and flight directors to react quickly and decisively in the event of an in-flight emergency. Lunney suggested that Bushuyev might want to comment on these topics, since the Panel had thus far only heard his own version.

Bushuyev prefaced his remarks by saying that safety had been a central concern of both sides since the very earliest days of the joint sessions. Through a series of detailed documents, the Soviet and American technical specialists had certified that their respective spacecraft were free from the hazards outlined by Nason. As for the ability of the crews and the flight directors to make command decisions in the event of an emergency, the Professor reminded the panel members of the extensive crew training in both flight procedures and language. The intercontrol center simulations, interpreters at the flight consoles, and visiting technical specialists in the two control centers were all for the purpose of providing split-second decision making on the ground as well. Given the experience with the crew and ground control training sessions to date, the Soviet director was convinced that by the time of the flight, the crews and the flight directors would be able to cope with any unforeseen circumstances. He added that his confidence was enhanced by his knowledge that every effort had been made to eliminate all possible sources of trouble. Lunney concurred and suggested that having worked together throughout most of the preparations for the flight the crews and flight directors would "understand each other's thinking" in the unlikely event that an emergency should require an immediate, on-the-spot decision during the mission.

Having had a chance to talk with Bushuyev and to watch the manner in which the technical directors worked together, the ASAP members were convinced that a two-nation partnership had indeed been worked out that was capable of conducting the first international manned space flight. They also began to understand Lunney's respect for the Soviet team. What they might not have fully appreciated, however, was the manner in which Lunney had handled Soviet concerns over issues that reflected the safety of Apollo. Safety was a full time interest of both teams, and there had been times when the Soviets had expressed concern about the manner in which Apollo was to be flown during the joint phase of the mission. The shoe could be on the other fellow's foot.

SOVIET WORRIES ABOUT APOLLO MINUS X THRUSTERS

According to the Americans, there was nothing to worry about. As Apollo approached Soyuz, the attitude control motors used to brake the craft would not send exhaust far enough to burn the thermal insulating blanket that protected Soyuz from the heat of the sun, nor would the Soyuz radio antennas be affected. But the Soviets were worried that the plume of the thrusters might hit their craft if the astronauts forgot to shut down those $-X$ engines* after capture by the docking gear. As in the case of the American worries over Soyuz, the home team could not quite see what all the fuss was about. Nevertheless, they had to attend to their visitors' unease.[63]

Ed Smith, whose Working Group 2 had to deal with such problems, traced the origin of the Soviet interest in the possible impingement of the $-X$ thrusters on Soyuz to a Skylab movie that Max Faget had taken to the U.S.S.R. to show at a gathering of space scientists. In this movie, the Skylab parasol fluttered in reaction to firings of the Apollo thrusters as the docking approach was executed. Subsequently, at the August-September 1974 sessions in Moscow, Vladimir Timchenko, Soviet chairman of Group 1, asked his American colleagues if these thruster firings could be expected to have any effect on Soyuz. Timchenko's concern was associated less with the possibility of damage to Soyuz than that the control system firings in the vacuum of space would upset the attitude of Soyuz and cause the ship to deviate from the reference attitude it needed to maintain during the docking maneuvers.

Richard Haken, a contract employee working with Group 2, said that he would pull together all the data JSC had concerning the expected lengths of time for $-X$ thruster firings when Apollo was both approaching and docking with Soyuz. He gave his findings to the Soviets during the November 1974 Houston meeting, and they took them home to study. Following their analysis, they sent a specialist, B. P. Skotnikov, to Houston in December with a Working Group 1 delegation to work with Smith, Haken, Steven Pollock, and Roscoe Lee on the possibility that the Apollo control system firings could create disturbance torques that would upset Soyuz. They concluded in the negative:

> At this meeting, both sides have presented and discussed the materials on the evaluation of the disturbance forces and torques, which affect the Soyuz spacecraft during operation of the Apollo RCS jets when docking. . . . As described in USA WG2-051, during normal docking, the Apollo RCS jets work in pulses. It was noted that beginning with a distance of six meters during docking, Apollo RCS jet pulses of 0.5 sec and larger are extremely rare and their repetition extremely remote. Both sides concluded that when the Apollo RCS jets work in pulses, small disturbance torques exist which do not cause deviation in the attitude of the Soyuz spacecraft.[64]

*Minus X engines are the forward firing thrusters used to brake or slow down the Apollo spacecraft.

But the Apollo thruster question did not end in November. At the January-February meetings, the Soviet delegation included A. G. Reshetin, an aerothermal expert who wanted to discuss the thermal impingement of the thruster firings on the surface of Soyuz. These talks covered the entire approach and docking sequence. From these reconsiderations of the $-X$ thruster firings, the Americans came to understand that the basic worry the Soviets had was not how long the reaction control system (RCS) jets might be fired during approach but just what guaranteed that they would be shut off after capture by the docking system. The Americans said that one had to rely upon the crewmen to throw the switches that would inhibit further operation of the thrusters. This reliance on men bothered the Soviets, who would have preferred to have those engines controlled by an automatic system.

The "RCS impingement problem," as it became known, bounced around during the January-February discussions until it promised to become a real issue of more magnitude than seemed justified in the minds of the Americans. To the Soviets, it continued to be a worrisome topic that needed further explanation and a definitive resolution. They wanted the Americans to state in their flight plan that the Apollo crew "shall not" use the $-X$ thrusters within 10 meters of Soyuz. Lunney had to step in and take a firm hand because the entire discussion was getting out of hand and no resolution appeared in sight.

Following a frank meeting on 30 January in which Lunney and his team discussed the meeting's progress to date, Ed Smith gave the Professor the run down on a procedure that would ensure that the $-X$ engines would not be fired after capture was made by the docking system. As Smith explained it, when the Apollo docking system captured Soyuz, an indicator light would appear in Apollo, and the Command Module Pilot would call out "contact" to the Commander, who would cease forward translation. At that point, the Commander would switch control of the RCS engines from the stabilization and control system to a second system controlled by the command module computer in a free mode (CMC-free), which would operate only upon a manual command given through the hand controller. While the RCS system was in this dormant condition, the commander would reach up and turn off the four RCS automatic select switches that control the forward firing thrusters. Finally, the computer would take over again, correcting the pitch and yaw of the two spacecraft as needed. The entire process would take only a matter of seconds.

When Smith completed his description, there followed a 40-minute discussion, and the process was described again and again. Bushuyev had a number of questions. V. P. Legostayev explained the process in Russian to Petrov and the Professor, while sketches were passed back and forth across

RCS impingement problem, January 1975

Soviet concern that the Soyuz spacecraft would be endangered by the exhaust from Apollo's reaction control system (RCS) received much attention during the meetings in Houston in January-February 1975. In one of the Working Groups, flight director Alexei Yeliseyev (left) listens as Pete Frank (gesturing with pen) explains the safeguards provided in the firing sequence of the RCS. Also listening (left to right) are Gene Cernan (obscured by Yeliseyev), Yu. S. Denisov, N. Latter, R. D. White, Frank, V. P. Legostayev, V. S. Syromyatnikov, and B. V. Nikitin.

At left above with arm extended, Yuri Zonov asks Lunney a question about RCS impingement. Finally, Professor Bushuyev (right, above), still not convinced, asks to see the astronauts throw the RCS isolation switches in the command module simulator to observe how much time it takes to disarm the −X thrusters. Left, all are smiling on the way back from the simulators where Lunney, Bushuyev, and others have watched the RCS exercise performed to everyone's satisfaction.

the table. With Yuri Zonov and Alex Tatistcheff interpreting, Lunney tried to answer the Soviets' several questions. The scene was hectic, and in the end Bushuyev still had some doubt in his mind about relying upon the crew to throw those important switches.

At this point, the whole "RCS impingement" issue was becoming an emotional subject. Some of the astronauts were openly upset that the

Soviets would question their training and discipline when it came to executing the proper sequence of actions. While Legostayev understood the Americans' explanation, Yeliseyev, the flight director, was still not convinced. According to Zonov, "Yeliseyev was the big skeptic." The issue was at last resolved on 5 February when Petrov and Yeliseyev climbed into the Apollo command module simulator and went through the procedures from the calling out of "capture" to throwing the proper switches. Once they saw how it worked, the troublesome issue seemed to go away. It also appeared that Petrov was growing tired of this recurring topic, and being personally convinced that there was no real problem he told his people to drop it. At the ASAP briefing the next morning, Bushuyev indicated that there had been some concern about the possibility of RCS impingement but that it had been the subject of enough discussion.

The RCS controversy demonstrated the different approaches to rendezvousing and docking spacecraft—automatic versus pilot controlled—and while it had been a real worry for many of the specialists, Lunney had not let it upset him. It was a problem for which there was a technical solution. Once his men told him how they intended to handle the matter, he explained it to the Soviets and invited them to the simulator where they could see that the proposed solution was indeed satisfactory. It was all part of a day's work, and there was no need to become emotional. On 31 January, when the issue was still pending, Lunney was asked how the meeting was going. He responded, "This . . . is a piece of cake"; he wished that he could get away from the Grumman building, where the talks were being held, and go back across NASA Road 1 to clean up the paper work on his desk. When asked about the RCS situation, he smiled with a characteristic twinkle in his eyes and said, "*Nyet problém.*" Then he added, "If you think that is a hot issue, you should have seen some of our earlier go-arounds." Chomping down on his cigar, he went in search of a cup of coffee.[65]

FINAL ROUND OF CREW TRAINING

While the technical specialists debated the readiness of the spacecraft, eight cosmonauts arrived in Washington on 7 February to begin their final training session in the United States. After being met in the nation's capital by Tom Stafford, the men flew to Kennedy Space Center (KSC) the next morning. Joined there by the other ASTP astronauts, the two flight teams spent all of Saturday at the launch complex. They were given briefings on the operations conducted at the facilities, and they had an opportunity to see the nearly flight-ready CSM 111 and docking module, which were in the Vehicle Assembly Building along with the launch vehicle.[66]

```
                         ASTP Flight Menu

    Alexei A. Leonov-SC              Valeriy N. Kubasov-FE

       Potato Soup                      Seafood Mushroom Soup
       Beef Steak                       Beef Steak
       Rye Bread                        Rye Bread
       Cheese Spread                    Cheese Spread
       Almonds                          Almonds
       Strawberries                     Strawberries
       Tea w/Lemon & Sugar              Tea w/Lemon & Sugar
```

On Sunday, the 14 men visited Disney World. Like other tourists, the cosmonauts and astronauts rode a number of the rides in the amusement park, including one that featured miniature spaceships. After a trip on a Mississippi river boat and handshakes with Donald Duck and a space-suited Mickey Mouse, they left for the Kennedy Space Center, as Br'er Bear waved good-bye to the visitors.[67]

After a second day at the Cape, the crews flew to Houston. Tuesday morning was occupied by a welcoming ceremony and then briefings on the joint scientific activities planned for the flight. During the afternoon, the crews were given an update on contingency plans covering possible emergency situations.[68] The two teams practiced all the joint flight activities in the Apollo, Soyuz, and docking module mockups, using the latest version of the onboard flight documents. While the prime crews did their walk-through, the backup crews practiced in the simulator. Later, as the prime crews rehearsed communication techniques, their backups practiced the joint activities. This training pattern continued into the weekend. After taking Sunday off, the crews got back to work on Monday, rehearsing the joint activities step by step until Friday the 28th.[69]

Each member of the prime crew completed 61.5 hours of language training during the February sessions. While the greater emphasis was placed on the joint phase, the men also found time to meet with the press and work out in the gym. Leonov and Kubasov tried out the American meals they would eat during their visit to Apollo.[70] Their individual menus looked like the sample in the box at the top of the page. But the visiting Americans might find something different awaiting them aboard Soyuz. (See box below.)

```
                         ASTP Flight Menu

    Alexei A. Leonov-SC              Valeriy N. Kubasov-FE

       Potato Soup                      Seafood Mushroom Soup
       Beef Steak                       Beef Steak
       Rye Bread                        Rye Bread
       Cheese Spread                    Cheese Spread
       Almonds                          Almonds
       Strawberries                     Strawberries
       Tea w/Lemon & Sugar              Tea w/Lemon & Sugar
```

A month and a half later, the crews met again, this time in the Soviet Union. From 14-30 April, the astronauts followed a pattern of activities similar to their Houston training. They practiced transfer and other joint phases of the flight in the Soyuz mockups and worked on numerous contingency situations in the Soyuz simulator. Radio communications skills were polished while "flying" the rendezvous and docking simulators. Two of the highlights of the trip for the Americans were the visit to the mission control center at Kaliningrad on 19 April and the journey to Baykonur Cosmodrome on the 28th. At the Soviet launch center, the U.S. team saw the actual flight hardware as well as the primary launch pad, which was about 2 kilometers distant. When they returned to the States in May, the U.S. crew told the press about what they had seen.[71]

The airfield they had flown into served Leninsk, the modern city of 50 000 built in the desert of Kazakhstan to provide living quarters and logistical support to the space projects at Baykonur. While they did not get exact information on the size of the center, Stafford's visual impression was one of vastness. On their evening flight back to Moscow, the crews saw the lights on launch pads and related complexes for more than 15 minutes, and according to Stafford, "that makes Cape Kennedy look very small."

At one point in the press session, Jules Bergman of ABC News broke in. It had been one of his "pet peeves for years" that the Soviets kept calling their launch complex the *Baykonur* Cosmodrome.

> Baykonur, if you'll look on the coordinates, is 135 miles [217 km] away or something. Tyuratam may only be a railhead, but it is the Tyuratam Launch Complex. They call it Baykonur, I know.... I'm going to call it Tyuratam. ABC is going to call it Tyuratam. SAC [Strategic Air Command] calls it Tyuratam. Can we once and for all straighten that out and arrive at a ... name for it, Tom?[72]

Although Bergman thought that calling the facility Baykonur was like referring to KSC as the "Tampa Space Port," as he put it, Stafford told him that Tyuratam was only "a little bitty old city" that butts up to the new city called Leninsk. Many reporters thought that the Soviets were calling it Baykonur to hide its true location, but Slayton told them that if they really wanted to use the name the Soviets commonly used, they would have to say Baykonur. Indeed, it appeared that the entire area was called Baykonur, much as Texans would talk about the Panhandle or the Gulf Coast.*

Another source of questions from the reporters was the aborted Soyuz launch of 5 April. All the American crewmembers repeated the assurances

---

*With an area of 2 756 000 square kilometers, Kazakhstan is the largest of the union republics, second only to the Russian Soviet Federative Socialist Republic. Even Texans would be staggered by the size, since it is nearly four time larger than the Lone Star State.

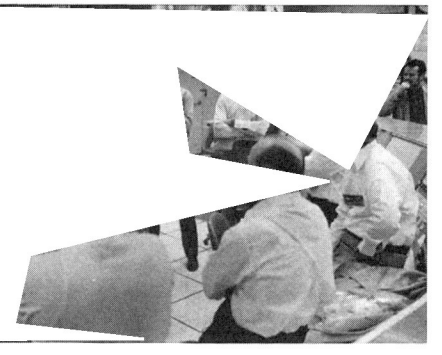

Above, ASTP crewmen are briefed on Apollo-Saturn automatic checkout equipment during a three-day inspection tour at the Kennedy Space Center, 8-10 February 1975. Right, Alexei Leonov enters the Apollo command module being readied for the joint mission. Below, Tom Stafford playfully tweaks Mickey Mouse's nose at Disney World, Florida. Fellow sightseers are (left to right) V. N. Kubasov, Deke Slayton, Vance Brand, A. A. Leonov, and V. A. Shatalov.

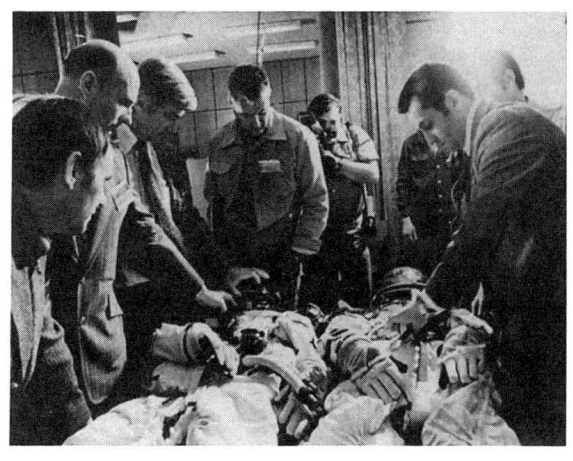

April 1975 American ASTP crews tour the Soviet ssion Control Center at Kaliningrad (above) during e final pre-flight training session in the Soviet ion. At right, Cosmonaut Aleksandr Ivanchenkov plains the operation of the Soviet space suit to tronauts Bean, Stafford, Cernan, and Overmyer ring the crew visit to Baykonur (Soviet Academy Sciences photos).

given earlier by Glynn Lunney that the upper stage failure reported by the Soviets would have no effect on ASTP. The two technical directors had briefly discussed the matter on 8 April over the telephone. The launch vehicle that had failed was one of the earlier models of the basic booster. For ASTP, a newer vehicle with a higher thrust capacity for weight would be used, and that booster had a successful launch history. Bushuyev promised Lunney full details on the "Soyuz anomaly." The astronauts were not worried about its impact on the flight; after all, the Soviets had a second spacecraft and launch vehicle set aside in the unlikely event of a repetition of the problem.[73]

Nick Chriss of the *Los Angeles Times* asked Stafford if he would be "satisfied with the type of automatic abort system they have, since the two [cosmonauts] almost landed in China." The astronaut replied that it had performed as it was supposed to; it had saved the cosmonauts' lives. If an American crew ever had to use their abort system, it might put them down in the "mid-Atlantic or the far-Atlantic Ocean," but wherever it put them the important thing is for the system to work. Stafford and Brand once again said that they were genuinely satisfied with the flightworthiness of Soyuz.[74]

The American crew was not particularly concerned about the 5 April Soyuz "anomaly," but Senator Proxmire was. Three days after the failure, he had requested "that the Central Intelligence Agency (CIA) make a safety assessment of the Soviet manned space technology in view of the failure of another Soviet space mission . . . and the pending U.S.-U.S.S.R. Apollo-Soyuz joint project." He told fellow senators that the "in launch failure . . . reinforces my deep concern that the upcoming . . . experiment may be dangerous to American astronauts." He did note that NASA claimed that the mission would be as safe as any other flight in the Apollo program and that they had produced a mass of studies to back up this assessment. But the senator from Wisconsin disagreed with their conclusions, saying "The history of the Soviet manned program shows an appalling lack of consistency. As soon as one severe problem is solved another occurs." Proxmire presented a box score of Soviet failures:

> Since April of 1967, the U.S.S.R. has conducted 18 manned Soyuz flights. Of these, two have been catastrophic failures with loss of four lives. In addition, two other flights, Soyuz 10 and Soyuz 15 have had docking problems and cannot be considered successes. Most recently was the launch failure of what would have been Soyuz 18. Thus, five out of 18 Soyuz flights have been marred by some sort of failure.[75]

Given this "poor track record," Proxmire wanted the CIA to investigate the Soviet program and report to the HUD, Space, and Science subcommittee well in advance of the July launches.

PREPARING FOR THE MISSION

As Senator Proxmire worried about the safety of Soyuz, the astronauts returned to their daily training routine. Stafford, Slayton, and Brand participated in the second session of a three-part series of intercontrol center simulations. On the morning of 13 May, 1 hour before the projected launch time of Soyuz, the simulations began. For over 25 hours into a simulated mission, the Soviet and American ground controllers and flight crews rehearsed both launches and made a number of the scheduled maneuvers. Beginning on the 15th, the two teams conducted a 56-hour continuous simulation that covered the period from 47 hours, 10 minutes, to 103 hours ground elapsed time following the Soyuz launch; rendezvous, docking, crew transfers, undocking, second docking, and final separation were rehearsed. Four days later, on the 20th; a 9-hour rerun of the rendezvous and docking exercise was performed. This training gave the control center personnel and the crews another chance to check out plans covering emergencies that might arise during the flight.[76]

Before the final simulation—29 June to 1 July—a major meeting took place in Moscow. Coming at the end of the last plenary gathering of the Working Groups, the Joint Flight Readiness Review chaired by V. A. Kotelnikov and George Low represented the final preflight evaluation. Once Kotelnikov and Low agreed that the spacecraft were ready to be flown, the two teams could attend to the final details—preparing the launch vehicles, as the clock ticked off the remaining hours to launch. Crew training would continue up to the very last day, the real test of their efforts coming on the 15th of July.

X

Final Examinations

In May 1975, George Low traveled once again to the Soviet Union, this time to inspect the Soviet spacecraft and to jointly chair the Flight Readiness Review (FRR). As with the Mid-Term Review, Low's correspondence prior to the meeting had been with Soviet Academy President Keldysh. When the NASA delegation arrived in Moscow, Petrov told Low that Keldysh's health had taken another turn for the worse; he was in the hospital. Three days later Petrov gave Low the news—Keldysh had decided to step down from his post. Vladimir Aleksandrovich Kotelnikov, as acting President, would supervise the FRR for the Soviet side. Kotelnikov, who had a good command of English, at 63 was well known throughout the U.S.S.R. for his textbooks in the field of radio and electronics engineering. While saddened at the news of Keldysh's poor health, Low knew he could work with his successor. But prior to the review, Low and his colleagues visited Baykonur.

Final checkout of the American communications equipment and the docking target alignment tests had been carried out at the Baykonur Cosmodrome during mid-May. This last major activity involved 16 Americans and their Soviet colleagues under the direction of R. H. Dietz and B. V. Nikitin. They had finished their work on the 17th, ahead of schedule, in time for Low; Arnold Frutkin; Glynn Lunney; John F. Yardley, the Associate Administrator for Manned Space Flight; and Walter J. Kapryan, the Director of Launch Operations at Kennedy Space Center (KSC), to visit the cosmodrome on an inspection and orientation tour. Low and his group left Moscow's Vnukovo airport on the afternoon of the 18th. After an evening's stay at the Cosmonauts' Hotel in Leninsk, the five Americans and Professor Bushuyev set out for the launch pad in the van usually reserved for transporting the cosmonauts. Their ride took them through the launch site industrial area to the launch stand.

At the launch pad, they stopped first at a small monument commemorating *Sputnik I*, which had been launched from this stand on 4 October 1957. Low noted in his trip report:

> [This same] pad was used for Sputnik I, for Gagarin's flight, and will be used for one of the ASTP birds. We asked how many launch vehicles had gone off

this pad and got two different answers: one being 100 and the second being 300. It was well preserved and painted, and apparently had been repainted prior to our visit. The basic sequencing is all mechanical. The vehicle is not held down but is guided by various arms which are part of the stand. It flies out of the launch pad without holddown, and the arms and booms which support the launch vehicle fall back under counter balance and the force of gravity. Various platforms underneath the launch vehicle are moveable and apparently collapse in a certain way so that they can all be rolled underneath the pad.[1]

Low also reported that it was very windy at Baykonur. He had been told "that the temperatures in the summertime go to 40°C and in the winter to -40°C." The Soviets indicated that a minimum of work was done out of doors in the winter months.

While at the launch stand, the Soviets and Americans discussed various aspects of launching spacecraft. Bushuyev and Dmitri Bolshakov, the director of the Baykonur Cosmodrome, were interested to learn why the U.S. launch vehicles were held down for a short period of time after the engines were ignited. The Americans explained that this ensured smooth combustion and thrust buildup. Only after the engines were running satisfactorily were the launch vehicles released. Low noted, "apparently [the Soviets] measure the thrust buildup curve for all 20 engines and can shut down during the buildup until just before lift-off." Since they did not need a hold down system, they avoided this complex procedure. The Soviet launch vehicle was also "slightly more efficient . . . from the point of view of fuel consumption," and the Soviets said that they had never lost a launch vehicle as a result of improper thrust buildup.[2] From the launch pad, the Americans were taken to the industrial area where the spacecraft were readied for their flights.

Since the industrial sector was only a short distance from the pad and since there were homes and a hotel there, Low asked if this area was evacuated during launches. Bolshakov responded that it was cleared just before a flight. After the Americans were escorted into a huge building that housed the equipment used to check out spacecraft systems, they visited the Soviet equivalent of the KSC vehicle assembly building. Along one side of the building, the prime launch vehicle rested horizontally. The two spacecraft (prime and backup), which the astronauts had examined in April and which Dietz' team had checked out earlier that month, were also housed here. Low recorded that he "spent considerable time walking around the launch vehicle"; he "asked a lot of questions all of which were answered." He had been told earlier that "the same [kind of] launch vehicle had been used since October 1957 and the first Sputnik launch." Low concluded that the Soviets periodically introduced "block changes" into the launch vehicles either to enhance reliability or to replace obsolescent component parts. He continued his report:

> We were told that the ASTP launch vehicle is one in a series of which more than 10 have already been flown. The April 5 launch was conducted with a launch vehicle left over from the previous series. The failure of the April 5 launch was explained to us again in detail, and our people appear to be satisfied with the explanation and with the fact that the changes made in the ASTP launch vehicle should prevent this kind of failure. The launch vehicle (all three stages) was on the right hand side of the center aisle. On the left side were the two spacecraft and the spacecraft shrouds. The launch vehicle was horizontal; the spacecraft were vertical. In order to place the spacecraft inside the shrouds, they are tipped to a horizontal position and cantilevered horizontally from the launch stand. The shroud is then slid over the spacecraft and attached to the bottom ring. At some point in the process, the spacecraft is then taken out on a railroad car and fueled. Then the launch vehicle and spacecraft are mated in a horizontal position on the car which ultimately takes them to the launch pad.[3]

The Americans spent the remainder of their visit to Baykonur attending a briefing on the communications electromagnetic compatibility tests just

Above, George Low (right) visits the Soviet monument commemorating the launching of Sputnik I *in October 1957. Others present are, left to right, W. J. Kapryan, G. S. Lunney, and A. W. Frutkin. Inside the Manned Spacecraft Assembly Building at Baykonur (above, right), George Low (center), Professor Bushuyev, and the Soviet interpreter listen to a briefing on the Soyuz launch vehicle. The first-stage engine nozzles are visible at the far right. Both ASTP Soyuz spacecraft are being readied for the joint mission (right). The extended solar panels will be folded back so that the protective launch shroud (white cylinder to the left of the far craft) can enclose the spacecraft (Soviet Academy of Sciences photos).*

completed, touring the Korolev and Gagarin cottages at the launch site, and visiting Leninsk. During the stop at the Korolev house, the Americans got a rare insight into Bushuyev's past. Throughout ASTP, the Soviets had given the Americans little information about their personal backgrounds in the space program. But when the U.S. team visited Korolev's cottage, "Bushuyev told us that he spent much time there with Korolev and apparently stayed there on several occasions," Low noted. On further questioning, Bushuyev told them that he had started working with Korolev right after World War II on the launch vehicle for Sputnik and on the spacecraft, too. Since then, he said, he had concentrated mostly on spacecraft. The Americans and their Soviet hosts then returned to Moscow for the FRR.[4]

REVIEWING FLIGHT READINESS

On 22 May, Low and Kotelnikov chaired the joint Flight Readiness Review at the Presidium of the Soviet Academy of Sciences. Reviewing the extensive preparations and testing that had taken place since the Mid-Term Review held in October 1973, the FRR was patterned after those traditionally conducted by NASA before all U.S. space flights. This formal management evaluation was designed "to assure that all appropriate steps [had] been taken by both sides to verify that the critical equipment and operations of each side [had] been planned or manufactured to meet the IED/ASTP Documentation Requirements."[5] Nearly all the joint pre-flight activities had been completed in time for the FRR. Concurrent with the launch site activities, representatives of Working Groups 2, 3, 4, and 5 had put the finishing touches on their pre-flight preparations, and the Technical Directors had cleared up their last minute questions of mutual interest.

Lunney and Bushuyev began the 5-hour review with technical histories of their respective spacecraft. In addition to comments about the readiness of ASTP hardware, the Technical Directors indicated that 133 documents had been negotiated and signed. While some flight-related documents were to be updated prior to the mission, only the post-mission report remained to be prepared. During their hour-presentation on Working Group 0 activities, Low raised a few questions. He asked Bushuyev if there had been any hardware anomalies in either of the unmanned ASTP test flights or in *Soyuz 16* that would require changes in the ASTP flight hardware. Bushuyev indicated that there had been only two minor problems—the cabin cooling system had directed too much cool air on the cosmonauts' feet, and the crew had had some minor difficulties with the food. Neither of these problems had required hardware changes. Bushuyev reported no difficulties with ASTP hardware during the test missions.[6]

FINAL EXAMINATIONS

With the completion of their presentations, the Technical Directors turned the proceedings over to the Working Group chairmen. During the course of the review, fifteen Soviets and twelve Americans at the main conference table followed the presentations in specially prepared notebooks, which contained briefing charts and bilingual illustrations. The Working Group presentations were made by the co-chairmen who had not reported during the Mid-Term Review—V. A. Timchenko, V. P. Legostayev, R. D. White, B. V. Nikitin, and W. W. Guy.[7]

MISSION OPERATING PLANS—REVIEW

Timchenko, speaking for Working Group 1, addressed that team's four major areas of responsibility—flight operations, operations training, experiments, and spacecraft compatibility. Planning the flight operations had been an exhausting and time consuming exercise, for besides planning for the projected 15 July launch date, the flight planners had had to map out alternative flight plans for a series of launch dates so as to be prepared for a postponement in case of equipment failures or weather problems. Mission planning analysts led by Kenneth A. Young and Oleg Georgiyevich Sytin had to consider a host of variable factors with each subsequent launch date. Lighting conditions at the Soviet and American launch and recovery sites constrained their planning considerably. Experiments keyed to the position of the sun or other stellar objects had to be juggled around in each flight plan to make certain that they would take place at the precise moment and place required by the experiment plan. Each alternate launch date also required its own tailored flight plan and trajectory computations, as well as documents verified in both Russian and English.

Timchenko reported that the process of planning for prime and alternate missions was completed. ASTP 40 301, "Joint Crew Activities Plan," reflected their work. Furthermore, the other groups had verified these flight plans for compatibility; training exercises in the mockups and simulators had disclosed no difficulties in flying the mission as outlined. Timchenko concluded that there were no unresolved questions relating to flight procedures or the mission timeline.[8]

He then turned to discussion of the control centers interaction plan—how the two centers in Moscow and Houston would operate during the mission. Over the many months of negotiations, teams under M. P. Frank and F. C. Littleton and A. S. Yeliseyev and Timchenko had codified several key agreements concerning control of the mission. Flight operations were to be directed by a flight director in each control center, with each side having basic responsibility for its own spacecraft and crew. These men would converse with the crews through the spacecraft communicator and with each other through the Joint Flight Directors and their interpreters. Under both normal and emergency conditions, these interpreters would play a key role in the management of the mission. Once the mission was underway, the burden of the responsibility would be on the shoulders of the flight directors and the Joint Flight Directors. Lunney and Bushuyev would act in a liaison and advisory role. To ensure the prompt resolution of technical questions that might arise during the flight, each side was to appoint a group of visiting specialists (the "consultative group," as the Soviets called these teams) to be present in a support staff room near the other country's control center.

George Low asked Timchenko about contingency planning. Should an emergency call for a deviation from the established flight plan, who would make the decision about the proper corrective action to be taken? Timchenko replied that the flight director would make the decision with preference being given to a solution based upon procedures that had been worked out before the flight. Low then inquired as to which side would make the decision about an in-flight emergency. In the case of a problem involving crew safety, Timchenko answered, the country whose men were in danger could take unilateral action. For example, the endangered crew could call for an undocking, which would be evaluated by that side's flight director, who would notify the other crew and ground controllers of his decision through the Joint Flight Director. Bushuyev interjected at this point that there were plans for a number of specific types of emergencies, the so-called "examined contingencies." Low probed deeper and asked Timchenko what would happen in a case where there were no communications with the ground. The Soviet group chairman responded that such possibilities were specifically addressed in the "Flight Plan Guidelines," ASTP 40 300, and the "Contingency Plan," ASTP 40 500. Going still further, Low inquired what would happen in the event of an "unexamined contingency."

Flight Readiness Review, Moscow, May 1975. Seated together, the five Soviet Working Group chairmen: left to right, B. V. Nikitin, Yu. S. Dolgopolov, V. S. Syromyatnikov, V. P. Legostayev, and V. A. Timchenko.

Timchenko said that the crews had been trained to make joint decisions on their own if necessary. Bushuyev added that each commander had the prime responsibility for his craft. Should a problem arise in Apollo, Stafford would have the responsibility to solve it; in Soyuz, such a decision would be Leonov's burden.[9]

Pursuing the issue of command further, Timchenko indicated that one country's spacecraft could communicate through the other country's ground stations to its own control center. This arrangement especially broadened the amount of contact time Moscow control would have with Soyuz since the ASTP trajectory took the craft on a path away from many of the Soviet ground tracking stations. In addition, communications between the centers would consist of ten voice channels, two Teletype channels, two television channels, as well as channels for retransmitting communications with the crews and for transmitting facsimiles of document pages or computer printouts. This complex system, worked out by the subgroup on Intercontrol Center Coordination led by John H. Temple and Viktor Dmitriyevich Blagov, had been tested in December 1974 and in March and May 1975. These tests indicated that the system and the bilingual personnel assigned to work on both sides as interpreters could work satisfactorily under normal and emergency situations.[10]

In his report on crew and ground support personnel training, Timchenko summarized the joint training sessions. While the crew sessions had received considerable publicity, the equally important work of the control center personnel had not. Teams of flight controllers, visiting specialists, communications technicians, and interpreters had worked in both the Moscow and Houston control centers for ten-day familiarization exercises. The American controllers completed their training in Moscow on 27 September 1974, and the Soviets finished their studies on 6 November 1974 in Houston. These sessions had been followed by joint simulations, which not only provided an evaluation of the communications but also gave all parties an opportunity to work together in a condition similar to that of the mission. Problems and equipment failures were introduced by the training leaders to give the flight control teams experience in coping with emergency situations. Timchenko indicated that the crews and flight controllers had successfully completed their training and appeared to be ready for the flight.[11]

In the final part of his report, Timchenko summarized the preparations made for conducting the five joint experiments. The requirements for each of these (microbial exchange, zone-forming fungi, furnace systems, artificial solar eclipse, and ultraviolet absorption) were documented in separate interacting equipment documents, and the operating procedures and plans for each had been incorporated into the appropriate onboard instructions documents. Those procedures and plans had been verified and practiced by the crews in the mockups and simulators. Only the ultraviolet absorption experiment had required further work to perfect the flight maneuvers associated with it.[12] Timchenko indicated that all the major tasks of Working Group 1 had been completed. The only remaining work related to solving some communications difficulties identified during the simulation and to conducting the June joint simulation. Otherwise, he reported that Group 1's personnel were ready to carry out their part of the mission. Having completed his remarks, he turned the meeting over to V. P. Legostayev, who addressed Working Group 2's preparations.

GUIDANCE AND CONTROL–REVIEW

Beginning with a report on the ASTP docking targets, Legostayev indicated that the primary target had been mounted on Soyuz and its proper alignment had been verified by Soviet and American specialists during the joint preparations at Baykonur during May. These checkout procedures involved the use of an American-designed fixture mounted on the face of the Soviet docking gear. A contingency target made from three fixed metal plates was also installed on Soyuz, in case the folded primary target failed to erect. In either event, the Apollo commander would have a target on which to sight during the final phase of docking.[13] After discussing the orientation lights and the optical tracking hardware, Legostayev turned to the subject of control systems.

The Soviet Working Group 2 chairman noted that extensive studies had been made of the control system from operational and safety standpoints.

Docking target alignment fixture.

FINAL EXAMINATIONS

The joint team had agreed to two major limitations on the use of the Apollo control system. First, to prevent undue stress from being placed on the Soyuz solar panels, the American agreed to use only two of the four roll jets to rotate the spacecraft during the docked phase of the flight. The second agreed restriction had to do with shutting down the Apollo forward-firing reaction control system (RCS) engines. As agreed during the final meeting in Houston, to protect Soyuz from the RCS plume, the forward-firing engines would be shut off within 2 seconds after docking system capture. No forward firing would be allowed during the docked part of the flight operations.[14]

The basic safety of the Apollo and Soyuz propulsion and control systems was documented in reports on each spacecraft, which provided a functional description of those systems and of how they operated. The Soviet and American specialists concluded that the control and propulsion systems of both spacecraft operated in a non-hazardous manner and posed no dangers to the crews during docked operations. Finally, Legostayev told the FRR Board that there were no outstanding issues; from Group 2's point of view all was in readiness for the flight.[15] After a few quick questions from John Yardley and Academician Petrov, the FRR attendees adjourned for lunch.

DOCKING SYSTEM–REVIEW

When the meeting resumed at 2:15, Bob White spoke for Group 3. After reviewing the major elements of the docking hardware, he summarized the recent test history of that equipment. Following the fourteen-week mate and dynamic development tests conducted in 1973, three more major test activities had been carried out successfully. The first of these, the mate and dynamic qualification testing, had been done in Houston (1 July-5 September 1974) to certify physical, functional, and operational compatibility, to verify the integrity of the docking systems under maximum docking loads; and to observe the operation of the guide ring and capture hooks on the active system. In two parts, the examination had covered the mate check and the functional test of the hardware when fit together under ambient temperatures. Subsequently, the hardware was placed on the dynamic docking vehicle simulator to test it at high (70°C) and low (-40°C) temperatures. The docking systems were run together in a conglomerate of experimental dockings to determine the operation of the systems under both normal and worst-possible circumstances. Roll, pitch, yaw, and axial misalignments were combined with various initial contact velocities to determine how this equipment, built to the same specifications as the flight hardware, could be expected to function in space.

Only one problem developed during these tests. When the U.S.S.R. system went through the active phase of the cold temperature test, the resulting data did not agree with that obtained either during the earlier development tests or with the Soviet mathematical analysis. When the hot temperature test was run, the testers noticed oil leaking from the attenuator rods. Upon examination, they discovered that an anti-corrosion grease placed in the hollow rods to prevent rusting had not been removed. The grease had gummed up some operating parts. The second Soviet qualification unit did not have the grease problem, nor did the flight units back in Moscow. Once the errant lubricant was removed from the unit in Houston, it functioned satisfactorily. Otherwise, the U.S. and U.S.S.R. systems performed as expected.[16]

When he turned to the pre-flight compatibility verification test of the actual flight docking systems, White noted one major alteration that had been made during the test. The Americans had altered the alignment pins and sockets on their docking gear. Group 3 had recognized a potential problem with the existing alignment pins and sockets in October 1974 during the acceptance testing of the docking system at Rockwell International. This exercise, designed to prove the flight hardware was acceptable prior to the government's paying for it, tested the docking systems in a special horizontal fixture, which took into account the absence of earth's gravity in space. Guide rails with roller bearings kept the extended active guide and ring from drooping from its own weight. These guide rails were supposed to align the two docking gear, but there was a slight misalignment. When the two systems were brought together and capture was achieved, the active system could not fully retract to achieve structural latching. A gap of about 6 millimeters existed, a small distance but enough to prevent a transfer in space. Once the Rockwell engineers discovered that the guide rails were not aligning the two systems properly, they adjusted them and tried the exercise again. The second time there was no problem; there had been no difficulty with the prime flight docking system.[17]

But this experience at Rockwell left Glynn Lunney with a nagging concern. He decided that the issue required further analysis. Meanwhile two of the docking systems were shipped to Moscow for the compatibility verification tests. Bob White and his eleven technical specialists followed. Once Rockwell completed their analytical work on the so-called friction lock problem, Lunney called White to discuss the results. In Moscow, White and his colleagues, the "Dirty Dozen" as they called themselves, disagreed with Rockwell's findings, which indicated that friction lock was a problem that could influence the conduct of the flight. White argued that Rockwell's assumptions were too severe. The circumstances as described by Earl

A NASA inspector observes mating of American Docking Systems 4 and 5 on the horizontal test fixture at Rockwell International factory in Downey, California.

V. S. Syromyatnikov, R. D. White, E N. Harrin, and Professor Bushuyev, in the latter's Moscow office, discuss the need to change the alignment pin and socket in the American docking system on the telephone with Glynn Lunney in Houston, November 1974. In December Ray Larson and C. E. Kindelberger of Rockwell International (right) fly to Moscow and install the modified alignment pin.

Holman, who conducted the analysis, were unlikely to occur in space. White also opposed changing the flight hardware at that very late date because the alterations were likely to create real problems while attempting to solve a possible one. Both sides had agreed that following the qualifications tests the design of the "interfacing hardware" (i.e., the docking system) would be frozen. No further changes should be made.

Nevertheless, Lunney, in consultation with NASA Headquarters and Rockwell, decided that the pins and sockets had to be altered. The agency saw the following hazard:

> The potential pin and socket binding problem of both docking systems can lead to the stall of the guide ring drive in an active USA docking which can produce an overload in one of the three cables retracting the ring, which in

turn can lead to its failure. Failure of a guide ring retract cable will result in the inability to further use the US docking systems either in the active or passive mode and, as a consequence, the failure to complete the basic purpose of the flight.[18]

As a consequence, Lunney telexed Professor Bushuyev and explained the problem to him.[19] (See box on facing page.)

While Rockwell manufactured the new pins and sockets, Bob White discussed the problem with V. S. Syromyatnikov and Professor Bushuyev. White believed that "Vladimir was sympathetic to our problem and agreed that the change was necessary."[20] Without Syromyatnikov's understanding, it would have been very difficult to sell Bushuyev on the alteration. The only constraint the Soviets placed upon the Americans was that the U.S.S.R. pins and sockets would remain unchanged.

> The USSR docking system guide pin and socket are installed from the back side of the structural ring front flange. In addition, the socket is installed before the docking system differential drive assembly is installed. Therefore, the socket removal and installation requires substantial docking system disassembly and readjustment of the kinematic coupling of six attenuator rods with the differential drive assembly which is a laborious operation.
>
> Considering that all three USSR flight docking systems are assembled and in readiness for the Preflight Mate Test . . . the docking system rework is impossible without a review of the schedule of preparing the spacecraft for flight and launch date itself.[21]

White and Syromyatnikov agreed that it would be sufficient to change only the U.S. pin and socket. This understanding was officially recognized by Lunney and V. A. Timchenko in their telephone conversation of 19 November.[22]

To accomplish the modification in the shortest time, White and Syromyatnikov modified the test plans and worked out procedures to test the new components while waiting for Arnold D. Aldrich, Lunney's Deputy, and Ray F. Larson, Rockwell's command and service module manager, to arrive. Once in Moscow, Aldrich worked with the Soviets to draft minutes covering the changes and establishing a test plan for an additional examination of the modified system in January 1975 at Downey. Ray Larson and Earl Holman worked with the "Dirty Dozen" to complete the changes. The American Group 3 delegation divided into two groups; the "Mod Squad" made the alterations on one docking system, while the "Test Team" continued testing the other system. Whereas Rockwell had predicted that it would take four days to complete the modifications, the specialists actually needed only 4 to 5 hours for each docking system. Total time lost

TO:

PROFESSOR K. BUSHUYEV
ACADEMY OF SCIENCES OF THE USSR
14 LENINSKI PR
MOSCOW, V-71, USSR

UNCLAS IN REPLY TO PA-LSN-213-74

MR. ALDRICH MENTIONED TO YOU DURING THE NOVEMBER 5 TELECON THAT WE
WERE SOMEWHAT CONCERNED ABOUT THE POSSIBILITY THAT UNDER CERTAIN
CONDITIONS IT MIGHT BE POSSIBLE FOR THE CURRENT DOCKING SYSTEM
ALIGNMENT PIN AND SOCKET DESIGN TO BIND AND PREVENT FINAL
RETRACTION. WE HAVE JUST COMPLETED A SERIES OF TESTS AND HAVE
CONCLUDED THAT BINDING OF A U.S. PIN IN A U.S. SOCKET CAN OCCUR UNDER
THE FOLLOWING CONDITIONS: A SIDE LOAD ON THE PIN ON THE ORDER OF 300
LBS [136 kg] OR GREATER WILL CAUSE BINDING WHEN A HESITATION AND
RETRACTION OF THE PIN FROM THE SOCKET OF 0.020 INCHES [0.5 mm] OCCURS
JUST AS THE PIN IS ENTERING THE CYLINDRICAL PORTION OF THE SOCKET. THE
HESITATION AND REENTERING OF THE PIN CAUSES THE SOCKET TO ROTATE
TO THE HARD STOP POSITION.

OUR ANALYSIS INDICATES THAT PREDICTED MISALIGNMENTS AND THERMAL
CONDITIONS COULD PRODUCE WORST CASE IN-FLIGHT SIDE LOADS OF 250-450 LBS
[113-204 kg]. WE ALSO FEEL THAT THE DYNAMIC ENVIRONMENT IS SUCH THAT
A SLIGHT HESITATION AND RETRACTION OF THE PIN DURING DOCKING SYSTEM
RETRACTION IS POSSIBLE, AS ARE MANY OTHER DYNAMIC EFFECTS.

BASED UPON THESE TESTS, WE ARE INCREASINGLY CONCERNED THAT THE
CURRENT PIN AND SOCKET DESIGN IS INADEQUATE, AND ARE SERIOUSLY
CONSIDERING THE POSSIBILITY OF CHANGING THE U.S. HARDWARE CURRENTLY
IN MOSCOW. WE ARE IN THE PROCESS OF BUILDING THE PARTS NECESSARY TO
CHANGE BOTH THE U.S. PIN AND THE U.S. SOCKET ON DOCKING SYSTEMS 5 AND
7. THIS HARDWARE WILL BE BROUGHT TO MOSCOW BY MR. ALDRICH ON
NOVEMBER 18, ALONG WITH INSTALLATION TOOLING AND PROCEDURES.

THE OPTIMUM SOLUTION TO THE PROBLEM, HOWEVER, MAY BE TO REPLACE BOTH
THE U.S. AND U.S.S.R. SOCKETS WITH A NEW NON-ROTATING SOCKET WHICH HAS
A SHORTENED CYLINDRICAL SECTION, AND NOT CHANGE THE U.S. PIN. FOR THIS
REASON, WE RECOMMEND THAT YOU PROCEED TO FABRICATE A NEW SOCKET.
THIS WOULD ALLOW YOU TO BE PREPARED SHOULD WE DECIDE A CHANGE IS
REQUIRED. MR. WHITE HAD THE PROPOSED SOCKET DESIGN. I HAVE ASKED MR.
WHITE TO DISCUSS THIS PROBLEM WITH DR. SYROMYATNIKOV, AND WOULD
APPRECIATE ANY INFORMATION OR TEST HISTORY WHICH YOU HAVE THAT MAY
RELATE TO THIS PROBLEM. THESE DATA SHOULD BE PROVIDED TO MR. WHITE.

I PLAN TO TALK TO MR WHITE AGAIN ON TUESDAY, NOVEMBER 12, AT 5:00 PM
MOSCOW TIME

I BELIEVE THAT WE MUST REACH A FINAL DECISION ON THIS QUESTION AT OUR
NOVEMBER 19 TELECON.

PLEASE PROVIDE MR. WHITE WITH A COPY OF THIS MESSAGE.

GLYNN S LUNNEY
MANAGER
APOLLO SPACECRAFT PROGRAM   NOV 8 1974

from the compatibility test was two days. Looking back, Bob White commented, "It's amazing how much work you can do when you have no other choice but to get it done." The Soviets were astounded, too![23]

After the Moscow trials, joint pin and socket interaction verification tests using a Soviet and American docking system were conducted 16-23 January 1975 at Downey. This exercise confirmed that the alterations made to the American pin and socket eliminated any possibility of friction lock. Over a two-year period, Working Group 3 had conducted six major joint tests involving over thirty-three weeks of activity.[24]

COMMUNICATIONS AND TRACKING—REVIEW

After White's 40-minute presentation at the FRR, Boris Nikitin gave a 10-minute summary of Group 4's activities. Although he did not dwell on them, this group also had conducted an important series of tests since the Mid-Term Review. First, they had to establish that the spacecraft-to-space-craft radio and cable intercommunication systems would work without any interference from internal or external power sources. In addition, they had to verify the safety of the pyrotechnics used in the two spacecraft—for example, the explosive bolts reserved for emergency undocking. To prove that the radio waves from neither craft—especially from the powerful Apollo high gain antenna used to communicate with the ATS-6 satellite—could detonate those pyrotechnic components, Working Group 4 had directed a series of radiofrequency radiation experiments. All of these tests had been favorably concluded. Just prior to the Flight Readiness Review, R. H. Dietz and his fifteen teammates had participated in the checkout of the American communication and ranging equipment that had been installed into the prime and backup Soyuz spacecraft and in electromagnetic compatibility tests of American equipment scheduled to be transferred into Soyuz during the mission.

Every spacecraft had its own electromagnetic environment created by the sum of all the electronic and electric components onboard. Just as an electric drill may affect television reception or as a citizen's band transmitter may affect FM radio reception, so too may the energy radiating from switches, fan motors, or power cables interfere with television cameras or communications gear. Electromagnetic compatibility (EMC) was not generally a problem when one party developed its own equipment for use in its own spacecraft. But the possibility of a problem with electromagnetic interference might arise when equipment from one electromagnetic environment was transferred to another. During January and February 1975, Soviet specialists had accompanied their counterparts to the Kennedy Space Center to check out their television camera inside the command and docking modules for EMC. And during May, the Americans had taken their television camera, motion picture camera, headsets, microphones, and speaker boxes to Baykonur to determine if the electromagnetic environment of Soyuz interfered with their performance. The test team found that all the American equipment operated satisfactorily. Although the Soviet half of Group 4 had experienced some difficulty meeting project deadlines before the Mid-Term Review, those problems had been resolved. Nikitin could report at the FRR that all their joint work had been completed.[25]

Academician Petrov had a few questions for Nikitin. His main concern was radio receiver interference on the Soviets' 121.75-megahertz frequency. During *Soyuz 16,* the crew reported receiving broadcasts from commercial aviation sources transmitting on that frequency. Petrov had asked the Americans to help them get the international radio users to vacate that channel during the mission, but NASA had decided not to take such an action. So when Petrov asked Nikitin what plans had been made to deal with such interference if it developed during the flight, the Soviet chairman said that they would just try to identify the source and ask the transmittor to suspend broadcasts during the remainder of the mission. After some further discussion on this point, Walt Guy reported on Working Group 5.[26]

LIFE SUPPORT AND CREW TRANSFER–REVIEW

Group 5 had conducted a series of environmental control tests since the Mid-Term Review to determine the flight readiness of the Apollo docking module and Soyuz. Equally significant was the work done since October 1973 to ensure the non-flammability of American and Soviet equipment that was to be transferred from one spacecraft to another. This topic had not been addressed in any detail until the Mid-Term Review. Since the Soviets used an 80 percent nitrogen/20 percent oxygen atmosphere, spacecraft flammability was not as severe a worry as it was for the Americans in their nearly pure oxygen atmosphere spacecraft. Walt Guy, in looking back on this topic, commented:

> We had seen movies in which they wore what appeared to be woolen clothes and fur hats, so we didn't feel that they had addressed the question of flammability. At one point, we considered putting all transferred equipment—such as space suits—into our fire proof bags. After EVA went away, our concern became one of not introducing materials into each other's spacecraft that could cause a fire. Our safety people were still concerned that their spacecraft might be on the lucky side instead of the safe side. Obviously, we flew a lot of missions before our Apollo disaster, which proved that we were more lucky that safe. There was a lot of concern about the basic design of the Soyuz from a safety point of view. Lunney got the Soviets to agree to

a certification document for the non-flammability of each piece of transferred equipment. When we eliminated the EVA and reduced much of the equipment to be transferred, the list became much shorter; we were able to consolidate all those documents into a single document.[27]

After considerable discussion, the Soviets agreed to use the American flammability test procedures to determine the safety of their equipment. One key point dealt with the cosmonauts' flight suits. Since the Americans could not let the Soviet crew enter Apollo wearing wool or cotton clothing, they volunteered to give the Soviets enough material to manufacture new suits. But Lavrov declined the offer, saying that the Soviets intended to develop a flameproof material of their own. After several experiments, Lavrov's team produced a cloth that Walt Guy noted was superior in its self-extinguishing characteristics to the material used by the Apollo crew. In a pure oxygen environment, the Soviet cloth, called Lola, would self-extinguish, whereas the American material tended to burn very slowly. During the development of their fabric, the Soviets had brought successive samples of the material to the U.S. for the Johnson Space Center (JSC) specialists to test. Lavrov was proud of the work that his Group 5 people had done, and he had used the samples to demonstrate their progress to Guy and his colleagues.

Once they got involved in the fire safety topic, the Soviets subjected nearly all the items they planned to transfer to rigorous testing. On the American side, the NASA team used four methods to determine the flame-proof nature of their materials. In addition to testing, they used analysis, similarity, and waiver. Seventeen items of American equipment to be transferred to Soyuz were certified by analysis to be safe by virtue of the materials from which they had been fabricated, for example, sunglasses, wrist watches, writing instruments, sliderules, and the like. Six other pieces of equipment, such as speaker boxes, had been approved for flight by determining that they were similar to hardware previously tested and found safe. Only four articles were certified using waivers. Walt Guy could tell the FRR Board that Working Group 5 had no open items. All equipment to be transferred had been cleared for fire safety.[28]

Summarizing the Working Group reports, the Technical Directors indicated that all project milestones had been completed as scheduled. The two teams had finished their detailed review of joint flight safety issues and had prepared safety assessment reports to clarify the safety of selected design areas. Lunney and Bushuyev listed areas in which work remained to be completed:

JUNE MCC [Mission Control Center] SIMULATION
FINAL UPDATE OF ON-BOARD DOCUMENTS

PREPARATION FOR JOINT SCIENTIFIC EXPERIMENTS
ANALYSIS OF FAILURE OF SENSOR INDICATING UV [ultraviolet]
   RETROREFLECTOR OPEN
POST FLIGHT REPORT
PUBLIC INFORMATION ITEMS[29]

George Low recorded in his notes on the FRR that "there were no serious open issues, and it was quite clear that this review was considered to be a formality by the ... Soviet side."[30] R. H. Dietz noted that during the morning session of the FRR Low was the only person to extensively question the Directors and chairmen. Since the Deputy Administrator himself remarked about this after the lunch break, Petrov asked a few questions and Kotelnikov asked one when the meeting resumed. Low speculated that prior to the FRR the Soviets had satisfied themselves internally as to the readiness of the two sides for the mission. For his own part, Low felt that all his questions had been "well answered by the working group co-chairmen from both sides." Still, he was a little uneasy about the possibility of clear cut decisionmaking in the event of an emergency. In his trip report, he noted:

> My remaining concern after this FRR has to do with command and authority of command, particularly in contingency situations. At no time is there a single commander in space nor is there a single flight director on the ground who is in charge. The project had tried to accommodate this situation by trying to anticipate all possible contingencies. I asked what would happen in the event of an unanticipated contingency or in case there is a difference in interpretation of whether or not a contingency exists. Although these questions were answered rather forcefully, I am still not convinced that this is not a potential problem area.[31]

At the end of the review, Low and Kotelnikov signed a protocol indicating that "the Apollo-Soyuz Test Project is proceeding in accordance with the agreed schedule and ready to proceed toward the launching, planned for July 15, 1975."[32]

REPORTING ON THE FRR

After Low returned to Washington, Administrator Fletcher reported on ASTP flight readiness to President Gerald R. Ford, a leading supporter of the project, and to Senator Proxmire, the major critic. Fletcher's letter to the President was short and cordial. He noted that ASTP was on schedule and expressed his hope that Ford would take an active part in the last Apollo launch. "We believe your personal involvement would further demonstrate this country's commitment to increasing cooperation with other nations."

Should his schedule preclude attendance at the Apollo launch, Fletcher suggested that the President might want to speak with the crews during the joint phase of the mission.[33]

In writing to Senator Proxmire, Fletcher forwarded him a full explanation on the Soviets' 5 April Soyuz launch failure. During the course of the May meeting in Moscow, Glynn Lunney and Robert O. Aller, Chet Lee's Deputy, had been given a detailed briefing by Professor Bushuyev on the Soyuz launch abort. Since Proxmire had expressed a desire to be kept informed of all developments possibly affecting the safety of ASTP, Fletcher enclosed a summary of the findings. He told the Senator that NASA had reviewed all the Soviet data in detail and had concluded that the failure would not affect the safety of the Apollo crew.

Professor Bushuyev had told the Americans that several minutes after lift-off, when the central sustainer core of the launch vehicle was supposed to separate from the third stage, a sequencer relay failed and permitted some pyrotechnic latches to fire prematurely. This disabled three other pyrobolts and prevented the complete release of the sustainer core. Since the third-stage engine had been ignited, the pyro failure caused the vehicle to stray from its path. The abort sequence was automatically initiated when the spacecraft reached a 10-degree deviation from the programmed flight path. In quick succession, the third stage engines were shut down, the spacecraft was separated from the lower stage, and the retrorockets were employed to ensure the proper trajectory for landing. At the time of the abort, Soyuz had reached an altitude of 180 kilometers, traveling at about 5.5 kilometers per second. Lazarev and Makarov—veterans of *Soyuz 12*—experienced *g* forces equivalent to nearly 14 times those on earth as they descended. Their landing site was 1800 kilometers downrange from the launch pad, covered with waist-deep snow.

In his briefing to Lunney and Aller, Bushuyev noted that there were two basic differences in the launch vehicle that failed and the ones assigned to ASTP. A new type of relay was being used, and the pyro lock circuitry had been changed to prevent a premature firing of the explosive bolts. These modifications, which made the asymmetric separation as experienced in the 5 April flight impossible, had been included in a series of launch vehicles prior to the failure. That updated group of boosters had been flown ten times, including the two unmanned ASTP precursor missions and *Soyuz 16*. NASA was convinced that the aborted April launch did not pose a hazard to the American crew of Apollo-Soyuz. Furthermore, the agency was satisfied that this type of failure would not occur on 15 July. But should something prevent the successful launch of the prime Soyuz, the Soviets would have a second launch vehicle, spacecraft, and crew ready to count down. Despite

Senator Proxmire's concerns, the people at NASA expected to meet the Soviets in orbit.[34]

PUBLIC AFFAIRS PREPARES FOR THE FLIGHT

While the Senator from Wisconsin pondered the safety of the joint mission, Soviet and American negotiators were completing Part II of the Public Information Plan. In the months that had passed since their October 1973 meeting in Moscow, the public affairs specialists had met many times to hammer out an agreement about flight-related activities. Central to all these discussions was the American insistence on live in-flight television coverage of ASTP. Negotiations of the television agreements were conducted at two levels—managerial and technical. While John Donnelly and Bob Shafer worked with I. P. Rumyantsev and V. S. Vereshchetin in an attempt to reach an accord on policy, several other Americans worked with the Soviet technical representative, Vladimir Aleksandrovich Denisenko. The U.S. television team was led by Jack King, Bennett W. James, and Gene Cernan. While King acted as policy coordinator, Ben James oversaw the requirements public affairs had for television and Gene Cernan managed the technical team, implementing the hardware and mission planning aspects of onboard television. The task was a large one, but it was not limited to in-house considerations. External to the space agency, for instance, NASA had to make provisions for the American networks to place a pool television production trailer on the recovery ship. Once Shafer and Donnelly discovered that the Soviets planned to cover their recovery live, arrangements had to be made to broadcast from the U.S.S. *New Orleans.* Then there was the question of the exchange and conversion of American and Soviet television signals during the mission. The agreement to exchange television was merely the first step. NASA and the Soviet Academy had to arrange to convert the signals so they would be compatible with each other's system at the Raistings television ground station operated by the Postal Department of the Federal Republic of Germany. And finally, the European Broadcast Union was wired into the circuit so that continent could also watch the joint mission.

To get ASTP television pictures into millions of homes across the globe was a complex task. Realizing this, Bob Shafer had begun the discussions of ASTP television planning in August 1972, with a proposed scenario for mission video coverage. Up to that point, dialogue on onboard television had related principally to the desire to include cameras in the command and docking modules. Once it had been agreed that live television would be broadcast from the spacecraft, Shafer composed a new scenario describing how best to put it to use.[35] After nearly a year of only limited Headquarters Public Affairs participation in the work on ASTP television, Donnelly and Shafer had advised George Low that the mission "would be flown in the dark" if he did not take some action to guarantee proper planning.[36] On 31 August 1973, Low wrote a memo to Chet Lee, noting that "in preparing for Skylab, we had a great deal of last-minute confusion because the planning for television coverage had not been properly taken into account in the overall Skylab mission planning." Since Low believed it essential to have "highly professional TV coverage of significant ASTP events," he asked Lee to coordinate with Donnelly and Shafer, letting them know at an early date:

1. The goals and objectives of ASTP TV.
2. The planned hardware implementation to meet these goals.
3. The planned programming implementation to meet these goals.
4. Key milestones in meeting the objectives.
5. A listing of responsible individuals who will make it happen.[37]

To smooth over possible intra-agency friction, Low had indicated in the same memo that "this [mission] has to be a joint effort with the Office of Public Affairs, with that office being responsible for programming requirements and for signing off on the hardware implementation."[38]

But friction did exist. On 14 September 1973, Shafer addressed a memo to Lee that read, "unequivocally and for the record, no one in Public Affairs drafted Dr. Low's memo to you, proposed its contents, suggested the language, or in any other way assisted in its preparation." Shafer believed it imperative that Public Affairs and Manned Space Flight proceed with their work without any misunderstandings over the agency's commitment to television as represented by the Low memorandum. "I would not think the management conviction it demonstrates should be at issue, particularly in connection with ASTP, but if that is the case perhaps you should discuss your concern directly with him."[39] These early differences were caused in part by the failure to understand some of the technical problems associated with providing television from such a low earth orbit. Lunney had advised Lee in September that the Apollo Office had been planning to use the Skylab type of video recorders because time for broadcasting live pictures was so limited—17.8 percent of each orbit. The argument over large-scale, live television coverage remained an academic debate until early October when the Office of Manned Space Flight (OMSF) gave its final approval for use of the ATS-F relay satellite to enhance all ASTP communications.[40]

On 2 October 1973, Lee briefed the House Manned Space Flight Subcommittee on the advantages of using the Applications Technology Satellite. He noted that ATS-F would permit direct communication with the ground for 48 minutes of each 88-minute orbit, an increase of 33 minutes per revolution over reliance only on ground station signal time. Dale Myers reported to Administrator Fletcher on 2 November on the status of ATS-F:

> It was determined that the ASTP vehicle would accommodate the additional weight and instrumentation and funds were identified for incorporating the necessary instrumentation. Concurrently, discussions were held with the ATS-F Program Office, which established an acceptable plan to both OMSF and the Office of Applications for the use of the satellite. During the period of negotiations with the ATS-F Program Office, a tentative approval by the ASTP Program Office was given to JSC for expenditure of about 10% of the necessary funds in order to proceed with the necessary engineering details. Shortly after receipt of the Office of Applications' support commitment letter of August 14, 1971, the ASTP Program Office approval was given to JSC to expend up to about $2.1M for the necessary modifications.[41]

Myers also told Fletcher that the Office of Tracking and Data Acquisition was working on both the hardware and the diplomatic aspects of placing a special ATS-F antenna at the Madrid Tracking Station.

Once all the television hardware elements had been identified, work began on preparing the equipment for the mission. A key to the success of this effort was Chet Lee's decision to ask Lunney to appoint a single individual to be responsible for television communications.[42] Lunney's response "as a result of the very high priority placed on television during ASTP" was to establish "a special TV planning team to coordinate all of the necessary activities to assure the best television we can have." Gene Cernan was designated chairman of this planning team, and he was directed to begin "a regular series of meetings to cover all aspects including policies and requirements, the hardware implementation (ground and air) and the plans for training and inflight use." At the same time, Samuel Sanborn was given responsibility for the technical aspects of preparing the television hardware.[43] In the 18 months between December 1973 and the Moscow FRR, a sizable team worked the technical issues associated with onboard television.[44]

By early 1974, the NASA television preparations were well on their way. The next task at hand was to obtain Soviet agreement on Part II of the information plan. Shafer told Donnelly:

> The agreement is not only necessary, but urgently so, because neither their working group members nor ours can proceed much beyond the present status until the television requirements can be discussed as bilateral, rather than unilateral, considerations. Mission planning is moving ahead rather rapidly, and it will soon be virtually impossible to rework all of the technical issues involved in meeting those requirements.[45]

The first detailed discussions of Part II were held in Moscow at the end of March 1974. A second negotiating session spanned the April meeting in Houston.[46] In March, Vereshchetin told the Americans that Part I had obviously been based on a NASA proposal. The Soviets, he said, were glad that Donnelly, Shafer, and King had shown the way. But Part I sounded just like an American document translated into Russian. As a consequence, Vereshchetin wanted Part II to sound more Russian. Donnelly and Shafer agreed to this consideration as long as the language reflected policies that NASA could accept. The task of reworking Vereshchetin's proposed Part II fell to Jack King, who had a new draft completed by April. Negotiations at that meeting were slow and difficult, but before the Soviets departed from Houston, a basic agreement had been established. Donnelly indicated to Low that substantial progress had been made in the areas of real-time television exchange, news personnel accreditation, creation of mission press centers, press kits, and the like. After four months of feverish activity in Houston and Washington, the Americans went to Moscow in September 1974 to conclude agreement on Part II.[47]

Press reaction to the information accord was mixed. United Press International noted that "Russia has agreed to distribute live television coverage of the launch of two Soyuz cosmonauts and full radio communications during their joint orbital flight with an American Apollo." The UPI wire story indicated that this was the first time live television and in-flight radio communications of a Soviet space flight would be released to the West. John Donnelly, in an interview with the UPI correspondent, said the information agreement called for the video broadcasts to begin as Leonov and Kubasov boarded their spacecraft about 2 hours before launch, followed by a live picture of the lift-off. The latter would be not only a first for Western viewers but also a unique event for Soviet citizens, who heretofore had seen only video replays.[48]

Despite the public affairs accomplishment, *Aviation Week,* among others, was critical of the American space agency because it had not held out for media access to the Soviet launch site. *Aviation Week's* editors stated their feelings bluntly:

> U.S. space negotiators have retreated another step in efforts to provide open access to the Apollo-Soyuz Test Project mission. With little protest from NASA officials, the Soviets have all but killed any prospects for U.S. or other Western press representatives to be present at Tyuratam during the Soyuz launch or at the Kalinin control center during the flight.[49]

Associated Press President and General Manager Wes Gallagher made a formal complaint to NASA about being excluded from the Soviet centers. Administrator Fletcher responded strongly to this criticism in a letter to Gallagher:

> The public affairs agreement between NASA and the Soviet Academy of Sciences provides for the most complete, comprehensive release ever to the

U.S. news media of real-time information related to a Soviet space mission. It provides among other things for the exchange and release of live airborne and ground-based television; for the transmission to and release by our control center of air-to-ground commentary between the Soviet control center and its spacecraft; for a running description by a Soviet commentator of mission events as they occur; for the operation of a press center to which U.S. correspondents will be registered to cover the mission; and for the exchange between press centers of public affairs officers and interpreters to assist the press in its coverage of the activities as they take place. All of these are firsts for the Russians.[50]

Fletcher added that something must have become "garbled somewhere along the way since the ASTP public affairs agreement in no way limits, restricts or excludes the American press from Baikonur." The public information agreement related only to joint activities, while recognizing the right of each side to make decisions about independent activities, such as the Soyuz launch, in accordance with its own obligations and traditions.

NASA's Administrator did not want to take any action that might compromise the agency's policy of running an open program. "While the Soviets have held steadfast to their right to refuse to admit the U.S. press to Baikonur," Fletcher saw no reason why NASA should retaliate by excluding Soviet newsmen from the American launch site. "It would compromise our own open-program principles without changing theirs." Fletcher believed that upon reflection Gallagher would "agree that under no circumstances should we compromise our policy to parallel or conform more closely to another system."[51]

Part II of the Public Information Plan provided a framework for mission and post-mission press activities. But during the nine months between the signing of Part II and the launch, the agreement was fleshed out somewhat. This work included developing detailed television transmissions, as well as preparing for the onboard press conference and determining when and where symbolic activities (exchanging flags, signing flight records, etc.) would take place. By the time of the Moscow Flight Readiness Review, all but a few minor questions had been resolved. Final ratification of the updated version of Part II was signed on 10 July, five days before the launch.[52] Public Affairs was ready for the mission.

NEW WORRIES

On 2 July, Senator Proxmire voiced another objection to the joint flight. He made public the testimony of a top Central Intelligence Agency official who raised questions about the ability of the Soviets to control two space shots at one time—ASTP and *Soyuz 18/Salyut 4,* which had been launched on 24 May 1975. Proxmire had part of Carl Ducket's testimony declassified so he could release it to the press. The Senator's news release read in part:

> During hearings before the HUD and Independent Agencies Subcommittee on June 4, the CIA Deputy Director for Science and Technology, Carl Ducket, stated, "I do not think they (the USSR) are in good shape to handle two missions at once from the command point of view."
>
> . . .
>
> "This warning from the nation's top scientific intelligence expert should not be taken lightly," the Senator said.
>
> The Soviet Union has announced that the two Russian cosmonauts already in space in the Salyut space lab will not be brought back to Earth before the July 15th launch of the joint US-USSR space mission.
>
> In view of the potential hazards that already exist during the joint mission, the added complexity of having two space missions going at once should be avoided at all costs.
>
> Soviet communications capabilities and central management facilities are greatly inferior to those of the U.S. Having two missions in space at once, including one involving two spacecraft of different nations, is complex enough to warrant concern that the ASTP mission may not get the full support it needs to be successful.
>
> Particularly troublesome is the potential for inadequate command and control should one or the other mission encounter difficulty.[53]

Proxmire urged NASA to postpone the ASTP launch until the Soviets brought *Soyuz 18* home. He said that it would be "a simple matter to de-orbit the two cosmonauts. . . . Then the joint mission could proceed without concern over this particular problem."[54] Administrator Fletcher responded on 3 July to Proxmire's request to postpone the launch.[55] "Although the Soviets have not made any official announcements with respect to their plans for the Salyut mission," Fletcher told the Senator, the Soviet press on 27 June had quoted Leonov as saying that the Salyut mission would continue during ASTP. Since the final full-scale simulation for ASTP had involved the two countries' control centers, Glynn Lunney had used that occasion to discuss the multiple flight control matter with Bushuyev.

The Professor indicated that there had been no final decision on the length of the *Soyuz 18/Salyut 4* mission.[56] During their conversation, Bushuyev assured Lunney that should the two missions overlap, the Soviets would use two separate ground control teams and control centers for the two missions. ASTP would be directed from the center at Kaliningrad, while *Soyuz 18/Salyut 4* would be conducted by the center that had been used prior to *Soyuz 12.* The Professor also told his American counterpart that the

ASTP mission had been assigned priority if the two sets of space vehicles should pass simultaneously within the same zone of coverage of a U.S.S.R. tracking station. This, of course, would be highly unlikely because ASTP and Salyut had distinctly separate flight paths. In fact, NASA's tracking specialists had made independent calculations that indicated that the two Soviet missions would be in communication with the same U.S.S.R. ground station only twice during the ASTP flight—and then only for intervals of about 0.5 and 1.5 minutes. Administrator Fletcher told Proxmire that based upon the data available and the nature of the two missions, "NASA has concluded that the *Soyuz 18/Salyut 4* mission does not constitute a hazard to ASTP and that there is no reason to delay the launch of ASTP if the Salyut mission is still in operation."[57]

Senator Proxmire, however, would not let the issue die. After inserting anti-ASTP articles in the *Congressional Record* on 11 July, he leveled another blast at the joint flight on the 14th, the eve of the launch.[58] Citing CIA data, the Wisconsin senator noted that:

> the Soviets have encountered severe problems in space and their technology is inferior to that of the U.S. in almost every category.
> —the Soyuz rendezvous and docking system has failed almost half the time
> —the current level of Soviet preparation still is below that of the US
> —the threat of a minor fire poses a moderate risk to the ASTP while a major fire is much less likely
> —Soviet communications are not up to the quality of US communications
> —Cosmonaut training and ground control crew proficiency are inferior to that of US counterparts.
> —There has been some technology flow to the Soviet Union as a result of the ASTP. Future joint missions would pose more of a potential for technology drain
> —the primary advantage to the USSR from the ASTP has been in observing US management and program operational techniques
> —the Soviet lunar program has produced a string of failures.
> In summary, the US has a significant technological lead over the USSR in the following areas: communications, management and quality control, handling of emergency situations, launch coordination and procedures, computerized functions, capability for inflight mission changes, space medicine, and crew training.[59]

Looking back on the Senator's remarks, American ASTP Commander Tom Stafford said that this was the first time that Proxmire had been worried about aerospace safety. Stafford had seen the Soviet flight hardware and had worked with the Soviet crews. And he was ready to fly. Stafford believed that Proxmire was simply opposed to space flight in general.

Whatever the sources of his concern, NASA did not share them. Nearly everyone was ready for the launch, and the space agency personnel had said so at the Headquarters Flight Readiness Review on 12 June.[60]

REVIEWING APOLLO READINESS

Within NASA, a series of additional reviews were conducted before 15 July. The Headquarters FRR was held at the Kennedy Space Center on 12 June, when representatives from the Johnson, Marshall, and Kennedy Space Centers gathered at the Cape to report on their respective pre-flight preparations. Glynn Lunney led off by summarizing the Moscow FRR and showing filmed highlights of the NASA team's visit to Baykonur Cosmodrome. The Technical Director said that he had been given full details on the April aborted Soyuz launch and explained the steps the Soviets had taken to prevent recurrence of the failure. M. P. Frank described the ASTP mission profile and enumerated the activities scheduled for the nine-day flight, and Arnold Aldrich listed the technical reviews to which the spacecraft had been subjected. Lunney's assistant went on to indicate that there were no spacecraft hardware "issues"—problems to be resolved prior to launch—with CSM 111, SLA-18, DM-2, or DS-5 that might interfere with an on-time lift-off. Gary A. Coultas of the Apollo Project Engineering Office presented a similar evaluation of government furnished equipment (the color television subsystem, photo/optical equipment, space suits, and rescue equipment). Safety assessments for the spacecraft and experiment materials were delivered by Bobby J. Miller of the JSC Safety Office. After further remarks from Houston personnel on flight rules, mission control center readiness, flight controller, and crew preparedness, Ellery B. May, Saturn Program Manager at Marshall, reported on the Saturn IB launch vehicle.[61]

SA-210 had been built in 1967 and stored since then. By the date of the launch, several of its components would be nine years old. Since some of the materials used in fabricating the booster were subject to possible deterioration from aging, periodic inspections had been made to monitor the condition of its various components. May noted that during the course of one such routine inspection, cracks had been discovered in two of the mounting points for the large tail fins of the first stage. After an intensive study of the fins and the cracks (caused by stress corrosion), Marshall and Headquarters engineers decided to replace all eight of the first stage fins. Subsequent analysis and monitoring of the hardware indicated that this action had corrected the problem, and it appeared as a resolved anomaly in the FRR.* At the conclusion of his presentation, May stated that all

*A chronology of SA-210 related events, including the stress corrosion problem, is presented in appendix F.

Inside the high bay of the Vehicle Assembly Building at the Kennedy Space Center, work progresses on the replacement of the first-stage fins of ASTP's Saturn IB launch vehicle, March 1975. On 24 March the Saturn-Apollo 210 launch vehicle and spacecraft were moved out to the launch pad. A special lightning mast is atop the service tower.

Marshall offices and contractors had been polled; they agreed that SA-210 was ready to fly.

KSC personnel provided data on the launch center's preparations for the flight. One of their major concerns was the possibility of thunderstorms and lightning strikes before and at the time of the launch. July was the worst time of the year for both at Cape Canaveral. William H. Rock, Manager for Sciences, Applications, and ASTP, spoke first for the launch team. In addition to covering all the ASTP-related modifications to the launch pad and control center, he gave some historical background on the ASTP lightning protection system installed on the mobile launcher. A lightning strike had long been a worry, one that had been reinforced by the twin bolts that had struck *Apollo 12.* To combat the effects of such a strike, the KSC team had installed a larger lightning rod atop the launch tower. This 25.6-meter fiberglass mast was designed so that the ground wires would not come any closer than 15 meters to the mobile launcher structure, thus eliminating the arcing of electrical current from the wires to the structure of the spacecraft. Rock noted that since 9 May four lightning strikes had been recorded; none had posed a threat to the hardware.*[62]

*The lengthy efforts related to lightning protection are documented in note 62.

Although the lightning hazard appeared to have been minimized by the new arrestor, the possibility of rain storms and high winds still concerned the mission planners. Jesse R. Gulick, KSC staff meteorologist, reported on the weather prospects. Starting with a recapitulation of last year's storm patterns, he said that, according to current mission rules, "we could not have launched on one-third of the days, that several such days could occur in a row . . . [but] there was no time last year when we could not have launched within a period of four days." The thunderstorm probability for 15 July 1975 at the proposed lift-off hour was 23 percent. Probability of a tropical storm or hurricane winds affecting the KSC area was less than 3 percent. The mixed forecast, some good news and some bad, led to the crossing of fingers. While everyone hoped that they would not be needed, the mission planners were relieved that they had spent so much time working out alternative launch dates—just in case.[63]

Captain Lee took the floor after statements from KSC, Goddard (Spacecraft Tracking and Data Network), and Department of Defense personnel. He canvassed the Center Directors for comments, and they all remarked favorably on the preparations, expressing their confidence that the mission would be a success. George Low and John Yardley echoed these sentiments. Low said that he was especially impressed by all the effort and detail that had gone into building and checking out the ASTP hardware. Attention to detail had been one of the hallmarks of the Apollo and Skylab Programs, but he sensed a special feeling of pride behind the hard work and devotion of the ASTP team. Low believed that their energies would help assure a favorable outcome for the first international space flight. He asked them to continue their hard work until the mission was completed and the crews safely aboard the recovery ship. At the close of the FRR, Lee told the launch crews to continue with their preparations.[64]

COUNTDOWN TO LAUNCH

During the days remaining till the 15th of July, Lunney and Bushuyev kept in touch by telex and telephone. On 23 June, Lunney sent the "Launch minus 21 days" report to the Professor:

SPACECRAFT:	S-IB launch vehicle stage fuel (RP-1) loading was completed satisfactorily on July 23, 1975. Countdown demonstration test preparations have started and the nine-day test will start at 0700 EDT on June 25, 1975.
GROUND SYSTEMS:	ATS-6 satellite testing is complete. All MCC-H network interface testing is complete except for S-band tracking test to be conducted on June 25, 1975.[65]

FINAL EXAMINATIONS

Three days later, Lunney and Bushuyev discussed a variety of topics, including the joint control center simulations scheduled for the end of the month and some public affairs questions.[66] Meanwhile, the astronauts continued their training.

At 2:50 on the afternoon of the 24th, the prime crew began a three-week preflight medical isolation program—the "Flight Crew Health Stabilization Plan." Stafford, Slayton, and Brand were limited to specific working and training areas at JSC, and only previously screened personnel could come in contact with them. These "primary contacts" were required to wear surgical masks when in the presence of the crew. No one, especially the three astronauts, wanted a change in crew assignments because of the sniffles or any other common illness. In their off-duty hours, the prime three were quartered in mobile homes near the astronauts' gym in the northeast corner of the space center. In addition to further practice in the simulators, they continued their work on Russian.[67] (See table X-1.)

Table X-1. *ASTP Crew Training Summary as of 15 July 1975*

Training activities	Hours accomplished		
	Stafford	Brand	Slayton
Briefing/reviews:			
Command and service module	26.4	55.6	60.3
Docking module	8.0	21.5	24.9
Launch vehicle	2.6	3.0	2.6
Experiments	94.8	95.8	100.3
Flight plan/checklist	13.5	38.0	19.5
Mission technique/rules	29.5	61.0	23.5
Soyuz	2.0	2.0	2.0
Systems Training:			
Transfer procedures	17.5	27.0	17.5
Crew systems	18.5	27.5	22.5
TV	2.5	5.0	5.5
Photo	12.5	12.5	11.0
Experiments	81.3	82.3	117.5
Stowage	1.0	2.0	7.0
Bench checks	12.0	12.0	12.0
Egress/fire	13.0	13.0	13.0
Spacecraft test	106.5	98.5	174.5
Morehead planetarium	0.0	0.0	12.5
Medical	33.5	22.5	31.5
Simulators:			
Command module simulator/docking module simulator	428.0	474.2	549.3
Command module procedures simulator	32.5	0.0	56.0
Russian language	1016.5	923.5	1077.5
Joint crew activities	737.6	812.3	735.1
Total hours	2689.7	2789.2	3075.5

On 25 June, the Countdown Demonstration Test (CDDT) began at 7:00 a.m. After participating in the "Joint Orbital Operations Simulation" between the Houston and Moscow control centers, which began early on Sunday the 29th and continued for 56 hours, the prime crew departed Houston on 2 July for KSC, where they would take part in the manned portion of the CDDT. During one phase of the test that included a simulated ignition and lift-off, Vance Brand's space suit leaked, as it had during a high altitude test run. This time the problem was traced—to one of the pressure-sealing slide fasteners. A minor modification to this "sophisticated zipper" fixed the leak, and technicians also altered Stafford's and Slayton's suits as a precautionary measure. As time ticked away toward the hour of launch, the tempo of activities quickened.[68]

After a one-day holiday and an additional check of Brand's suit, the prime crew flew their T-38s back to Ellington Air Force Base on Saturday, 5 July. That afternoon, they reviewed the flight data file, and on Sunday they took to their T-38s again and later practiced Russian. As last minute checks went on at the Cape, the crew spent a busy week in the command module simulator. Rendezvous was practiced, with and without systems failures thrown in by their instructors, and solutions to possible docking malfunctions were studied and worked out in the simulator. On 13 July, the crewmen once again climbed into their jets and departed for Florida.[69]

Reports from Baykonur Cosmodrome indicated that the Soviet crewmembers were also in quarantine. After talking to Stafford on the 10th, Leonov and Kubasov continued reviewing their flight plans at the Soviet launch center. While they studied, the launch crews readied the two space vehicles. The prime spacecraft and booster were on the pad and fueling began on the 11th. Two days later, the second Soyuz and launch vehicle were transported to the secondary pad some 20 kilometers distant. All preparations were on schedule, and Soviet mission control advised the Americans that the tracking stations and tracking ships *Akademik Sergei Korolev* and *Kosmonavt Yuri Gagarin* were ready for the flight.[70]

Since the personnel of the Soviet and American control centers were scheduled to start round-the-clock duty scheduled on 14 July, 24 hours before the launch of Soyuz, the teams of visiting specialists had to be in place ahead of time. A group of 15 specialists and interpreters from the Soviet Union had arrived in Houston on the 8th.* NASA's flight control

*O. I. Babkov headed the delegation, which included S. G. Grishin, Deputy Director of MCC-M for Information; V. V. Illarionov, CapCom; A. S. Korolev, onboard systems specialist; V. V. Kudryavtsev, docking systems specialist; V. K. Novikov, life support system specialists; I. P. Shmyglevskiy, control system specialist; V. I. Staroverov, flight controller; V. D. Yastrebov, trajectory specialist; G. I. Kharitonov, Intercosmos; B. S. Kunashev, Space Studies Institute; and interpreters, B. P. Artemov, Y. N. Sergeyeva, and O. G. Yavorskaya.

The Soyuz launch vehicle and ASTP spacecraft are transported to the launch site by railroad flat car (Soviet Academy of Sciences photo).

specialists led by Charles R. Lewis arrived at Moscow Sheremetyevo Airport on the 12th, where they were met by V. G. Kravets, V. D. Blagov, and S. P. Tsybin, members of the Soviet ground control teams.* With everyone in place, Chet Lee conducted the "L – 2 Day Review" to determine whether all systems were set for launch.[71] Gulick could give a reasonably favorable weather forecast—broken clouds, wind out of the east southeast at 10 knots, a temperature of about 29°C, with about a 70-percent change of thunderstorms. All stations reported that they were ready to go.

---

*C. R. Lewis headed the delegation, which included R. F. Overmyer, Capcom; R. L. Haken, control system specialist; R. W. Becker, trajectory specialist; J. A. Kamman, guidance and navigation specialist; J. S. McLendon, electrical and instrumentation specialist; R. L. Grafe, ECS specialist; J. E. Riley, F. G. Williams, and A. P. Alibrando, public information; and interpreters J. O. Glikman, I. A. Mamantov, A. Rodzianko, T. Krivosheim, and K. Javorskaya.

XI

Come Fly with Us

15 JULY 1975–LAUNCH

At 5:50 a.m.,* Glynn Lunney entered the second floor Mission Operations Control Room (MOCR), located in the Mission Control Center-Houston (MCC-H). All smiles, with just traces of sleepiness in his eyes, he spoke to several of the men who were already on duty at their flight consoles. As he fitted his headset, Lunney nodded a good morning to R. Terry White, who for this shift was the "Voice of Apollo Control." Alex Tatistcheff, Lunney's interpreter, arrived at five past six, just as the Soviet launch complex was first shown on the television monitors and on the large eidophor picture screen in the MOCR. By 6:55 when the first televised public information release was transmitted from Moscow Mission Control Center (MCC-M), George Low and Chris Kraft had joined the growing number of people in the Houston control room.[1]

Reports from Baykonur indicated that the weather was perfect for the launch—clear skies, light winds, and hot July sunshine. With the crew on board and 45 minutes remaining until lift-off, the ground team removed the semicircular halves of the service structure. Soyuz 19 sat poised for the launch. In Houston, Ross Lavroff interpreted the commentary as it was broadcast from MCC-M in Kaliningrad:

> This is the Soviet Mission Control Center. Moscow time is 15 hours, 15 minutes. Everything is ready at the Cosmodrome for the launch of the Soviet spacecraft Soyuz. Five minutes remaining for launch. Onboard systems are now under onboard control. The right control board ... opposite the commander's couch is now turned on. The cosmonauts have strapped themselves in and reported that they are ready. They have lowered their face plates. The key for launch has been inserted. . . . The crew is ready for launch.[2]

Five minutes later, the fueling tower was removed, and the command was given for launch. "Ignition. The engines are powered up. The launch; the

---

*Unless otherwise indicated, all times given are central daylight (CDT) or Soyuz ground elapsed time (SGET). For Moscow time add 8 hours to CDT; for Greenwich mean time (GMT) add 5 hours to CDT.

booster is off. Moscow time 15 hours, 20 minutes, 10 seconds. The flight is proceeding normally." At 120 seconds into the flight, the strap-on booster units of the first stage were separated. Then at 160 seconds, the emergency abort system was jettisoned, followed by the separation of the launch shroud and the firing of the second-stage engines. Third-stage ignition took place at 270 seconds, orbital insertion at 530 seconds. The third stage was shut down, and the antennas and solar panels were extended. Kubasov asked the ground, "How do you read?" MCC-M responded that they heard them well. The initial orbital parameters were 220.8 by 185.07 kilometers, at the desired inclination of 51.80°, while the period of the first orbit was 88.6 minutes. There were smiles in Moscow and in Houston.[3]

Max Faget, who was seated in the viewing room overlooking the MOCR, expressed the feeling of most of the American flight team. "It's our turn to hit the ball. Now we've got to get into orbit." Early evening at Baykonur was mid-afternoon in Moscow and early morning in Florida and Texas. While the American crew slept, Chet Lee, Launch Director Walt Kapryan, and Kennedy Space Center (KSC) Director Lee Scherer monitored the continuing preparation of SA-210. At the time of the Soyuz lift-off, liquid oxygen was flowing into the tanks of the Apollo launch vehicle at a fast fill rate of 4543 liters per minute. After the U.S.S.R. launch, Lee, Kapryan, and Scherer got a briefing on the predicted weather conditions for the afternoon—there were thunderstorms in the vicinity of the Cape, but they were not expected to affect the American lift-off. In Houston, Lunney called Professor Bushuyev to congratulate him on the success of the Soyuz launch and to advise him that the countdown was proceeding on schedule with the best weather forecast in months. Bushuyev reported in turn that the orbit of Soyuz was within 2 or 3 kilometers of the desired figures.[4]

Stafford, Slayton, and Brand were awakened at 9:10. While they were having their final medical examination, the team that assists the crew at the launch site set out for the spacecraft. Following their visit with the doctors, the astronauts sat down to the traditional pre-flight breakfast of steak and scrambled eggs. As they ate, they watched a video replay of the Soyuz launch. Robert Crippen, the Backup Command Module Pilot, meanwhile began the final preparation of the command and service module (CSM) cockpit in anticipation of the crew's arrival. Once they completed their breakfast, the three men went to the suit room in the Manned Spacecraft Operations Building and donned their space suits. At 11:37, accompanied by John Young, Chief of the Astronaut Office, they rode down the elevator and boarded their van for the 25-minute ride to the launch pad.[5]

With the assistance of their suit technicians, the crew arrived at Pad 39B, where they made their way by elevator to the 100-meter level of the mobile launch tower. Once there, they crossed over swing arm number 9 and entered the White Room surrounding the spacecraft. Stafford was the first into the cockpit, where he moved into the left couch, assisted from the inside by Crippen, who also connected Stafford's electrical, oxygen, and communications umbilicals. Slayton was next, and Crippen went through the same procedures after he was seated in the right-hand couch. Brand was last. When Crippen completed his check of Brand's fittings, he removed the protective covering from the crewman's helmet, as he had for the other two. At 12:02, Stafford called to the test conductor Clarence Chauvin, "Looks like it's a good day to fly."[6]

Crippen slid down under the center couch and crawled out the hatch above Brand's head. After some additional checks, the CSM hatch was closed at approximately 12:22. As the first live launch pad color television pictures of the interior of the CSM were broadcast to the world, the crew began to run through the final checklist. Stafford asked Karol "Bo" Bobko, the Spacecraft Communicator (CapCom) at 1:10, "Are you giving us the countdown in English or Russian today?" Bobko responded, "Oh, I figured I'd give it in English." In Moscow, the Soviet flight director was reminding Leonov and Kubasov that the Apollo lift-off was set for 10:50 Moscow time (2:50 CDT). At T minus 7 minutes, 52 seconds, the Apollo crewmembers finished their checkout of some 556 switches, 40 event indicators, and 71 lights on the console. Stafford told Bobko to tell Soyuz to get ready for them. "We'll be up there shortly."[7]

After the final minutes of waiting, at 2:49:50, the now famous count backwards from 10 began. "10, 9, 8, 7, 6, 5, 4, 3, 2, engine sequence start, 1, 0, launch. . . . We have liftoff. Moving out, clear the tower." Above the roar of the first-stage engines, Stafford reported that the ride had been a little shaky at lift-off, but now it was "smooth as silk." Fifty seconds into the flight, the acceleration force equaled 2 gs, twice the gravitational force normally experienced on earth. At 124 seconds, the crewmen were experiencing 4 gs as they dropped off the first stage and continued their journey under the power of the S-IVB stage. Fifty-two seconds later, they jettisoned the launch escape tower, and Stafford remarked, "Tower jett. There she goes! . . . Adios. . . . At 4:40, back to one g acceleration and looking good."

Dick Truly, CapCom:	Apollo, Houston. At 5 minutes you're GO.
Stafford:	Roger. 5 minutes. Looks good onboard, Dick. And we've got a beautiful sight.
Truly:	Roger. Wish I could see it.
Stafford:	Roger.
Slayton:	Man, I tell you, this is worth waiting 16 years for.

Brand:	Got a beautiful ocean out . . . here, Dick.
Truly:	Roger, I believe all that.
Stafford:	Okay, at 5:30, onboard trajectory looks beautiful.
Truly:	Roger. Concur, Tom. You're right on the money.[8]

On the ground, Ed Smith and R. H. Dietz with grins on their faces echoed the same thoughts when they said, "We've got a ball game!" The rendezvous chase was on. Apollo had achieved orbital insertion at 2:59:55.5 central daylight time. Brand exclaimed, "Miy nakhoditsya na orbite!"*

Stafford notified Houston at 3:55 p.m. that the crew was preparing to execute the transposition, docking, and extraction maneuver in 2 minutes. As a preliminary to removing the docking module (DM) from the spacecraft lunar module adapter (SLA) truss assembly, the CSM was separated from the S-IVB stage, and as the CSM moved away from the adapter section, the panels of the SLA were explosively jettisoned. In bringing the spacecraft about to face the docking module, the crew encountered its first minor problem of the flight. When Stafford looked through his alignment sight (COAS) at the Saturn IVB and docking module, the attitude was such that all he could see was the glare from the sunlit earth. At first he thought that the light illuminating the cross hairs in his sight had burned out. But when he put his hand in front of the COAS, Stafford reported that he could see the green reticle. Swearing under his breath, he knew that he would just have to wait until the two craft were positioned differently. Stafford moved the CSM toward the S-IVB and docking module until about only 10 meters separated them. Watching the stand-off cross on the docking module truss in the S-IVB stage, the Apollo crew assumed a stationkeeping status. Slowly the target vehicle appeared to move toward the earth's horizon. Stafford squinted and leaned his head to one side so he could see the reticle. "Finally when I got it in line," he later recounted, "I could just tell my general attitude and moved in." Despite the problems, Stafford's docking was perfect. He had aligned the two spacecraft to within a hundredth of a degree, the best alignment ever achieved with the Apollo docking system. By the time he had lined up his target, Apollo had passed out of radio contact with the ground.[9]

When Apollo re-established communications over Rosman, North Carolina, Stafford told Truly that they had achieved a real hard docking with the DM; all hatches were locked. The commander was happy to have this first docking completed. He later recalled that given the past problems with the Apollo probe and drogue, he had really been "sweating out" this exercise. Once it was over, he looked forward to meeting Soyuz. The new

*"We are in orbit!"

docking mechanism was a pilot's dream, and he knew that he could fly it in for a smooth docking.[10]

During a subsequent 5-minute pass over the tracking ship USNS *Vanguard,* the crewmembers advised Houston that they had completed the extraction of the docking module. The spacecraft, configured as it would be for the meeting with the Soviet craft, was now in an orbit 173.3 by 154.7 kilometers with an orbital period of 87 minutes, 39 seconds, and an orbital velocity of 7820 meters per second. Additional maneuvers would bring Apollo and Soyuz into the proper orbital relationship for rendezvous. Apollo's orbit was circularized at 167.4 by 164.7 kilometers at 6:35. From this orbit, the first Apollo phasing maneuver was executed at 8:28 to provide the proper catch-up rate, so that docking with Soyuz could occur on the 36th Soviet revolution. This 20.5-meter-per-second change placed Apollo in a 233- by 169-kilometer path. The next phase and plane correction maneuver of 2.7 meters per second was scheduled for the 16th revolution.[11]

In the midst of this precision flying, there were some lighter moments. At 6:10, Brand asked Truly to tell the launch crew at the Cape that they permitted a stowaway to board the spacecraft. "We found a super Florida mosquito flying around here a few minutes ago." Slayton said that he planned to feed it to the fish that they were carrying onboard if he could catch it, and Brand wanted to bring it back and give it astronaut wings.* These transmissions were conducted through the *ATS 6* satellite. While that particular communications satellite had been an unknown quantity throughout much of the mission planning, it was working very satisfactorily.

Placed in a geosynchronous orbit at 42 596 kilometers on 30 May 1974, *ATS 6* had remained at a fixed point over the Galapagos Islands, permitting educational television transmissions to remote areas at relatively low costs. Following transmission experiments to Appalachia, the Rocky Mountains, and Alaska, the satellite on command from the ground moved to a new position over Africa, where it was to be used for a year-long educational experiment in India. It reached its present location, 35° east longitude on the equator, on 2 July, in time for the ASTP team to borrow its communications channels for the joint flight. Broadcasting through the spacecraft tracking and data network station at Buitrago, Spain, the Apollo crew and the team in Houston were able to talk and transmit data for 55 minutes of each 87-minute revolution. This three-fold increase in communications, impossible without *ATS 6,* made all the hard work and worry about its success worthwhile.[12]

Later in the evening, after a cabin overheating problem had been solved, Brand asked Karol Bobko, who had relieved Truly as CapCom, about the

---

*After flying about for several hours, the mosquito was never seen again. Apparently, it died in the reduced pressure pure oxygen of the CSM.

15 July 1975
Baykonur, U.S.S.R.

Cosmonauts Alexei Leonov and Valeriy Kubasov (above) speak to reporters, then board their spacecraft for the launch and flight to position for rendezvous by Apollo. Below, the launch of their craft, Soyuz 19.

With Soyuz successfully in orbit, Astronauts Vance Brand and Tom Stafford (above, left), followed by Deke Slayton, arrive at the 100-meter level of the Saturn launch tower where their Apollo spacecraft awaits them. Final countdown is smooth, and the launch (right) is on time.

...e in orbit, the Apollo spacecraft separated from ...S-IVB stage, then doubled back to dock with the ...king module (DM) poised on the S-IVB (water-...r conception by Paul Fjeld). After collecting the ..., Apollo fired up for rendezvous with Soyuz. ...first problem of the flight occurred when Brand ...nd he could not remove the docking probe from ...tunnel entrance to the DM. Troubleshooters ...nd that an improperly installed pyrotechnic ...nector (shown in this Rockwell photo taken be-...e acceptance of the probe) had eluded Rockwell ...NASA inspectors.

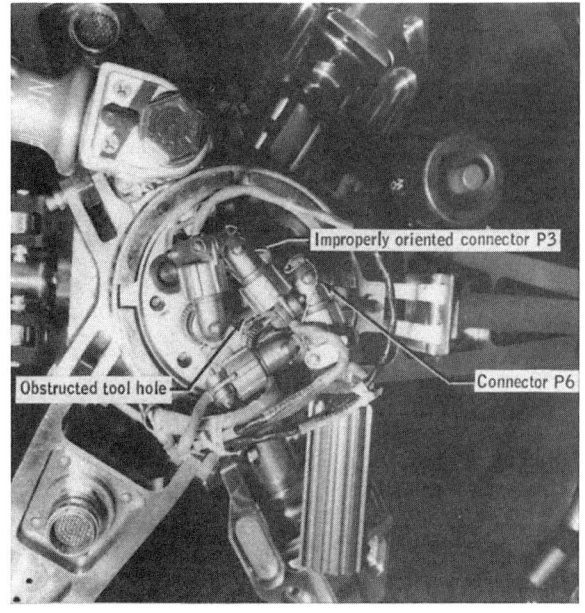

status of Soyuz. Bobko reported that Leonov and Kubasov were asleep and that to this point in the flight their only problem was a television camera that refused to work. He told Vance that they had tried without success to repair it but they planned to work on it some more after their sleep period. Apollo meal time arrived at 10:06, and Slayton coined a new space phrase for eating when he indicated to Houston that he and his crewmates were in the "food intake mode."[13]

A problem later that night, however, caused some concern both on the ground and in the spacecraft. Bobko had wished the crew good night in Russian, and they were supposed to be bedding down for a rest period, when at 22 minutes past midnight Stafford called to the ground. Brand had attempted to remove the probe assembly from the tunnel between the CSM and the DM so that he could open the hatch and store overnight a freezer* in the passageway, but he found that he could not insert the tool that unlocked and collapsed the probe. Brand went on to explain the difficulty:

> **Brand:** Okay, Bo. Everything in the probe removal checklist on the cue card ... has been going great up through step 11. Step 12 is "Capture latch release, tool 7." You insert it in the pyro cover. You turn it 180 degrees clockwise to release the capture latches. Well, here's where the problem is, and let me explain it to you.... do you have somebody there that knows the probe that can listen?
>
> **Bobko:** Roger. Go ahead.
>
> **Brand:** Okay, as I look in the back of the ... pyro cover, I'm looking with my flashlight through the hole where I insert this tool, and there's something behind the pyro cover that's preventing me from putting this tool all the way in.... it's actually one of the pyro connectors.... this tool has to go down through the pyro cover in between ... some pyro connectors. But one of these pyro connectors has rotated such that it's in the way....[14]

Neil B. Hutchinson, flight director at the time the probe problem was discovered, later told press representatives that the ground team and the crew had discussed the difficulty for about 18 minutes. Their first decision had been to forget transferring the freezer into the tunnel and just have the crew close the hatch and go to sleep. But when Brand tried to close the hatch, he discovered that the partially removed probe assembly prevented him from doing so. Since the three men were already past their sleep time and the open hatch did not pose a hazard, the two teams—ground and

---

*The cryogenic freezer for the electrophoresis experiment was cooled by liquid nitrogen. Since it released nitrogen into the cabin, the flight plan called for it to be stored in the DM-CSM tunnel overnight.

space–agreed to postpone any further work on the probe until morning. As a precaution, the crew raised by a very slight amount the cabin pressure, which provided additional oxygen to compensate for the nitrogen that was boiling off the freezer. The crew went to sleep, and Hutchinson went to a 3:15 a.m. news briefing.[15]

In his explanation, the flight director indicated that the problem was not serious, just an annoyance. In the morning, the crew would have to run back through the 11 steps to re-engage the probe in its fully locked position. Then one of the men would have to remove the pyro cover, straighten out the misaligned pyro cap, go through the 11 disassembly steps, and on the 12th insert the key and unlock the capture latches. Afterwards the removal of the probe would follow according to the original plan. When asked if this was the same type of problem encountered in *Apollo 14*, Hutchinson answered that although this was the same probe assembly as that used in *Apollo 14*, the difficulty was an entirely different one.[16] Everyone had to wait until morning to determine if the solution would be as simple as anticipated.

16 JULY–CHASE

While the Apollo crew slept, Leonov and Kubasov were awakened in the early morning hours of the 16th and were advised by Moscow control of the Apollo probe difficulty. The American ground team was still refining its solution to the problem. Besides exchanging greetings with the Soviet crew aboard *Salyut 4* through the Moscow center, the ASTP cosmonauts continued to attempt repairs on their troublesome black and white television system. Following instructions from the ground, the Soyuz crew went as far as to attempt a repair involving cutting away some of the lining of the spacecraft so they could gain access to a television wiring junction box. This unorthodox in-flight repair procedure failed, and the black and white system never did work. This failure upset some Americans, notably Bob Shafer, because this system's absence meant that there would be no pictures of Apollo during the flight. While some of the NASA team groused about this turn of events, the Soyuz crewmembers prepared for the circularization manuever that would bring their spacecraft into a 225- by 225-kilometer orbit.[17] As they were executing that maneuver, the Apollo crew was awakened to the rock sounds of Chicago's "Good Morning Sunshine."

Medical reports and breakfast filled the first minutes of the Apollo crew's morning activities. With the exception of some minor frustrations like the slow functioning urine dump system and some spilled strawberry juice, everything was proceeding satisfactorily. CapCom Crippen advised the crew that Soyuz had completed its circularization maneuver and was "in orbit waiting for you." Truly replaced Crippen and gave Brand the latest information on how to remove the probe. As they were disassembling the back end of the probe, Stafford commented, "Dick, it wouldn't be a normal flight if we didn't have our little probe problems."[18]

Stafford came back on the air-to-ground communications loop at 9:55 a.m. to tell Houston that the probe was out. With that "glitch" solved, the crewmembers could return their attention to the flight plan. Preparation for televising pictures from the cabin and checking out the docking module were the next activities on the list. As they worked through their schedule, the Soviet crewmembers were transmitting their first television pictures with their color camera. Talking to the Soviet flight director, V. A. Dzanibekov, Leonov gave the folks at home a commentary on their first 28 hours in space and then conversed with Klimuk and Sevastyanov, who had been aboard the space station Salyut since 24 May. Sevastyanov commented that the ASTP crews had a very responsible task and that a large portion of the world's population was watching and listening to their progress. Referring to the seven men now in space, two aboard Salyut and the five involved in ASTP, Klimuk said, "these are the magnificent seven." With pleasantries concluded, the Soviet crews returned to their respective duties. Leonov and Kubasov began lowering the pressure of their ship to 500 mm of Hg in preparation for the docking.[19]

Aboard Apollo, Stafford, Slayton, and Brand were settling into the routine of flight. Their day was filled with independent experiments (electrophoresis, helium glow, and earth observation)* and collecting biomedical data. During the earth observation pass, Stafford told Bobko to inform Farouk El-Baz, the principal investigator for that experiment, that at ASTP altitude one could see far more detail than in Project Gemini, where Stafford and Cernan had flown at a higher altitude (+60 kilometers). The ground reported to the crew that the medical information received from the exercise period was very good. To round out its other activities, the crew made another course change at 3:18 p.m. In anticipation of their big day on the 17th, the Apollo team bedded down a few minutes after eight, and the Soviet crew had been resting since about 2:50 that afternoon. Throughout their "night," the spacecraft were coming closer together as Apollo closed the gap between them by about 255 kilometers per revolution.[20]

17 JULY—RENDEZVOUS

Roused at 3:07 a.m. by an alarm and warning signal from the guidance system, the crewmembers decided to stay awake after determining that the

*Details concerning the experiments are summarized in appendix E.

warning was a false alarm. That morning Slayton observed a grass fire in Africa, and Stafford saw a forest fire atop a mountain in the U.S.S.R. Slayton commented that things looked just the same as in an airplane at 12 000 meters. At 7:56, 5 minutes after completing another maneuver to bring the craft into better attitude for rendezvous, the Apollo crew attempted radio contact with Soyuz. Brand reported at 8:00 that he had sighted Soyuz in his sextant. "He's just a speck right now."[21]

Voice contact between the two ships was established 5 minutes later. Speaking in Russian, Slayton called, "Soyuz, Apollo. How do you read me?" Kubasov answered in English, "Very well. Hello everybody."

Slayton:	Hello, Valeriy. How are you. Good day, Valeriy.
Kubasov:	How are you? Good day.
Slayton:	Excellent. . . . I'm very happy. Good morning.
Leonov:	Apollo, Soyuz. How do you read me?
Slayton:	Alexey, I hear you excellently. How do you read me?
Leonov:	I read you loud and clear.
Slayton:	Good.[22]

Thirty-two minutes later at Slayton's signal, Kubasov turned on the range tone transfer assembly to establish ranging between the ships. The gap had been reduced to 222 kilometers. At 9:12, Apollo had changed its path again when the crew executed a coelliptic maneuver that sent the craft into a 210-by 209-kilometer orbit. Apollo was spiraling outward relative to the earth to overtake the Soviet ship.

A 0.9-second terminal phase engine burn at 10:17 brought Apollo within 35 kilometers, and the crew began to slow the spacecraft as it continued on the circular orbit that would intersect that of the Soyuz. CapCom Truly advised Stafford at 10:46, "I've got two messages for you: Moscow is go for docking; Houston is go for docking, it's up to you guys. Have fun." Immediately, Stafford called out to Leonov, "Half a mile, Alexey." Leonov replied. "Roger, 800 meters."[23] In accordance with the flight plan, the Soyuz crew had moved back into the descent vehicle and closed the hatch between them and the orbital module. Inside Apollo, the men had closed the CSM and DM hatches preparatory to docking. At a command from Stafford, Leonov performed a 60° roll maneuver to give Soyuz the proper orientation relative to Apollo for the final approach. On the television monitors in Houston and Moscow, Soyuz was seen as a brilliant green against the deep black of space as the onboard camera recorded the final approach.

Visitors had begun to gather in the MOCR viewing room about 2 hours before the docking. Among the early arrivals were General Samuel C. Phillips, former Apollo Program Director; Astronauts Scott, Allen, Garriott, McCandless, Musgrave, and Schweickart; and Captain Jacques Cousteau.

Just before 10:00, Dr. and Mrs. Fletcher, accompanied by John Young, escorted Ambassador Anatoliy Fedorovich Dobrynin and his wife into the viewing room. Other guests included Elmer S. Groo, Associate Administrator for Center Operations, and his wife; the Gilruths; D. C. Cheatham; D. C. Wade; and C. C. Johnson. As Apollo silently closed the remaining gap, the MOCR and viewing area grew quiet. Only the air-to-air and air-to-ground transmissions broke the spell.

Leonov called out as the two ships came together. "Tom, please don't forget about your engine." This reference to the $-X$ thrusters made Stafford and many of those on the ground who knew the story chuckle (see chap. IX). Stafford called out the range, "less than five meters distance. Three meters. One meter. Contact." The hydraulic attenuators absorbed the force of the impact, and Leonov called out, "We have capture, . . . okay, Soyuz and Apollo are shaking hands now." It was 11:10 in Houston. Stafford retracted the guide ring, actuated the structural latches, and compressed the seals. In Russian he said, "Tell Professor Bushuyev it was a soft docking." "Well done, Tom," congratulated Leonov, "It was a good show. We're looking forward now to shaking hands with you on . . . board Soyuz."[24]

The chase of Soyuz by Apollo had ended in a flawless docking. Stafford later recalled, "Later that night, we checked the alignment and noticed that the center of the COAS was sitting right on the center of a bolt that held the center of the target in for Soyuz." That is dead center. A feeling of relief and exultation swept the control center in Houston. Lunney with a cigar in hand called Professor Bushuyev. Watching each other on their television monitors, the Technical Directors smiled as they exchanged congratulations, while both crews went through pressure integrity checks on their craft. When Slayton opened the hatch into the docking module, he caught the strong scent of burned glue. This news dampened spirits on the ground for a short time. As a precaution, Vance Brand donned his oxygen mask, and Stafford advised Leonov: "Soyuz, this is Apollo. Now we have . . . a little problem. I think we have somewhat of a bad atmosphere here. I think soon that we will no longer have any problems."[25] While his Russian might not have won any prizes, the Soviet commander got Stafford's message. Once the odor dissipated and the ground crews decided that they could not discover any danger in this unexpected development, the crews continued the procedures leading to the opening of the hatches between the spacecraft.

Prior to that first handshake in space, Viktor Balashov, a noted Soviet television announcer, read a message from Leonid Ilyich Brezhnev over the air-to-ground link:

> To the cosmonauts Alexey Leonov, Valeriy Kubasov, Thomas Stafford, Vance Brand, Donald Slayton. Speaking on behalf of the Soviet people, and

for myself, I congratulate you on this memorable event. . . . The whole world is watching with rapt attention and admiration your joint activities in fulfillment of the complicated program of scientific experiments. The successful docking had confirmed the correctness of the technical decisions developed and realized by means of cooperative friendship between the Soviet and American scientists, designers and cosmonauts. One can say that the Soyuz Apollo is a forerunner of future international orbital stations.

Brezhnev's remarks continued, noting that "the detente and positive changes in the Soviet-American relations have made possible the first international spaceflight." He saw new possibilities for cooperation in the future and gave his best wishes to the crews.[26]

Stafford and Slayton meanwhile had entered the docking module and closed behind them the hatch (no. 2) leading to the CSM. They raised the pressure from 255 to 490 millimeters by adding nitrogen to the previously 78 percent oxygen atmosphere. In Soyuz, the crew had reduced the cabin pressure to 500 millimeters before the docking. The pressure in the tunnel between the docking module hatch (no. 3) and the Soyuz hatch (no. 4) had been raised from zero to equal that of the docking module. Leonov and Kubasov were the first to open the hatch leading to the international greeting. During the transfer that was to follow, the pressure in the DM and Soyuz would be the same—510 millimeters.

Then at 2:17:26 p.m. on the 17th of July, Stafford opened hatch number no. 3, which led into the Soyuz orbital module. With applause from the control centers in the background, Stafford looked into the Soviet craft and, seeing all their umbilicals and communications cables floating about, said, "Looks like they['ve] got a few snakes in there, too." Then he called out, "Alexey. Our viewers are here. Come over here, please." High above the French city of Metz, the two commanders shook hands.* Their dialogue was broken—part personal, part technical. They appeared to accept their amazing technical accomplishment with the same nonchalance that had characterized their practice sessions in the ground simulators. There were no grand speeches, just a friendly greeting from men who seemed to have done this every day of their lives. In the background was a handlettered sign in English—"Welcome aboard Soyuz."[27]

When they talked later with President Ford, however, the crews appeared somewhat less at ease. Ford had watched the Soyuz launch two days earlier in the State Department auditorium with Ambassador Dobrynin and Administrator Fletcher, while Mrs. Dobrynin interpreted for them. Keenly interested in the ASTP flight, Ford had wanted an opportunity to

*5°47'37" E and 49°10'12" N.

speak with the crews. Dennis Williams, the information officer attached to the International Affairs Office at NASA, had drafted a series of possible questions for the White House that could be asked of each crewman. Neither Williams nor the mission control team in Houston expected Ford to use all the questions, but that is exactly what he did. The crew, who had been advised the night before of the conversation, were taken by surprise when the President, watching the men on a television monitor in the Oval Office, talked for 9 minutes instead of the scheduled 5. He asked a barrage of questions that sent the crews scrambling to trade off their three flight helmets to they could respond to him. But despite the confusion, Ford and the five space men seemed to enjoy the chat. Ford began:

> Gentlemen, let me call you to express my very great admiration for your hard work, your total dedication in preparing for this first joint flight. All of us here in ... the United States send to you our very warmest congratulations for your successful rendezvous and for your docking and we wish you the very best for a successful completion of the remainder of your mission.

Stressing the same themes of cooperation as had Brezhnev, Ford pointed out that it had "taken us many years to open this door to useful cooperation in space between our two countries." When he asked Stafford whether he thought the new docking system would be suitable for use in future international manned space flights, the Apollo commander responded, "Yes, sir, Mr. President, I sure do. Out of the three docking systems I've used, this was the smoothest one so far. It worked beautifully." Ford spoke in turn to Leonov, Slayton, Brand, and Kubasov. The President asked Slayton, "as the world's oldest space rookie, do you have any advice for young people who hope to fly on future space missions?" Slayton responded that the best advice he could give was "decide what you want to do and then ... never give up until you've done it." To Ford's question about space food, Kubasov noted that the meals were different than the one the crews had shared with the President, especially since there was neither seafood nor beer available during the flight. In signing off, the President wished the men a "soft landing."[28]

Next Stafford, Slayton, Leonov, and Kubasov made a symbolic exchange of gifts, while Brand remained in the command module monitoring the American craft and waiting for his turn to visit Soyuz. Stafford speaking first, said:

> Alexey, Valeriy. Permit me, in the name of my government and the American people, to present you with 5 flags for your government and the people of the Soviet Union. May our joint work in space serve for the benefit of all countries and peoples on the Earth.[29]

Leonov thanked Stafford for "these very valuable presents" and in return gave Soviet flags to the Americans. During succeeding transfers, other symbolic items would be exchanged. Apollo would return a United Nations flag launched in Soyuz, and the two crews would sign the *Fédération Aeronautique Internationale* certificates for the official record books.[30]

The four men settled down to their first joint space banquet. On the ground, too, some people went in search of a snack. John Young escorted the Fletchers, the Dobrynins, and the Groos to a third floor snack bar in the Houston control center. Over ice cream bars and coffee, they discussed the events of the day. The Ambassador asked Fletcher why the ships had docked a little early, and the NASA Administrator indicated that they were so well lined up that there was no reason not to complete the docking. Fletcher told Dobrynin that the crews had not known until late the preceding night that they would be speaking directly with Mr. Ford. After a few good-hearted comments about the President's tendency toward long-windedness, the Americans bid farewell to the Dobrynins, who left for Washington.

Glynn Lunney and Chet Lee met with representatives from the press late on the afternoon of the 17th to comment on the status of the meeting in space. Lunney said that those who had seen him in similar "change of shift briefings" in the past had seen a busy flight director with a dozen or so pages of notes. On this particular day, he had not taken many notes; he had mainly sat in the control center "watching the Flight Directors and the rest of the team work." He continued:

> I would like to say that I've enjoyed today one hell of a lot. I have talked a number of times to the man on the other side of the ocean, Professor Konstantin Bushuyev, who's my counterpart and Director of the ASTP program for the Soviet Union and I could tell from the sound of his voice that he's enjoying the day as much as I am. . . .[31]

With his characteristic good humor, Lunney fielded a number of questions from the media representatives—the glue smell had not posed a problem; the crews had not talked much during their meal because "their mothers told them not to"; and there had been a scramble for headsets because no one had anticipated the President's desire to ask questions of all five men. Technically, diplomatically, and socially, the 17th had been a good day.

Stafford and Slayton said good-bye to Leonov and Kubasov at 5:47 and floated back through the tunnel into the docking module. Stafford returned to the command module, while Slayton closed the DM hatch. In Soyuz, the Soviets were securing their hatch, also. During the ensuing pressure integrity check, a possible leak through hatch nos. 3 or 4 was detected by the Soyuz monitoring equipment. This apparent flow of gas between the two hatches, while not serious, caused the crews to get to sleep a little later than planned.

Finally, by 7:36, the Apollo crewmen had bid the ground good night and were beginning to settle down.[32]

18 JULY—TRANSFERS

Awakened by "Midnight in Moscow," the Americans began their fourth work day in orbit at 2:00 a.m. Houston time. While the crews had slept, the two ground teams—in Houston led by Walt Guy and V. K. Novikov—had been watching the pressure levels of both ships and had conferred about the leak between the hatches. They had concluded that after the two hatches were closed and the pressure had been reduced to 260 millimeters the gases trapped between them heated up. The pressure sensing devices could not distinguish between the expanding gases and a leak. Neil Hutchinson commented on working with Soviet Flight Director Vadim Kravets, whom he had never met:

> the hatch integrity check ... involved me getting on the loop and talking to my counterpart who happened to be Kravets ... the answers were all forthcoming in a timely fashion and very professionally done.... I think the one thing, as I sit back and look at it now that makes me wonder; I wish there was another one of these flights. We've gone to all this trouble to learn how to work with those people. It's like going to the moon once and never going back. 90 per cent of the battle is over with ... getting all the firsts done.... I could run another Apollo Soyuz or another joint anything with a heck of a lot less fuss than it took to get this one going.[33]

Though some of the worry in both Houston and Moscow had been in vain, the two teams had confirmed that they could work together in analyzing an unforeseen problem.

With breakfast behind them and their early morning activities completed, Kubasov and Brand conducted a broadcast session from "your Soviet/American TV center in space," as Kubasov called it. In giving his tour of Soyuz, the Soviet flight engineer pointed out what various instruments were for and televised a picture of Brand in "the kitchen" (the food preparation station) warming up lunch. Stafford reciprocated by giving Leonov and the Soviet viewers a Russian language tour of the command module. Despite some problems with communications to the ground, the space television production was just one more unique aspect of the joint mission. Appearing casually simple from the perspective of the home viewer, these broadcasts had required hours of negotiation and planning, just as all other aspects of the flight had. Soviet viewers were particularly enthralled by the live coverage of the mission, but many Americans seemed to accept shows from 225 kilometers up as commonplace.[34]

Kubasov later gave an English language travelogue as the two craft passed over the U.S.S.R. "Dear American TV people," he began. "It would be wrong to ask which country's more beautiful. It would be right to say there is nothing more beautiful than our blue planet." After explaining that he would be giving a description of "what flows below the spacecraft," Kubasov continued:

> Our spacecraft, Soyuz, is approaching the USSR territory. Our country occupies one-sixth of the Earth's surface. Its population is over 250 million people. It consists of 15 Union Republics. The biggest is the Russian Federal Republic with the population of 135 million people.... At the moment we are flying over the place where Volgograd city is. It was called Stalingrad before. In winter 1942-43, German fascist troops were defeated by the Soviet Army here....

With the television camera still trained out the port of the orbital module, Leonov continued to describe the panorama. In the command module with Stafford and Slayton, the Soviet commander spoke of the Ural Mountains, and he pointed out the area below in Kazakhstan from which they had been launched three days before. Toward the end of the 10-minute commentary, Brand added some remarks about the countryside he could see from his vantage point and concluded, "as you can tell, Soviets very much remember the war 30 years ago. Fortunately, we've come a long way since then...."[35]

Fifteen minutes later at about 8:20, Brand and Kubasov began filming some science demonstrations that could later be used in science classrooms back on earth to demonstrate the effects of zero gravity on various items. Originally proposed by Marshall Space Flight Center, Kubasov became very enthusiastic about the idea of such demonstrations, which were similar in concept to those filmed during Skylab. As a result, he suggested simple illustrations of basic principles of physics, such as the gyroscope, to be recorded during the flight. Brand narrated the film in English, and Kubasov gave the Russian commentary. Literally nowhere on earth could a classroom instructor duplicate the experiments, not to mention having such celebrities give the explanations.

During this second transfer, Brand had lunched in Soyuz and Leonov in Apollo. At 10:43, Brand returned to Apollo, and Stafford and Leonov moved into Soyuz. Kubasov then transferred into the command module in this exacting cosmic ballet. With each movement of the crewmembers, the atmospheric composition of Soyuz had to be checked to make certain that not too much nitrogen had been removed. Once everyone was in place, the hatches between the orbital and docking modules were closed as a further step toward maintaining the proper cabin atmospheres. The highlight of the third transfer—the space-to-ground press conference—was about to begin.

17-19 July 1975, Soyuz and Apollo Meet in Space

Left, Soyuz waits as Apollo (right) maneuvers closer (below) and docks. In the docked position the view out Apollo's window is along part of the docking module to the spherical bulk of Soyuz, one of whose solar cell arrays extends out parallel to the white earth horizon.

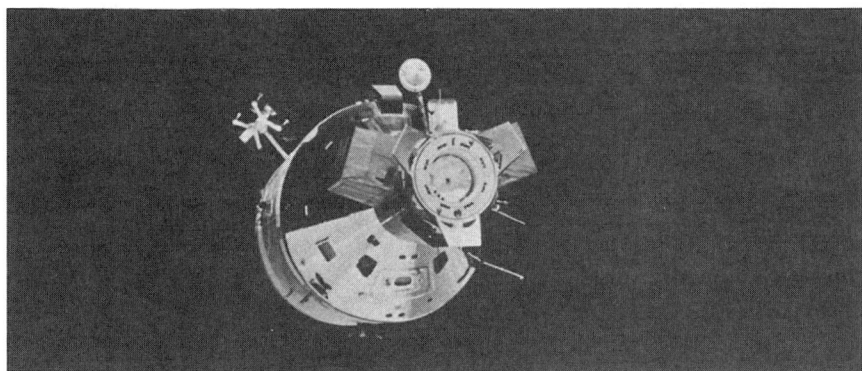

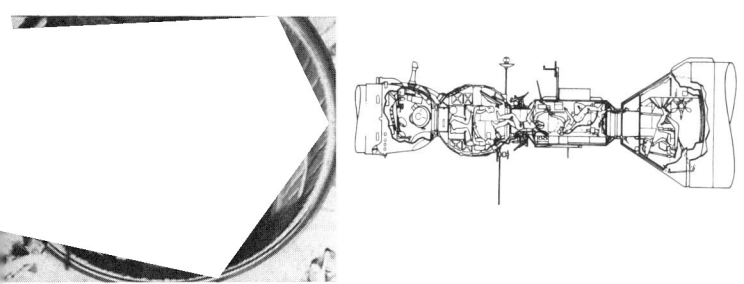 

Above left, Cosmonaut Leonov and Astronaut Stafford meet in the docking module; above, a cutaway drawing of the two spacecraft in docked position; above right, Leonov displays his sketch of Stafford; center, Stafford snaps photo of himself, Leonov, and Slayton literally putting their heads together; below, left, in the hatchway between the Soyuz orbital module and the descent module, Kubasov is at work as Leonov checks the flight plan; below, right, Brand monitors the Apollo controls while his fellow crewmen visit Soyuz.

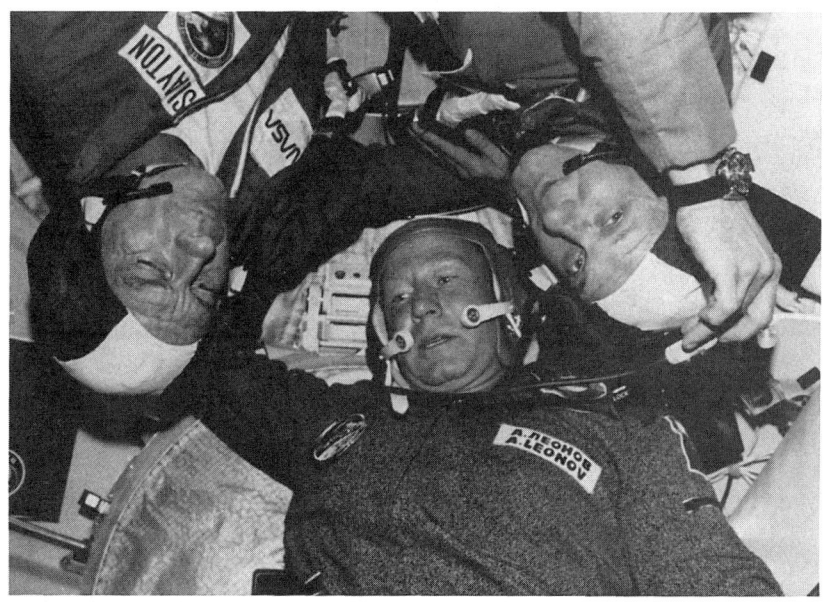

On the ground in the control center in Houston during the Apollo-Soyuz flight, Cernan, Lunney, and Tastistcheff follow mission progress (left); in Moscow (right), Viktor Legostayev (foreground) and Alexei Yeliseyev do the same (Soviet Academy of Sciences photo).

Other mission activities in Houston: Soviet technical specialists V. V. Illarionov (above, foreground), O. I. Babkov, and V. I. Staroverov monitor the flight plan; at left, Working Group Chairmen Guy (foreground), White, and Dietz jointly monitor the mission; below, Robert Shafer (center) sits listening to Harold Stall discuss the mission's TV; bottom left, Administrator Fletcher directs Soviet Ambassador Anatoliy Dobrinin's attending to the large eidophor display screen, while technical assistant Leonard Nicholson stands by; bottom right, cheering the Apollo splashdown are controllers Don Puddy (left), Frank Littleton, M. P. Frank, Neil Hutchinson, and Gene Kranz.

COME FLY WITH US

Having collected questions in advance from news people in Moscow and Houston, Valeriy Vasil'yevich Illarionov of the visiting specialists team and Karol Bobko read the questions to the crews from Houston. The queries and the responses were friendly, in the spirit of the mission. Stafford began by saying that it had been a very rewarding two days in space. He felt that the success of the mission was the result of "the determination, the cooperation, and the efforts by the governments of the two countries, by the managers, engineers, and all the workers involved." When he first opened the hatch to greet Leonov and Kubasov, he had a couple of thoughts that he was unable to express at the time. He believed that when they opened those hatches in space, they were opening the possibility of a new era on earth. "I would have said," in Russian, "we were opening back on Earth a new era in the history of man." He noted that just how far that new era would go would depend upon "the determination, the commitments, and the faith of both countries and of the world." The "climate of detente and a developing cooperation between our countries" has made this mission possible, Leonov added.

Because of his participation in the first space welding in 1969, Kubasov was asked about materials processing in space. Kubasov believed that one of the future benefits of space programs would be the development of better and different alloys resulting from space processing. "It seems to me that the time will come when space will have whole plants, factories, for the production of new materials and new substances with new qualities, which could be . . . made only in space." Linked to that question was one from Moscow addressed to Stafford about the justification of spending money on space programs when there were so many problems in the world that needed solving.

Stafford noted that this was not a new question. He certainly believed that the costs would be repaid by the long term benefits. Science and applications were the likely areas of payoff, but the uplift to the human spirit was also implicit in his words and those of his colleagues. All the men agreed that they preferred news of peace and tranquility, and Kubasov especially hoped that all children would have a future filled with peace, so that they would never have to know what it was like to lose parents or loved ones in a war. On a lighter note, when a Soviet reporter asked Leonov to transmit a sketch "that would depict the meaning, the essence of the joint mission," Leonov and Stafford held up two flags, one from the United States and one from the Soviet Union—although backwards, the message was clear enough. Leonov then went on to show the television audience a number of sketches that he had drawn—"Here's a whole cosmic portrait gallery."

The best lines of the press conference came later. When asked how he liked the American food, Leonov diplomatically answered, "I liked the way it was prepared, its freshness."

> But as an old philospher says, the best part of a good dinner is not what you eat, but with whom you eat. Today I have dinner together with my very good friends Tom Stafford and Deke Slayton because it was the best part of my dinner.

Slayton was asked how the experience of space flight compared to all the stories he had been told over the years. He said that he did not think he had discovered anything new.

> We've had the same kind of problems up here that people have complained about since MR-3. . . . Not enough space, and a little congestion to the time line, difficulty in keeping up with things. It's a lot slower getting things done up here than you realize when you're down there in one-g. . . . In some respects, it's easier because weighty things are easier to move around, but, on the other hand, everything just tends to take off if you let go of it. . . . it's been a great experience. I don't think there's any way anybody can express how beautiful it is up here.

Looking to the future, Leonov was convinced that mankind was just at "the beginning of a great journey into outer space." As with the other ASTP crewmen, he hoped to have a chance to fly again. Stafford agreed and said that he would like to fly on one of the early Shuttle missions. "And I would hope that if Alexey would have a vehicle developed by [his] country that we could fly . . . in a joint mission." Not to be outdone, Leonov added, "I would always like to fly with friends . . . whom one trusts and with whom it is not dull to work. . . ."[36]

The crews returned to other items on their flight plan. Slayton, as part of the earth observation experiment (MA-136), took photos of ocean currents off the Yucatan Peninsula and in the Florida straits. He also tried to observe the red tide phenomena—marine micro-organisms that cause the water to appear red—off the coast of Tampa and in the vicinity of Cape Cod. But this visual exercise was not completed because of cloud cover. Brand's travelogue of the East Coast of the U.S. was likewise hindered by the clouds, but he gave the narration anyway, describing the climate and flora of Florida, North Carolina, Virginia, Washington, the Middle Atlantic states, and New England. As the ships passed over Massachusetts, Brand noted that Robert H. Goddard had launched the world's first liquid fueled rocket from that state on 16 March 1926.[37]

Leonov narrated the events of the fourth transfer as he saw them. He stressed the large amount of work they had to accomplish during the joint phase of the mission, including five bilateral experiments. Although this "saturated program" seemed at times to be more than the five men could handle, they managed to complete all their tasks. Slayton, Brand, and Kubasov assembled the two halves of a medallion commemorating the flight, and then they exchanged tree seeds. As Slayton juggled television equipment, Stafford and Leonov bid their final farewell. All these exercises climaxed one of the most complex television scenarios ever conceived and executed.

Tom Stafford shook hands with Leonov and Kubasov, bidding them farewell at about 3:49 in the afternoon. He then moved back into the docking module, and the space men closed the hatches for the last time at 4:00. Once the checklist for securing the hatches and executing the pressure integrity check of the seals was completed, the crews set about routine housekeeping chores—stowing equipment and making certain that all was in readiness for their next meal. For the statistically minded, the records indicate that Stafford spent 7 hours, 10 minutes aboard Soyuz, Brand 6:30, and Slayton 1:35. Leonov was on the American side for 5 hours, 43 minutes, while Kubasov spent 4:57 in the command and docking modules. To those at work in space and on the ground, it seemed longer.

Before finishing all the items on their pre-sleep checklist, the Americans paused to listen to the news and sports as read by CapCom Truly. Included in his report was mention of an American home exhibit that had just opened to enthusiastic crowds in Moscow. Called "Technology in the American Home," the display was designed to give Soviet citizens an idea of the gadgetry available to the American homemaker. While no one commented on the fact, it was just such an exhibit that had sparked the Nixon-Khrushchev debate in 1959. In 16 years' time, the international scene seemed to have changed dramatically.

Although the crew signed off for the evening on schedule at 7:20, they spent an uneasy first few hours. In addition to being very tired from the activities of their fourth day in space, they were jangled awake an hour later by a master alarm that reported a reduction in docking module oxygen pressure. This problem was no real hazard, and it was quickly solved by an increased flow of oxygen into the DM, but it kept the crew from getting all the sleep for which they had been scheduled. When wake-up time came at 3:13 on the morning of the 19th, the crew failed to hear the musical strains of "Tenderness" as sung by the Soviet female artist Maya Kristalinskaya, with which the ground team had hoped to gently waken them. But 15 minutes later, they were awake and ready to begin their fifth day. Next door, beyond hatches three and four, Leonov and Kubasov were getting prepared, too.

19 JULY—EXERCISES

During day five of the flight, the crews concentrated on docking exercises and experiments that involved the two ships in the undocked mode. During the interval between the first undocking and the second docking, the Apollo crew placed its craft between Soyuz and the sun so that the diameter of the service module formed a disk which blocked out the sun. This artificial solar eclipse, as viewed from Soyuz, permitted Leonov and Kubasov to photograph the solar corona. Ground-based observations were conducted simultaneously, so that the Soviet astronomer G. M. Nikolsky could compare views of the solar phenomena with and without the interference of the earth's atmosphere. Skylab had provided a long term look at the corona, and the ASTP data would give scientists an opportunity to compare findings made a year and a half later. This "artificial solar eclipse" (MA-148) experiment would be the last American chance for such information gathering until the Shuttle era.

Another major experiment, "ultraviolet absorption" (MA-059), was an effort to more precisely determine the quantities of atomic oxygen and atomic nitrogen existing at such altitudes as the one in which Apollo and Soyuz were orbiting. Again this information could not readily be obtained from ground-based observations because of the intervening layers of atmosphere. Apollo, flying out of plane around Soyuz, first at 150 meters, then at 500 meters, and finally in plane at 1000 meters, projected monochromatic laserlike beams of light to retroreflectors mounted on Soyuz. When the beams were reflected back to Apollo, they were received by a spectrometer, which recorded the wavelength of the light. Subsequent analysis of these data would yield information on the quantities of oxygen and nitrogen. Some very precise flying was called for in these experiments.

After being docked for nearly 44 hours, Apollo and Soyuz had parted for the first time at 7:12 a.m. while out of contact with the ground. Slayton advised Bobko after radio contact was re-established that they had undocked without incident and were stationkeeping at a range of 50 meters. Meanwhile, Soyuz had extended the guide ring on its docking system in order to test the Soviet mechanism in the active configuration. Once they completed the solar eclipse experiment, with Slayton at the controls, Apollo moved towards Soyuz for the second docking. As he did, Stafford called out to the ground, "Okay, Houston, Deke's having the same problem with the COAS washout that I had." As Slayton explained it, he could see Soyuz and the target initially when they were against the dark sky, but at "about 100 meters or so, it went against the earth background and zap. Man, I didn't have anything." Although worried that he might run over Soyuz, he pressed on with the docking "by the seat of the pants and I guess I got a little closer than they or the ground anticipated."[38] There was too much light flowing into the optical alignment sight for Slayton to get a good view of the docking target. Contact with Soyuz came at 7:33:39, and Leonov advised the Americans that he was beginning to retract his side of the docking assembly.

As viewed via Apollo television, this docking looked as if it had been harder than the first, and the two ships continued to sway after capture had been completed. Slayton, speaking in a debriefing, later said:

> The docking was normal, you guys gave me contact as usual and then I gave it thrusting. The only thing that happened then was they seemed to torque off. I was surprised at the angle they banged off there after we had contact.[39]

Despite this oscillation, the Soyuz system aligned the two craft and a proper retraction was completed. Subsequently, there was some discussion of this docking, and the Soviet docking specialist Syromyatnikov was at first worried that an unnecessary strain might have been placed on the Soyuz gear. Bob White said that analysis of the telemetry data indicated that Slayton had inadvertently fired the roll thrusters for approximately 3 seconds after contact, and that this sideways force caused the craft to oscillate after the docking systems were locked and rigid.

But even with the extra thrusting, the second docking was within the limits of safety established for the docking system. Slayton's docking took place at a forward velocity of 0.18 meter per second versus 0.25 meter per second for Stafford's docking, but the difference lay in the inadvertent thrusting. Momentarily an issue, the extra motion of Slayton's try was not a serious concern after all the data had been evaluated. Even Syromyatnikov had to concede that "the mechanism functioned well under unfavorable conditions." It was a case of things looking worse than they really were. In the end, the incident only demonstrated the reliability and hardiness of the new docking system.[40]

It was 10:27 when Apollo and Soyuz undocked for the second and final time. This 4-minute exercise was conducted by Leonov, since it was a Soyuz active undocking. Slayton then moved his ship to a stationkeeping distance, about 40 meters away. As he did, Leonov opened the retroreflector covers so that the ultraviolet absorption (UVA) experiment could be performed. A difficult series of maneuvers were called for in this test. As Soyuz continued its circular orbit, Slayton took Apollo out of plane with Soyuz and oriented his craft so that its nose was pointed at the reflector on the side of the other ship. Orbiting sideways in this configuration, Slayton flew Apollo in a small arc from the front of Soyuz to the rear of that ship while the spectrometer gathered the reflected beams. On the 150-meter phase of the experiment, light from a Soyuz port led to a misalignment of the spectrometer, but on the 500-meter pass excellent data were received; on the 1000-meter pass satisfactory results were also obtained.

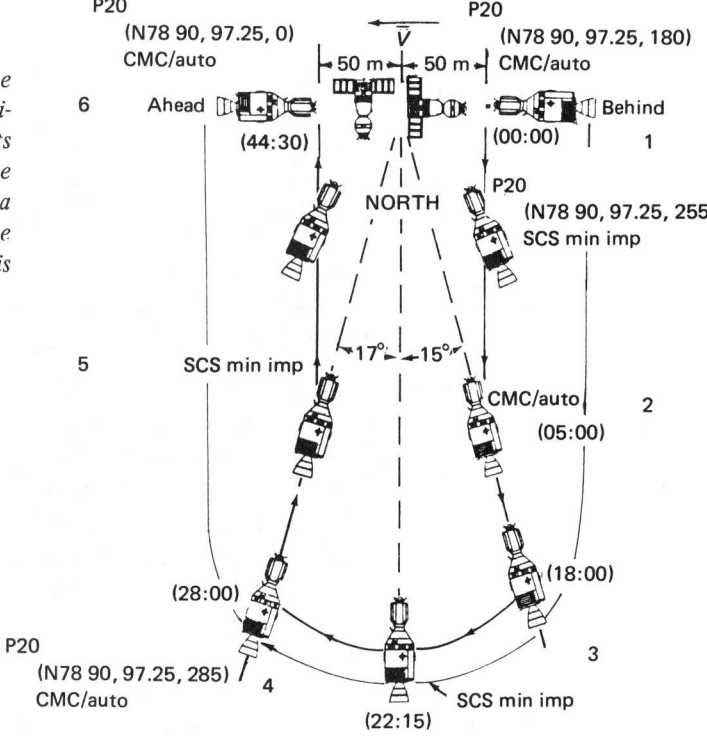

Diagram of the flight plan for the 500-meter ultraviolet absorption experiment. Starting from behind Soyuz in its orbital path, Apollo swings out of plane and around the Soviet craft for a 10-minute data take. At the end of the 44-minute exercise, the American ship is ahead of Soyuz.

After nearly 3 hours of tough flying, Bobko congratulated the crew. "You people flew it fine." Slayton responded:

> Okay. Great, Bo. And you can thank ol' Roger Burke, Steve Grega, and Bob Anderson, down there, that everything came off right. 'Cause they sure did all the work to make it go.[41]

The three men Slayton mentioned had spent hours in the simulators working out the procedures to fly this complicated maneuver. Burke, who had worked with developing flight procedures for years, felt that this was one of the hardest experiments a crew had ever been called on to do, especially since the flight plan for it had continued to evolve until a couple of days before launch.[42] Slayton later noted that it had taken all three Apollo crewmen to complete the ultraviolet absorption experiment. "I was doing the flying, Vance was running the computer and we had Tom down in the equipment bay opening and closing doors, turning on sensors and so forth. So, it was a busy time for all of us." He indicated that the maneuvers were difficult because orbital mechanics came into play as they tried to fly around Soyuz. When the Apollo crew changed the velocity of their craft, they also affected its orbit. They would have no difficulties if they had had unlimited fuel resources, but being out of plane and playing orbital mechanics with "a very limited fuel budget ... made it a great challenge."[43] Stafford added that the thruster firings had to be timed because the onboard accelerometers could not measure the changes in velocity.[44]

Apollo performed a separation maneuver at 1:42 to prevent re-contact with Soyuz, placing the American craft in a 217- by 219-kilometer orbit. With all the joint flight activities completed, the ships were going their separate ways. Soyuz was below and moving ahead of Apollo at a rate of 6 to 8 kilometers per orbit. Leonov and Kubasov prepared to go to sleep, but the American crew had several hours of work scheduled in their crowded flight plan after their mid-afternoon meal before they could settle down for a rest period. The fifth day of ASTP—the second of joint activities—had been a success, and everyone in the Moscow and Houston control centers was pleased that all had gone so well.

20 JULY—INDEPENDENT ACTIVITIES

Kubasov and Leonov began their sixth day in space at 1:10 a.m. while their American friends slept. They had breakfast and carried out a series of activities that included earth and solar photography and recording data photographically on the joint zone-forming fungi experiment and other unilateral experiments. Leonov also ran through a simulation of the deorbit procedures—orientation, retrofire, and the deployment of the parachutes. At mid-afternoon Moscow time (sunup in Houston), the Soviet space travelers gave a television broadcast to their viewers at home. Afterwards, they continued their experiments and preparations for their re-entry 24 hours later.[45]

Houston control tried for a second morning to wake the crew with "Tenderness" in Russian. This time they succeeded, and the men began their sixth day at 1:54 a.m. In addition to a day-long earth observation, which they started before breakfast, they concentrated on experiments during their first independent day in orbit. Included in the flight plan were experiments in the multipurpose furnace (MA-010), extreme ultraviolet surveying (MA-083), crystal growth (MA-085), and helium glow (MA-088). In the midst of their work during an *ATS 6* communication session, CapCom Crippen gave them a news report.

Crippen included a special item in his report. "Six years ago today at 3:17:40 central daylight time we landed on the Moon. At 9:56, that's when Neil said his famous words about 'small step for man, giant leap for mankind.' " Stafford responded, "Roger. Remember it well."

Slayton:	Say, what day of the week is this, incidentally?
Crippen:	This happens to be Sunday.
Brand:	[garbled] . . . our day off.

Crippen:	Oh, yeah. We'll get them off after you guys get back. Y'all . . . are certainly not getting a day off today.
Brand:	We're not complaining.[46]

While there was still much to do, the pressure of the first days of the mission was gone, and the crew was settling down to the routine. The sixth day of the ASTP flight was noticeably void of the drama that had been associated with the joint activity.

21 JULY–FAREWELL

Leonov and Kubasov had signed off the air shortly after 1:37 (9:37 in Moscow) on the afternoon of the 20th, after stowing all of the returnable items in the descent module. Following a rest period of nearly 10 hours, the Soyuz crewmen advised the ground that they were awake and that all systems were normal. After exchanging flight data and receiving a weather report, they ate breakfast and donned their space suits. Their pressure integrity check, conducted at about 3:30 a.m. indicated that their suits were functioning normally. Leonov and Kubasov ran through their re-entry checklist. Moscow Mission Control gave the following announcement:

> The Mission Control Center's calculated the descent-orbit data. This data has been entered into the program computer . . . the crew is monitoring the orientation and also the transmission of information. The deorbit data . . . is the following . . . the braking pulse to shift the spacecraft from Earth to a

At left, ASTP Cosmonauts Leonov and Kubasov (speaking) are interviewed by Soviet newsmen shortly after their safe landing in Kazakhstan. Below, Kubasov autographs the side of the Soyuz descent vehicle (Soviet Academy of Sciences photos).

descent trajectory will be 120 meters per second. This braking pulse . . . will operate for 194.9 seconds at the altitude of 214 kilometers and 13 hours and 9 minutes.[47]

Throughout the Soviet Union, crowds gathered in homes and in stores with televisions to watch the rare treat of a live broadcast of a Soyuz recovery. At Houston, a few hardy souls, in addition to the ground control team, were up to witness this early morning event.

The deorbit burn came exactly on time (5:09 in Houston), and the Soyuz crew notified Moscow that the retro-engine had fired for the calculated period and had been turned off at 5:13:38. Separation of the orbital and descent modules came 9 minutes later. Leonov advised the ground that the gravitational forces had built up, passed, and were less than he had anticipated. A task force of Soviet helicopters and ground-based personnel moved into the landing area. All in all, this formidable armada of trucks and aircraft was about equal in number to the size of the sea-based team that would later greet Apollo.

Soviet air rescue pilots began receiving radio signals from the spacecraft at approximately 5:40, and almost simultaneously helicopter-borne television cameras began transmitting pictures of the descent. As Soyuz floated downward, Walter Cronkite, in search of commentary on the event, noted for his viewers that the color quality of the pictures was not very good. But good or bad, they were extraordinary! Within a few feet of the ground, the automatically fired landing rockets slowed the "thumpdown" of the descent vehicle. A cloud of dust caused by the braking rockets of Soyuz engulfed the craft and caused momentary anxiety for those viewers who did not understand its meaning. Three minutes after landing, at 5:51, a slightly shaky Kubasov was the first to exit. Leonov and his flight engineer smiled broadly and waved to photographers on the scene. Houston Mission Control reported: "We're just looking at the TV here and see that Soyuz has landed safely, and Alexey and Valeriy were outside of the spacecraft and seem to be in good health."[48] Stafford asked Houston to give the Soviets their best and to say that he was glad to hear that everything went well. For the remaining three and a half days, Stafford, Slayton, and Brand would concentrate on their experiments, but in many respects the saga of Apollo and Soyuz had come to an end.

22-23 JULY—EXPERIMENTS

Some minor experiment hardware problems developed during the final days of the mission, but for the most part the crewmembers worked through their flight plan—which included 23 independent experiments—with few difficulties. CSM 111 was truly the best—as well as being the last—Apollo to fly. After a relatively quiet day of work on the 22d, the major part of the next day was devoted to preparing for and conducting the doppler tracking experiment (MA-089). Paired with the geodynamics experiment (MA-128), these investigations were designed to verify which of two techniques would be best suited for studying plate tectonics (movements of the earth's substrata) from earth orbit. Where the geodynamics experiment utilized Apollo and *ATS 6* in an attempt to measure these movements (the so-called low-high approach), the doppler tracking experiment involved the use of two satellites in low earth orbit (the low-low approach) to measure the existence of "mass anomalies" greater than 200 kilometers in size. When the jettisoned docking module and the CSM were separated by 300 kilometers, they would theoretically have their orbits affected by the greater gravitational forces exerted by these mass anomalies. As their orbits were perturbed, the radio signals transmitted from one to another would correspondingly be affected.

Prior to releasing the docking module on its separate journey, the crew had participated in a second press conference from space. During that 32-minute session, the crewmen were asked to philosophize about the future of manned space flight in general and upon such diverse topics as trips to Mars and their own participation in the Shuttle program. Their answers were filled with optimism and good humor. Deke Slayton's statement that he had done nothing in space that his 91-year old aunt could not have done sent reporters scrambling to find out her name (Mrs. Sadie Link) so they could meet their deadlines. Following the press period, CapCom Crippen told the crew, "you guys did a great job there. Professional as always." He also gave them the news that Leonov had been promoted from colonel to major general.

With congratulations over, Stafford told the ground, "Now, back to work." After donning their space suits, the crew vented the command module tunnel and at 2:41 jettisoned the docking module. Filled with all their trash and used equipment that need not be returned, the DM tumbled into space at exactly the proper rate. Stafford and his team then executed their separation maneuver so that they could take the necessary doppler measurements. The docking module would continue on its way until it re-entered the earth's atmosphere and burned up in August 1975.[49]

24 JULY—LAST SPLASH

Approximately 24 hours after they parted from the docking module, Stafford, Slayton, and Brand began their journey homeward. On the ground, the flight control team played Jerry Jeff Walker's "Redneck Mother" to wake the crew. With a cheery "Good morning, gents. Party's over. Time to come home," CapCom Crippen told them to rise and shine. At half past seven, the crew started preparing for its mid-afternoon deorbit. As the men rubbed the sleep from their eyes, ate breakfast, and gathered data for the medical doctors on the ground, Crippen read them the news for the last time, news most of which the Apollo crew was making. The newspapers said that Slayton would fly again and that Stafford was still undecided about the future. Would it be NASA, the Air Force, industry, or politics? "That last option is sure out. I'll clue you, ol' buddy," was the General's response. Crippen gave a favorable weather forecast for the prime recovery area— visibility 16 kilometers, winds at 17 knots, scattered cloud cover at 600 meters, and wave height 1.1 meters.[50]

CSM deorbit came at 3:37:47, or about 13 seconds ahead of schedule. Six and a half minutes later, the command module was separated from the service module. As the reentry vehicle descended, Slayton and Brand commented on the buildup of gravity forces and the fireball that flared up as the heatshield pressed against the earth's atmosphere. At 4:18:24, Apollo splashed down about 7300 meters from the recovery ship *New Orleans*. Houston control was filled with smiling faces and cigar smoke. Unknown at that time to the celebrants was the fact that the crew had inhaled nitrogen tetroxide fumes during the descent.

The descent phase had gone without incident until about 15 000 meters. In the days that followed the recovery, the story of the failure to actuate the Earth landing system (ELS) was told and retold several times by Glynn Lunney, John Young, and others. Vance Brand presented his version during the crew technical debriefing. When the CM reached an altitude of 9144 meters, two earth landing switches that permitted the apex cover to be jettisoned at 7310 meters were normally armed. The drogue parachutes would then be released, followed by the main chutes. Commenting on the descent, Brand said that as Stafford read steps from the Entry Checklist he threw the proper switches. There was quite a bit of noise in the cabin from the command module's thrusters and the passage of the craft through the atmosphere.

At 30K [9144 meters], normally we arm the ELS AUTO, ELS LOGIC, that didn't get done. Probably due to a combination of circumstances. I didn't hear it called out, maybe it wasn't called out. Any case 30K to 24K [9144-7315 meters] we passed through that regime very quickly. I looked at the altimeter at 24K, and didn't see the expected apex cover come off. Didn't see the drogues come out. So, I think at about 23K, I hit the two manual switches. One for the apex cover and also, the one for drogues. They came out. That same instant the cabin seemed to flood with a noxious gas, very high concentration it seemed to us. Tom said he could see it. I don't remember for sure now, if I was seeing it, but I certainly knew it was there. I was feeling it and smelling it. It irritated the skin a little bit, and the eyes a little bit, and, of course, you could smell it. We started coughing. About that time, we armed the automatic system, the ELS. . . .[51]

The manual deployment of the drogue chutes caused the CM to sway, and the reaction control system thrusters worked vigorously to counteract that motion. When the crew finally armed the automatic ELS 30 seconds later, the thruster action terminated.

During that 30 seconds, the cabin was flooded with a mixture of unignited propellant and oxidizers from the thrusters. Prior to drogue deployment, the cabin pressure relief valve had opened automatically, and in addition to drawing in fresh air it also brought in unwanted gases being expelled from the roll thrusters located about 0.6 meter from the relief valve. Brand manually deployed the main parachutes at about 2700 meters, and despite the gas fumes in the cabin, the crewmembers continued to work through their checklist as best they could. Due to severe coughing and intercom noise, they had difficulty talking to one another and to the ground.

Following a normal but hard splashdown, the command module flipped over, leaving the three men hanging upside down in their couches from harnesses. Brand, who was coughing the most because he was closest to the

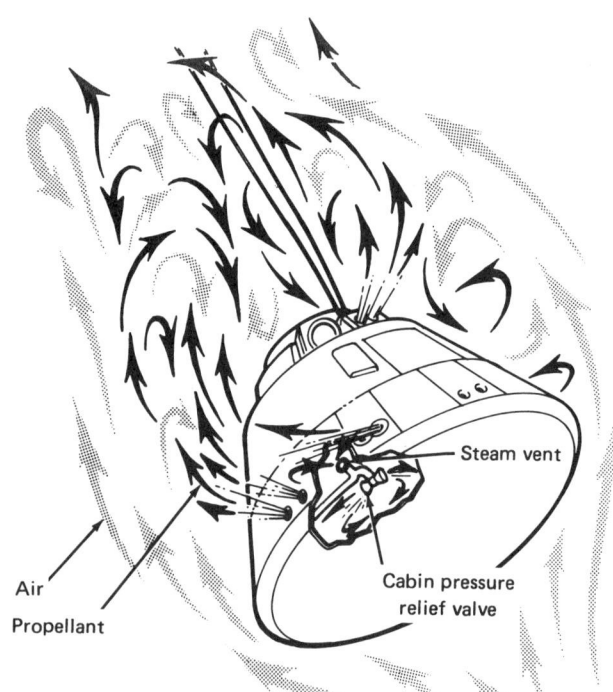

Diagram of air and propellant flow around the Apollo command module during descent through the atmosphere. Note propellant gases being drawn into the steam vent.

Air

Propellant

Steam vent

Cabin pressure relief valve

steam duct opening, saw that Slayton was feeling nauseous and reminded Stafford to get their oxygen masks. The commander recalled:

> For some reason, I was more tolerant to [the bad atmosphere], and I just thought get those damn masks. I said don't fall down into the tunnel. I came loose and . . . had to crawl . . . and bend over to get the masks. . . . I knew that I had a toxic hypoxia . . . and I started to grunt-breathe to make sure I got pressure in my lungs to keep my head clear. I looked over at Vance and he was just hanging in his straps. He was unconscious.[52]

After Stafford secured the oxygen mask over Brand's face and held it there, he began to come around. Once the entire crew was breathing pure oxygen, Brand actuated the uprighting system. When the command module was upright in the water, Stafford opened the vent valve, and with the in-rush of air the remaining fumes disappeared.[53]

Failure to throw the ELS switches led to an unanticipated two-week hospital stay for the crew in Honolulu. For Slayton, it also meant the discovery of a small lesion on his left lung and an exploratory operation that indicated it was a non-malignant tumor. After a short convalescence, Slayton joined the other four ASTP flyers for two tours, one of the Soviet Union and one of the United States. Despite a grueling month on the road, neither Slayton nor his team mates seemed any the worse for wear, and the warm public reception wherever they went seemed to indicate that the unfortunate accident at the end of the flight had not detracted from the basic success of the Apollo-Soyuz Test Project.[54] Rivalry had produced the first manned space flights in the early 1960s. But that sense of conflict had been overcome with the creation of an international test project. Ironically, this first joint flight also marked the end of an era. NASA's manned space program had seen its last splashdown. Apollo would fly no more.

Epilogue

While the American ASTP crew toured the Soviet Union and the United States with their Soviet counterparts, major changes were occurring at NASA Headquarters and the Johnson Space Center. In October, the interior walls of the Apollo Program Office in Houston were quite literally moved around to create the Space Shuttle Payload Integration and Development Office. Glynn Lunney, who on the last day of Apollo's flight had been put in charge of managing the Space Shuttle cargoes, had told Boris Artemov, Bushuyev's interpreter, during a telecon on 29 October, "Don't mind the banging, Boris, they're just tearing down the building." Shifting walls were indicative of the changes sweeping the halls at Johnson Space Center (JSC).[1]

With the splashdown of Apollo, a major chapter in the history of NASA had come to a close. All three generations of American spacecraft—Mercury, Gemini, and Apollo—had been single-flight vehicles. In these essentially experimental craft, the NASA team had mastered the problems of orbital and cislunar flight. Knowing that truly economical space flight would be possible only when the same spacecraft could be flown many times, NASA had begun the search for a reusable vehicle in the late 1960s. The Space Shuttle grew out of that quest. Consisting of three major elements—an orbiter, an external fuel tank, and solid-rocket, strap-on boosters—Shuttle was designed for a crew of four and up to six payload specialists. With a payload bay 18 meters in length by 4.5 meters in diameter, Shuttle would have the capacity to carry a 30 000-kilogram cargo. Initially, the orbiter would be able to stay in space for seven days at a time; later that period would be expanded to 30 days.

Those who had been responsible for the Apollo-Soyuz Test Project in Houston were given new assignments related to Shuttle. Arnold Aldrich, Lunney's deputy during the mission, was placed in charge of program assessment in the Shuttle Program Office. Bob White and Frank Littleton went to work for him, evaluating the management aspects of that effort. Ed Smith turned his full attention to Shuttle simulation planning, which had received only part of his time during the ASTP years. Pete Frank, as Chief of the Flight Control Division, devoted his time to Shuttle flight control problems. R. H. Dietz divided his energies between Shuttle payload communications questions and feasibility studies of a large solar power station satellite. Walt Guy also looked toward the future; his concern was new environmental control systems on the Shuttle orbiter.

Stafford, Slayton, and Brand, recently a crew, went their separate ways. In November 1975, Stafford left NASA to resume his career with the Air Force. With a second star on his shoulder, he assumed command of the Air Force Flight Test Center, Edwards Air Force Base, California. He was gone from JSC but not totally out of the picture. Shuttle would make its first approach and landing tests (ALT), after being carried aloft by a modified Boeing 747, on the dry lake bed at Edwards. Slayton, the director of ALT for NASA, would be visiting his alma mater—the test pilot school at Edwards, from which so many of the astronauts had graduated—to oversee those first unpowered glide flights. Brand, working in the Astronaut Office at JSC, had the responsibility for developing flight techniques for Shuttle, especially in entry and landing.

There was to be a hiatus in American manned space flight, but the pause should not be all that long. The approach and landing tests, begun in 1977, are to study the glide characteristics of the new orbiter. The first orbital flight test is set for 1979, and six developmental flights are on the drawing boards for mid-1980. Then Shuttle would begin regular and frequent operations, promising to become the DC-3 of outer space.

When Professor Bushuyev and his colleagues arrived in Houston for their final ASTP visit on 10 November 1975, many of the Shuttle changes were already visible at the space center. But the question on everyone's mind was, "What next with the Soviets?" Since the October 1973 meeting in Moscow, the Soviets had been deferring on the future systems aspect of the space cooperation agreement. Low, Lunney, and other Americans had continued to prod the Soviets about their plans for joint activities after ASTP, and each time the Soviets had asked the U.S. team to wait until after the joint flight. Bushuyev had told Lunney repeatedly that he did not have the personnel required to both prepare for ASTP and discuss future activities. So the talks that had begun as an effort to explore joint missions with future generations of spacecraft remained incomplete, despite recommendations from each Working Group concerning future operations based upon the lessons learned from ASTP.[2]

So how do we judge the success of the joint project? Evaluation of ASTP within the large context of continued cooperation between the United States and the Soviet Union will have to wait. Certainly, we can say that ASTP had a political dimension, one that reflected the improved relationship between the two countries that Presidents Nixon and Ford and Secretary of State Kissinger were seeking. But for now, the mission can be judged only upon its merits as a test flight. During the joint activity, the television media

Apollo and Soyuz in the docked configuration on display in the Smithsonian Institution's new National Air and Space Museum, August 1976, now a part of the history of flight.

presented a favorable, sometimes glowing, commentary on the live show from space, but several newspaper journalists were critical of what they termed "a costly space circus."[3] Robert B. Hotz, editor-in-chief of *Aviation Week and Space Technology*, editorialized:

> the real tragedy for this country was the decision to put its scarce space dollars into the political fanfare of Apollo-Soyuz. . . .
>
> Now that it is over, it is apparent that the decision to fly Apollo-Soyuz, instead of another Skylab or whatever else could yield a good return on the Apollo investment already made, was as foolish and feckless as those other facets of the Nixon-Kissinger detente—the SALT talks, the trade deals and that great treaty that brought peace to Vietnam.[4]

This catchy, facile opinion was one widely held by many American journalists. As with so many aspects of American national policy, NASA's programs had always reflected the current environment of foreign affairs. Apollo, which had begun as a response from the Kennedy administration to the technological competition initiated by the Soviets in 1957, had been converted by NASA Administrators Paine and Fletcher into a means of cooperation with the Soviets. The joint flight could be seen as a part of detente, but the people at NASA saw it as much more.

On the most pragmatic level, ASTP gave the NASA team an opportunity to stay in the manned space flight business between the splashdown of *Skylab 4* on 8 February 1974 and the first orbital flight test of the Shuttle orbiter. Considerable thought by NASA planners had been given to flying the backup Skylab workshop, but this effort was abandoned in mid-1971 because it would have been too expensive; a duplicate Skylab would have drained scarce Shuttle funds. ASTP, on the other hand, gave the agency an opportunity to evaluate new hardware and flight techniques and the chance to carry a modest package of new or updated scientific experiments. Candidly, Chris Kraft thought that ASTP had been good for the American manned space program—good for morale, and it kept the flight team working. In addition, it was "a very big first step to international space flight cooperation."[5]

But could ASTP be equated with the seemingly endless Strategic Arms Limitations Talks (SALT)? Was it little more than the "great wheat deal in the sky?" Those who worked with the joint project did not think so. Unlike the arms talks, ASTP had a specific goal and a precise timetable. Once NASA and the Soviet Academy of Sciences agreed to fly in July 1975, the technological imperatives inherent in getting hardware ready for flight created an inner determinism within the project that helped to eliminate the possibility of either country stalling for political reasons. In the SALT negotiations, goals were less clearly defined and there was no deadline. While SALT participants continued to talk, the ASTP team brought their project to completion. The next steps in space cooperation, like the progress of the arms limitation discussions, would depend upon the international climate. Though ASTP had been a unique project, future cooperation, like SALT, was anchored in politics.

In April 1976, Tom Stafford noted that the Soviet and American space teams had met all their joint goals—they had designed, developed, and produced the hardware and systems whereby two spacecraft from different traditions could be joined together in space. "Where both systems were completely separate before," Stafford said, "we got together and worked [the differences] out. . . . the political implications were [such] that we could work in good faith." Stafford underscored good faith as "the key to something this technically difficult."[6] Glynn Lunney agreed with this observation. The real breakthrough made in ASTP was in bringing together teams from the U.S. and U.S.S.R. to "implement, design, test and finally fly a project of this complexity." ASTP had been a big job. "Perhaps we've gotten a bit blase about it . . . but we [had] an awful lot of hardware that [had] to work well,"[7] Lunney added.

Director Kraft pointed out that far from being a giveaway project, as many had claimed ASTP to have been, NASA had discovered many things about the Soviet space program that the American agency otherwise probably would not have learned. While he conceded that some of this information could have been ferreted out if there had been a reason to do so, both sides had been too busy with their own projects to study in any depth the other's efforts. As the Americans and the Soviets worked together, they learned just how differently they had approached various aspects of manned space flight. Designer Caldwell Johnson, who had retired in 1974, commented on the prevailing Soviet policy of flying unmanned spacecraft to test out their systems. NASA had always built elaborate facilities on the ground to simulate the space environment. Each side preferred the approach to which it had become accustomed, and Johnson could not say in absolute terms which was the best.[8]

Stafford, Kraft, Lunney, and Johnson saw this adherence to tradition as the basic reason not to be concerned about the transfer of technological concepts or secrets to the Soviet Union. In terms of the pace at which aerospace technology developed, Apollo equipment was already old hat when the last flight thundered off the launch pad. There was really little to worry about when the Americans loaned an Apollo transceiver to the Soviets, since that piece of equipment was being replaced in Shuttle by newer transceivers. Even if the Soviets had taken the transceiver apart—and there was no evidence that they ever tried—without the manufacturing capacity to make the components, looking inside would have been akin to trying to assemble a solid black jigsaw puzzle.

Chris Kraft did see one area in which the Soviets might possibly have learned something from NASA that could benefit their space program. "I think they learned the large amount of complexity we go into to build our space vehicles . . . they learned generally how we go about manufacturing a space vehicle . . . [but] above all, [they] found out how we *manage* programs." Management was the key lesson that the Soviets could have learned from NASA. Still, Kraft was not certain that even after having been exposed to the process the Soviets understood how the Americans laid out their programs—how the agency projected what it was going to do in a milestone schedule; how the agency forced its personnel to manage resources as well as hardware; or how the agency integrated operational planning with the design and manufacture of equipment. "I doubt [if] they could take what we do and apply it to their way of doing business," he added. Stafford agreed: "the only thing they could have learned from us was management," but this lesson would have no significant impact on the Soviet space program, based upon the limited insights he had been able to gain about their managerial organization.[9]

While there was general belief within NASA that ASTP had been successful, there was uncertainty about what if anything would happen next with the Soviet Academy. During the winter of 1975-1976, the American

Government's attitude toward detente changed dramatically with the Soviet-Cuban involvement in Angola. As detente disappeared from the foreign policy vocabulary, Chris Kraft reflected upon the meaning of these changes for international cooperation in space. "I guess that you would conjecture that this whole business of the tightening of the belt on both sides relative to each other's exploits in the world of foreign policy these days is certainly bound to rub off on these kinds of negotiations . . . unfortunate, but a fact of life."[10] But Kraft was hopeful that ASTP was not the end of cooperation. He thought that the United States and NASA needed to "continue rubbing elbows with the Russians in a technical space flight sense. And I hope that we can develop a continuing rapport with those people . . . setting goals . . . between ourselves, that we both want to meet, and then working towards them, even if they are long range." Kraft went on:

> Now that doesn't mean that we have got to fly in the same spacecraft . . . together, but if we have a cooperative attitude . . . and maybe plan some of our work together, I think [it] will lead to a quicker approach to the solution of problems; that would be very beneficial to the world, and certainly has got to be beneficial politically.[11]

George Low, who left NASA in the summer of 1976 to become president of Rensselaer Polytechnic Institute, also maintained that Apollo-Soyuz had been a success. Looking back on the project, he believed that it had established a solid technical and managerial foundation upon which subsequent joint ventures could be built. Low also understood that cooperation was important for two reasons. First, space exploration was too costly for the Americans and Soviets to continue indefinitely their duplicative efforts. Second, he said, "We live in a rather dangerous world. Anything that we can do to make it a little less dangerous is worth doing. I think that ASTP was one of those things."[12]

In the Apollo-Soyuz Test Project, the Soviets and Americans had done the dramatic. Once they had proven that they could work together, they needed to develop other meaningful activities. There were indications that this could be done. Over a two-year period, 1974-1976, NASA scientists had worked with Soviet counterparts to develop a package of four biological experiments that were flown on the Soviet satellite *Cosmos 782* (U.S.S.R./U.S. Biosatellite program).[13] In mid-summer 1976, the three-volume *Foundations of Space Biology and Medicine,* first discussed during the 1964 Dryden-Blagonravov talks, was finally distributed in separate English and Russian editions as a joint publication of the Soviet Academy of Sciences and the National Aeronautics and Space Administration.[14] In addition to the dramatic, the two sides were beginning to cooperate on more everyday activities. Still, the future was uncertain.

EPILOGUE

Would the past be the prologue? Which past? The twelve years of competition or the five years of ASTP? In looking at the dozen years that preceded the joint flight, one would not likely have predicted such a cooperative venture. But single-minded individuals in the United States and the Soviet Union had pursued the goal of the docking mission and secured it. Looking toward the future, members of NASA's ASTP team could only hope that their efforts would lead to further cooperation and that the era of rivalry and competition would not return. But they knew from the moment that Apollo splashed down that the decision—to cooperate or to compete—was not theirs to make. They could only hope.

Source Notes

PROLOGUE

1. Thomas O. Paine, "Man's Future in Space," 1972 Tizard Memorial Lecture, Westminster School, London, 14 Mar. 1972, p. 4; and Paine, "... For all Mankind: Space Progress to the Year 2000" (typescript, NASA HQ History Office Archives), p. 5-2.

2. Paine to Edward C. Ezell, 12 July 1974; and interview, Paine-Eugene M. Emme, 3 Aug. 1970.

3. Interview, Paine-Emme, 3 Aug. 1970.

4. Interview, Paine-Emme, 3 Sept. 1970.

5. Ibid.

6. Paine to Anatoliy Arkadyevich Blagonravov, 30 Apr. 1969, with enclosure: *Opportunities for Participation in Space Flight Investigations,* NASA-NHB 8030.1 (Washington, 1967).

7. Ibid. According to Arnold W. Frutkin, "Whenever we invite proposals for experiments on our spacecraft by U.S. or foreign scientists, we now include the Soviet Union as a routine matter and invite them to submit proposals along with the others," NASA News Release, HQ, "Background Press Briefing, U.S. and USSR Cooperation in Space," 13 Oct. 1970, p. 3. "Proposals for flight investigations from scientists outside the U.S. should be sent first to the official national space agency in the scientist's country. After review, this organization will then forward the endorsed proposal to NASA where it will go through the same evaluation and selection as a US-originated proposal," NASA, *Opportunities for Participation in Space Flight Investigations,* p. III-3.

8. Letter, Paine to Blagonravov, 29 May 1969.

9. TWX, Blagonravov to Paine, 10 July 1969.

10. NASA, *Astronautics and Aeronautics, 1969: Chronology on Science, Technology, and Policy,* NASA SP-4014 (Washington, 1970), p. 233.

11. U.S. Congress, Senate, Committee on Aeronautical and Space Sciences, *NASA Authorization for Fiscal Year 1971: Hearings on S. 3374,* 91st Cong., 2d sess., 1970, p. 1038. See also Blagonravov, "Apollo 11 and the Soviet Lunar Programme," *Spaceflight* 11 (Dec. 1969): 414-416.

12. NASA, *Astronautics and Aeronautics, 1969,* p. 233.

13. Mstislav Vsevolodovich Keldysh to Paine, message [July 1969].

14. Paine to Keldysh, 21 Aug. 1969. The Viking launch was subsequently slipped to 1975 because of budgetary restrictions, NASA, *Astronautics and Aeronautics, 1970,* NASA SP-4015 (Washington, 1971), p. 11.

15. Keldysh to Paine, 5 Sept. 1969, as telegraphically transmitted by the American Embassy in Moscow, 9 Sept. 1969.

16. Paine to Keldysh, 15 Sept. 1969.

17. Keldysh to Paine, 12 Dec. 1969.

18. Paine to Keldysh, 10 Oct. 1969.

19. Space Task Group, *The Post-Apollo Space Program: Directions for the Future,* Space Task Group Report to the President, Sept. 1969, p. 6.

20. Ibid., p. 7.

21. Paine to Keldysh, 10 Oct. 1969.

22. Foy D. Kohler, memo for record, "Memorandum of Conversation," Dec. 1969, as cited in Dodd L. Harvey and Linda C. Ciccoritti, *U.S.-Soviet Space Cooperation* (Coral Gables, Fla., 1974), p. 270; and Richard LeBaron, "U.S.-U.S.S.R. Space Cooperation," Oct. 1973 (typescript, NASA HQ History Office), p. 17. LeBaron cites Kohler's memo as a State Dept. memorandum. See also Keldysh to Paine, 12 Dec. 1969.

23. LeBaron, "U.S.-U.S.S.R. Space Cooperation," pp. 18-21; Robert F. Packard to Staff Committee, Space Task Group, "International Implications of the Space Program for the Next Decade," 4 June 1969; and William P. Rogers to Richard M. Nixon, "International Space Cooperation," 14 Mar. 1969, with attachment, "New Initiatives in Space Cooperation," 10 Mar. 1969.

24. Paine to Keldysh, 20 Feb. 1970.

25. Paine to Ezell, 12 July 1974.

26. Frutkin, memo for record, "Paine-Blagonravov Meeting 4/24," 12 May 1970, p. 1.

27. Paine to Ezell, 12 July 1974.

28. Frutkin, memo for record, 12 May 1970, p. 2.

29. NASA, *Astronautics and Aeronautics: 1970,* p. 176.

30. George M. Low, "Notes Concerning Trip to the Soviet Union, May 19-24, 1970," p. 3.

31. Philip Handler to Ezell, 9 Oct. 1974.

32. Ibid.

33. Handler to Paine, 28 May 1970; Handler to Ezell, 9 Oct. 1974; and Handler, "Trip Report" [n.d.].

34. Handler to Paine, 28 May 1970.

35. Handler to Paine, 29 July 1970; and Handler to Ezell, 9 Oct. 1974.

36. Handler to Keldysh, 29 July 1970, in which he emphasized that "Dr. Paine's communication quite properly constitutes the official invitation on the part of the Government of the United States to embark upon the negotiations you requested through your Embassy." Paine had sent another letter to Keldysh on 30 June 1970 saying that he planned to be in Europe during July and that perhaps he could meet with Keldysh. This letter got lost in the confusion caused by the Handler conversation with Keldysh as to which agency represented the U.S. space program—NAS or NASA; and Handler to Ezell, 1 and 19 Nov. 1974.

37. Paine to Keldysh, 31 July 1970.

38. Ibid. Paine sent Handler a handwritten note: "It is important in my view to keep our momentum in US-USSR space cooperation and to let Keldysh know that my leaving will not affect our position. Hence this note to him." Paine to Handler, 31 July 1970.

39. Paine to Ezell, 21 July 1974.

40. Paine to Keldysh, 4 Sept. 1970.

41. Keldysh to Paine, 11 Sept. 1970. Keldysh to Handler, 10 Sept. 1970 contained basically the same substance: "In consideration of the statement in your letter on July 29 of this year that T. Paine's proposal to conduct negotiations . . . constitutes in fact the official proposal on the part of the Government of the United States, we have sent to Mr. T. Paine a reply confirming our positive attitude toward that question."

42. Keldysh to Paine, 11 Sept. 1970.

43. Low to Keldysh, 25 Sept. 1970.

44. TWX, Keldysh to Low [n.d.], as cited in Low to Keldysh, 21 Oct. 1970.

I

1. Constance McLaughlin Green and Milton Lomask, *Vanguard: A History,* NASA SP-4202 (Washington, 1970), pp. 19-20; interview, Lloyd V. Berkner-Jay Holmes, 4 June 1959, pp. 22-24 and 26; and Berkner obituary, *New York Times,* 5 June 1967. For a summary of the IGY, see Walter Sullivan, *Assault on the Unknown: I.G.Y.* (New York, 1961); Hugh L. Dryden, "The International Geophysical Year: Man's Most Ambitious Study of His Environment," *National Geographic Magazine* 109 (Feb. 1956): 285-298; and Richard W. Porter, "International Cooperation in Space," in *Astronautical Engineering and Science from Peenemunde to Planetary Space,* Ernst Stuhlinger et al., ed. (New York, 1963), pp. 350-359.

2. Green and Lomask, *Vanguard,* pp. 22-23.

3. Ibid., p. 23.

4. Adam B. Ulam, *The Rivals: America and Russia since World War II* (New York, 1971), p. 382. Background on the differing interpretations of the Cold War and its origins can be found in Norman A. Graebner, ed., *The Cold War: Ideological Conflict or Power Struggle?* (Boston, 1963); Thomas G. Patterson, ed., *The Origins of the Cold War* (Lexington, Mass. [1970]); and Thomas G. Patterson, *Soviet-American Confrontation: Post War Reconstruction and the Origins of the Cold War* (Baltimore, 1974).

5. Green and Lomask, *Vanguard,* p. 33.

6. Ibid., pp. 31-32.

7. Richard S. Lewis, *Appointment on the Moon* (New York, 1969), p. 39; and Clifford C. Furnas, "Birthpangs of the First Satellite," *Research Trends* [Cornell Aeronautical Laboratory, Inc.] 18 (spring 1970): 15-18. Furnas, founder of Cornell Aeronautical Laboratories, was a member of the Ad Hoc Advisory Group on Special Capabilities, a euphemism for the satellite study committee established by the Department of Defense in May 1955. Other members of the committee were H. J. Stewart, R. R. McMath, C. Lauritsen, J. B. Rosser, R. W. Porter, G. H. Clement, and J. Kaplan.

8. "Mezhdunarodnii kongress astronavtov" [International congress of astronauts], *Pravda,* 5 Aug. 1955; a translated version of this Tass dispatch appears in F. J. Krieger, *Behind the Sputniks: A Survey of Soviet Space Sciences* (Washington, 1958), pp. 330-333.

9. Krieger, *Behind the Sputniks,* pp. 4-5, addresses the problem of interpreting Sedov's com-

SOURCE NOTES

ments: "Although the White House announcement on July 29, 1955—that the United States intended to launch an earth satellite sometime during the International Geophysical Year (1957-1958)—led to considerable speculation concerning the Soviet position and capability in this field of technology, the imperturbable Russians, as usual, did not commit themselves. . . .

A notable event occurred in the week following the White House announcement. The Sixth International Astronautical Congress sponsored by the International Astronautical Federation convened in Copenhagen, Denmark. It was notable because, unlike previous meetings, it was attended by two Soviet scientists, Academician L. I. Sedov, Chairman of the USSR Academy of Sciences Interdepartmental Commission on Interplanetary Communications, and Professor K. F. Ogorodnikov of the department of astronomy at Leningrad State University. . . .

The Russians were observers at the Congress and did not participate in any formal discussion of the papers. Sedov, however, did hold a press conference on August 2 at the Soviet Legation in Copenhagen, but unfortunately some of the statements attributed to him were garbled in the Western press. Three days later, on August 5, *Pravda* published an official version of the press conference.

For a comparison, see *New York Times,* 3 Aug. 1955; *New York Herald Tribune,* 3 Aug. 1955; Green and Lomask, *Vanguard,* p. 39; Evgeny Riabchikov, *Russians in Space* (Garden City, N.Y., 1971), p. 142; and Frederick C. Durant, III, "Impressions of the Sixth Astronautics Congress," *Jet Propulsion* 25 (Dec. 1955): 738-739. Sedov's impressions of the Aeronautical Congress appeared in *Pravda* on 26 Sept. 1955 and are reprinted in Krieger, *Behind the Sputniks,* pp. 112-115. See also Leonid Ivanovich Sedov, "O poletakh v mirovoe prostranstvo" [On flights into world space], *Pravda,* 26 Sept. 1955, a translation of which appears in Krieger, *Behind the Sputniks,* pp. 112-115.

10. Vernon Van Dyke, *Pride and Power: The Rationale of the Space Program* (Urbana, Ill. 1963), pp. 5-38 and 267-276; and John M. Logsdon, *The Decision to Go to the Moon: Project Apollo and the National Interest* (Cambridge, Mass., and London, 1970), pp. 40-62. For an inside view of post-Sputnik Washington, see Oliver M. Gale, "Post-Sputnik Washington from an Inside Office," *Cincinnati Historical Society Bulletin* 31 (winter 1973): 225-252.

11. National Advisory Committee for Aeronautics, Working Group on Vehicular Program, *A National Integrated Missile and Space Vehicle Development Program* (Washington, 1958), pp. 6-7. The background of the Stever Committee is presented in interview, H. Guyford Stever-Alex Roland and Eugene M. Emme, 4 Feb. 1974.

12. Loyd S. Swenson, Jr., James M. Grimwood, and Charles C. Alexander, *This New Ocean: A History of Project Mercury,* NASA SP-4201 (Washington, 1966); and interview, Gerald W. Siegel-Jay Holmes, 25 June 1968, which sheds light on the congressional scene. Siegel was counsel to the Preparedness Investigating Subcommittee and became the staff director of the Special Committee on Space and Aeronautics, which held the Senate hearings on the National Aeronautics and Space Act, May 1958. A longtime political adviser to Lyndon B. Johnson, he stresses the role the Senator from Texas played in the creation of a civilian space agency.

13. Robert L. Rosholt, *An Administrative History of NASA, 1958-1963,* NASA SP-4101 (Washington, 1966), pp. 8-13 and 34-36; Elisabeth A. Griffith, *The National Aeronautics and Space Act: A Study in the Development of Public Policy* (Washington, 1962); and Mary Stone Ambrose, "The National Space Program; Phase I: The Passage of the National Aeronautics and Space Act of 1958" (Master's thesis, American University, 1960).

14. Public Law 85-568, 72 Stat. 426.

15. Arnold W. Frutkin, *International Cooperation in Space* (Englewood Cliffs, N.J., 1965), p. 8.

16. Rosholt, *Administrative History of NASA,* p. 332; and T. Keith Glennan to James R. Killian, Special Asst. to the President for Science and Technology, 29 Oct. 1958.

17. McKinsey & Company, Inc., "Organizing Headquarters Functions: National Aeronautics and Space Administration," with letter of transmittal, 31 Dec. 1958, pp. 2-26, 2-27.

18. NASA News Release, HQ, "Henry E. Billingsley Named NASA's Director of the Office of International Cooperation," 6 Jan. 1959.

19. Frutkin, *International Cooperation in Space,* p. 19.

20. Ibid., p. 21.

21. Ibid., p. 22.

22. Ibid., pp. 28-84 describes U.S. efforts to create a basis for cooperation within the framework of NASA's programs.

23. Ibid., pp. 85-88.

24. *Public Papers of the Presidents of the*

United States, Dwight D. Eisenhower, 1958 (Washington, 1959), p. 82.

25. Quoted in Dodd L. Harvey and Linda C. Ciccoritti, *U.S.-Soviet Cooperation in Space* (Coral Gables, Fla., 1974), p. 16.

26. Frutkin, *International Cooperation in Space,* p. 142. For the subsequent exchange of letters between Eisenhower and the Soviet leaders see Harvey and Ciccoritti, *U.S.-Soviet Cooperation in Space,* pp. 17-22.

27. William Hines, "Soviet Space Scientists Tell Little of Ventures," *Washington Star,* 18 Nov. 1959.

28. George M. Low to Ezell, "Comments on December 1975 Draft of ASTP History," 29 Dec. 1975; Frutkin, *International Cooperation in Space,* p. 89; and U.S. Congress, Senate, Committee on Aeronautical and Space Sciences, *Soviet Space Programs: Organization, Plans, Goals, and International Implications,* 87th Cong., 2d sess., 1962, p. 179.

29. "Russians Unimpressed by Space Man Project," *Washington Star,* 23 Nov. 1959.

30. Ulam, *The Rivals,* p. 249.

31. Bela Kornitzer, *The Real Nixon: An Intimate Biography* (New York, 1960), pp. 297-310, gives material concerning the kitchen debate.

32. Strobe Talbot, tr. and ed., *Khruschev Remembers: The Last Testament* (Boston and Toronto, 1974), pp. 443-455.

33. NASA Hq News Release, "Memo to the Press," 5 May 1960; and David Wise and Thomas B. Ross, "The U-2 Affair: Memo to Press Hastily Drawn," *Washington Star,* 8 June 1962.

34. Louis Kraar, "Space Partnerships," *Wall Street Journal,* 26 Sept. 1960. *Krasnaya Zvezda* [Red Star], 23 July 1961, stated that *Tiros III* and *Midas III* were comparable to the U-2: "A spy is a spy, no matter what height it flies."

35. John F. Kennedy, "If the Soviets Control Space . . . They Can Control Earth," *Missiles and Rockets,* 10 Oct. 1960, pp. 12-13; and in the same issue, Clarke Newlon, "Kennedy's Stand on Defense and Space," p. 50.

36. Richard M. Nixon, "Nixon: Military Has Mission to 'Defend' Space," *Missiles and Rockets,* 31 Oct. 1960, pp. 10-11; in the same issue, "Candidates' Views Compared," p. 12; and in the same issue, Clarke Newlon, "Nixon Drops Party Line on Space," p. 50.

37. Frutkin, *International Cooperation in Space,* pp. 89-91.

38. *Public Papers of the Presidents of the*

United States, Dwight D. Eisenhower, 1960-1961 (Washington, 1961), pp. 714-715.

39. Ad Hoc Committee on Space, Jerome B. Wiesner, Chairman, "Report to the President-Elect of the Ad Hoc Committee on Space," 12 Jan. 1961; Swenson, Grimwood, and Alexander, *This New Ocean,* pp. 304-306; W. H. Lawrence, "Kennedy Warned of Space Setback," *New York Times,* 12 Jan. 1961; "Excerpts from Task Force's Report to Kennedy on U.S. Position in Space Race," *New York Times,* 12 Jan. 1961; and James Baar, "Space Shake-up Coming," *Missiles and Rockets,* 16 Jan. 1961, pp. 11-12.

40. *Public Papers of the Presidents of the United States, John F. Kennedy, 1961* (Washington, 1962), pp. 1-2, 26-27, and 93-94.

41. The differing views can be seen by comparing Theodore C. Sorenson, *Kennedy* (New York, 1965), p. 524, with Logsdon, *Decision to Go to the Moon,* p. 93.

42. Interview, Hugh L. Dryden-Jay Holmes, 26 Mar. 1964.

43. Logsdon, *Decision to Go to the Moon,* pp. 111-112.

44. Sorenson, *Kennedy,* p. 524.

45. Logsdon, *Decision to Go to the Moon,* pp. 111-112.

46. Ibid., p. 126.

47. *Public Papers of John F. Kennedy, 1961,* pp. 405-406.

48. U.S. Congress, Senate, Committee on Aeronautics and Space Sciences, *NASA Authorization for Fiscal Year 1962, Hearings on H.R. 6874,* 87th Cong., 1st sess., 1961, p. 155.

49. TWX, Kennedy to Nikita Sergeyevich Khrushchev, 13 Feb. 1961, as printed in U.S. Congress, Senate, Committee on Aeronautical and Space Sciences, *Documents on International Aspects of the Exploration and Use of Outer Space, 1954-1962,* 1963, p. 190.

50. TWX, Khrushchev to Kennedy, 15 Feb. 1961, ibid.

51. Sedov, "Sovetskii Soyuz-pioner v osvoenii kosmosa" [The Soviet Union—pioneer in space], *Pravda,* 9 May 1961.

52. William Beller, "AIAC Stressing Peace," *Missiles and Rockets,* 9 Oct. 1961, p. 14.

53. Dept. of State, Memo of conversation, "Vienna Meeting between the President and Chairman Khrushchev," 3 June 1961 [John F. Kennedy Library]; Pierre Salinger, *With Kennedy* (Garden City, New York, 1966), p. 178.

54. Salinger, *With Kennedy,* pp. 189-196; and Ulam, *The Rivals,* pp. 322-323.

55. Harrison E. Salisbury, "World Scientists Map Coordinations," *New York Times,* 8 Sept. 1961; and Salisbury, "Space Proposals for World Near," *New York Times,* 9 Sept. 1961.

56. For the tenor of the time, see Harry Schwartz, "Khrushchev Presses Hard to Force Settlement on His Terms," *New York Times,* 11 Sept. 1961; John W. Finney, "U.S. Tests to Preserve Lead over Soviets," *New York Times,* 11 Sept. 1961; and Richard Lowenthal, "Negotiating with Russia—What's the Use," *New York Times Magazine,* 11 Sept. 1961, pp. 21 and 116-117.

57. John W. Finney, "Soviet Block Boycotts U.S. Weather Satellite Symposium," *New York Times,* 15 Nov. 1961.

58. David Halberstam, *The Best and the Brightest* (Greenwich, Conn., 1973), p. 82.

II

1. "Programme of the Communist Party of the Soviet Union Adopted October 31, 1961," in B. Dmytryshyn, *A Concise History: USSR* (New York, 1965), p. 479.

2. The Soviets also approved the General Assembly Resolution 1721 (XVI) on 20 Dec. 1961, which laid down the general ground rules for cooperative activities in space. The text of the resolution is given in U.S. Congress, Senate, Committee on Aeronautical and Space Sciences, *International Cooperation and Organization for Outer Space,* 89th Cong., 1st sess., 1965, pp. 201-203. The events leading to the passage of this resolution are described in U.S. Congress, Senate, Committee on Aeronautical and Space Sciences, *Soviet Space Programs: Organization, Plans, Goals, and International Implications,* 87th Cong., 2d sess., 1962, pp. 163-173; and Arnold W. Frutkin, *International Cooperation in Space* (Englewood Cliffs, N.J., 1965), pp. 91-142, 142-145 ff.

3. "Text of Gherman Titov's October 26 Speech," *22d CPSU Congress,* 15: 56.

4. "SShA-vtoraya strana, poslavshaya cheloveka v kosmos, polet Dzhona Glenna" [USA—second country to send a man into space, pilot John Glenn], *Izvestiya,* 22 Feb. 1962.

5. Nikita Sergeyevich Khrushchev to John F. Kennedy, 21 Feb. 1962, as printed in U.S. Congress, Senate, Committee on Aeronautical and Space Sciences, *Documents on International Aspects of the Exploration and Use of Outer Space, 1954-1962,* 88th Cong., 1st sess., 1963, p. 232.

6. Kennedy to Khrushchev, 21 Feb. 1962, as printed in Committee on Aeronautical and Space Sciences, *Documents on International Aspects of Space,* p. 233.

7. *Public Papers of the Presidents of the United States, John F. Kennedy, 1962* (Washington, 1963), pp. 151-152 and 157-158; and Theodore C. Sorenson, *Kennedy* (New York, 1965), p. 529.

8. *Public Papers of John F. Kennedy, 1962,* pp. 157-158.

9. Interview, Hugh L. Dryden-Arnold W. Frutkin, Walter D. Sohier, and Eugene M. Emme, 26 Mar. 1964, p. 20.

10. Ibid., pp. 20-21; and A. W. Frutkin comments on draft history, 12 Feb. 1975.

11. "Address by the Director of the Office of International Programs, National Aeronautics and Space Administration [Frutkin], on International Cooperation in the Exploration of Space, February 16, 1960," as printed in Committee on Aeronautical and Space Sciences, *Documents on International Aspects of Space,* pp. 168-175. For an alternate and critical view, see Eugene B. Skolnikoff, *Science, Technology and American Foreign Policy* (Cambridge, Mass., 1967), pp. 32-36.

12. Frutkin, *International Cooperation in Space,* p. 93. For full text of the Kennedy letter, see *Public Papers of John F. Kennedy, 1962,* pp. 244-245; and Committee on Aeronautical and Space Sciences, *Documents on International Aspects of Space,* pp. 242-244.

13. *Public Papers of John F. Kennedy, 1962,* pp. 244-245.

14. Philip J. Farley to George W. Ball, memo, "Designation of Technical Representatives for U.S.-Soviet Space Cooperation Talks," 9 Mar. 1962; Ball to Kennedy, memo, "Designation of Technical Representatives for U.S.-Soviet Space Cooperation Talks," 16 Mar. 1962; interview, Dryden-Frutkin, Sohier and Emme, 26 Mar. 1964, p. 22; Thomas J. Hamilton, "U.N. Space Panel Hears U.S. Urge Cooperation," *New York Times,* 20 Mar. 1962; and "Dryden Hard to Fool in Science or Politics," *Washington Star,* 22 Mar. 1962.

15. Khrushchev to Kennedy, 20 Mar. 1962, as printed in Committee on Aeronautical and Space Sciences, *Documents on International Aspects of Space,* pp. 248-251.

16. Ibid., p. 250. For further comment, see

R. Cargill Hall, "Rescue and Return of Astronauts on Earth and in Outer Space," *American Journal of International Law* 63 (Apr. 1969): 197-210.

17. Khrushchev to Kennedy, 20 Mar. 1962, pp. 250-251.

18. Ibid., pp.251-252.

19. *Public Papers of John F. Kennedy, 1962,* p. 264.

20. See "Poslanie N. S. Khrushcheva Prezidenty SShA Dzh. Kennedi" [Message of N. S. Khrushchev to President of the USA John Kennedy], *Pravda,* 22 Mar. 1962; and "Kosmicheskie isspedovaniya-na sluzhbu delu mira, poslanie N. S. Khrushcheva Prezidenty SShA Dzh. Kennedi" [Space investigation–into the affairs of the world, message of N. S. Khrushchev to President of the USA John Kennedy], *Izvestiya,* 22 Mar. 1962. The American newspapers treated the exchange of letters at some length. Milton Besser, "Soviet Says It Would Help Set up Satellite Communications System," *Washington Post,* 21 Mar. 1962; Thomas J. Hamilton, "Soviet Promises Space Data to U.N.," *New York Times,* 21 Mar. 1962; "Khrushchev Accepts Bid for Cooperation in Space," *New York Times,* 22 Mar. 1962; "Russia Agrees to Plan for Joint Space Ventures," *Wall Street Journal,* 22 Mar. 1962; Carroll Kilpatrick, "U.S. and Soviet Move Toward Joint Space Use," *Washington Post,* 22 Mar. 1962; and "Joint Space Efforts," editorial, *Washington Post,* 24 Mar. 1962.

21. Dryden, "Preliminary Summary Report: U.S.-Soviet Space Cooperation Talks, New York, N.Y., March 27, 28, 30, 1962," 30 Mar. 1962; Frutkin, *International Cooperation in Space,* p. 95; and TWX, Adlai E. Stevenson to Secretary of State, 21 Mar. 1962. Stevenson reported the Soviets felt that the press treatment of the talks must be limited. The Soviets had "urged that if [the] talks are to be fruitful, they cannot be conducted in a goldfish bowl."

22. Dryden, "Preliminary Summary Report, U.S.-Soviet Space Cooperation Talks," p. 1. The NASA position papers were attached to the Dryden report: "Tentative Basis for Further Discussions of Meteorological Satellite Cooperation," "Tentative Basis for Further Discussions of Cooperation in Data Acquisition," and "Tentative Basis for Further Discussions of Mapping the Earth's Magnetic Field."

23. Ibid.

24. Frutkin, *International Cooperation in Space,* p. 95.

25. To follow the dialogue between the U.S. and the U.S.S.R. on the subject of "weather" and "propaganda" balloons, see in the *Department of State Bulletin* the following: correspondence and DoD press release, 8 Jan. 1956, p. 293; "Transcript of Secretary Dulles' News Conference," 20 Feb. 1956, pp. 280-282; "Correspondence with U.S.S.R. Concerning Weather Balloons," 20 Feb. 1956, pp. 293-295; "U.S. Restates Position on Weather Balloons," 12 Mar. 1956, pp. 426-428; text of 5 Sept. 1958 U.S. note, 29 Sept. 1958, p. 504; and "U.S. Replies to Soviet Note on Balloons" (10 Nov. 1958), pp. 739-740. Also, see "Balloons over the Red World," *America; National Catholic Weekly Review,* 18 Feb. 1956, p. 547; "The Russians, Too," *Newsweek,* 12 Mar. 1956, pp. 33-34; "Freedom Balloons Aimed" *Science Newsletter,* 25 Aug. 1951, p. 124; and "Iron Curtain Balloons," *Flying,* Dec. 1955, p. 79.

26. Frutkin, "Topical Summary of Bilateral Discussion with the Soviet Union, March 27-30, 1962," 1 May 1962, as cited in Dodd L. Harvey and Linda C. Ciccoritti, *U.S.-Soviet Cooperation in Space* (Coral Gables, Fla., 1974), p. 94; [Frutkin], "Status of US/USSR Bilateral Space Talks," 21 Apr. 1962; [Robert F. Packard, memo for record, "Meeting with Under Secretary McGhee Concerning US-USSR Cooperation in Outer Space Activities," 24 Apr. 1962; and NASA, *Astronautics and Aeronautics, 1963: Chronology on Science, Technology, and Policy,* NASA SP-4004 (Washington, 1964), p. 219. On 29 May 1963, a U.N. subcommittee on space addressed the U.N. Committee on the Peaceful Uses of Outer Space with the following message: the "urgency and importance of the problem of preventing potentially harmful interference with peaceful uses of outer space" cannot be overemphasized. This warning, referring to the USAF-sponsored Project West Ford communications experiment, had been initiated by the Soviet delegate Anatoliy A. Blagonravov. He denounced the experiment as a danger to other space studies, including flights by manned satellites. This charge was denied by Homer E. Newell, Jr., a U.S. representative.

27. Dryden to T. Keith Glennan, 26 Apr. 1962, as cited in Harvey and Ciccoritti, *U.S.-Soviet Cooperation in Space,* p. 94; and Dryden to James A. Van Allen, 20 Apr. 1962.

28. Howard Simons, "U.S.-Russia Open Talks on Co-operation in Space," *Washington Post,* 28 Mar. 1962; Lawrence O'Kane, "U.S. and Soviet Start Space Talks," *New York Times,* 28 Mar.

SOURCE NOTES

1962; and "New Parley Slated on Space Research," *New York Times,* 31 Mar. 1962. Dryden commented privately on the drafting of this statement. The Americans "proposed to list the three specific projects on which agreement seemed possible but the Soviet delegation wished either to include all projects mentioned in the letters plus the reconnaissance satellite item or none. . . ." Nevertheless, Dryden considered the Soviet attitude businesslike and a "good sign," since "Blagonravov stated that he favored the negotiation of agreements for those projects on which we can agree as agreement is reached rather than attempting to cover all projects in a single negotiation. Such a procedure appears to dispose of the reconnaissance satellite pledge as a precondition for agreements and is favorable to a fruitful outcome of the negotiations." Dryden, "Preliminary Summary Report, U.S.-Soviet Space Cooperation Talks."

29. "Chelovechestvo raskroet tainy kosmosa" [Mankind gets back secrets from space], *Pravda,* 10 Apr. 1962; and "Blizkie prostopi vselennoy, interv'y c Yu. Gagarinim i G. Titovim" [Near Space is Universal, an Interview with Yu. Gagarin and G. Titov], *Izvestiya,* 12 Apr. 1962.

30. "Rech'tovarishcha M. V. Keldysha" [Speech by comrade M. V. Keldysh], *Pravda,* 13 Apr. 1962.

31. "Khrushchev Drops Summit Pressure," *New York Times,* 25 Apr. 1962.

32. Interview, Margaret Chase Smith over radio, 1 Apr. 1962, as cited in U.S. Congress, House of Representatives, Committee on Science and Astronautics, *Astronautical and Aeronautical Events of 1962, Report,* 88th Cong., 1st sess., 12 June 1963, p. 46.

33. U.S. Congress, House of Representatives, Committee on Science and Astronautics, George P. Miller, "Press Release," 23 Feb. 1962.

34. William E. Minshall of Ohio reprinted the results of a poll of his constituents. Of some 20 000 respondents, 47 percent were opposed and 13.4 percent had no opinion, U.S. Congress, House of Representatives, *Congressional Record,* 87th Cong., 1st sess., vol. 108 (18 Apr. 1962), p. A3035.

35. Committee on Science and Astronautics, *Astronautical and Aeronautical Events of 1962,* p. 74.

36. Dryden to Glennan, 28 Apr. 1963, indicates that Dryden did not expect the meetings to resume in Washington but in Geneva in May.

37. Interview, Dryden-Frutkin, Sohier, and Emme, 26 Mar. 1964, p. 23.

38. Frutkin, *International Cooperation in Space,* p. 96. The full text of the "Bilateral Space Agreement between the US and the USSR," together with letters of transmittal and news releases, are presented in *Department of State Bulletin,* 24 Dec. 1962, pp. 962-965. The agreement took shape much as Dean Rusk had predicted, Rusk to Kennedy, memo, 15 May 1962:

> . . . the Soviets prefer to develop such arrangements on a step-by-step basis, not on the basis of an overall formal agreement between the two governments. Further, the Soviets are apparently interested in working primarily within multilateral programs (i.e., those of the World Meteorological Organization and the International Telecommunications Union), but on the basis of prior US-USSR agreement. It appears unlikely that significant joint effort in outer space activities will develop in the near term, but there is a prospect that the Soviets will agree to some modest cooperation in the form of coordinated satellite launch schedules, compatible instrumentation and some additional exchange of technical information.

39. "U.S.-Russian Pact on Weather Probes Drafted in Geneva," *New York Times,* 9 June 1962; "Joint Communique on US-USSR Talks," 8 June 1962; and Packard to E. C. Welsh, H. L. Dryden, J. B. Wiesner, W. P. Bundy, F. W. Reichelderfer, and H. Scoville, memo, "Meeting with Under Secretary McGhee Concerning U.S.-USSR Cooperation in Outer Space Activities," 12 June 1962, with attachments, "Dryden-Blagonravov Memorandum," 8 June 1962, and "Joint Communique." The Americans participating in the Geneva talks who had not been in New York City were Furnas, Wexler, Heppner, and Valdes. The Soviet delegation consisted of Blagonravov, Barinov, Stashevsky, Klokov, Kalinin, and Bugaev.

40. James E. Webb to Keldysh, 30 Oct. 1962; "Bilateral Space Agreement between the US and the USSR," *Department of State Bulletin,* 24 Dec. 1962, pp. 964-965; and John W. Finney, "U.S. Prods Soviet on Space Accord," *New York Times,* 20 Sept. 1962.

41. Considering the political climate, Dryden had little difficulty in setting up the next meeting with Blagonravov, Dryden to Blagonravov, 11 Dec. 1963; and McGeorge Bundy to George C. McGhee, memo, "Bilateral Cooperation with the USSR in Outer Space Activities," 10 Dec. 1963. In Blagonravov to Dryden, 7 Jan. 1963, the

Soviet representative asked that the meeting be scheduled for March rather than January as proposed by Dryden. In Dryden to Blagonravov, 21 Jan. 1963, Dryden agreed to the postponement but requested that a larger number of technical experts be present so the talks could be "more substantial and expeditious." See also Dryden to Donald F. Hornig, 21 Jan. 1963.

42. Richard J. H. Barnes, Acting Director, International Programs, to Webb and Robert C. Seamans, Jr., memo, "US-USSR Bilaterals," 1 Nov. 1962. Barnes commented on Webb's letter to Keldysh of 30 Oct.:

> State understands that the Webb reply to Keldysh was sent by international registered mail Tuesday evening, October 30 and that Dr. Dryden was undertaking last night to notify Congressmen Miller and Fulton, and Senators Smith, Cannon, and Kerr, of the exchange of correspondence and the White House embargo on publicity. Both actions had been authorized by Undersecretary McGhee, and State had urged that the Congressional Committee members be informed yesterday because of the news stories and yesterday's editorial in the Washington Post.

See also "Light in Space?" editorial, *Washington Post,* 31 Oct. 1962; and John W. Finney, "Space Pact Nearer for U.S. and Russia," *New York Times,* 30 Oct. 1962.

43. NASA News Release 62-257, "US-USSR Join in Outer Space Program," 5 Dec. 1962.

44. Tass International Service, 8 Dec. 1962.

45. "First Memorandum of Understanding to Implement the Bilateral Space Agreement of June 8, 1962, between the Academy of Sciences of the USSR and the National Aeronautics and Space Administration of the US," together with news releases and correspondence, is reproduced in *Department of State Bulletin,* 9 Sept. 1963, pp. 404-410.

46. Frutkin, *International Cooperation in Space,* pp. 97-105; [Joint U.S.-USSR news release], "Progress Made in US-USSR Space Talks," 20 Mar. 1963; and "U.S.-Soviet Agree to Program for Weather Probes in Space," *New York Times,* 21 Mar. 1963.

47. NASA News Release, HQ [unnumbered], "News Media Briefing: Joint US-USSR Talks on Cooperative Space Research Projects Held in Rome, Italy," 25 Mar. 1963, pp. 8-9.

48. U.S. Congress, Senate, Committee on Aeronautical and Space Sciences, *NASA Authorization for Fiscal Year 1964: Hearings on S. 1245, Pt. 1.,* 88th Cong., 1st sess., 1963, pp. 33-34; and

John W. Finney, "Conflicts Peril Accord on Space," *New York Times,* 28 Apr. 1963.

49. Committee on Aeronautical and Space Sciences, *NASA Authorization for Fiscal Year 1964,* p. 35.

50. Frutkin, *International Cooperation in Space,* p. 12.

51. Ibid., pp. 14-15.

52. "Kommunike o podnisanii drugimi gosudarstvami dogovora o zapreshchenii ispitanii yadernoga oruzhiya v atmosfere, v kosmicheckom prostranstve i pod vodoy" [Communique about other countries signing the agreement on forbidding the testing of nuclear weapons in the atmosphere and under water], *Pravda,* 17 Aug. 1963; "V imya progressa, sovetsko-amerikanskoe sotrudnichestvo v mirnom osvoenii kosmosa" [In the name of progress, Soviet-American collaboration in the peaceful use of space], *Izvestiya,* 17 Aug. 1963; and Robert C. Toth, "U.S. and Russia Agree to Share Satellite Data." *New York Times,* 17 Aug. 1963.

53. Sir Bernard Lovell to Dryden, 23 July 1963. This letter and the response from Webb are reprinted in full by Frutkin, *International Cooperation in Space,* pp.127-131.

54. Frutkin, *International Cooperation in Space,* p. 129.

55. Ibid.

56. During congressional testimony, Dryden said, "the Russians are proposing an international forum of scientists to discuss our program not theirs," U.S. Congress, House of Representatives, Committee on Appropriations, Subcommittee on Independent Offices, *Independent Offices Appropriations for 1964: Hearings, Pt. 3,* 88th Cong., 1st sess., 1963, p. 105.

57. *Public Papers of the Presidents of the United States, John F. Kennedy, 1963* (Washington, 1964), pp. 567-568. Expanded discussion of the Lovell letter and its impact is provided by Frutkin, *International Cooperation in Space,* pp. 105-111, and Harvey and Ciccoritti, *U.S.-Soviet Cooperation in Space,* pp. 112-119.

58. *Public Papers of John F. Kennedy, 1963,* pp. 695-696.

59. Ibid; and Julian Scheer, memo for record, 29 Oct. 1963, with distributed attachments designed to guide discussion on the value of Project Apollo: Attachment A—NASA response to UPI story that Russia had "withdrawn" from the "moon race," 26 Oct. 1963; Attachment B— NASA subsequent response to queries from news

media; Attachment C—Presidential State of the Union speech, 25 May 1961; Attachment D—Webb speech excerpt; Attachment E—List of reasons why the U.S. has mounted broad-based program as outlined in recent presentation.

60. Dryden, memo for record, 17 Sept. 1963, reprinted in part by Harvey and Ciccoritti, *U.S.-Soviet Cooperation in Space,* pp. 118-119.

61. John W. Finney, "U.S. Aide Rebuffs Soviets' Moon Bid," *New York Times,* 18 Sept. 1963; and Howard Simons, "Soviet Interest in U.S. Space Ties Seen Growing," *Washington Post,* 18 Sept. 1963.

62. Bundy to Kennedy, memo, "Your 11 a.m. Appointment with Jim Webb," 18 Sept. 1963.

63. Interview, James E. Webb, 19 Sept. 1972, as cited in Harvey and Ciccoritti, *U.S.-Soviet Cooperation in Space,* p. 122; and Webb, "Leadership Evaluation in Large Scale Efforts," paper presented to the General Accounting Office Fifty Year Anniversary Meeting [n.d.], pp. 14-15.

64. Interview, Webb, 19 Sept. 1972; and Webb to Ezell, [May 1975].

65. Interview, Dryden-Frutkin, Sohier, and Emme, 26 Mar. 1964, p. 25. Schlesinger's role is related in A. M. Schlesinger, Jr., *A Thousand Days: John F. Kennedy in the White House* (Boston and Cambridge, Mass., 1965), pp. 918-921.

66. Interview, Webb, 19 Sept. 1972.

67. *Za Rubezhom* [Abroad], 28 Sept. 1963, as cited in "Russian Says Moon Shot Idea of President Is Premature," *Washington Post,* 29 Sept. 1963. The Walter Lippman column, "Today and Tomorrow: Purifying the Moon Project," had been published in the American papers on 24 Sept. 1963 and was reprinted in Moscow as "Trezvii podkhod" [Sober approach], *Pravda,* 2 Oct. 1963.

68. Harvey and Ciccoritti, *U.S.-Soviet Cooperation in Space,* pp. 124-126.

69. A sample of the responses is as follows: Howard Simons, "Opinion Divided Here on Joint Moon Shot Plan"; "Russian News Reports Delete Moon Trip Plan"; "Goldwater Criticizes Moon Plan"; "A Lofty Appeal," editorial, *Washington Post,* 21 Sept. 1963; Thomas J. Hamilton, "Kennedy Asks Joint Moon Flight by U.S. and Soviets as Peace Step; Urges New Accords in U.N. Speech"; and John W. Finney, "Washington Surprised at Retreat from Insistence That U.S. Reach Moon First," *New York Times,* 21 Sept.

1963. A quick analysis of the Kennedy proposal was prepared for the RAND corporation by Alton Frye, *The Proposal for a Joint Lunar Expedition: Background and Prospects,* report no. P-2808 (Santa Monica, 1964).

70. Warren Burkett, "J. F. K. Offer to Cooperate 'No Surprise,' " *Houston Chronicle,* 21 Sept. 1963.

71. "Combined U.S.-Russian Space Problems Feared," *Houston Chronicle,* 19 Sept. 1963; and Burkett, "J. F. K. Offer." Gilruth was speaking at the Goddard Memorial Dinner where he was being honored with the Dr. Robert H. Goddard Memorial Trophy.

72. *Space Business Daily,* 7 Nov. 1963, p. 217 summarizes Khrushchev's reported position and commented editorially on its significance:

> Soviet Premier Nikita Khrushchev has made it emphatically clear that the USSR has neither "deferred," postponed, or "withdrawn" its competitive lunar landing program. Rather, he says his country will launch a man to the Moon when all preparations have been completed that will ensure his safety.
>
> In making the announcement he chided United States speculation that the Soviet Union has changed its lunar landing plans for economic reasons. "In regard to the question of whether we have given up our lunar project. You're the ones who said that."
>
> Khrushchev's remarks, hopefully, coming at a time when the memories are still fresh, will be a warning to members of the general press and many members of our national leadership that inaccurate translation, quotation and interpretation or analysis of the antagonists proclamations on the still very technical and complex arena of space can afford a very embarrassing psychological victory to those antagonists. If members of the general press had resorted to the expediency of consulting with our own national space leadership, for instance, Dr. Edward Welsh of the National Space Council (a technical advisor more than a political appointment), it would not have appeared that our entire national space program was a tail wagged by Premier Khrushchev.
>
> This past year has seen an excess of spur-of-the-moment interpretations reflected in assaults upon the whole concept of the national space program and the age-old philosophy of national competition to a point where there has been a weakening of our nation's determination at a time when the antagonist is demonstrating a continuing space technology leadership. At the risk of being repetitive, it should be recorded that many of our general press have a history of placing our space leaders on trial by the proclamations from an audience either completely ignorant of the technology of space

and its implications, or from foreign or antagonistic onlookers.

73. Public Law 88-215, *An act making appropriations . . . for the fiscal year ending June 30, 1964, . . . ,* 88th Cong., 1st sess., 1963, p. 16. The rumor that the Soviets had withdrawn from the "moon race" had led to substantial cuts in the NASA budget. Kennedy Administration efforts to restore part or all of the $600 million were unsuccessful.

74. *Public Papers of the Presidents of the United States, Lyndon B. Johnson, 1963-1964 I* (Washington, 1964), pp. 72-73. Kennedy had begun to publicly back away from his proposal; see Kennedy to Albert Thomas, 23 Sept. 1963, in which he said, "In my judgment, therefore, our renewed and extended purpose of cooperation, so far from offering any excuse for slackening or weakness in our space effort, is one reason the more for moving ahead with the great program to which we have been committed as a country for more than two years."

75. Louis B. Fleming, "Adlai Renews Proposal for Joint Trip to Moon," *Washington Post,* 3 Dec. 1963; and Kathleen Teltsch, "U.S. Renews Call to Soviet to Join in Moon Venture," *New York Times,* 3 Dec. 1963. NASA and the White House continued to study the topic of cooperation internally. See Kennedy to Webb, "Cooperation with the USSR on Outer Space Matters," National Security Action Memorandum No. 271, 12 Nov. 1963; and Webb to U. Alexis Johnson, 18 Dec. 1963, which included a NASA position paper, "US-USSR Cooperation in Space Research Programs," which had been developed by Frutkin's office in response to the President's memo of 12 Nov.

76. Blagonravov to Dryden, 14 Jan. 1964. "At present we are completing the preparation of the first stage of observations of the satellite during the period of its inflation. Our stations, which are located within the zone of visibility, will observe the moment of inflation by visual and photographic means only. . . . For the purpose of facilitating the successful conduct of this work we should like to ask you to issue instructions that we be informed of the moment of the launch."

77. NASA News Release, HQ, 64-11, "NASA to Launch Second Echo Communications Satellite," 21 Jan. 1964; Standard Packaging Corp., National Metallizing Division, news release, "Aluminum Coated Space Package Launched Is Historic Feat," 9 Aug. 1960; Mission Operation

Report No. S-622-64-03, "Echo C Project," 20 Jan. 1964; and Bill Becker, "Echo 2 Is Orbited; Soviets to Aid Tests," *New York Times,* 26 Jan. 1964.

78. NASA News Release, HQ, 64-21, "Echo II Monitored by USSR," 27 Jan. 1964; "Russians Are Tracking Echo 2 in Joint Experiment with U.S.," *New York Times,* 27 Jan. 1964; "Observations of the Echo-II," *Krasnaya Zvezda,* 28 Jan. 1964; and "Mirror in the Cosmos (Echo-2)," *Izvestiya,* 28 Jan. 1964.

79. Blagonravov outlined the program of radio communications experiments in a letter, Blagonravov to Dryden, 27 Jan. 1964; Homer Newell summarized the *Echo II* experiment results in U.S. Congress, House of Representatives, Committee on Science and Astronautics, Subcommittee on Space Science and Applications, *1966 NASA Authorizations: Hearings on H.R. 3730, No. 2, Pt. 3,* 89th Cong., 1st sess., 1965, pp. 987-991; H. L. Baker, Project Manager, Echo II, to Harry J. Goett, memo, "A Quick Look Evaluation of USSR Optical Data as Submitted by Professor Massevich," 9 Apr. 1964; and Newell to Goett, memo, "Report on the Echo II Experiments with US-UK-USSR," 8 June 1964.

80. U.S. Congress, Senate, Committee on Aeronautical and Space Sciences, *NASA Authorization for Fiscal Year 1965: Hearings on S. 2446, Pt. 2,* 88th Cong., 2nd sess., 1964, pp. 358-359.

81. U.S. Congress, Senate, Committee on Aeronautical and Space Sciences, *NASA Authorization for Fiscal Year 1966: Hearings on S. 927, Pt. 1,* 89th Cong., 1st sess., 1965, pp. 60-61.

82. Webb, "U.S.-Soviet Space Capabilities," speech, Missouri Cotton Producers Association, Sikeston, Mo., 30 May 1964; Mission Operation Report No. S-622-64-03, "Echo II-Post Launch Report No. 1," 27 Jan. 1964; and Mission Operation Report No. S-622-64-03, "Echo II-Post Launch Report No. 2," 31 Mar. 1964.

83. Dryden to Blagonravov, 26 Mar. 1964; Dryden to Blagonravov, 1 Apr. 1964; TWX, Blagonravov to Dryden, 6 May 1964; "Suggestion for Private Meeting (D/B)" [n.d.] ; TWX, Frutkin to State Department, 11 May 1964; and Leonard Jaffee to Dryden, memo, "Proposed Second Stage of US-USSR Echo II Experiments," 19 May 1964.

84. "Second Memorandum of Understanding to Implement the Bilateral Space Agreement of June 8, 1962, between the Academy of Sciences

of the USSR and the National Aeronautics and Space Administration of the US," 6 June 1964; "Protocol for the Establishment of a Direct Communications Link between the World Meteorological Center in Moscow and Washington in Accordance with the Bilateral Agreement on Outer Space Dated June 8, 1962, between the Academy of Sciences of the USSR and the National Aeronautics and Space Administration of the USA," 6 June 1964; and NASA News Release, HQ [unnumbered], "News Conference on Implementation of U.S.-U.S.S.R. Bilateral Space Agreement," 8 June 1964.

85. NASA News Release, "News Conference on Implementation," 8 June 1964.

86. Ibid.; "Mirnoe ispol'zovanie kosmosa" [Peaceful use of space], *Pravda,* 10 June 1964; and Blagonravov, "Collaboration between the USSR and the United States in Space Research," *Vestnik Akademii Nauk SSSR* [Herald of the Academy of Sciences USSR], No. 10 (1964) (JPRS Translation No. 28,890), pp. 82-84. Blagonravov summarized the joint talks through the summer of 1964 and presented the Soviet view.

87. An official report on the status of the Soviet-American meteorological exchange was presented by the U.S. Weather Bureau, as cited in Subcommittee on Space Sciences and Applications, *1966 NASA Authorization,* p. 900.

88. U.S., Congress, House of Representatives, Committee on Appropriations, Subcommittee on Independent Offices, *Independent Offices Appropriations for 1966: Hearings, Pt. 2,* 89th Cong., 1st sess., 1965, p. 1007. For summary of Dryden's last meeting with Blagonravov, see "Memorandum of Conversation between Dr. Hugh L. Dryden, Deputy Administrator, NASA, and Academician A. A. Blagonravov, USSR Academy of Sciences, Held May 14, 1965, at Mar del Plata, Argentina, 2-3:15 PM"; and diary note, Frutkin, "Notes on US/USSR Bilateral and Soviet Participation in COSPAR Meeting, May 1965, Mar del Plata, Argentina," 15 May 1965.

89. Richard K. Smith, comp. and ed., *The Hugh L. Dryden Papers, 1898-1965: A Preliminary Catalogue of the Basic Collection* (Baltimore, 1974), p. 32.

90. U.S. Congress, Senate, Committee on Aeronautical and Space Sciences, *NASA Authorization for Fiscal Year 1969, Hearings on S.2918, Pt. 1,* 90th Cong., 2d sess., 1969, p. 58; and A. J. Dessler, "Discontent of Space-Science Community," 30 Oct. 1969.

III

1. Frederick C. Durant, III, "Space Flight Needs Only Money, Time," *Aviation Week,* 27 Sept. 1954, p. 46. On 14 July 1952, the executive committee of the National Advisory Committee for Aeronautics passed a resolution that "NACA devote modest efforts to problems of unmanned and manned flights at altitudes from 50 miles to infinity and at speeds from Mach 10 to escape from the earth's gravity." NACA to High Speed Flight Research Station, "Discussion of Report on Problems of High Speed, High Altitude Flight, and Consideration of Possible Changes to the X-2 Airplane to Extend Its Speed and Altitude Range," 30 July 1953, which contains the NACA directive.

2. Joe Reddy, memos for record, " 'Man in Space': Production Story," and "#20 Disneyland-TV 'Man in Space'," Walt Disney Productions synopsis and background, 1955. Von Braun, Haber, and Ley had long been advocates of space flight, and as early as 1952 they had contributed articles to a *Collier's* symposium entitled "Man will conquer space soon." The articles included Wernher von Braun, "Crossing the Last Frontier," pp. 24-29 and 72-74; Willy Ley, "A Station in Space," pp. 30-31; Fred L. Whipple, "The Heavens Open," pp. 32-33; Joseph Kaplan, "This Side of Infinity," p. 34; Heinz Haber, "Can We Survive in Space," pp. 35 and 65-67; and Oscar Schachter, "Who Owns the Universe?" pp. 36 and 70-71, *Collier's,* 22 Mar. 1952.

3. Kyril Feodorovich Ogorodnikov to Durant, 23 Sept. 1955; Leonid Ivanovich Sedov to Durant, 24 Sept. 1955; Durant, "Impressions of the Sixth Astronuatical Congress," *Jet Propulsion* 25 (Dec. 1955): 738; and interview (via telephone), Durant-Ezell, 13 Dec. 1974. Based upon his conversations with the two Soviet scientists, Durant conjectured that the then anonymous Soviet "Chief Designer of Spacecraft," S. P. Korolev, was to be the intermediary who would present *Man in Space* to the Soviet political and technical leadership.

4. Maxime A. Faget, *Manned Space Flight* (New York, 1965), p. 8.

5. Loyd S. Swenson, Jr., James M. Grimwood, and Charles C. Alexander, *This New Ocean: A History of Project Mercury,* NASA SP-4201 (Washington, 1966), pp. 39-46, summarize the studies of multiple *g* loads as they relate to Project Mercury; and S. P. Umansky, *Chelovek*

na kosmicheskoy orbite (Moscow, 1974), pp. 31-32 (available in translation as *Man in Space Orbit,* NASA Technical Translation F-15973).

6. Swenson, Grimwood, and Alexander, *This New Ocean,* pp. 36-39, summarize the American studies relating to weightlessness; "Problemy nevesomosti" [The problems of weightlessness], *Nauka i Zhizn'* 22 (Dec. 1955): 17-20; translated in F. J. Krieger, *Behind the Sputniks: A Survey of Soviet Space Science* (Washington, 1958), pp. 127-133.

7. David G. Simons, "Use of V-2 Rocket to Convey Primate to Upper Atmosphere," Air Force Technical Report 5821, Air Material Command, Wright-Patterson AFB, Ohio, May 1949; J. P. Henry, E. R. Ballinger, P. J. Maher, and D. G. Simons, "Animal Studies of the Subgravity State during Rocket Flight," *The Journal of Aviation Medicine* 23 (Oct. 1952): 421-432; D. G. Simons, "Review of Biological Effects of Subgravity and Weightlessness," *Jet Propulsion* 25 (May 1955): 209-211; and Don Bailer, " 'Alas, Poor Yorick'," *MSC Space News Roundup* [reprint, courtesy of Aerojet-General], 22 Aug. 1959, p. 6.

8. Evgeny Riabchikov, *Russians in Space,* Nikolai P. Kamanin, ed., Guy Daniels, trans. (Garden City, N.Y., 1971), pp. 140-142 and 149-151; A. V. Pokrovskii, "Comment se comportent les animaux a 100 km d'altitude" [How animals behave at an altitude of 100 km], *Etudes Sovietiques,* no. 106 (Jan. 1957): 65-70; translated in Krieger, *Behind the Sputniks,* pp. 156-163. Umansky, *Chelovek na kosmicheskoy orbite,* pp. 33-40, cites six articles on weightlessness that appeared in Soviet journals between 1970 and 1972. Based on Riabchikov's comments (p. 141), Pokrovskii appears to have been in charge of canine experimentation in the Soviet space program. A general view of the Soviet use of animals is presented in the Novosti Press Agency pamphlet, *Animals, Pioneers of Outer Space; Physiological Experiments with Animals Flying in Geophysical Rockets; Biological Studies during Space Flights* (Moscow, n.d.).

9. Umansky, *Chelovek na kosmicheskoy orbite,* p. 49.

10. Ibid., pp. 50-54; Faget, *Manned Space Flight,* pp. 98-100; Swenson, Grimwood, and Alexander, *This New Ocean,* pp. 231 and 558, note 21; and Eugene B. Konecci, "Soviet Bio-astronautics–1964," paper, National Space Club, Washington, 15 Dec. 1964, pp. 4-7. Konecci summarizes the comparative advantages and dis-advantages of sealevel and 258-mm-Hg cabin atmospheres.

11. Umansky, *Chelovek na kosmicheskoy orbite,* pp. 50-51; and Faget, *Manned Space Flight,* pp. 100-102.

12. Swenson, Grimwood, and Alexander, *This New Ocean,* pp. 225-233, discuss development of the Mercury environmental control system; Frank H. Samonski, Jr., *Technical History of the Environmental Control System for Project Mercury,* NASA Technical Note D-4126 (Langley, Va., 1967); and interview (via telephone) Samonski-Ezell, 21 Jan. 1975.

13. P. T. Astashenkov, *Akademik S. P. Korolev* (Moscow, 1969) (available in translation as *Academician S. P. Korolev, Biography,* Foreign Technology Division edited translation HC-23-542-70, pp. 185-186); and Konstantin Petrovich Feoktistov, "Razvitie sovetskikh pilotruemuikh kosmicheskikh korablei," *Aviatsiya i Kosmonavtika,* no. 11 (1971): 36-37 (available in translation as "Development of Soviet Manned Spacecraft," National Lending Library for Science and Technology, Boston Spa, Yorkshire, England, and available from NASA as N73-15876). Feoktistov stated the following reasoning for adoption of the sphere: "The aerodynamic characteristics of the sphere, the drag coefficient and the position of the centre of masses were well known for the entire velocity range (from the first cosmic [i.e., orbital velocity] down to subcosmic velocity). In addition, the problem of maintaining stability of movement of a spherical vehicle in the atmosphere could be easily solved by just shifting the gravity centre of the vehicle off the centre of the sphere. This provides for the static stability and, as revealed by computations, for good dynamics of vehicle movements around the centre of masses even in the case of arbitrary orientation of the vehicle prior to re-entry and in descent when controls are no longer available."

14. Astashenkov, *Academician S. P. Korolev, Biography,* pp. 185-186; Hartley A. Soulé to James M. Grimwood, 29 Aug. 1965; Swenson, Grimwood, and Alexander, *This New Ocean,* pp. 71-72; and Ames Aeronautical Laboratory, "Preliminary Investigation of a New Airplane for Exploring the Problems of Efficient Hypersonic Flight," 18 Jan. 1957. In appendix B of the Ames report, there is a description of a proposed 1.5-meter spherical ballistic spacecraft, pp. 30-31.

15. U.S.S.R. Academy of Sciences, comp.,

SOURCE NOTES

Kosmicheskiy korabl' Vostok (Moscow, 1969); available in translation as *The Spaceship "Vostok,"* Foreign Technology Division edited translation HT-23-705-70, pp. 5-6; and Leonid Vladimirov, *The Russian Space Bluff,* David Floyd, trans. (London, 1971), pp. 89-91. Vladimirov indicates that a Peter Dolgov was killed when his space suit was ripped during a test of the ejection system. "Korolyov's [Korolev's] reaction to Dolgov's death was to take a number of urgent and clever measures. First he had the exit hatch made larger. Secondly, he increased to two seconds the interval between shooting off the hatch and the operation of the ejector mechanism."

16. H. Julian Allen, "Hypersonic Flight and the Reentry Problem," *Journal of the Aeronautical Sciences* 25 (Apr. 1958): 217-230; Alfred J. Eggers, Jr., "Performance of Long Range Hypervelocity Vehicles," *Jet Propulsion* 27 (Nov. 1957): 1147-1151; and Swenson, Grimwood, and Alexander, *This New Ocean,* pp. 55-82. The authors of *This New Ocean* describe the background of NACA and Air Force research into the problem of reentry vehicle design; also see William M. Bland, Jr., "Project Mercury," in *The History of Rocket Technology Essays on Research, Development, and Utility,* Eugene M. Emme, ed. (Detroit, 1964), pp. 214-215.

17. Swenson, Grimwood, and Alexander, *This New Ocean,* pp. 68-69.

18. Robert R. Gilruth, "Memoir: From Wallops Island to Mercury; 1945-1958," paper, Sixth International History of Astronautics Symposium, Vienna, Austria, 13 Oct. 1972, pp. 31-32.

19. Swenson, Grimwood, and Alexander, *This New Ocean,* p. 86; Grimwood, *Project Mercury: A Chronology,* NASA SP-4001 (Washington, 1963), p. 17; "How Mercury Capsule Design Evolved," *Aviation Week,* 21 Sept. 1959, pp. 52-53, 55, and 57; and David A. Anderton, "How Mercury Capsule Design Evolved," *Aviation Week,* 22 May 1961, pp. 50-71 passim.

20. Faget, Benjamin J. Garland, and James J. Buglia, "Preliminary Studies of Manned Satellites—Wingless Configuration: Nonlifting," in "NACA Conference on High-Speed Aerodynamics, Ames Aeronautical Laboratory, Moffett Field, Calif., Mar. 18, 19, and 20, 1958: A Compilation of Papers Presented," pp. 9-34, reissued as NASA Technical Note D-1254 (Langley, Va., 1962).

21. Grimwood, *Project Mercury: A Chronology,* pp. 19-24; Gilruth, "Memoir: From Wallops Island to Mercury," pp. 34-37.

22. Gilruth, "Memoir: From Wallops Island to Mercury," p. 37.

23. Feoktistov, "Razvitie sovetskikh pilotruemuikh kosmicheskikh korablei," p. 37; Astashenkov, *Academician S. P. Korolev, Biography,* pp. 187-190; Riabchikov, *Russians in Space,* p. 155; and Vladimirov, *The Russian Space Bluff,* p. 89. Vladimirov indicates that one of the more serious problems encountered by the Soviet team was the heart attack suffered by Korolev on 3 Dec. 1960.

24. Feoktistov, "Razvitie sovetskikh pilotruemuikh kosmicheskikh korablei," p. 37.

25. Ibid.

26. Swenson, Grimwood, and Alexander, *This New Ocean,* pp. 341-511, summarize the operational phase of Project Mercury.

27. Barton C. Hacker and Grimwood, *On the Shoulders of Titans: A History of Project Gemini,* NASA SP-4203 (Washington, 1977), p. xvi.

28. Ibid.; Hacker, "The Idea of Rendezvous: From Space Station to Orbital Operations in Space-Travel Thought, 1895-1951," *Technology and Culture* 15 (July 1974): 373-388; and John M. Logsdon, "Selecting the Way to the Moon: The Choice of the Lunar Orbital Rendezvous Mode," *Aerospace Historian* 18 (June 1971): 63-70.

29. James A. Chamberlin, "Project Gemini Design Integration," Lecture 36 in a series on engineering design and operation of manned spacecraft, presented during summer, 1963, at the Manned Spacecraft Center and to graduate classes at Louisiana State University, the University of Houston, and Rice University. The series was later edited and published as chapter 35 in Paul E. Purser, Faget, and Norman F. Smith, eds., *Manned Spacecraft: Engineering Design and Operations* (New York, 1964), pp. 365-374.

30. Hacker and Grimwood, *On the Shoulders of Titans,* p. xvi.

31. Ibid., p. xvii.

32. The Soviets are relatively vague in their descriptions of Voskhod and its development. See Astashenkov, *Academician S. P. Korolev, Biography,* pp. 226-230; and Riabchikov, *Russians in Space,* pp. 207-211. Vladimirov, *The Russian Space Bluff,* pp. 123-127, argues that Khrushchev wanted a space mission that would surpass the accomplishments promised by

371

Gemini. Equally questioning of the design merits of Voskhod are James E. Oberg, "The Voskhod Program: Khrushchev's Folly!" *Spaceflight* 16 (Apr. 1974): 145-149; and Peter Sullivan, "The Voskhod Spacecraft," *Spaceflight* 16 (Nov. 1974): 405-409.

33. Apparently, the Soviets employed the Venus upper stage to launch the Voskhod. Sullivan, "The Voskhod Spacecraft," pp. 407-408, speculates that Korolev and his colleagues had to extemporize because of the tight schedule imposed upon them:

> The information submitted to the FAI (Federation Aeronautique Internationale) stated that the launch vehicle for Voskhod I consisted of a seven engine launcher compared with a six engine launch vehicle for Vostok. From photographs we know that in each case five of the engines refer to the bottom central sustainer surrounded by four boosters (the 1½ stage booster) and the term "engine" means an independent unit. . . . above the basic 1½ stage booster was a long stage, followed by the standard short final stage as used on the Vostok. The reason for this inefficient set up resulted from the speed with which the Voskhod programme was conceived and had to be executed to be effective.
>
> At the time the only trustworthy extra stage that could be man-rated was too powerful and the Vostok launch vehicle had been stretched to its limit and was not capable of launching heavier assembly. . . . Instead of replacing the final stage by simply the longer, more powerful stage, the final stage was retained as dead weight to lower the altitude that was attained. Even so, it still resulted in a much higher altitude than any of the Vostok or forthcoming Soyuz missions.

34. Vladimirov, *The Russian Space Bluff,* pp. 125-126; and statement, Hugh L. Dryden to PAO (dictated over telephone), 12 Oct. 1964.

35. Sullivan, "The Voskhod Spacecraft," pp. 405-406; and interview, Willard M. Taub-Ezell, 28 Feb. 1975.

36. Riabchikov, *Russians in Space,* p. 208; and Astashenkov, *Academician S. P. Korolev, Biography,* pp. 227-228.

37. U.S. Congress, Senate, Committee on Aeronautical and Space Sciences, *Soviet Space Programs, 1966-1970; Goals and Purposes, Organizations, Resources, Facilities and Hardware, Manned and Unmanned Flight Programs, Bioastronautics, Civil and Military Applications, Projections of Future Plans, Attitudes Toward International Cooperation and Space Law; Staff Report,* 92d Cong., 1st sess., 1971, p. 186.

38. Riabchikov, *Russians in Space,* pp. 210-211. Also see memo, M. Scott Carpenter to Gilruth et al., "Cosmonaut Training," 24 Nov. 1964.

39. There has been considerable speculation as to the cause of *Cosmos 57's* disintegration; e.g., William J. Normyle, "Cosmos 57 Believed Destroyed by Soviets," *Aviation Week and Space Technology,* 12 Apr. 1965, p. 34.

40. U.S. Congress, Senate, Committee on Aeronautical and Space Sciences, *Soviet Space Programs, 1962-1965; Goals and Purposes, Achievements, Plans, and International Implications; Staff Report,* 89th Cong., 2d sess., 1966, pp. 206-208. Vladimirov, *The Russian Space Bluff,* p. 140, states that it was L. A. Voskresensky's idea not to pressurize the Voskhod cabin, but to use instead "a light tube . . . attached to the hatch of the space-craft to form an exit chamber. . . ."

41. Committee on Aeronautical and Space Sciences, *Soviet Space Programs, 1962-65,* p. 207; and Peter L. Smolders, *Soviets in Space; The Story of the Salyut and the Soviet Approach to Present and Future Space Travel* (London, 1973), pp. 144-145.

42. Hacker and Grimwood, *On the Shoulders of Titans,* p. 235.

43. Ibid., p. 248.

44. Ibid., p. 275.

45. TWX, Rhett Turnipseed to NASA, Houston, "Text of an Interview by an *Izvesti[y]a* Correspondent with the Soviet Cosmonaut Pavel Romanovich Popovich [21 Dec. 1965]," 29 Dec. 1965.

46. "Gemini 7/6 Astronaut Post Flight Press Conference," 30 Dec. tape 8, p. 2; and Grimwood and Ivan D. Ertel, "Project Gemini," *Southwestern Historical Quarterly* 81 (Jan. 1968): 407.

47. Hacker and Grimwood, *On the Shoulders of Titans,* p. 289.

48. Ibid., pp. 338-339.

49. Vladimirov, *The Russian Space Bluff,* pp. 136-137, 140-141, and 145.

50. The "moon" or "space race" has been a topic of continuing debate and a subject of considerable speculation. A sample of views are included here in the absence of a definitive Soviet statement. The Novosti Press book by Riabchikov, *Russians in Space* (1971), does not address the space race question but indicates that the Soviets were concentrating on earth-orbital missions that would lead to the development of a

space station. This thesis is reemphasized in the 1973 edition of Smolders, *Soviets in Space.* The Soviet emigre Vladimirov wrote *The Russian Space Bluff* to argue that the limited technical capability of the Soviet space program could not have possibly sent men to the moon and that the whole program was inspired by Khrushchev's desire to gain a propaganda advantage over the U.S. Nicholas Danilov in *The Kremlin and the Cosmos* (New York, 1972) suggests that after Khrushchev's ouster there was a retreat from the competitive posture and that the Soviet leadership opted instead for a two-part space program—automatic spacecraft for lunar and planetary exploration (Luna and Venera probes) and manned earth orbital missions (Soyuz and Salyut).

51. A. Yu. Dmitriyev et al., *Ot komicheskikh korabley-k orbital'nymn stantsiyam,* 2d ed. (Moscow, 1961), pp. 24-25, (available in translation as *From Spaceships to Orbiting Stations,* NASA Technical Translation F-812); and interview, Faget-Ertel and Grimwood, 15 Dec. 1969. Faget commented that:

> One of the things we kind of set as a policy in all our design studies was adequate amount of volume inside the command module. About this time one of the other things that was being studied was the possibility of a two compartment vehicle. . . . Now in order to have enough volume, of course, they had to make the thing bigger which meant we had to carry along a lot of extra heat protection systems, so it seemed a very attractive thing to do to divide that volume in two pieces and we had for a long while a command module and a mission module, the mission module being where everybody was supposed to do their business. This started off to be a very attractive idea but as we went through their studies it became clear that less and less things were going on in that mission module, and every thing that was vital for one reason or another . . . also was vital during entry so you either did it twice, once in the mission module and again in the command module, or you did it once in the command module. So it seems that the mission module was turning out to provide nothing but extra room. There were no systems and no particular activity that anyone really wanted to carry out in the mission module other than to stretch out and perhaps get a little sleep. The consequence of this was that it didn't look like it was worthwhile to have a mission module. So in the final analysis we ended up with a single cabin version. You might have noticed that the Russians ended up going into something very close to our two-compartment vehicle that we were considering at that time. I don't know where they got their ideas, but it might

have been from us because we made no secret of these considerations.

52. Dmitriyev et al., *Ot komicheskikh,* p. 26.

53. Ibid., pp. 26-28; and Smolders, *Soviets in Space,* pp. 151 and 154-155.

54. Smolders, *Soviets in Space,* pp. 157-159.

55. Memo, Julian Scheer to HQ Program and Staff Offices, 24 Apr. 1967; and NASA News Release, HQ [unnumbered], "Russian Accident Statement," 24 Apr. 1967.

56. Smolders, *Soviets in Space,* p. 160: and NASA, Apollo 204 Review Board, "Report of Apollo 204 Review Board to the Administrator, National Aeronautics and Space Administration," 5 Apr. 1967. Uri Marinin, "Where Does Danger Lurk?" *Space World* D-5-41 (May 1967): 43-44, presents a Soviet commentary on the Apollo 204 fire and the dangers inherent in a 100-percent oxygen system.

57. The details of the Apollo spacecraft story will be documented in Courtney G. Brooks, Grimwood, and Swenson, "Chariots for Apollo: A History of the Lunar Spacecraft," in process. Until that official history is available, there are four very useful chronologies: Ertel and Mary Louise Morse, *The Apollo Spacecraft: A Chronology, Volume I, through November 7, 1962,* NASA SP-4009 (Washington, 1969); Morse and Jean Kernahan Bays, *The Apollo Spacecraft: A Chronology, Volume II, November 8, 1962-September 30, 1964,* NASA SP-4009 (Washington, 1973); Ertel and Brooks, *The Apollo Spacecraft: A Chronology, Volume III, October 1, 1964-January 20, 1966,* NASA SP-4009; and Ertel and Roland W. Newkirk, with Brooks, *The Apollo Spacecraft: A Chronology, Volume IV, January 21, 1966-March 4, 1974,* NASA SP-4009.

58. Ertel and Morse, *Apollo Chronology, Vol. I,* pp. 106, 128, 135, and 168; and Morse and Bays, *Apollo Chronology, Vol. II,* p. 5.

59. Viktor P. Legostayev and B. V. Raushenbakh, "Avtomaticheskaya sborka v kosmose," paper, 19th Congress of the IAF, New York, Dec. 1968 (available in translation as "Automatic Assembly in Space," NASA Technical Translation F 12, 113). Legostayev and Raushenbakh presented an analysis of an automatic rendezvous system of the type used on *Cosmos 186, 188, 212,* and *213.* Committee on Aeronautical and Space Sciences, *Soviet Space Programs, 1966-70,* pp. 230-234; Smolders, *Soviets in Space,* pp. 162-168; and Georigy Ivanovich Petrov, ed., *Osvoenie kosmicheskogo prostranstva v SSSR; ofitsian'nye soobsh cheniya TASS i materialy*

tsentrol'noi pechati oktyabr', 1967-1970 gg (Moscow 1971) (available in translation as *Conquest of Outer Space in the USSR; Official Announcements by Tass and Material Published in the National Press from October 1967 to 1970,* NASA Technical Translation F-725 [New Delhi, 1973], pp. 15-70).

60. The difference in views as to the goals of the Beregovoy mission is illustrated by Smolders, *Soviets in Space,* pp. 163-164; and Committee on Aeronautical and Space Sciences, *Soviet Space Programs, 1966-1970,* p. 233.

61. "Columbuses of Space," *New York Times,* 22 Dec. 1968; "The Christmas Journey," *Washington Post,* 22 Dec. 1968; NASA *Astronautics and Aeronautics, 1968,* NASA SP-4010 (Washington, 1969), p. 325.

62. "Apollo 8 Called Key Flight Space Program," *Baltimore Sun,* 24 Nov. 1972.

63. "Soviet Scientist Hails Apollo 'Courage' and Skill," *New York Times,* 31 Dec. 1968; "Soviet Cautious on Moon Flights," *Baltimore Sun,* 31 Dec. 1968; and Boris Nikolaevich Petrov, "O polete Apollona-8" [On the flight of Apollo-8], *Pravda,* 30 Dec. 1968.

IV

1. Donald R. Morris to George M. Low, memo, 21 Sept. 1970.

2. Dale D. Myers to Philip E. Culbertson, note, 19 Aug. 1970.

3. Leroy Roberts to Eldon W. Hall, memo, "US/USSR Space Cooperation," 31 Aug. 1970; and Roberts, "International Cooperation in Space," 31 Aug. 1970.

4. Roberts to Hall, memo, 31 Aug. 1970.

5. Ibid.

6. Culbertson to Hall, note, Aug. 1970.

7. Roberts to R. Cerrato, memo, "Cooperative Space Activity," 3 Sept. 1970; and Roberts to Douglas R. Lord, William D. Green, Jr., and Richard J. Allen, "Cooperative Space Activity," 4 Sept. 1970.

8. Roberts to Jack C. Heberlig and Willard M. Taub, memo, "Cooperative Space Activities," 3 Sept. 1970. Background correspondence between Paine and Keldysh had been transmitted to Heberlig by Roberts on 26 Aug. 1970.

9. William C. Schneider to Myers, memo, "International Cooperation in the Skylab Program," 4 Sept. 1970.

10. Thomas O. Paine to Mstislav Vsevolodovich Keldysh, 4 Sept. 1970; and Paine to Keldysh, 31 July 1970.

11. Myers to Paine, memo, draft (prepared by Roberts), "US/USSR Space Cooperation," 10 Sept. 1970; and Lord to Roberts, memo, "Feasibility of Compatible US and USSR Docking Systems," 14 Sept. 1970.

12. "Station No. 1," *Newsweek,* 27 Jan. 1969, pp. 93-94. This article was passed around within OMSF, because it described the differences between Soviet and American spacecraft.

13. Myers to Paine, memo, draft, 10 Sept. 1970, with enclosures.

14. Charles W. Mathews to Myers, note, 15 Sept. 1970.

15. Myers to Low, memo, "US/USSR Space Cooperation," 17 Sept. 1970.

16. Keldysh to Paine, 11 Sept. 1970; Keldysh to Philip Handler, 10 Sept. 1970; and Lord to Low, memo, 21 Sept. 1970.

17. Low to Keldysh, 25 Sept. 1970.

18. Roberts, note for record [handwritten chronology of events, n.d.]; Mathews to Arnold W. Frutkin, memo, "Suggested Agenda Items," 24 Sept. 1970; and interview, Low-Edward C. Ezell, 30 Apr. 1975.

19. Myers to Frutkin, memo, "OMSF Participation in International Meeting," 5 Oct. 1970; and Mathews to Hall, note, 5 Oct. 1970.

20. Interview, Robert R. Gilruth-Ezell, 25 Mar. 1975; and Mathews to Frutkin, memo, "US/USSR Space Cooperation," 9 Oct. 1970.

21. Interview, Glynn S. Lunney-Ezell, 23 July 1974; interview, Caldwell C. Johnson-Ezell, 27 Mar. 1975; and interview (via telephone), George B. Hardy-Ezell, 4 Apr. 1975.

22. NASA News Release, HQ [unnumbered], "Background Press Briefings; U.S. and USSR Cooperation in Space," 13 Oct. 1970, p. 5; and NASA News Release, HQ, 70-173, "U.S.-Soviet Meeting," 12 Oct. 1970.

23. Walter Sullivan, "U.S.-Soviet Space Docking Is Said To Be under Study," *New York Times,* 8 Oct. 1970; "Russians to Join Talks on Space Rescue Plan," *Washington Evening Star,* 8 Oct. 1970; and Howard Benedict, "Joint Space Rescue Symposium Slated," *Denver Post,* 8 Oct. 1970. Space rescue was a subject of considerable interest in the late 1960s, as exemplified by R. Cargill Hall in "Rescue and Return of Astronauts on Earth and in Outer Space," *American Journal of International Law* 63 (Apr. 1969): 197-210. Hall points out the need for hardware compatibility (p. 208).

24. NASA News Release, "Background Press Briefings; U.S. and USSR Cooperation in Space," 13 Oct. 1970, p. 6.

25. Ibid., pp. 8-9.

26. For the press comment, see Albert Sehls-tedt, Jr., "U.S., Soviet to Meet on Space," *Baltimore Sun*, 13 Oct. 1970; John Noble Wilford, "5 NASA Officials to Visit Moscow," *New York Times*, 13 Oct. 1970; "Space Linkup Far in Future," *Washington Post*, 14 Oct. 1970; "NASA to Discuss Docking with Russians," *Washington Evening Star*, 14 Oct. 1970; "Co-operation in Space," *Houston Post*, 14 Oct. 1970; and D. J. R. Bruckner, "Space Efforts Could Take Giant Leap through International Cooperation," *Los Angeles Times*, 16 Oct. 1970.

27. Interview, Gilruth-Ezell, 25 Mar. 1975; and interview, Johnson-Ezell, 27 Mar. 1975.

28. Interview, Gilruth-Ezell, 25 Mar. 1975.

29. Interview, Johnson-Ezell, 27 Mar. 1975.

30. Interview, Gilruth-Ezell, 2 Mar. 1975; and debriefing tape, made by Gilruth, Lunney, Johnson, and Hardy, following their return from Moscow [n.d.].

31. Hedrick Smith, *The Russians* (New York, 1976), p. 106.

32. Gilruth et al., debriefing tape; and Lunney to distribution, memo, "Trip Report—Delegation to Moscow To Discuss Possible Compatibility in Docking," 5 Nov. 1970.

33. Lunney to distribution, memo, 5 Nov. 1970; Gilruth et al., debriefing tape; and Vladimir Aleksandrovich Shatalov; "Komandnia rubka 'Soyuza'" [Command module of the Soyuz], *Aviatsiya i Kosmonovtika* (Oct. 1970), pp. 34-36, describes the onboard systems of the command module and illustrates the control displays.

34. Gilruth et al., debriefing tape.

35. Lunney to distribution, memo, 5 Nov. 1970; and Vu-graphs from presentation by Lunney, Moscow, 26 Oct. 1970.

36. Lunney to distribution, memo, 5 Nov. 1970; and Viktor Pavlovich Legostayev and Boris V. Raushenbakh, "Avtomaticheskaya sborka v kosmose," paper presented at the 19th Congress of the IAF, New York, Dec. 1968 (available in translation as "Automatic Assembly in Space," NASA Technical Translation F12, 113).

37. Johnson, "Fundamentals of Gemini and Apollo Docking and Crew Transfer and Docking Concepts for Future Spacecraft," with Vu-graphs, 20 Oct. 1970, p. 4.

38. Ibid., p. 5.

39. Ibid.

40. North American Aviation Corp., "Apollo Monthly Progress Report," SID 62-300-4, 31 May 1962, p. 66; North American Rockwell Corp., Space Division, Kenneth A. Bloom and George E. Campbell, "The Apollo Docking System," SD 69-42, Aug. 1969, p. 1; and NASA, MSC, "ASPO Status Report," 4 Dec. 1963.

41. NASA, MSC, Advanced Spacecraft Technology Division, Johnson, "A Docking Gear Concept for AAP and Advanced Spacecraft," May 1967, pp. 2-3.

42. Interview, Johnson-Ezell, 27 Mar. 1975.

43. Frutkin sent a copy of Syromyatnikov's paper to Gilruth, see note, Frutkin to Gilruth, 3 Dec. 1970. Vladimir Sergeyevich Syromyatnikov, "Docking-Mechanism Attenuator with Electro-Mechanical Damper," in *Fifth Aerospace Mechanisms Symposium: Proceedings of a Conference Held at Goddard Space Flight Center, Greenbelt, Maryland, June 15-16, 1970*, NASA SP-282 (Washington, 1971), pp. 43-48. Following his paper, Syromyatnikov showed a movie, "Docking of the *Soyuz IV* and *V*," and gave a narrative description of slides made from the article "Linked Soyuz Spacecraft Shown by Soviets at Japan's Expo 70," *Aviation Week & Space Technology*, 27 Apr. 1970, pp. 70-74. "Discussion: Soviet Spacecraft" is presented on pp. 49-57 of the *Proceedings*. Bloom and Campbell presented "Apollo Docking System" at the same symposium (pp. 3-8 of the *Proceedings)*. This talk was a variation of their presentation cited in note 40.

44. Lunney to distribution, memo, 5 Nov. 1970. On 28 October, the Soviets gave the NASA representatives four papers: "Kratkoye opisaniye stykovochnogo ustroystva kosmicheskikh korabley tipa 'Soyuz'" [A short description of the docking device of "Soyuz" type spacecraft]; "Kratkoye opisaniye stykovochnogo ustroystva kosmicheskikh korabley tipa 'Soyuz' (S vnyutrennim perekhodom)" [A short description of the docking device of "Soyuz" type spacecraft (with internal passage)]; "Opisaniye printsipial'noy skhemy sblizheniya i prichalivaniya kosmicheskikh korabley tipa 'Soyuz'" [A description of the conceptual arrangement for the rendezvous and docking of "Soyuz" type spacecraft]; and "Kratkoye opisaniye radioapparatury sblizheniya kosmicheskikh korabley tipa 'Soyuz'" [A short description of the radar approach equipment of "Soyuz" type spacecraft].

45. Lunney to distribution, memo, 15 Nov. 1970; and Gilruth et al., debriefing tape.

46. Gilruth et al., debriefing tape.

47. Ibid.

48. Ibid.

49. Ibid.; and Keldysh to Low, 19 Oct. 1970.

50. Johnson, "Initial Efforts toward the Development of Compatible Rendezvous and Docking Hardware and Software for USSR and US Spacecraft," 3 Nov. 1970. Johnson sent with this an undated note to Gilruth saying, "I wrote the attached only as a guide to feasibility studies that I propose to begin within Spacecraft Design Office."

51. Johnson, "Initial Efforts," 3 Nov. 1970.

52. Low to Henry A. Kissinger, 29 Oct. 1970; Low to U. Alexis Johnson, 30 Oct. 1970; Low to Keldysh, 5 Nov. 1970; Keldysh to Low, 2 Dec. 1970; TWX, Robert F. Freitag to Frank A. Bogart, "US-USSR Agreements and Studies," 10 Nov. 1970; René A. Berglund to Gilruth, memo, "Status of USSR/USA Docking System Activities," 9 Dec. 1970; and interview (via telephone), Shirley Malloy-Ezell, 22 May 1975.

53. Keldysh to Low, 19 Oct. 1970; Low to Keldysh, 24 Nov. 1970; Keldysh to Low, 4 Dec. 1970; and Low to Edward E. David, Jr., Science Adviser to the President, 2 Dec. 1970.

V

1. Mstislav Vsevolodovich Keldysh to George M. Low, 19 Oct. 1970; and Thomas O. Paine to Keldysh, 15 Sept. 1970.

2. Low to Keldysh, 24 Nov. 1970; Keldysh to Low, 4 Dec. 1970; and TWX, Keldysh to Low, 30 Dec. 1970.

3. Low, "Notes on Trip to the Soviet Union," 15-22 Jan. 1971.

4. Harry Schwartz, "Threats and Bombs—A Nasty Phase for the Two Nations," New York Times, 10 Jan. 1970; David A. Andelman, "Dangerous Campaign to Harass Russians," New York Times, 17 Jan. 1971; "Soviet Union—Limited Leniency," Time, 11 Jan. 1971, pp. 19-21; "Lapel Diplomacy," Time, 18 Jan. 1971, p. 27; and "The Private Jewish War on Russia," Time, 25 Jan. 1971, pp. 18 and 21.

5. Low, "Notes on Trip to the Soviet Union," 15-22 Jan. 1971.

6. Interview, Low-Edward C. Ezell, 30 Apr. 1975.

7. Robert R. Gilruth to Boris Nikolaevich Petrov, 23 Nov. 1970, with enclosures ("Operations and Functions of the Apollo Guidance Computer During Rendezvous"; "A Summary Description of the Apollo Command and Service Module Telecommunications System"; "A Summary Description of the Apollo Docking System"; "The Apollo Radar Systems"; and "A Summary Description of the Apollo Command Module Environmental Control System"); Gilruth to Dale D. Myers, 23 Nov. 1970; Petrov to Gilruth, 16 Dec. 1970, with enclosures ("Kratkoye opisaniye radioapparatury sblizheniya kosmicheskikh korabley tipa 'Soyuz' " [A brief description of the radio equipment used for rendezvous by Soyuz type spacecraft]; "Radiotelefonnaya svaz' mezhdu pilotiruyemymi korablyami tipa 'Soyuz' " [Radio-telephone communications between manned spacecraft of the Soyuz type], and "Spravochnyye dannyye po parametram atmosfery zhilykh otsekov korabley tipa 'Soyuz' " [Reference data on the parameters of the atmosphere in the living compartments of Soyuz type spacecraft]); and Keldysh to Low, 4 Dec. 1970.

8. Interview, Arnold W. Frutkin-Ezell, 5 May 1975.

9. Low to Henry A. Kissinger, 29 Oct. 1970; and Low to U. Alexis Johnson, 30 Oct. 1970.

10. Interview, Low-Ezell, 30 Apr. 1975; and Low, "Notes on Trip to the Soviet Union," 15-22 Jan. 1971.

11. Low, "Notes on Trip to the Soviet Union," 15-22 Jan. 1971; Peter Reich, "Lunokhod Revives Debate on Manned vs. Robot Explorers," Chicago Today, 28 Dec. 1970; John Noble Wilford, "Exotic Fragments Found in Apollo Lunar Sample, New York Times, 11 Jan. 1971; Thomas O'Toole, "Soviet Scientist Details Plans for Lunar Robots," Washington Post, 15 Jan. 1971; and Alexandr Pavlovich Vinogradov, "Preliminary Data on Lunar Ground Brought to Earth by Automatic Probe 'Luna-16'," paper presented at the Second Lunar Science Conference, Houston, Tex., 11-14 Jan. 1971.

12. Low, "Notes on Trip to the Soviet Union," 15-22 Jan. 1971; and Low, "Speech Delivered before the National Space Club," Washington, 26 Jan. 1971.

13. NASA News Release, HQ, 71-57, "U.S.-Soviet Agreement," 31 Mar. 1971; and NASA News Release, HQ, 71-9, "U.S.-USSR Space Meeting," 21 Jan. 1971.

14. Low, "Notes on Trip to the Soviet Union," 15-22 Jan. 1971.

15. Don Kirkman, "Soviet Invites Space Talks," Washington Daily News, 6 Jan. 1971; and Bernard Gwertzman, "U.S. and Russians Reach Moon Pact," New York Times, 22 Jan. 1971.

16. Low, "Notes on Trip to the Soviet

Union," 15-22 Jan. 1971; and interview, Low-Ezell, 30 Apr. 1975.

17. NASA, MSC, Clarke Covington, "USSR/US Docking Studies," 5 Jan. 1971; René A. Berglund to Gilruth, memo, "Status of USSR/USA Docking System Activities," 17 Dec. 1970; NASA, MSC, Covington, "USSR/US Docking Study Review," 21 Dec. 1970 (draft of 5 Jan. presentation); and interview, Covington-Ezell, 3 Apr. 1975.

18. Covington, "USSR/US Docking Study Review."

19. Interview, Covington-Ezell, 3 Apr. 1975.

20. Ibid.; interview, Walter W. Guy and James R. Jaax-Ezell, 17 May 1974; and interview, Guy and Jaax-Ezell, 7 July 1974. See also NASA, MSC, NASA Technical Brief, B72-10690, "An Efficient Prebreathing Apparatus for Humans during Decompression," Dec. 1972.

21. Interview, Covington-Ezell, 3 Apr. 1975; and Covington, "USSR/US Docking Study Review."

22. NASA, MSC, "Preliminary Rendezvous and Docking Requirements for United States Spacecraft," 2 Feb. 1971. This document was compiled with the help of many people at MSC under the editorial supervision of Leonard S. Nicholson.

23. NASA, MSC, "A Concept for a Union of Soviet Socialists Republics/United States of America Rendezvous and Docking Mission," 2 Feb. 1971.

24. Gilruth to Petrov, 17 Feb. 1971.

25. Ibid.; and Gilruth to Myers, 12 Feb. 1971.

26. NASA, MSC, "A Synopsis of the Russian/American Docking Systems Activities," 8 Feb. 1971.

27. Briefing Vu-graphs from "Planning for USA/USSR Preliminary Mating" [n.d.]; Eldon W. Hall to Philip Culbertson, memo, "Weekly Activities Report," 25 Mar. 1971; NASA, MSC, "E&D Weekly Activity Report," 6-12 Mar. 1971, 20-26 Mar. 1971, and 3-9 Apr. 1971; and James L. Roberts to Glynn S. Lunney, memo, "International Rendezvous and Docking Activity," 20 Aug. 1971.

28. Petrov to Gilruth, 15 Mar. 1971. The seven papers that constituted the Soviet technical requirements were as follows: "Tekhnicheskiye trebovaniya k sistemam i oborudovaniyu kosmicheskikh korabley i stantsiy obespechivayushchim ikh sblizheniye i stykovku" [Specifications for rendezvous and docking systems and equipment for spacecraft and space stations], "Predlozheniya po sistemam doordinat dlya razrobotki metodov i sredstv sblizheniya i stykovki kosmicheskikh korabley i stantsiy" [Proposed coordinate systems for developing methods and equipment for use in the rendezvous and docking of spacecraft and space stations], "Tekhnicheskiye trebovaniya na razrabotku stykovochnogo ustroystva dlya kosmicheskikh korabley i stantsiy, obespechivayushchego ikh stykovku" [Specifications for the development of a docking mechanism for docking spacecraft and space stations], "Tekhnicheskiye trebovaniya k atmosfere obitayemykh otsekov, sposobam otkrytiya vkhodnykh lyukov, pnevmogidroelcktroraz'yeman mezhdu skafandrami kosmonavtov i bortom kosmicheskikh korabley ili stantsiy" [Technical requirements for cabin atmospheres, means of opening entry hatches, pneumo-hydro-electrical connections between cosmonauts' suits and equipment aboard space vehicles and stations], "Technicheskiye trebovaniya k radio-apparatus dlya svyazi mezhdu ekipazhami kosmicheskikh korabley i stantsiy" [Technical requirements for radio equipment for communications between crews of space vehicles and stations], "Ogranicheniya po raspolozheniyu elementov konstruktsii i oborudovaniya pri sblizhenii i stykovke kosmicheskikh korabley i stantsiy" [Restrictions on the placement of structural elements and equipment in rendezvous and docking of space vehicles and stations], and "Technicheskiye po dinamike stykovki kosmicheskikh korabley i stantsiy" [Technical requirements in regard to the docking of space vehicles and stations].

29. Gilruth to Petrov, 9 Apr. 1971; and Lunney to Caldwell C. Johnson, memo, "Preparations for Mid-May Soviet Talks," 9 Apr. 1971.

30. TWX, Jacob D. Beam, Ambassador, to Secretary of State, "Space Cooperation: Local Expenses for Exchanges," 19 Mar. 1971.

31. Vu-graph briefing charts from "Dr. Gilruth/Mr. Frutkin Meeting, May 7, 1971, USSR Visit" [n.d.]; and TWX, Beam to Secretary of State, "Space Docking Meeting," 7 May 1971: "Vereshchetin informed science officer that delegation must postpone until sometime in mid-June. Letter from Keldysh on whole range of cooperation to Low to be delivered to science officer Monday."

32. TWX, Petrov to Gilruth, 8 May 1971; interview, Nicholson-Ezell, 16 July 1974; and interview, Nicholson-Ezell, 6 June 1975.

33. Berglund to Gilruth, note, 14 May 1971;

Gilruth to Petrov [not sent]; "Soviet-U.S. Parley on Space Cancelled," *New York Times,* 23 May 1971; and Jack Hartsfield, "Apollo-Soyuz Link-up Proposal Gets Nowhere," *Huntsville Times,* 4 May 1971.

34. Theodore Shabad, "A Long Manned Orbital Flight Is Predicted by Soviet Official," *New York Times,* 15 Apr. 1971; and Shabad, "A Decade after Man's First Space Flight; Soviet Sets New Goals," *New York Times,* 13 Apr. 1971.

35. "Polet 'Soyuza 10' " [Flight of "Soyuz 10"], *Priroda* [Nature], No. 6 (1971), pp. 2-3 (available as NASA Technical Translation F 13, 931).

36. Tass International Service radio broadcast transcript in English, "Soviet Union FBIS," 26 Apr. 1971, p. L5. For a fuller description of the docking process, see T. Borisov, "A Meeting in Space," *Trud,* 25 Apr. 1971. For American news media comments, see "Salyut Boosts Self to Higher Orbit," *Aviation Week and Space Technology,* 3 May 1971, pp. 14-15, which claims the mission was "aborted." Also see "Soyuz Review Leaves Puzzles Unresolved," *Aviation Week and Space Technology,* 10 May 1971, p. 19.

37. TWX, Petrov to Gilruth, 21 May 1971; TWX, Gilruth to Petrov, 1 June 1971; and TWX, Secretary of State to American Embassy, Moscow, "Meeting of Compatible Docking Working Group," 1 June 1971.

38. Roberts, "Minutes of Working Group Meeting on US/USSR Compatible Docking Mechanisms Conducted at MSC on June 21-25, 1971" [n.d.; handwritten notes].

39. Ibid.; and NASA, MSC, "Orientation Meeting," 21 June 1971.

40. Roberts, "Minutes of Working Group Meeting on US/USSR Compatible Docking Mechanism"; "Russians 'Fly' Moonship Simulators at Texas Center," *New York Times,* 22 June 1971; and Jim Maloney, "Space Docking Talks Progress," *Houston Post,* 23 June 1971.

41. NASA, MSC, "United States Summary Presentation," 21 June 1971 (with English and Russian texts); and "Data Provided to Soviet Delegation" [n.d.]. That data included the following: NASA, MSC, Bedford F. Cockrell, "Proposed Coordinate Systems for International Rendezvous and Docking of Spacecraft," 5 May 1971 (MSC-04245 and MSC Internal Note No. 71-FM-157); Jaax, "A Description of Cabin Atmosphere, Environmental Control and Life Support, Crew Transfer and Airlock Systems for Future Space Vehicles," 2 June 1971; "U.S. Proposed Items for Discussion; Working Group #1," June 1971; and MSC comments on the Soviet papers "Specifications for Rendezvous and Docking Systems and Equipment for Spacecraft and Space Stations," "Technical Requirements for Radio Equipment for Communication between Crews of Space Vehicles and Stations," and "Specifications for the Development of a Docking Mechanism for Docking Spacecraft and Space Stations."

42. Roberts, "Notes from Summary Presentation by USSR, " 22 June 1971.

43. Ibid.; and Roberts, "Minutes of Working Group Meeting on US/USSR Compatible Docking Mechanism"; and interview, Johnson-Ezell, 27 Mar. 1975.

44. Interview, Covington-Ezell, 3 Apr. 1975; interview, Nicholson-Ezell, 16 July 1974; Roberts, "Minutes of Working Group Meeting on Compatible Docking Mechanism"; and Roberts, "Notes from Summary Presentation by USSR," 22 June 1971.

45. Jack C. Waite to Gilruth, memo, "Comments Regarding Soviet Delegation Visit to MSC," 7 July 1971.

46. MSC environmental control system specialists gave the Soviets a copy of NASA, MSC, Jaax, "A Description of Cabin Atmosphere, Environmental Control and Life Support, Crew Transfer and Airlock Systems for Future Space Vehicles," 2 June 1971, which expanded upon the materials presented by the Soviets earlier that March.

47. "Summary of Results, Attachment A, Working Group No. 1, Minutes of Meetings," 22-23 June 1971.

48. "Summary of Results, Attachment B, Minutes of the Meeting of Working Group No. 2," 22-25 June 1971; and "Summary of Results, Attachment C, Minutes of Meeting of Working Group No. 3," 22-25 June 1971.

49. "Summary of Results," 22-25 June 1971, pp. 1-3; NASA News Release, MSC, 71-43, 25 June 1971; Petrov to Gilruth, 29 June 1971; "Russia, U.S. Exchange Space-Tragedy Notes," *Baltimore Sun,* 20 July 1971; Jim Maloney, "U.S., Russia Pass up Joint Press Talks," *Houston Post,* 25 June 1971; and Thomas O'Toole, " '74 Linkup by Soviets, U.S. Hinted," *Washington Post,* 26 June 1971.

50. Interview, Frutkin-Ezell, 5 May 1975; and Frutkin to Ezell, memo, 12 Feb. 1976.

51. NASA Press Conference, MSC, "Summary of U.S. & U.S.S.R. Joint Docking Space Systems Discussions," 28 June 1971.

52. Ibid.; note, George B. Hardy to Berglund, 30 June 1971; and Waite to Gilruth, memo, "USSR Delegation Visit," 30 June 1971.

53. NASA Press Conference, MSC, "Summary of U.S. & U.S.S.R. Joint Docking Space Systems Discussions," 28 June 1971; Jim Maloney, "Joint Space Effort Forecast for 70s," *Houston Post,* 29 June 1971; Bruce Hicks, "Soviet Talks on Linkups in Space Landed," *Houston Chronicle,* 29 June 1971; and Robert C. Cowen, "NASA Finds Soviets Sincere in Space-docking Talks," *Christian Science Monitor,* 29 June 1971.

54. Library of Congress, Federal Research Division, "Soyuz-11 Triumph and Tragedy," S & T News Feature, Item No. 454 [n.d.].

55. V. Golobachev and T. Borisov, "Duty Carried Out to the End," *Trud,* 2 July 1971.

56. "Cause Sought in Soyuz Tragedy," *Aviation Week and Space Technology,* 5 July 1971, pp. 12-15; and Peter Smolders, *Soviets in Space* (Guildford and London, 1973), pp. 246-247.

57. "3 Cosmonauts in Space Lab Found Dead after Recovery," *Washington Post,* 30 June 1971; Arthur J. Snider, "Has Man Reached His Space Limit," *Washington Evening Star,* 30 June 1971; Don Kirkman, "Soyuz Air Leak Blamed," *Washington Daily News,* 1 July 1971; Robert C. Cowen, "Soviet Loss Underscores Space Dangers," *Christian Science Monitor,* 1 July 1971; Thomas O'Toole, "Soyuz 11 Deaths Assessed," *Washington Post,* 1 July 1971; Jonathan Spivak, "Deaths of Cosmonauts Are Unlikely To Delay U.S. Space Program," *Wall Street Journal,* 1 July 1971; Charlotte Saikowski, "Space Tragedy Probed," *Christian Science Monitor,* 1 July 1971; and interview, Charles A. Berry-W. David Compton, 10 Apr. 1975. See also NASA Press Conference, MSC, "Statement on the Soyuz 11 Flight," 30 June 1971.

58. TWX, George G. Coletto to all Station Directors, MSFN, 30 June 1971.

59. Stuart Auerbach, "Cosmonaut Deaths Laid to Faulty Hatch," *Washington Post,* 3 July 1971. Low offered the following statement on 30 June: "The death of the three Soviet Cosmonauts is a terrible tragedy. They were pioneers in their achievements in space—in establishing the first manned space station. Our hearts go out to their families and to their colleagues."

60. Petrov, "On the Threshold of New Achievements," *Pravda,* 4 July 1971; "Salute Missions to Go On," *Washington Post,* 5 July 1971; Walter Sullivan, "Tragedy: When the Hatch Was Opened the Men Were Dead," *New York Times,* 4 July 1971; and Bernard Gwertzman, "Soviet Space Scientist States Salyut Program Will Continue," *New York Times,* 5 July 1971.

61. TWX, Goddard Network Operations Control Center to all Station Directors, 13 July 1971; Bernard Gwertzman, "Cause Confirmed in 3 Soyuz Deaths," *New York Times,* 12 July 1971; and Dean Mills, "Russia Blames the Soyuz Deaths on Failure of Seal," *Baltimore Sun,* 12 July 1971.

62. Petrov to Gilruth [n.d.; with enclosure, letter, Yu. P. Khomenko to Gilruth, 17 July 1971]; and "Russia, U.S. Exchange Space-Tragedy Notes," *Baltimore Sun,* 20 July 1971.

63. "Apollo's Safeguards Are Emphasized by U.S. Space Experts," *Chicago Tribune,* 20 July 1971; and Thomas O'Toole, "Apollo 15 Crewmen To Suit Up To Avert Soyuz 11 Disaster," *Washington Post,* 20 July 1971.

64. NASA News Release [unnumbered], "FY 1971 Interim Operating Plan News Conference," 2 Sept. 1970; and "New Year Prospects Glum in Aerospace Industry," *San Diego Union,* 3 Jan. 1971.

65. NASA, MSC, "Apollo 14 Mission Report" MSC-04112, May 1971; and NASA, MSC, "Apollo 14 Mission Anomaly Report No. 1; Failure to Achieve Docking Probe Capture Latch Engagement," MSC-05101, Oct. 1971.

66. Ronald Kotulak, "Manned Moon Flight Program Nears End," *Chicago Tribune,* 14 Feb. 1971; NASA, MSC, "Apollo 15 Mission Report," MSC-05161, Dec. 1971; "America's Future in Space," *Washington Post,* 16 Aug. 1971; Walter Sullivan, "Apollo 15: New Clues from the Men on the Moon," *New York Times,* 8 Aug. 1971; and Thomas O'Toole, "Our Surprising Moon," *Washington Post,* 8 Aug. 1971.

67. "New Director of Space Agency: James Chipman Fletcher," *New York Times,* 2 Mar. 1971.

68. Thomas O'Toole, "NASA Nominee Favors Cooperation with Russia" *Washington Post,* 11 Mar. 1971; and U.S. Congress, Senate, Committee on Aeronautical and Space Sciences, *Nomination of Dr. James C. Fletcher to be Administrator of National Aeronautics and Space Administration,* 92d Cong., 1st sess., 1971, p. 14.

69. John Noble Wilford, "U.S., Soviet Weigh

Space Linkup," *New York Times,* 18 July 1971; NASA, MSC, "Post Skylab Mission: Summary Report," 17 Mar. 1971 (enclosure to letter, Gilruth to Myers, 25 Mar. 1971); Maxime A. Faget to distribution, memo, "Post-Skylab Mission Study," 30 Apr. 1971; Berglund to distribution, memo, "Post-Skylab Mission Study," 14 May 1971; Gilruth to Myers, 25 Aug. 1971; NASA News Release Apollo 15 PC22, "Space Shuttle Briefing, Kennedy Space Center," 24 July 1971; and interview, Christopher C. Kraft-Ezell, 29 Mar. 1976.

70. John Noble Wilford, "U.S., Soviet Weigh Space Linkup," *New York Times,* 18 July 1971; and interview, Kraft-Ezell, 29 Mar. 1976.

71. "Informal Notes for Glynn Lunney, Subject: Results of the Briefing to MSC Management on Common Docking," draft, 13 July 1971; NASA, MSC, "CSM/Salyut Program Briefing," 12 July 1971; NASA, MSC, "E&D Weekly Activity Report," 3-9 July 1971; and Caldwell C. Johnson to Covington et al., memo, "CSM/Salyut Mission," 15 July 1971.

72. Johnson to Covington et al., memo, "CSM/Salyut Mission," 15 July 1971.

73. D. A. Nebrig to R. C. Lashbrook, "Contract NAS 9-150, Statement of Work for International Rendezvous and Docking Mission," 29 July 1971; and Berglund to distribution, memo, "International Rendezvous and Docking Missions Statement of Work," 28 July 1971. The "Statement of Work" was attached to both documents. The CCA#4162;500-300, dated 29 July 1971, was a change to contract NAS 9-150, 21 Dec. 1961.

74. NASA, MSC, "Statement of Work: International Rendezvous and Docking Mission," 28 July 1971, pp. 2-3.

75. Berglund to distribution, memo, "International Rendezvous and Docking Mission Study Team Staff Meeting of August 4, 1971," 4 Aug. 1971; and MSC Announcement 71-123, "Establishment of a Study Task Team," 23 Aug. 1971.

76. Berglund to distribution, memo, "International Rendezvous and Docking Mission Study Team Staff Meeting of August 16, 1971," 18 Aug. 1971, together with enclosure, "Docking Module Study Plan," 14 Aug. 1971.

77. Interview, Covington-Ezell, 3 Apr. 1975.

78. "Minutes of International Rendezvous and Docking Mission, ECS Meeting," 24 Aug. 1971; and R. T. Everline to Berglund, memo, "IRDM Docking Module Meeting," 26 Aug. 1971. At this latter meeting, held on 24 Aug.,

Covington presented his briefing of the 16th again, this time for North American.

79. Interview, Covington-Ezell, 3 Apr. 1975; NASA, MSC, "Docking Module Design Study," 29 Sept. 1971 (revised 5 Oct. 1971).

80. NASA, MSC, "Docking Module Design Study," 29 Sept. 1971.

81. Ibid.

82. Lunney to Gilruth, memo, "Status Report," 30 Aug. 1971.

83. Berglund to distribution, memo, "Minutes of the NR International Rendezvous and Docking Mission Study Status Review," 25 Aug. 1971; Vu-graphs from "International Rendezvous and Docking Mission Status Review," Aug. 24, 1971; North American Rockwell, Space Division, "International Rendezvous and Docking Mission First Status Review," AP71-19, 24 Aug. 1971; NASA, MSC, "E&D Weekly Activity Report," 21-27 Aug. 1971; and Lunney to Gilruth, memo, "Status Report," 30 Aug. 1971.

84. A standard method of including new data in a contracted study was the issuance of Document Change Requests (DCRs). Between 5 Aug. and 9 Nov., 60 DCRs were made to the document, NASA, MSC, "International Rendezvous and Docking Mission Study Guidelines and Constraints Document," MSC-04750, 5 Aug. 1971, which outlined the scope of the missions and concepts to be considered in the IRDM study. For example, see letter, D. A. Nebrig to R. C. Lashbrook, "Contract NAS 9-150, Guidelines and Constraints Approved for Inclusion in MSC-04750 (International Rendezvous and Docking Mission Study)," 9 Nov. 1971, which enclosed DCRs 48-60.

85. Berglund to distribution, memo, "International Rendezvous and Docking Mission Study Team Staff Meeting of August 30, 1971," 31 Aug. 1971; Berglund to distribution, memo, "International Rendezvous and Docking Mission Staff Meeting of September 13, 1971," 15 Sept. 1971; and North American Rockwell, Space Division, "International Rendezvous and Docking Mission, Second Status Review," AP71-21, 29 Sept. 1971.

86. North American Rockwell, Space Division, "International Rendezvous and Docking Mission Final Briefing," AP71-23, 16 Nov. 1971.

87. Ibid.; also sent with North American's briefing were the following: "IRDM Programmatic Considerations Summary (Briefing)"; "Layouts of DM, SLA Truss"; "IRDM Science Supplement (Briefing)," AP71-23-2, 16 Nov.

1971; "Total Stowage Plan/List (Briefing/List)";
"Devel/Cert Test/Anal Requirements (Write up)";
"MU/Trainer Requirements (Write Up)"; Manufacturing Engineering Producibility Study on
International Docking Module," MPA 35005,
10 Nov. 1971; and "International Docking Systems Development Plan," SID 71-684, 5 Nov.
1971. North American Rockwell subsequently
submitted a formal report entitled "International
Rendezvous and Docking Mission," SD71-700,
with Addendum, Dec. 1971. Also see Lashbrook
to A. H. Atkinson, "Transmittal of International
Rendezvous and Docking Mission (IRDM) Report," 14 Dec. 1971; and Berglund to distribution, memo, "North American Rockwell (NR)
International Rendezvous and Docking Mission
(IRDM) Final Report," 17 Dec. 1971.

VI

1. NASA, MSC, "A Docking Mechanism for
Apollo/Salyut-Type Spacecraft," 17 Nov. 1971;
and Robert R. Gilruth to Arnold W. Frutkin, 29
July 1971, asking transmittal of letter, Glynn S.
Lunney to Konstantin Davydovich Bushuyev, 3
Aug. 1971. Before this letter was sent, Frutkin
asked for Chuck Mathews' concurrence. This
formality was subsequently dropped as Lunney's
authority broadened and efforts were made to
speed communication. See J. Leroy Roberts to
Charles W. Mathews, note, 2 Aug. 1971, with
Mathews' concurrence dated 3 Aug.

2. Gilruth to Frutkin, 16 Aug. 1971, asking
transmittal of letter, Lunney to Bushuyev, undated, with the following enclosures: NASA,
MSC, Bidford F. Cockrell, "Coordinate Systems
Standards for International Rendezvous and
Docking of Spacecraft," MSC Internal Note No.
71-FM-312 (MSC-04746), 9 Aug. 1971; "Recommendations for a Communication Channel between the USA and USSR Mission Control
Centers to Support International Manned Spacecraft Rendezvous and Docking" [n.d.]; NASA,
MSC, "Technical Requirements for Compatible
USA and USSR Docking Systems," 6 Aug. 1971;
and NASA, MSC, "Several Concepts of Communications and Tracking Systems for US/USSR
Compatibility Study," 13 Aug. 1971. See also
Gilruth to Frutkin, 20 Sept. 1971, asking transmittal of letter, Lunney to Bushuyev [n.d.], with
separate agendas for all three Working Groups
attached. René Berglund had been concerned
about the need to differentiate between the near
and far term in June. See René A. Berglund to

Gilruth, memo, "Notes on the June 21-25 Soviet
Visit," 30 June 1971.

3. Bushuyev to Lunney, 8 Oct. 1971, with
two enclosures: "Predlozheniya po znacheniyam
parametrov sistem upravleniya, radionavedeniya i
svazi, obespechivayushchikh sblizheniye i stykovku kosmicheskikh korabley i stantsiy SSR i
SShA" [Proposed value of parameters for control, radio guidance, and communication system
ensuring the rendezvous and docking of USSR
and USA spacecraft and stations] and "Tekhnicheskiye trebovaniya k atmosfere obitayemykh
otsekov, sposobam perekhoda, agregatam i sistemam, neobkhodimyye dlya obespecheniya pere-
khods ekipazhey posle stykovki kosmicheskikh
korabley ili stantsiy SSR i SShA" [Specifications
for crew compartment atmosphere, transfer
methods, and units and systems needed to
provide for the transfer of crews after USSR and
USA spacecraft and space stations have docked].

4. Bushuyev to Lunney, 28 Oct. 1971; and
Lunney to Bushuyev [n.d.].

5. Lunney to distribution, memo, "Organization of Material for US/USSR Activities," 8 Sept.
1971.

6. Lunney to distribution, memo, "Review of
Material for US/USSR Meeting," 22 Sept. 1971;
NASA, MSC, "E&D Weekly Activity Report,"
2-8 Oct. 1971; NASA, MSC, "E&D Weekly
Activity Report," 16-22 Oct. 1971, p. 11; NASA,
MSC, "E&D Weekly Activity Report," 6-12 Nov.
1971, p. 11; and [MSC], "Review of Material for
Next Meeting with USSR on Compatibility of
Rendezvous and Docking," 10 Nov. 1971.

7. [MSC], "Review of Material for Next
Meeting with USSR," 10 Nov. 1971.

8. Ibid.; and interview, Caldwell C. Johnson-
Edward C. Ezell, 27 Mar. 1975.

9. [MSC], "Review of Material for Next
Meeting with USSR," 10 Nov. 1971.

10. Lunney to distribution, memo, "Trip
Report on Visit to Moscow on Compatible
Rendezvous and Docking" [n.d.].

11. The two American papers were NASA,
MSC, "U.S. Summary of Possible Apollo-Salyut
Test Mission," Nov. 1971; and NASA, MSC,
"U.S. Summary of Present Status of Technical
Requirements," Nov. 1971.

12. Lunney to distribution, memo, "Trip
Report" [n.d.].

13. Ibid.

14. MSC, "Review of Material for Next Meeting with USSR," 10 Nov. 1971.

15. Interview, Gilruth-Ezell, 25 Mar. 1975;

and "Summary of Results," 29 Nov.-6 Dec. 1971.

16. "Summary of Results," 29 Nov.-6 Dec. 1971.

17. "Minutes of Meetings, Working Group No. 1," 30 Nov.-6 Dec. 1971 [retyped 30 Dec. 1971].

18. "Minutes of the Working Group No. 2," 29 Nov.-7 Dec. 1971, with appendices.

19. Johnson to Gilruth, memo, "Miscellaneous Engineering Information Pertinent to CSM/Salyut Docking," 30 June 1971.

20. NASA, MSC, "A Docking Mechanism for Apollo/Salyut-Type Spacecraft," 17 Nov. 1971; interview, William K. Creasy-Ezell, 7 July 1975; and Johnson to Lunney, memo, "Documents and Visual Aids for Moscow Meeting," 16 Nov. 1971.

21. Johnson to Lunney, memo, "IRDM Docking Mechanism, Concept Verification Study, January through May 1972," 16 Dec. 1971; [V. S. Syromyatnikov], "Printsipal'naya konstruktivniaya skhema stukovochnogo ustroistva periferiinogo androginnogo tipa" [Design concept of a docking mechanism of the peripherial and androgynous type], 2 Dec. 1971; and R. W. Kubicki to W. W. Petynia and Johnson, memo, "International Docking System Design Review," 26 Oct. 1971, which presents an in-house systems engineering view of the hydraulic attenuators.

22. Interview, Creasy-Ezell, 7 July 1975.

23. Johnson to Lunney, memo, "IRDM Docking Mechanism, Concept Verification Study, January through May 1972," 16 Dec. 1971.

24. "Appendix C, Working Group No. 3, Minutes of Meetings on Assuring the Compatibility of the Docking Systems and Tunnel," 29 Nov.-6 Dec. 1971.

25. Johnson to Lunney, memo, "IRDM Docking Mechanism, Concept Verification Study," 16 Dec. 1971.

26. Ibid.

27. Lunney to distribution, memo, "Schedule of Work as a Result of the Third Meeting on International Compatibility of Rendezvous and Docking," 16 Dec. 1971. Lunney had taken drafts of the Summary of Results and Working Group minutes with him to Moscow in November. See also Dale D. Myers to Lunney and Frutkin, memo, "Sample Summary of Results," 22 Nov. 1971.

28. Myers to George M. Low, memo, "Need for FY '73 Funding for Post Skylab CSM Mission," 22 Oct. 1971; letter, Myers to Gilruth, 16 Sept. 1971; Myers to Rocco A. Petrone,

memo, "Excess Apollo Flight Hardware," 29 Oct. 1971; Gilruth to Myers, 25 Mar. 1971, with NASA, MSC, "Post Skylab Missions Summary Report," 17 Mar. 1971 enclosed; and Gilruth to Myers, 25 Aug. 1971. This latter letter from Gilruth had argued for completion of 115 and 115A for the IRDM mission and provided specific cost figures. In the final months of 1971, several briefings were held at MSC on the subject of how best to use the remaining CSMs. See [MSC], "CSM Utilization Briefing," 28 Oct. 1971; NASA, MSC, "Utilization of Apollo Hardware Between Skylab Period and Shuttle Availability," 15 Nov. 1971 and, as revised, 7 Dec. 1971.

29. William C. Schneider to Myers, memo, "Docking Module," 24 Sept. 1971; and Myers to Gilruth, 22 Nov. 1971.

30. Schneider to Myers, memo, "Docking Module," 24 Sept. 1971.

31. Myers to Schneider, note, 28 Sept. 1971.

32. Gilruth to Myers, 13 Dec. 1971.

33. "Summary of Results," 29 Nov.-6 Dec. 1971; and Myers to Lunney and Frutkin, memo, "Sample Summary of Results," 22 Nov. 1971.

34. Myers to Gilruth, 14 Dec. 1971, with enclosure, NASA, MSC, "International Rendezvous and Docking Mission Program Plan," 21 Sept. 1971.

35. A sample of the documentation includes [MSC], "International Rendezvous and Docking Mission Contracting Situation," 23 Dec. 1971; [MSC], "Cost Estimate for USA/USSR Docking Mission," 7 Jan. 1972; [MSC], "Cost Assumptions," 27 Jan. 1972; [MSC], "Cost Assumptions," 28 Jan. 1972; [MSC], "Cost Assumptions," 1 Feb. 1972; [MSC], "CSM/AMDS Status Briefing," 1 Feb. 1972; [MSC], "Program Options," 17 Feb. 1972; NASA, MSC, "CSM/AMDS Planning Briefing," 8 Mar. 1972; NASA, MSC, "Residual Apollo Hardware Status," 21 Mar. 1972; NASA, MSC, "Residual Apollo Hardware Status," 27 Mar. 1972; and NASA, MSC, "NR Sustaining: Currently Negotiated Manpower NR Recommended Sustaining," 14 Mar. 1972.

36. Myers, memo for record, "Compatible Rendezvous and Docking Study and Potential Flight Test," 29 Mar. 1972.

37. NASA News Release, MSC, 72-15, 14 Jan. 1972; and Carol H. Sweeny to distribution, memo, "Agreements and Action Items from January 11-12, 1972 Meeting [OMSF]," 31 Jan. 1972.

38. Christopher C. Kraft to Frutkin, 19 Jan.

1972, asking transmittal of letter, Lunney to Bushuyev [27 Jan. 1972], with enclosure, "Docking Mechanism Subjects for Discussion." See also Johnson to Lunney, memo, "Preparations for Possible Telecons and Meetings with Soviets," 11 Feb. 1972.

39. TWX, Jacob D. Beam to Lunney, "NASA/Interkosmos Conference Call," 1 Mar. 1972; and Johnson to Lunney, memo, "Initial Telecon with Soviets to Discuss Technical Aspects of Contemplated CSM/Salyut Mission," 3 Mar. 1972.

40. Johnson to distribution, memo, "Transcript of Telecon between MSC, Houston, and Intercosmos, Moscow, 2 March 1972," 22 Mar. 1972, enclosing a copy of the transcript; [Soviet Academy of Sciences], "Summary of the March 2, 1972, Telephone Conversation between the NASA and the Academy of Sciences" [n.d.] ; and Roberts, notes, "Conference Call to Moscow" [n.d.].

41. Kraft to Frutkin, 22 Mar. 1972, asking transmittal of letter, Lunney to Bushuyev, 15 Mar. 1972.

42. Interview, Creasy-Ezell, 7 July 1975; and "Minutes of Meeting on Assuring Compatibility of Docking Systems and Tunnels," 3 Apr. 1972.

43. Interview, Low-Ezell, 30 Apr. 1975; [Low], "Visit to Moscow, April 1972, to Discuss Compatible Docking Systems for US and USSR Manned Spacecraft," 4-6 Apr. 1972; Lunney to Bushuyev [n.d.], with enclosure, "Apollo/Salyut Test Mission Considerations," 23 Mar. 1972; and Kraft to Frutkin, 4 Apr. 1972, asking transmittal of letter, Lunney to Bushuyev [n.d.], enclosing, NASA, MSC, "Project Technical Proposal for an Apollo/Salyut Test Mission" [n.d.] and NASA, MSC, "Proposed Project Schedule Document," ASTM 30 000, 3 Mar. 1972.

44. Interview, Low-Ezell, 30 Apr. 1975; [Low], "Visit to Moscow," 4-6 Apr. 1972.

45. John Noble Wilford, "U.S.-Soviet Accord in Sight on a Joint Space Mission," *New York Times,* 2 Apr. 1972.

46. Interview, Low-Ezell, 30 Apr. 1975; and [Low], "Visit to Moscow," 4-6 Apr. 1972. Kaiser had written articles on the docking talks in the past, a fact that brought little comfort to Low. Robert G. Kaiser, "U.S., Soviet Space Link-up Seen Near," *Washington Post,* 4 Dec. 1971.

47. Interview, Lunney-Ezell, 23 July 1974.

48. Ibid.; [Low], "Visit to Moscow," 4-6 Apr. 1972; and TWX, Secretary of State to Science Attaché, American Embassy, Moscow, "US/USSR Rendezvous and Docking Summary of Results," 25 Feb. 1972.

49. "Apollo/Salyut Test Mission Considerations," 23 Mar. 1972.

50. Interview, Low-Ezell, 30 Apr. 1975; and [Low], "Visit to Moscow," 4-6 Apr. 1972.

51. "Summary of Results," 4-6 Apr. 1972.

52. Interview, Low-Ezell, 30 Apr. 1975; and [Low], "Visit to Moscow," 4-6 Apr. 1972.

53. TWX, Lunney to Bushuyev, 10 Apr. 1972.

54. Lunney to Kraft, memo, "Minutes of the Apollo/Salyut Test Mission Telecon Held April 14, 1972," 26 Apr. 1972; and Donald C. Cheatham to Lunney, memo, "Telephone Conference with USSR Working Group No. 2 on April 14, 1972," 19 Apr. 1972.

55. "Working Group No. 2, May 11-17–Moscow 1972, Minutes of Meeting on Assuring Compatibility of Rendezvous and Docking Systems of USA/USSR Spacecraft," 17 May 1972; and interview, R. H. Dietz-Ezell, 28 June 1974.

56. Jack T. McClanahan to Robert N. Lindley, memo, "Voice Communication Frequency Assignments–US/USSR Cooperative Mission," 15 Feb. 1972; C. C. Kraft to D. D. Myers, 22 Mar. 1972; D. D. Myers to C. C. Kraft, memo, "Frequency Assignment and Rendezvous and Tracking System," 3 Apr. 1972; "USA/USSR Voice Communications System Frequency Selection Briefing Presented to M/Mr. Myers at Headquarters, April 5, 1972 by EG/D. C. Cheatham," 5 Apr. 1972; Leroy Roberts, "Minutes of Meeting–Voice Communication Frequency Assignment Meeting," 5 Apr. 1972; D. C. Cheatham to Robert A. Gardiner and G. S. Lunney, memo, "USA/USSR Voice Communications System Frequency Selection Meeting with Mr. Dale Myers, M/Associate Administrator Manned Space Flight," 6 Apr. 1972; and D. D. Myers to distribution, memo, "Frequency Assignment," 8 May 1972.

57. "Working Group No. 2 Minutes of Meeting," 17 May 1972, with "Appendix: The List of Documents Exchanged by the Sides at the Working Group No. 2 Meeting of May 11-17 1972"; and interview, Reinhold H. Dietz-Ezell, 30 July 1975.

58. Interview, Cheatham-Ezell, 24 July 1975.

59. "U.S.-Soviet Space Feat Likely by '75," *Baltimore Sun,* 5 Apr. 1972; Thomas O'Toole, "U.S.-Soviet Joint Efforts in Space Seen," *Washington Post,* 6 Apr. 1972; Jim Maloney, "Joint U.S., Soviet Space Trip Likely," *Houston Post,* 5

383

Apr. 1972; Al Marsh, "Deke Learning Cosmonaut Talk," *Today,* 17 Apr. 1972; "Deke Slayton Studies Russian and Dreams of Space,"*New York Times,* 27 Apr. 1972; "Team Up with the Soviets? 'The Chances Are Quite Good,'" *U.S. News and World Report,* 8 May 1972; Thomas O'Toole, "Summit in Space: June 15, 1975," *Washington Post,* 7 May 1972; Nicholas C. Chriss, "Joint Mission; NASA, Soviet Togetherness: It's Far Out," *Los Angeles Times,* 5 May 1972; and Jonathan Spivak, "Ivan and John? The U.S. and Russia Seem Ready To Join Hands in Outer Space; Soviets Need the Technology, NASA Needs the Money: Going to Mars Together? Hooking Up Apollo to Salyut," *Wall Street Journal,* 16 May 1972.

60. Interview, Low-Ezell, 30 Apr. 1975; and [Low], "Visit to Moscow," 4-6 Apr. 1972.

61. Ibid.

62. A press package released by NASA on 24 May 1972 included Richard T. Mittauer, "Note to Editors" [n.d.]; NASA News Release, HQ [unnumbered], "Text of US/USSR Space Agreement," 24 May 1972; "Statement by Dr. Fletcher," 24 May 1972; NASA News Release, HQ, 72-109, "US/USSR Rendezvous and Docking Agreement," 24 May 1972; NASA News Release, HQ [unnumbered], "Background on Rendezvous Results," 4-6 Apr. 1972; and "Summary of Results of the Low-Keldysh Agreements," 18-21 Jan. 1971. See also NASA press conference, HQ, "News Conference on US/USSR Rendezvous and Docking Agreement," 24 May 1972; and "White House Press Conference of the Vice President; Dr. James C. Fletcher, Administrator of NASA; Glynn S. Lunney, Assistant to the Manager for Operational, Experiment and Government Furnished Equipment, NASA; and Dr. Edward E. David, Jr., Science Adviser to the President," 24 May 1972.

63. Congressional Quarterly, Inc. (comp.), *Historic Documents 1972* (Washington, 1973), p. 440.

64. "Salt and the Moscow Summit, May 22-30, 1972," in Congressional Quarterly, Inc., comp., *Historic Documents, 1972* (Washington, 1973), pp. 431-463.

VII

1. Dale D. Myers to John P. Donnelly, memo, "Designation for Joint US/USSR Mission," 5 June 1972; and Donnelly to Myers, memo, "Project Designation," 30 June 1972.

2. Myers to Kurt H. Debus, Eberhard F. M. Rees, and Christopher C. Kraft, memo, "Apollo Soyuz Test Project," 13 June 1972; Philip E. Culbertson to Rocco A. Petrone, memo, "Transfer of Apollo Soyuz Project Responsibility," 9 June 1972; and MSC Announcement 72-31, "Key Personnel Assignment," 2 Mar. 1972.

3. Interview, Leonard S. Nicholson-Edward C. Ezell, 16 July 1974. Nicholson had discussed the issue of two-versus-one managers with Berglund prior to the latter's discussion with Kraft.

4. NASA News Release, MSC, 68-28, 3 Apr. 1968; Glynn S. Lunney, "Discussion of Several Problem Areas during the Apollo 12 Operation," paper presented to AIAA 7th Annual Meeting and Technical Display, Houston, Tex., 19-22 Oct. 1970 (A70-1260); and U.S. Congress, Senate, Committee on Aeronautical and Space Sciences, *Apollo 13 Mission; Hearings,* 24 Apr. 1970, 91st Cong., 2d sess. (Washington, 1970).

5. [NASA, MSC], "Proposed Operating Plan for US/USSR Meeting on the Apollo/Soyuz Test Project," July 1972.

6. Caldwell C. Johnson, "Working Group Meetings Nominal Procedures," 26 June 1972; Johnson, "Principal Events, July 5 thru July __," 26 June 1972; Johnson, "Working Group Procedures," 26 June 1972; Johnson, "Status Review," 28 June 1972; "Proposed Agenda for Joint Meeting of U.S./U.S.S.R. Working Groups for Compatible Means of Rendezvous and Docking" [n.d.]; "Outline of Discussion with Dr. Kraft (6/14/72) on Preparations for July Meeting with the USSR" [n.d.]; Glynn S. Lunney to Eziaslav Harrin, Tamara Holmes, Dmitri Arensburger, and Dmitry Zarechnak, memo, "Interpreters Assignments and Instruction," 5 July 1972; John W. King to Lunney, memo, "Apollo-Soyuz Test Project Meeting," 20 June 1972; and Lunney to distribution, memo, "Briefing on Apollo/Soyuz Test," 6 June 1972.

7. Lunney to distribution, memo, "Briefing on Apollo/Soyuz Test," 6 June 1972, with attached list of invitees.

8. Ibid.

9. Lunney and Bushuyev had their usual correspondence exchange before the July meeting; Konstantin Davydovich Bushuyev to Lunney, 7 June 1972; TWX, Lunney to Bushuyev, 15 June 1972; and TWX, Bushuyev to Lunney, 22 June 1972.

10. NASA decided to rely upon North American Rockwell to modify the CSM and build the related equipment needed for ASTP. The initial

IRDM study had been conducted under a contract change authorization to the original contract, issued 21 Dec. 1961. The work contemplated for the ASTP mission required a new contract, and drafting of a new Statement of Work (SOW) had started in early 1972. A preliminary version of that document, "Statement of Work for CSM/Advanced Mission Docking System," 28 Mar. 1972, was distributed throughout OMSF. See Culbertson to distribution, memo, "Preliminary Statement of Work for CSM/Advanced Missions Docking System," 12 Apr. 1972. ASPO had established an evaluation team to work with the proposed contractor in evaluating the contract proposal; see James A. McDivitt to distribution, memo, "Designation of Evaluation Team for ASTM (Apollo/Salyut Test Mission) Contract Proposal," 11 Apr. 1972. Meanwhile, Terrence Heil had prepared a procurement plan, "Development of Command and Service Module/Advanced Missions Docking Systems (CSM/AMDS)," on 2 Mar., which among other things contained a justification for a noncompetitive procurement. Culbertson described the reasons for selecting NAR in interview, Culbertson-Ezell, 5 May 1975. This procurement plan was approved by Kraft on 29 Mar. and forwarded to Headquarters. Following the May summit, the SOW was changed where necessary to reflect the shift from Salyut to Soyuz, and a letter contract was issued on 30 June. See letter contract, Heil to North American Rockwell Corp., Space Division, Contract NAS9-13100, 30 June 1972. This contract was accepted by NAR on 6 July and scheduled to run for 90 days; a definitive contract was to be negotiated by 29 Sept. Because there was a short delay and that definitive contract was not issued until 6 Oct., Kraft sought a 30-day extension on 22 Sept. See TWX, Kraft to Dale D. Myers, 22 Sept. 1972; and TWX, Myers to Kraft, 28 Sept. 1972. The definitive contract, issued on 6 Oct., was also numbered NAS9-13100.

11. [M. P. "Pete" Frank], "Working Group #1, Notes on Joint Meeting, July, 1972" [n.d.].

12. Reinhold H. Dietz to McDivitt, memo, "ASTP Working Group No. 4 Debriefing Notes, Joint Meetings, July 6-17, 1972," 28 July 1972, enclosing "Joint U.S.A./U.S.S.R. Meeting, July 6-17, 1972, Debriefing Notes, Working Group No. 4" [n.d.].

13. Interview, Lunney-Ezell, 23 July 1974.

14. Interview, Clarke Covington-Ezell, 3 Apr. 1975.

15. Ibid.; and NASA press conference, MSC, "US-USSR Apollo-Soyuz Test Project, Press Conference," 17 July 1972.

16. "Summary of Results," in "Apollo/Soyuz Test Project, Minutes, Fourth Joint Meeting, USSR Academy of Sciences and US National Aeronautics and Space Administration," 6-18 July 1972.

17. NASA Press Conference, MSC, "US-USSR Apollo-Soyuz Test Project, Press Conference," 17 July 1972.

18. "Minutes of Working Groups 1 and 5," 6-17 July 1972, in "Minutes of Joint Meeting."

19. Interview, Walter W. Guy and James R. Jaax-Ezell, 7 July 1974.

20. The document in question was "Description of the System Providing Gas Composition and Temperature Control of the Habitable Modules of a Soyuz-Type Spacecraft," 1972.

21. Ibid.; and memo, Robert E. Smylie to Lunney, "Summary of Working Group 5 Meetings in July," 31 July 1972, enclosing "Summary of US/USSR Meeting on ASTP," 7-16 July 1972.

22. Interview, Guy and Jaax-Ezell, 7 July 1974; interview, Guy-Ezell, 19 Aug. 1975; "Minutes of the Apollo Soyuz Test Project PDR No. 1 Board Meeting, July 13, 1972" [14 July 1972]; and [North American Rockwell Corp., Space Division], "Apollo/Soyuz Test Project Requirements & Implementation Concepts Review, PDR No. 1, MSC," AP72-14, 13 July 1972.

23. "Working Groups No. 2 and 4, Minutes on Apollo Soyuz Test Project," 6-17 July 1972, in "Minutes of Joint Meeting." Dietz summarized for internal distribution the technology transfer arguments in a document entitled "Technology Investigation—Apollo VHF Communications and Ranging Equipment" [n.d.], which appended NASA and RCA correspondence on the hardware in question. Sam Holt of RCA wrote Dietz on 18 Nov. 1971. "I have reviewed the information available to me in regard to availability of information on the key technology involved. . . . I feel that the information on the techniques involved . . . has been so widely circulated in the [U.S.] engineering community . . . that it now [is] public information."

24. Herbert E. Smith, Jr., to Lunney, memo, "Debriefing Activities and Future Plans, Status of Working Group No. 2," 27 July 1972.

25. Interview, Lunney-Ezell, 23 July 1974.

26. Owen G. Morris to distribution, memo, "ASTP #2 Board Meeting Minutes," 2 Aug. 1972; North American Rockwell Corp., Space

Division, "Apollo/Soyuz Test Project Requirements & Implementation Concepts Review, PDR No. 2, Downey," 27 July 1972; and Morris to distribution, memo, "ASTP Preliminary Design Review No. 3, Minutes," 30 Aug. 1972.

27. Lunney to distribution, memo, "Apollo Soyuz Test Project Staff Meeting," 26 July 1972. This memo is typical of those distributed prior to staff meetings; it contains an agenda and a distribution list.

28. Morris to distribution, memo, "ASTP Management Review," 20 Sept. 1972; and James C. Fletcher to Senator Clinton P. Anderson, 23 July 1972. The same letter was sent to Senators Curtis, Pastore, and Allott and Representatives Miller, Mogher, Boland, Jonas, Teague, Winn, Symington, and Frey. Petrone had been assigned responsibility as Program Director for ASTP in August. See NASA News Release, HQ, 72-174, "Petrone to Head ASTP," 21 Aug. 1972.

29. TWX, Lunney to Bushuyev, 5 Sept. 1972; Bushuyev to Lunney, 11 Sept. 1972; Lunney to Bushuyev, 12 Sept. 1972; TWX, Bushuyev to Lunney, 14 Sept. 1972; TWX, Lunney to Bushuyev, 15 Sept. 1972; and TWX, Lunney to Bushuyev, 19 Sept. 1972. George Low and Academician M. V. Keldysh had approved the results of the July meeting in an exchange of letters: Mstislav Vsevolodovich Keldysh to George M. Low, 9 Aug. 1972; and Low to Keldysh, 7 Sept. 1972.

30. "Minutes of the Joint Meeting of Working Group 1," 9-20 Oct. 1972, in "Apollo/Soyuz Test Project, Minutes, Fifth Joint Meeting, USSR Academy of Sciences and US National Aeronautics and Space Administration," 9-19 Oct. 1972.

31. Interview, Covington-Ezell, 3 Apr. 1975.

32. "Minutes of the Joint Meeting of Working Group 1," 9-20 Oct. 1972, in "Minutes of Joint Meeting."

33. Lunney, "Minutes, ASTP Staff Meeting, October 25, 1972," 30 Oct. 1972.

34. "Minutes of ASTP Working Group 2 Meeting," 9-20 Oct. 1972, in "Minutes of Joint ASTP Meeting," 16 Oct. 1972; and interview (via telephone), Smith-Ezell, 25 Aug. 1975.

35. Lunney, "Minutes, ASTP Staff Meeting, October 25, 1972," 30 Oct. 1972. Donald C. Wade expressed some unhappiness with the Soviet model in Wade to Lunney, memo, "Working Group #3 Debriefing Notes for the October 1972 Meeting in Moscow," 1 Nov. 1972.

36. "Apollo Soyuz Test Project, October 12 and 13, 1972, Preliminary Systems Review (Stage 1)" [13 Oct. 1972].

37. Ibid.

38. Wade to Lunney, memo, "Working Group #3 Debriefing Notes for the October 1972 Meeting in Moscow," 1 Nov. 1972.

39. "Minutes of Working Group No. 5," 9-20 Oct. 1972, in "Minutes of Joint Meeting"; and Lunney, "Minutes, ASTP Staff Meeting, October 25, 1972," 30 Oct. 1972.

40. Lunney, "Minutes, ASTP Staff Meeting, October 25, 1972," 30 Oct. 1972.

41. Ibid.; and "Minutes of Working Group No. 5," 9-20 Oct. 1972, in "Minutes of Joint Meeting."

42. Lunney, "Minutes, ASTP Staff Meeting, October 25, 1972," 30 Oct. 1972.

43. NASA, MSC, "Minutes of Meeting, ASTP CSM 111 Critical Design Review; DM-DS Preliminary Design Review," 8-10 Nov. 1972; and Morris to distribution, memo, "Apollo Soyuz Test Project (ASTP) Design Review," 2 Nov. 1972.

44. NASA, OMSF, "Manned Space Flight Management Council," 8 Nov. 1972.

45. Interview, Smith-Ezell, 2 Sept. 1975; and interview, A. Don Travis-Ezell, 2 Sept. 1975.

46. Smith to Lunney, memo, "Report of Working Group No. 2 Joint Meetings," 15 Dec. 1972, enclosing "Conduct and Results of ASTP Working Group No. 2 Meetings between NASA and the Soviet Academy of Sciences" [n.d.].

47. Interview, Smith-Ezell, 12 Feb. 1975.

48. Ibid.; and "Working Groups No. 2 and No. 4, Minutes of Meeting on Apollo/Soyuz Test Project," 24 Nov.-2 Dec. 1972.

49. "Working Groups No. 2 and No. 4, Minutes of Meeting on Apollo/Soyuz Test Project," 24 Nov.-2 Dec. 1972.

50. Letter, Lunney to Bushuyev [drafted 6 Dec. 1972].

51. Ibid.; and "Summary of ASTP Working Groups 2 & 4 Meeting, Houston, Texas, November 24-December 2, 1972" [n.d.].

52. Wade to Lunney, memo, "Working Group #3 Debriefing Notes for the December 7-15, 1972, Meeting in Moscow" [drafted 20 Dec. 1972]. Dr. W. R. Hawkins, who accompanied the Working Groups in Oct. 1972, said that the parasite encountered was diagnosed as *Giardia lamblia,* which produces intestinal distress and severe diarrhea. The U.S. Public Health Service concluded that the NASA team probably contracted the parasite in the drinking water during their visit to Leningrad; interview (via telephone), W. R. Hawkins-Ezell, 3 Sept. 1975.

53. Wade to Lunney, memo, "Working

Group #3 Debriefing Notes for the December 7-15, 1972, Meeting in Moscow" [drafted 20 Dec. 1972]; and "Apollo/Soyuz Test Project: Minutes of Meeting on Assuring Compatibility of Docking Systems, Working Group 3," 6-16 Dec. 1972.

54. Lunney to Kraft, memo, "Summary Report on Results of ASTP Working Group 3 Meeting in Moscow," 21 Dec. 1972, enclosing "Summary of ASTP Working Group 3 Meeting," 6-16 Dec. 1972.

55. Dodd L. Harvey and Linda C. Ciccoritti, *U.S.-Soviet Cooperation in Space* (Coral Gables, Fla., 1974), p. 246.

56. NASA Special Announcement, "Appointment of Program Director for the Apollo Soyuz Test Project Office of Manned Space Flight," 16 Jan. 1973; Keith Wible to James R. Elliott, memo, "Redesignation of Manned Space Flight Organizations," 15 Mar. 1973; and interview, Lunney-Ezell, 23 July 1974.

57. U.S. Congress, House of Representatives, Committee on Science and Astronautics, Subcommittee on Manned Space Flight, transcript, "Executive Session Briefing on Apollo/Soyuz Test Project," 2 Oct. 1973, p. 7; and Myers to Ezell, 3 Sept. 1975. NASA had been keeping the Manned Space Flight Committee abreast of ASTP developments to reduce their mistrust of the Soviets. See H. Dale Grubb to Fletcher, memo, "ASTP Informal Meeting," 8 Mar. 1973; Chester M. Lee to Bernard L. Johnson, memo, "Draft Responses to Congressman Fuqua," 7 May 1973; Willis H. Shapley to Myers, memo, 31 July 1973; and Myers to Olin E. Teague, 16 Aug. 1973.

58. Subcommittee on Manned Space Flight, transcript, "Executive Session Briefing on Apollo/Soyuz Test Project," 2 Oct. 1973, p. 7.

59. Ibid., p. 12.

60. David R. Scott to Lunney, memo, "ASTP Mission to Moscow, June-July 1973," 31 July 1973.

61. Lee to Fletcher and Low, memo, "US/ USSR July Working Group Meeting," 25 July 1973. Lunney had discussed the documentation issue with Bushuyev in several letters; e.g., Lunney to Lee, memo, "Transmittal of Letter to Moscow," 7 June 1973, asking transmittal of letter, Lunney to Bushuyev, 19 June 1973; and memo, Lunney to Lee, "Transmittal of Letter to Moscow," 24 Aug. 1973, asking transmittal of letter, Lunney to Bushuyev [n.d.].

62. Interview, Scott-Ezell, 21 Aug. 1974.

63. Lee to Fletcher and Low, memo, "US/

USSR July Working Group Meeting," 25 July 1973.

64. Ibid.

65. Ibid.

66. Subcommittee on Manned Space Flight, transcript, "Executive Session Briefing on Apollo/Soyuz Test Project," 2 Oct. 1973, pp. 17-18.

67. Ibid., pp. 33-34.

68. Ibid., pp. 37-38.

VIII

1. Glynn S. Lunney to Konstantin Davydovich Bushuyev [signed 24 Aug. 1973, but an earlier draft appears in the files dated 7 Aug. 1973 apparently prepared before Low's letter to Keldysh of 14 Aug. 1973].

2. Lunney to Bushuyev [24 Aug. 1973].

3. Ibid.

4. George M. Low to Mstislav Vsevolodovich Keldysh, 14 Aug. 1973. The two teams had discussed *Salyut 2* and *Soyuz 11* in some detail at the joint meetings in July 1973 in Houston, but Lee and others thought that additional information was necessary. See Chester M. Lee to Dale D. Myers, memo, "US/USSR July Working Group Meeting," 25 July 1973.

5. Keldysh to Low, 30 Aug. 1973.

6. TWX, Jack L. Tech to Arnold W. Frutkin, "NASA Delegation Visit to USSR," 4 Oct. 1973; TWX, Oscar E. Anderson, Jr., to Frutkin and Low, "NASA Delegation Visit to USSR," 5 Oct. 1973; TWX, Low to Keldysh, 10 Oct. 1973; and Michael E. DeBakey to Edward C. Ezell, 25 Feb. 1976.

7. Low, "Visit to Moscow, October 14-19, 1973," Dec. 1973; and interview (via telephone), Reinhold H. Dietz-Ezell, 18 Feb. 1976.

8. Low, "Visit to Moscow, October 14-19, 1973," Dec. 1973; and ASTP notebook, kept by Leonard S. Nicholson, for 1973.

9. Low, "Visit to Moscow, October 14-19, 1973," Dec. 1973; interview, Walter W. Guy-Ezell, 12 Sept. 1975; Thomas O'Toole, "Valve Mishap Blamed for Soyuz Deaths," *Washington Post,* 29 Oct. 1973; and John F. Yardley to Low, memo, *"Soyuz 11* Failure ," 3 Mar. 1975.

10. Donald C. Winston, "Soviet Space Center Being Expanded," *Aviation Week & Space Technology,* 25 June 1973, p. 18; and "Soviet Space: A Visit to Star City," *Time,* 9 July 1973, pp. 44 and 47.

11. Interview, Guy-Ezell, 12 Sept. 1975.

12. NASA, JSC, "Apollo Soyuz Test Project Presentation to Manned Space Flight Subcom-

mittee Staff," 15 Nov. 1973; "2-Day Soviet Flight," *Facts on File* 33 (30 Sept.-6 Oct. 1973): 814-815; Theodore Shabad, "Soviet Puts Soyuz 12, with 2 Aboard into Earth Orbit," *New York Times,* 28 Sept. 1973; and Murray Seeger, "Soviet Union Launches 2-Man Space Mission," *Los Angeles Times,* 28 Sept. 1973. To correct the problem encountered on *Soyuz 11,* the Soviets redesigned the valve and seal, improved the manual valve closing so that it took fewer turns of the handle to close it, reduced the power of the pyrotechnic bolts and replaced half of them with pyrotechnic/gas actuated latches, and provided for the crewmen to reenter in pressure suits.

13. Low, "Visit to Moscow, October 14-19, 1973," Dec. 1973.

14. Ibid.

15. Ibid.

16. Ibid.; and Low to Olin E. Teague, 31 Oct. 1973. M. Pete Frank also had a favorable evaluation of the Soviet facilities at Star City and Kaliningrad. In a memo, Frank to distribution, "October 1973 Working Group 1 Meeting," 31 Oct. 1973, he wrote the following about the control center:

> Tour of the Soviet Mission Control Center:
> The control center that will be used for the ASTP mission is located in a tightly secured complex northeast of Moscow. It is in an area called Kaliningrad just off the Yaroslavl highway. It took us about 40 minutes to reach the control center from the Rossiya Hotel. . . .
> The building in which the control center is contained was only one of many buildings inside this complex . . . surrounded by a high brick wall and . . . heavily guarded at the entrances. The buildings looked fairly new and were modern with very large glass windows. . . . The control center is nestled in among other buildings that have something to do with their space program.
> As far as general comments regarding the tour, it was a very detailed and comprehensive tour. I think the Soviets went out of their way to make the point that they were showing us everything. They were not holding anything back. They even showed us work areas (just office space) in the control center; and as we would go down the hall, they offered to open any of the doors. . . .
> The tour began with an initial briefing of the flight control and the operations aspects provided by Yeliseyev. The overall tour was conducted by Dr. Albert Melytsin who was called the Technical Director of the Mission Control Center. I have the feeling that he was responsible for the construction and operation of the facility. Yeliseyev's briefing covered two areas: the first was the organization of their flight control operation; the second was the flow of information from the tracking stations into the control center. . . .
> I was impressed by the quality of the equipment in the control center; I thought it was similar to ours. I was also impressed by the similarity of the flow diagrams of information from the tracking station to the control center. It seemed to include all the elements of our own system.
> The mission operations control room was quite large—it contained 16 two-man consoles, thus allowing them 32 flight controllers. I don't know how many they actually use for a mission, but there are capabilities for 32. In addition, there was a back row of consoles which were used for personnel such as the project technical director; this back row is also a work station for display controllers who control the main display boards in the front of the room. These large displays had a map with an orbit plotted on it with a computer driven indicator for the spacecraft position as it flew across the earth. There were digital indications of time and AOS and LOS times across the top of the screen; there were two large television screens about 10' × 10' on the right-hand side which could be configured for special displays; the consoles were equipped with a television display and communication panels. I think their communication capabilities were somewhat less than what we have, but there was some flexibility in that the flight director could call up people on individual basis or everybody at once to talk or listen. The television display system is capable of 100 different formats. These are changed from one mission to the next and can be selected by the local console operators simply by dialing up the proper number. The television system can display closed circuit views such as we have from the staff support rooms. It can also display digital data in real time from the computer system, and it can be used to display general information that is typed into a central display unit. It looked to be a very flexible system although I do not think it had anywhere near as much capability as our digital television system. In the back of the MOCR, the Russians have a balcony with several dozen seats which serve as a VIP viewing room; however, it is not glassed off (isolated) from the MOCR and it sits up at an elevated level. (It is interesting to note that although the equipment *appeared* to be very high quality, I had the definite feeling it did not have the performance capability we have.)
> Next the Soviets took us on a conducted tour around the building showing us the various staff support rooms in which the flight controller support teams function. They had similar television display capabilities and were able to communicate with their team leaders in the MOCR in a manner similar to the way we do.

The Russians showed us the teletype stations where messages are processed to send out to the remote sites; they showed us the telemetry ground stations and a room where a large number of chart recorders were used to "monitor data quality" as it comes in from the remote sites.

They showed us the computer facilities, which were very interesting. They had three main frames; each contained 16 memory drums. Each of these memory drums had a capacity of 32 thousand 48-bit words. . . .

Commanding to the Soyuz is not done from the control room; it is only accomplished at the remote sites. Of course, the commands to be sent are relayed to the site. (The remote sites are told what commands are to be sent but they cannot be sent directly from the control center.)

The large world map in the front of the control room showed the Soviet zones of coverage rather than tracking stations. This zone of coverage was from 25° east longitude to 150° east longitude and from approximately 38° north latitude to 53° north latitude. It was a rectangle on that Mercator projection map.

Another interesting comment was that all the voice tapes are saved until the mission is over, but once the mission is completed, these tapes are erased and used over again. Apparently, a permanent record of all the voice recordings is not made. They do record all the interior loops, loops between the control center and remote sites, as well as the air-to-ground; but these recordings are destroyed after the mission is completed.

I think that the control center has very recently been put into operation. The Soyuz 12 was the first manned mission that was flown from this control center; however, they did say that it had been used for unmanned missions prior to that. I would not be at all surprised that these were limited to Soyuz testing that had occurred just prior to the Soyuz 12 flight. I also had the impression that the control center was started approximately 3 years ago, although it may have been stated that it was completed approximately 3 years ago. It did not look that old to me.

The control center takes over control of the mission after the spacecraft is inserted into orbit, and the specific event that signals this is the separation of the spacecraft from the booster, that is T zero for the control center. Control during the launch phase up to separation of the spacecraft from the booster is maintained by a launch control facility which I assumed is located at the launch site. They stated that there is an automatic abort capability in the Soyuz as well as a manual abort capability and that this automatic capability is effective up until orbit insertion.

17. Letter, Low to Ezell, 15 Apr. 1976.

18. Ibid.

19. Robert J. Shafer to John P. Donnelly, memo, "IRDM," 19 May 1972, which responds to handwritten attachments to routing slip, Donnelly to Shafer, 2 May 1972.

20. Shafer to Donnelly, memo, "IRDM," 19 May 1972.

21. Interview, Donnelly and Shafer-Ezell, 26 and 28 Jan. 1976; interview, John E. Riley-Ezell, 10 Mar. 1976; and interview, John W. King-Ezell, 15 Mar. 1976.

22. [Riley, draft of ASTP Public Information Plan Part I], 10 Jan. 1973.

23. [Draft of ASTP Public Information Plan Part I], 15 Jan. 1973; and "Draft" [ASTP Public Information Plan], 1 Feb. 1973.

24. Routing slip, Richard Friedman to William J. O'Donnell, 13 Feb. 1973; and "R. Friedman: 2/13/73 Revisions" [ASTP Public Information Plan], 13 Feb. 1973.

25. Lee to Donnelly and Frutkin, memo, "Draft ASTP Information Plan Dated 13 February 1973," 2 Mar. 1973.

26. Ibid. Italics in the original.

27. Donnelly and Frutkin to Lee, memo, "Draft ASTP Information Plan," 9 Mar. 1973. This acceptable draft was signed by Donnelly, Frutkin, and Myers on 21 Mar. 1973 and by Low on 28 Mar. 1973.

28. "Apollo Soyuz Test Project, Minutes of Joint Meeting, USSR Academy of Sciences and US National Aeronautics and Space Administration," 9-20 July 1973; and Lunney to Bushuyev, 6 Sept. 1973.

29. Interview, Donnelly and Shafer-Ezell, 26 and 28 Jan. 1976.

30. Ibid.

31. Ibid.; Low, "Visit to Moscow, October 14-19, 1973," Dec. 1973; Vladimir Alexandrovich Kotelnikov to Low, 6 Nov. 1973; TWX, Henry A. Kissinger to American Embassy, Moscow, "Space Agreement: ASTP Information Plan," 29 Nov. 1973; Kotelnikov to Low, 29 Dec. 1973; and Lee to Lunney, memo, "PAO Plan Part I, Now in Effect," 15 Jan. 1974.

32. Interview, Donnelly and Shafer-Ezell, 26 and 28 Jan. 1976; and Ron Van Nostrand to Donnelly, 26 Nov. 1973.

33. Interview, Donnelly and Shafer-Ezell, 26 and 28 Jan. 1976; and Shafer [notes recorded during Moscow trip], 8-11 Oct. 1973.

34. Ibid.

35. "Apollo Soyuz Test Project, ASTP Public Information Plan Part I," ASTP 20 050, Part I, 12 Oct. 1973.

36. Shafer to Donnelly, memo, "ASTP Public

Information Plan," 23 Oct. 1973; and interview, Donnelly and Shafer-Ezell, 26 and 28 Jan. 1976.

37. "ASTP Information Plan Part I," 12 Oct. 1973, p. 6.

38. Shafer to Donnelly, memo, "ASTP Public Information Plan," 23 Oct. 1973.

39. "ASTP Information Plan Part I," 12 Oct. 1973, p. 5.

40. Two contemporary accounts of newsmen who have worked in the U.S.S.R. are contained in Robert G. Kaiser, *Russia: The People and the Power* (New York, 1976); and Hedrick Smith, *The Russians* (New York, 1976).

41. Riley to Donnelly, memo, "Soviet Documentary Photography during ASTP Crews' Visit to U.S.S.R.," 14 Feb. 1974.

42. Ibid.

43. Riley to Donnelly, memo, "Soviet Exclusion of Non-U.S. Western News Media at ASTP Crews News Conference in Star City," 14 Feb. 1973.

44. Letter, Low to Teague, 31 Oct. 1973.

45. Ibid.

46. "War Erupts in Middle East," *Facts on File* 33 (7-13 Oct. 1973): 833-838; and "Mideast War Mounts in Intensity," *Facts on File* 33 (14-20 Oct. 1973): 857-862.

47. James R. Jaax, comments on ASTP history draft, 19 Jan. 1976.

48. "Council of U.S. Academy of Sciences Expresses Concern to Soviet Counterparts over Sakharov Harrassment," *Science,* 21 Sept. 1973, pp. 1148-1149; "Soviet Academy Replies to NAS Defense of Sakharov," *Science,* 2 Nov. 1973, p. 1459; "Soviet Rebuts Americans on Sakharov," *New York Times,* 18 Oct. 1973; and "Soviet Letter on Sakharov," *New York Times,* 18 Oct. 1973. See also Smith, *The Russians,* pp. 439-445; and Kaiser, *Russia: the People and the Power,* pp. 419-428.

49. Low, "Visit to Moscow, October 14-19, 1973," Dec. 1973.

50. Hedrick Smith, "U.S. Space Team at Soviet Center," *New York Times,* 19 Oct. 1973.

IX

1. NASA News Release, MSC, 73-12, "ASTP Crew Named," 30 Jan. 1973; Loyd S. Swenson, Jr., James M. Grimwood, and Charles C. Alexander, *This New Ocean: A History of Project Mercury,* NASA SP-4201 (Washington, 1966), pp. 440-442; and NASA News Release [redistributed at JSC], "Excerpt from a Medical Briefing with Dr. Charles A. Berry, March 13, 1972, Discussing Donald K. Slayton's Heart Condition and His Return to Full Flight Status," 14 July 1975.

2. NASA News Release, MSC, 73-12, "ASTP Crew Named," 30 Jan. 1973.

3. Ibid.; U.S. Congress, House of Representatives, Committee on Science and Technology, *Astronauts and Cosmonauts: Biographical and Statistical Data,* 94th Cong., 1st sess. (Washington, 1975), p. 80; Harold M. Schmeck, Jr., "U.S. Selects Space Crew for Flight with Russians," *New York Times,* 31 Jan. 1973; and Thomas O'Toole, "Crew Picked for Joint Space Mission," *Washington Post,* 31 Jan. 1973.

4. NASA Press Conference, MSC, "Apollo Soyuz Test Project Prime Crew Press Conference," 1 Feb. 1973. Lunney and Lee had sought to coordinate the public announcement of the crewmembers with notification to the Soviets and to interested members of Congress. This required some careful timing. NASA did not want word to leak from Moscow about the American crew selection before all bases were touched in Washington. Chester M. Lee to Dale D. Myers, memo, "Naming of ASTP Astronauts," 30 Jan. 1973, together with draft TWX to Bushuyev, draft news release, and recommended list of congressmen and senators to be notified.

5. Ibid.

6. Ibid.

7. Committee on Science and Technology, *Astronauts and Cosmonauts,* pp. 123, 129, 131, 136, 140, 143, and 148-149; NASA News Release, JSC, 73-93, "ASTP Cosmonauts to Visit JSC," 6 July 1973; FBIS-Soviet, "Soviet Cosmonaut Crew Announced for Joint Space Program," from Moscow Tass International Service, 25 May 1973.

8. "Apollo/Soyuz Spotlighting Cooperation at Paris Air Show," *Aviation Week & Space Technology,* 28 May 1973, p. 14; and Shirley Malloy to George M. Low, 25 May 1973, message relaying comment from *Aviation Week* editor Robert Hotz: "Your Paris exhibit is the star of the show."

9. "Minutes of Discussion on Organization of Joint Exhibition at the 30th International Aviation and Space in Paris, May 1973," 14-15 Dec. 1972.

10. Letter, Glynn S. Lunney to Konstantin Davydovich Bushuyev, 20 Feb. 1973, with enclosure "Minutes of Paris Air Show Telephone Conversation, February 8, 1973."

SOURCE NOTES

11. "Apollo/Soyuz Spotlighting Cooperation at Paris Air Show," *Aviation Week & Space Technology,* 28 May 1973, p. 15.

12. Hotz, "Paris Vintage 1973," *Aviation Week & Space Technology,* 11 June 1973, p. 7; and interview, Charles A. Biggs-Edward C. Ezell, 28 Oct. 1975.

13. "Glenn, Titov Get Together," *Washington Star,* 3 May 1962; and "Transcript of News Conference of Titov and Glenn," *New York Times,* 4 May 1962.

14. "Astronauts Take Paris Spotlight," *New York Times,* 20 June 1965; and "Space Twins Steal Some Red Thunder," *Washington Daily News,* 19 June 1965.

15. "Russian Cosmonaut Greets Cooper and Conrad in Athens," *Washington Post,* 18 Sept. 1965.

16. "Space Rivals Drink Toast," *Baltimore Sun,* 27 May 1967.

17. Wade St. Clair to Julian F. Scheer, memo, "Report on NASA Participation at 1969 Paris Air Show," 12 June 1969. The USIA press attaché attended the meetings between the astronauts and cosmonauts and briefed the press on these sessions. For press reactions, see "U.S. and Russian Spacemen Meet," *Baltimore Sun,* 3 June 1969; and Clyde H. Farnsworth, "Astronauts Hold Paris Rendezvous," *New York Times,* 3 June 1969.

18. Frank Borman to Ezell, 20 Aug. 1975; "Borman Hopes for Joint Mission in Space with Soviets during '70s," *Washington Post,* 4 July 1969; and Bernard Gwertzman, "Podgorny Meets Borman, Voices Hope for Successful Moon Trip," *New York Times,* 10 July 1969.

19. "Cordial Cosmonauts Field Questions of U.S. Press," *Christian Science Monitor,* 28 Oct. 1969; and Donnie Radcliffe, "Cosmonauts Are Lionized," *Washington Evening Star,* 1 Nov. 1969.

20. Interview, David R. Scott-Ezell, 21 Aug. 1974; and Scott to Lunney, memo, "ASTP Mission to Moscow, June-July, 1973," 31 July 1973. For press reports on meetings between astronauts and cosmonauts, see "Admirers Mob Armstrong in Leningrad," *Baltimore Sun,* 26 May 1970; "Moscow 'Chill' Mars Visit by Armstrong," *Chicago Tribune,* 2 June 1970; "Soviet Spacecraft Studies Earth Resources," *New York Times,* 6 Oct. 1970; James Stanton, "Astronaut Calls Space Flights Key to Amity," *Philadelphia Evening Bulletin,* 5 Oct. 1970; "2 Cosmonauts Arrive Here on Goodwill Trip," *Washington Post,* 19 Oct. 1970; Paul W. Valentine, "2 Soviet Cosmonauts Orbit City on Goodwill Tour of U.S.," *Washington Post,* 20 Oct. 1970; "Astronauts All," *New York Times,* 3 June 1971; and Walter Sullivan, "58 Layers Found in Lunar Sample," *New York Times,* 21 Sept. 1971.

21. "Apollo Soyuz Test Project Crew and Ground Personnel Training Plan," ASTP 40 700, 26 Mar. 1973.

22. "Soviet Space: A visit to Star City," *Time,* 9 July 1973. John Noble Wilford of the *New York Times* had been one of the first Western news persons to visit Star City when he toured the center in March 1972 during his month sojourn in the U.S.S.R. His evaluation of the Soviet center and the pace of the Soviet program was similar to those in *Time* and *Aviation Week;* Wilford, "Friendly, Yes, but Trying to Be First," *New York Times,* 28 Mar. 1972.

23. Donald C. Winston, "Soviet Space Center Being Expanded," *Aviation Week & Space Technology,* 25 June 1973, p. 18.

24. "Minutes, Joint Meeting of Working Group 1," in "Apollo Soyuz Test Project, Minutes of Joint Meeting, USSR Academy of Sciences and US National Aeronautics and Space Administration," 9-20 July 1973; and interview, Mike S. Brzezinski-Ezell, 23 Sept. 1975.

25. "Minutes, Joint Meeting of Working Group 1," 9-20 July 1973; and NASA News Release, HQ [unnumbered], "Comminique on Results of Apollo Soyuz Test Project Meetings, July 8-20, 1973" [20 July 1973].

26. NASA Press Conference, JSC, "Apollo-Soyuz Test Project (ASTP) Technical Directors' and Prime Crew Press Conference," 20 July 1973; Wilford, "Astronauts from Soviet and U.S. Begin a Briefing in Houston for Joint Mission," *New York Times,* 10 July 1973; "Cosmonauts Hear Lectures," *Houston Post,* 11 July 1973; and Jack Waugh, "Building U.S.-Soviet Space Team," *Christian Science Monitor,* 19 July 1973.

27. "Visiting Cosmonauts Have a W. Texas Rural Look," *Houston Chronicle,* 19 July 1973.

28. Owen G. Morris to Christopher C. Kraft, memo, "Flight Crews for ASTP," 18 Sept. 1972; Morris to Kraft, memo, "ASTP Language Training," 7 Nov. 1972; Myers to Rocco A. Petrone, memo, "Astronaut Proficiency in the Russian Language," 17 Oct. 1972; Petrone to Meyers, memo, "Astronaut Proficiency in the Russian Language," 20 Nov. 1972; Donald K. Slayton to Morris, memo, "Joint US/USSR Crew Training,"

15 Dec. 1972; interview, Brzezinski-Ezell, 23 Sept. 1975; and interview, Nicholas Timacheff-Ezell, 1 Oct. 1974.

29. Morris to Kraft, memo, "Flight Crew for ASTP," 18 Sept. 1972.

30. The titles of the nine brochures all dated Oct. 1973 presented to the astronauts were as follows: "Soyuz" Mission Profile, Soyuz Spacecraft General Description, Androgynous Peripheral Docking System (APDS), Onboard Systems Manual Control General Concept, Thermal Control System (TCS), Atmospheric Composition Control and Integrity Check Systems General Concept, "Soyuz" Radio and TV Communications Systems Operation, Food and Water Systems, and Waste Management Systems.

31. "Minutes, US Astronaut Crews Familiarization Courses at the Y[u]. A Gagarin Cosmonaut Training Center" 19-30 Nov. 1973; Eugene A. Cernan to Lunney, memo, "ASTP Crew Visit to USSR," 13 Dec. 1973; Hedrick Smith, "U.S. Astronauts in Soviet to Train with Russians," *New York Times,* 20 Nov. 1973; Michael McGuire, "Joint Space Bid Off to Bumpy Start," *Chicago Tribune,* 20 Nov. 1973; and Vladimir A. Shatalov, "Na orbite sotrudnichestva" [On the orbit of cooperation], in *Soyuz i Apollon, rasskazivayut sovetskie uchenie inzheneri i kosmonavti—ychastniki sovmestnikh rabot s amerikanskimi spetsialistami* [Soyuz and Apollo, related by Soviet scientists, engineers and cosmonauts—participants of the joint work with American specialists], Konstantin D. Bushuyev, ed. (Moscow, 1976), pp. 199-214.

32. Interview, Robert D. White-Ezell, 30 Sept. 1975; "Apollo Soyuz Test Project, Results of Apollo Soyuz Docking Systems Development Tests," IED 50 013, 25 Dec. 1973; NASA, MSC, "E&D Weekly Activity Report," 9-15 Jan. 1974; and White to James M. Grimwood, memo, "Comments on E. C. Ezell's Draft Manuscript of the ASTP History," 6 Jan. 1976.

33. "U.S.A. Docking Module Environmental Control System Breadboard Testing Quick-Look Report," in "Minutes, Working Group 5," 14-25 Jan. 1974.

34. Data given by Brzezinski, 23 Sept. 1975; and interview (via telephone), Walter W. Guy-Ezell, 1 Oct. 1975.

35. Interview, Thomas P. Stafford-Ezell, 6 Apr. 1976.

36. "Integrated Testing of the Soyuz Life Support System," 11-22 Mar. 1974; "Apollo Soyuz Test Project, Minutes, Working Group 5," 11-22 Mar. 1974; and interview (via telephone), Guy-Ezell, 1 Oct. 1975.

37. NASA News Release, JSC, 74-64, "ASTP Working Groups to Meet at JSC," 10 Apr. 1974; Brzezinski, "ASTP Activity Planning Guide, April Cosmonaut Visit," 14-27 Apr. 1974; interview, Lonnie D. Cundieff-Ezell, 14 Aug. 1975; and "ASTP Teams Make Progress on Procedures for Joint Flight," *Aviation Week & Space Technology,* 6 May 1974. p. 18.

38. NASA Press Conference, JSC "ASTP Joint Crew Press Conference," 26 Apr. 1974.

39. Interview, Stafford-Ezell, 6 Apr. 1976; interview, Slayton-Ezell, 2 Mar. 1976; and interview, Anatole Forostenko-Ezell, 15 Mar. 1976.

40. NASA News Release, JSC, 74-183, "ASTP Docking Test Near Completion," 2 Aug. 1974; interview, Brzezinski-Ezell, 23 Sept. 1975; V. Sisnev, "SShA: Sovmestnii kosmicheskii eksperiment; ikh budet pyatero" [USA: joint experiment in space; there will be five of them], *Trud,* 1 Jan. 1975; and "Start nazachen na 15 yulya" [The launch is scheduled for 15 July], *Trud,* 1 Jan. 1975.

41. "Soyuz Gives Cosmonauts Little Control," *Aviation Week & Space Technology,* 21 Jan. 1974, p. 38; Theodore Shabad, "Soviet Puts Soyuz 12 with 2 Aboard, into Earth Orbit," *New York Times,* 28 Sept. 1973; James Oberg, "U.S.-Soviet Space Flight," *Los Angeles Times,* 30 Sept. 1973; "Nas planetoi sovetskii kosmicheskii korabl' v polete 'Soyuz-13' " [Above the planet, Soviet space ship, in the flight of "Soyuz-13"] *Izvestiya,* 19 Dec. 1973; "V polete–'Soyuz-13' " [In flight–"Soyuz-13"], *Pravda,* 19 Dec. 1973; A. Pokrovskii, " 'Soyuz-13': Lydi i sud'bi" ["Soyuz-13": people and fates], *Pravda,* 19 Dec. 1973; "Soyuz-13': Den' Vtoroy" ["Soyuz-13": the second day], *Pravda,* 20 Dec. 1973; "Smooth Sailing for Companions in Orbit," *Time,* 31 Dec. 1973, p. 41; B. Komovalov, "Zdravstvei, zemlya!" [Hello, earth!], *Izvestiya,* 27 Dec. 1973; "Soviet Cosmonauts Link Up to Salyut 3," *Newport Times Herald,* 4 July 1974; "Cosmonauts Star on Moscow TV, *Washington Star-News,* 14 July 1974; Malcolm Browne, "Russian Crew Ready to Return to Earth," *New York Times,* 19 July 1974; and Kenneth W. Gatland, "Salyut 3: Soviets Still Catching Up," *Christian Science Monitor,* 22 July 1974.

42. Ezell, "Notes on 0745 Director's Tag-up Meeting," 17 Apr. 1974.

43. Robert G. Kaiser, "Cosmonauts in Orbit; First in Two Years," *Washington Post,* 28 Sept. 1973.

44. William Proxmire to James C. Fletcher, 3 Sept. 1974. Harold M. Schmeck, Jr., "Proxmire, Citing Failures, Asks Study of U.S.-Soviet Space Plan," *New York Times,* 6 Sept. 1974; Richard D. Lyons, "Failures Mark Russian Space Program," *New York Times,* 26 Sept. 1974; "Cosmonauts Down Early but at Least Safely," *New York Times,* 1 Sept. 1974; "Soyuz Ends Flight; Salyut Linkup Fails," *Aviation Week & Space Technology,* 2 Sept. 1974, p. 23; Christopher S. Wren, "Soviet Astronauts Land Safely; Space Docking Apparently Fails," *New York Times,* 29 Aug. 1974; "Soyuz Ends Space Trip Early," *Washington Post,* 29 Aug. 1974; and Wren, "Two Soviet Astronauts in Good Health; First Night Landing Hailed in Moscow," *New York Times,* 30 Aug. 1974.

45. Lunney to Bushuyev, 3 Sept. 1974, enclosing "Minutes of the ASTP Telephone Conversation of August 27, 1974 [US Minutes]"; [Response to Query, HQ], "NASA's Reply to Proxmire Letter," 5 Sept. 1974; and Fletcher to Proxmire, 10 Sept. 1974.

46. "Summary of Results of the August-September 1974 Meeting of Specialists of the USA NASA and USSR Academy of Sciences on the Preparations for Conduct of the Test Flight of Apollo and Soyuz," 20 Sept. 1974, in "Apollo Soyuz Test Project Minutes of Joint Meeting, USSR Academy of Sciences and US National Aeronautics and Space Administration," 26 Aug.-20 Sept. 1974.

47. NASA Press Conference, JSC, "ASTP News Conference," 11 Sept. 1974.

48. NASA News Release, JSC, 74-272, "U.S. Tracking Soyuz 16," 3 Dec. 1974; "Minutes of the ASTP Telephone Conversation," 3 Dec. 1974; "Soyuz 16 Life Support Systems Operation," USSR WG5-035, 27 Jan. 1975; Elizabeth Pond, "Soviet Dress Rehearsal for U.S. Space Linkup," *Christian Science Monitor,* 3 Dec. 1974; Wren, "Soviet Orbits 2 in a Test for Joint Space Link-up," *New York Times,* 3 Dec. 1974; Soviets Orbit 2 Cosmonauts," *Washington Post,* 3 Dec. 1974; Wren, "Space Flight Test Pleases Soviet," *New York Times,* 8 Dec. 1974; A. Bessonov, "Provereno v kosmicheskom polete," *Novoe Vremya,* 13 Dec. 1974, pp. 6-7 (for translation, see the English version, "Tested in Space," *New Times,* Dec. 1974, pp. 6-7); and V.

Lesnikov, "Polet 'Soyuz A-16,' po sovmestnomu vapiantu" ["Soyuz-16," the joint version], *Aviatsiya i Kosmonovtika* (Jan. 1975), pp. 4-5.

49. "Minutes of Joint Meeting, USSR Academy of Sciences and US National Aeronautics and Space Administration," 26 Aug.-20 Sept. 1974.

50. Robert J. Shafer to John P. Donnelly, memo, "Advance Notification for Soyuz Launch," 7 Oct. 1974; and Shafer to Donnelly, memo, "Advance Notification for Soyuz Launch," 9 Oct. 1974.

51. TWX, Lunney to Bushuyev, 11 Oct. 1974; Gerald M. Truszynski to John F. Yardley, memo, "Procedure for Authorizing Tracking of Any ASTP Precursor Mission," 8 Nov. 1974; TWX, Lunney to Lee, "Tracking of Soyuz Flight," 8 Nov. 1974; and Yardley to Truszynski, memo, "Tracking of Soyuz Flight," 15 Nov. 1974.

52. Bushuyev to Lunney, 16 Dec. 1974, transmitting "Report on Telephone Exchanges during the Soyuz 16 Joint Tracking Experiment."

53. TWX, Lunney to Bushuyev, 11 Oct. 1974; "Minutes of Joint Meeting, USSR Academy of Sciences and US National Aeronautics and Space Administration," 20 Jan.-13 Feb. 1975; and "Soyuz 16 Life Support Systems Operation," USSR WG5-035, 27 Jan. 1975.

54. Transcription, report on *Soyuz 16* by Bushuyev, 31 Jan. 1975.

55. Ibid.

56. White to Grimwood, memo, "Comments on E. C. Ezell's Draft Manuscript," 6 Jan. 1976.

57. NASA Management Instruction 1156.14A, "Aerospace Safety Advisory Panel," 18 June 1973. Members of the panel and staff were Howard K. Nason, Charles D. Harrington, Frank C. Di Luzio, Herbert E. Grier, Lee R. Scherer, Henry Reining, Ian M. Ross, Warren D. Johnson, Bruce T. Lundin, Gilbert L. Roth, V. Eileen Evans, William A. Arazek, and Carl R. Praktish.

58. Aerospace Safety Advisory Panel, "Annual Report to the NASA Administrator, Part 1—Apollo Soyuz Test Project, Section 1— Observation and Conclusions," Feb. 1975; and Nason to Fletcher, 5 Feb. 1975.

59. Harrington, "Trip Report: Observation of Working Group No. 3 at Academy of Sciences, Moscow, USSR," 27 Nov. 1974.

60. Aerospace Safety Advisory Panel, "Annual Report," Feb. 1975.

61. Reporting on this part of the ASAP meeting with the Soviets is based on notes taken by Ezell during the discussions, 6 Feb. 1975. See interview, Alex Tatistcheff-Ezell, 6 Feb. 1975.

62. Bushuyev, ed., *Soyuz i Apollon, rasskazivayut sovetskie uchenie inzhineri i kosmonavti–ychastniki sovmestnikh rabot s amerikanskimi spetsialistami* [Soyuz and Apollo, related by Soviet scientists, engineers, and cosmonauts–participants of the joint work with American specialists] (Moscow, 1976), p. 28.

63. Interview, Herbert E. Smith, Richard Haken, Steven Pollock, and Roscoe Lee-Ezell, 12 Feb. 1975.

64. "Meeting Minutes, Joint Meeting of Working Group 1," 25 Nov. -20 Dec. 1974.

65. This description of the RCS impingement meeting is based upon notes taken by Ezell during that discussion and upon subsequent talks with Eugene A. Cernan, Lunney, and members of Working Group 2. Yuri Zonov's comment was made to Ezell after the trip to the simulator.

66. NASA News Release, JSC, 75-07, "Cosmonauts in U.S. for ASTP Training," 6 Feb. 1975; NASA, KSC, "News Clippings; Visit of Apollo Soyuz Test Project Cosmonauts and Astronauts," 8-10 Feb. 1975; and data supplied by Donalyn Epstein (handouts, briefing materials, and itineraries) from visit to KSC.

67. Viktor K. McElheny, "Soviet Astronauts Enjoy Flight into Fantasies of Disney World," *New York Times,* 10 Feb. 1975; and "Detente in Space," *New York Times,* 15 Feb. 1975.

68. "Status Report No. 1, Joint ASTP Crew Training," 12 Feb. 1975; and data provided by Brzezinski.

69. "Status Report No. 2, Joint ASTP Crew Training," 12 Feb. 1975; and "Status Report No. 7, Joint ASTP Crew Training," 24 Feb. 1975.

70. "Minutes of the Astronaut and Cosmonaut Joint Training Visit at the Johnson Space Center," 7-28 Feb. 1975; and TWX, Lunney to Bushuyev, 7 Jan. 1975.

71. "Minutes of the Training of Soviet and American ASTP Crews at Gagarin Cosmonaut Training Center," 14-30 Apr. 1975. In addition to the prime and backup crewmen, the four support crewmembers and Cernan went to the U.S.S.R. on this trip.

72. NASA Press Conference, JSC, "ASTP Crew Press Conference," 14 May 1975; "First Map of Baikonur Cosmodrome," *Defense/Space Daily,* 16 Apr. 1975; "New Map of Baikonur Cosmodrome," *Defense/Space Daily,* 17 Apr. 1975; Hal Piper, "Astronaut Describes Huge Soviet Center," *Baltimore Sun,* 15 May 1975; Thomas O'Toole, "Astronauts Tell of Huge Soviet Spaceport," *Washington Post,* 15 May 1975; and Ivan Borisenko, "The Baikonur Cosmodrome," *Sputnik,* (Apr. 1975); pp. 68-71. In F. Dennis Williams, memo to distribution, "Soviet Place Names for ASTP," 17 Apr. 1975, the Office of International Affairs took the following stance: *"Soviet launch site:* the Soyuz spacecraft will be launched from the Baykonur cosmodrome (also acceptable: 'Baykonur launch facility' or 'Baykonur launch complex' or 'Baykonur launch center') near the city of Tyuratam at about 40°N., 63°E. in the Kazakh Soviet Socialist Republic. Precise references should include both the name of the facility and the nearest city: That is, 'Baykonur launch complex near Tyuratam.' This (cf. Johnson Space Center, Houston) will avoid confusion that might arise from the fact that there is a well-known city called Baykonur (about 48°N., 66°E.) some 200 miles to the northeast of the launch complex. Since both Soviet and U.S. newsmen frequently use only 'Baykonur,' confusion about the physical location of the launch site might arise, a confusion which can best be avoided by use of the complete phrase: e.g., Baykonur launch facility near Tyuratam." In an interview, Reinhold H. Dietz-Ezell, 4 June 1975, Dietz emphasized the point that during the Working Group 4 tests at the launch site in May 1975 the Soviets used the work "Baykonur" to name the whole region surrounding Leninsk.

73. NASA News Release, JSC, 75-21, "USSR/USA Discussions on Soyuz Launch Failure," 8 Apr. 1975; "Minutes of the ASTP Telephone Conversation [US Minutes]," 8 Apr. 1975; and Epstein, "Non-Successful Launch of Soyuz, from Telecon, April 8, 1975," 9 June 1975.

74. NASA Press Conference, JSC, "ASTP Crew Press Conference," 14 May 1975; "Soyuz Off-Course, Mission Aborted," *Washington Post,* 7 Apr. 1975; James F. Clarity, "Soviet Space Shot with 2 Men Fails," *New York Times,* 7 Apr. 1975; Thomas O'Toole, "Soyuz Mishap May Peril Joint Project," *Washington Post,* 8 Apr. 1975; O'Toole, "Soyuz Failure Termed No Threat to Project," *Washington Post,* 9 Apr. 1975; and U.S. Congress, Senate, *Congressional Record,* 94th

Cong., 1st sess. (9 Apr. 1975), vol. 121, p. S5537.

75. News release, issued by the office of Senator William Proxmire, Wisconsin, 8 Apr. 1975. Bushuyev in *Soyuz i Apollon,* pp. 27-28, commented:

> Sometimes members of the American press gave the impression that they intentionally create a tense atmosphere with their questions on the safety of the joint flight. . . . Senator Proxmire also did not fail to add some fuel to the fire. He openly expressed concerns about the American astronauts' life during the joint flight. In the second half of 1974, . . . Proxmire wrote a special letter to NASA Administrator, James C. Fletcher, requesting that he personally check out all of the factors related with the Soyuz/Apollo project and take the necessary steps to guarantee the safety of American astronauts.
> To us, the participants of the project, it was perfectly clear that these artifically formed cautions had no foundation. The successful joint flight of the spacecraft Soyuz and Apollo confirmed this. . . . In spite of the concerns expressed by various prophets of doom on the subject of such 'adventure' the first international space flight of Soviet and American spacecraft was crowned with success.

76. For information on the first inter-control center simulations held in Mar. 1975, see Charles R. Lewis, memo to distribution, "Report on the March 1975 Joint Training Simulations," 7 Mar. 1975; and Linda N. Ezell, ASTP Narrative History Log Note 40, "Simulation Debriefing," 24 Mar. 1975. NASA News Release, JSC, 75-42, "Moscow/Houston Control Centers to Conduct Joint Apollo-Soyuz Simulations," 9 May 1975; and TWX, Lunney to Bushuyev, 3 June 1975.

X

1. George M. Low, "Notes from Visit to Soviet Union, May 17-23, 1975," 5 June 1975.
2. Ibid.
3. Ibid.
4. Ibid.
5. "U.S. Proposal for a Joint ASTP Flight Readiness Review" [n.d.].
6. ASTP notebook, kept by Leonard S. Nicholson, for May-Nov. 1975.
7. Low, "Notes from Visit to Soviet Union, May 17-23, 1975," 5 June 1975. According to data supplied by R. H. Dietz on 6 June 1975, the Soviet lineup at the main table was as follows: B. V. Nikitin, Yu. S. Dolgopolov, V. S. Syromyatnikov, V. P. Legostayev, V. A. Timchenko, A. A.

Leonov, Yu. V. Zonov, K. D. Bushuyev, V. A. Kotelnikov, B. N. Petrov, A. I. Tsarev, M. V. Sokolov, V. S. Vereshchetin, I. Rumyantsev, Tulin, and Sagdayev. The Americans present included C. M. Lee, H. E. Smith, R. O. Aller, A. W. Frutkin, A. B. Tatistcheff, G. M. Low, G. S. Lunney, J. F. Yardley, W. J. Kapryan, W. W. Guy, L. S. Nicholson, and R. D. White. See also "Apollo Soyuz Test Project, Flight Readiness Review, May 1975," 25 May 1975.

8. "Apollo Soyuz Test Project, Flight Readiness Review, May 1975," 25 May 1975, pp. WG-1-1 and WG-1-5; O. G. Sytin, "Ballistika EPAS" [ASTP trajectory questions], in *Soyuz i Apollon, rasskazivayut sovetskie uchenie inzhineri i kosmonavti—chastniki sovmestnikh rabot s amerikanskimi spetsialistami* [Soyuz and Apollo, related by Soviet scientists, engineers, and cosmonauts—participants of the joint work with American specialists], Konstantin D. Bushuyev, ed. (Moscow, 1976), pp. 77-99; and A. S. Yeliseyev and V. G. Kravets, "Upravlenie poletom" [Flight control], in *Soyuz i Apollon,* pp. 215-234.

9. ASTP notebook, kept by Nicholson, for May-Nov. 1975.

10. "Apollo Soyuz Test Project Flight Readiness Review, May 1975," 25 May 1975, pp. WG-1-12 and WG-1-13.

11. Ibid., pp. WG-1-22 and WG-1-26.

12. Ibid., pp. WG-1-27 to WG-1-40.

13. Ibid., p. WG-2-2. Results of the docking target verification tests were documented in "ASTP, Results of Verification Tests of Alinement Soyuz Docking Target," IED 50 203, 20 May 1975.

14. "Apollo Soyuz Test Project, Flight Readiness Review, May 1975," 25 May 1975, pp. WG-2-12 to WG-2-20.

15. Ibid., p. WG-2-21. The results of the safety assessments were presented in two documents: "ASTP Safety Assessment Report for Apollo Propulsion and Control Systems," ASTP 20 102, 3 May 1974, and "ASTP Safety Assessment Report for Soyuz Propulsion and Control Systems," ASTP 20 202.1, 1 May 1975. Soviet comments on guidance and control are contained in V. P. Legostayev, "Korabli idut na sblizhenie" [Spacecraft are approaching each other], in *Soyuz i Apollon,* pp. 100-115. The author emphasizes the superior automatic (or ground-controlled) guidance systems of the Soyuz and the changes that the Soviets had to make in their

spacecraft to accommodate the Apollo systems and configuration incompatibilities.

16. "Apollo Soyuz Test Project, Flight Readiness Review, May 1975," 25 May 1975, p. WG-3-13; and interview (via telephone), Robert D. White-Edward C. Ezell, 23 Mar. 1976.

17. "Apollo Soyuz Test Project, Flight Readiness Review, May 1975," 25 May 1975, pp. WG-3-21 to WG-3-29; interview, White-Ezell, 23 Mar. 1976; White to Glynn S. Lunney, memo, "Debriefing Notes Regarding Docking Systems Preflight Compatibility Vertification in Moscow," 13 Mar. 1975; and Ray F. Larson to James M. Grimwood, "Contractor Comments on Edward C. Ezell's Manuscript of the ASTP Narrative History for Review and Comment (NAS9-13972)," 13 Jan. 1976.

18. "ASTP Working Group 3 Minutes on the Question of Resolving the Problem of Possible Binding of the Docking System Guide Pin and Socket," 22 Nov. 1974.

19. TWX, Lunney to Bushuyev, 8 Nov. 1974.

20. Interview, White-Ezell, 23 Mar. 1976.

21. "ASTP Working Group 3 Minutes on the Question of Resolving the Problem of Possible Binding of the Docking System Guide Pin and Socket," 22 Nov. 1974.

22. "Minutes of the ASTP Telephone Conversation [U.S. Minutes]," 19 Nov. 1974; and "Report on the Telephone Conversation of November 19, 1974 between Drs. Timchenko and Lunney [U.S.S.R. Minutes]," [n.d.].

23. White to Lunney, memo, "Debriefing Notes," 13 Mar. 1975; and interview, White-Ezell, 23 Mar. 1976.

24. "ASTP Meeting Minutes, Joint Meeting of Working Group 3," 16 Jan.-12 Feb. 1975; "ASTP Report of USA/USSR Docking System Alignment Pin and Socket Verification Test," USA WG3-042, 27 Jan. 1975; and White to Grimwood, memo, "Comments on E. C. Ezell's Draft Manuscript of the ASTP History," 6 Jan. 1976. Also see V. S. Syromyatnikov, "Stikovka—eto uzhe sotrudnichestvo" [To dock is to cooperate], in Soyuz i Apollon, pp. 116-148. The technical explanations covering the American and Soviet docking systems are very elaborate, with considerable effort being made to use layman's terminology. Syromyatnikov points out that, while the Soviet system, its design, technology, and construction never once caused any problems in manufacture, testing, or the flight itself, the American docking sytem, "a

copy of an existing Soviet design," was a cause for concern on many occasions. "American technology just was not sufficiently developed" to build a system better than that.

25. "Apollo Soyuz Test Project, Flight Readiness Review, May 1975," 25 May 1975, pp. WG-4-1 to WG-4-53; interview, Reinhold H. Dietz-Ezell, 6 June 1975; interview (via telephone), Dietz-Ezell, 24 Mar. 1976; and B. V. Nikitin and B. F. Ryadinski, "Apollon, ya—Soyuz! Kak slishite?" [Apollo, this is Soyuz! How do you read?], in Soyuz i Apollon, pp. 149-165.

26. Interview, Dietz-Ezell, 18 Feb. 1976; interview; Dietz-Ezell, 24 Mar. 1976; and ASTP notebook, kept by Nicholson, for May-Nov. 1975.

27. Interview, Walter W. Guy and James R. Jaax-Ezell, 19 Jan. 1976.

28. Ibid.; "Apollo Soyuz Test Project Flight Readiness Review, May 1975," 25 May 1975, pp. WG-5-19 to WG-5-25; and I. V. Lavrov and Yu. S. Dolgopolov, "V poiskak obshcheiy atmosferi" [In search of a common atmosphere], in Soyuz i Apollon, pp. 166-179.

29. "Apollo Soyuz Test Project Flight Readiness Review, May 1975, 25 May 1975, p. WG-5-1.

30. Low, "Notes from Visit to Soviet Union, May 17-23, 1975," 5 June 1975.

31. Ibid.; and interview, Dietz-Ezell, 6 June 1975.

32. "Summary of the Joint Review of the Flight Readiness of the Apollo-Soyuz Project," 22 May 1975; "Space Officials Give Approval to Link-up of Apollo and Soyuz," New York Times, 23 May 1975; "All Systems Go," Washington Post, 23 May 1975; and "Apollo-Soyuz Flight is Okayed," Baltimore Sun, 23 May 1975.

33. James C. Fletcher to Gerald R. Ford, 29 May 1975.

34. Fletcher to William Proxmire, 4 June 1975, with enclosure, "Soyuz Launch Abort of April 5, 1975," 4 pp.; and Chester M. Lee, memo for record, "Information Regarding Soyuz 18," 18 Apr. 1975.

35. Robert J. Shafer to John P. Donnelly, memo, "Apollo/Soyuz Test Mission Television," 27 June 1972; Shafer to Donnelly, memo, "ASTP," 29 Aug. 1972; and TWX, Bushuyev to Lunney, 28 Mar. 1974.

36. Interview, Donnelly and Shafer-Ezell, 26 and 28 Jan. 1976.

37. Low to Lee, memo, "ASTP Television," 31 Aug. 1973.

38. Ibid.

39. Shafer to Lee, memo, "ASTP Television," 14 Sept. 1973; and Lunney to Lee, memo, "ASTP Television," 18 Sept. 1973.

40. Lunney to Lee, memo, "ASTP Television," 18 Sept. 1973; Lee to Shafer, memo, "Television Coverage During the ASTP Mission," 25 Oct. 1973; and Shafer to Lee, memo, "ASTP Television Coverage," 2 Nov. 1973.

41. Dale D. Myers to Fletcher, memo, "Summary of ASTP Actions Taken Regarding Use of the ATS-F Satellite," 2 Nov. 1973, with enclosure, "Chronology of ATS-F Coordination."

42. Lee to Low, memo, "ASTP Television Coverage" [12 Nov. 1973].

43. Lunney, memo to distribution, "ASTP Television," 14 Dec. 1973.

44. "ASTP TV Planning Team, Minutes," 8 Jan. 1974, 12 Feb. 1974, 1 May 1974, 4 June 1974, 10 July 1974, 6 Aug. 1974, 22 Aug. 1974, 10 Oct. 1974, 20 Feb. 1975, and 26 Mar. 1975.

45. Shafer to Donnelly, memo, "ASTP Television," 30 Jan. 1974. Cernan had a similar evaluation of the situation. See Eugene A. Cernan to Lunney, memo, "ASTP TV Status," 1 Feb. 1974.

46. Donnelly to Fletcher and Low, memo, "Moscow Trip Report," 28 Mar. 1974; Lee to Donnelly, memo, "ASTP Status Report," 7 June 1974; and Shafer to Donnelly, memo, "ASTP Public Information Plan, Part II–May 1974 U.S. Revisions," 10 June 1974.

47. "ASTP Public Information Plan–Part II," ASTP 20 050 Part II, 13 Sept. 1974.

48. UPI wire release, 9 Oct. 1974. The precise language of the agreement stated, "Beginning with the Soyuz launch and ending with Apollo recovery, inflight (on-board) information from both spacecraft, voice communications between the two spacecraft, US and USSR ground communications to and from both spacecraft, and ground-based and on-board television will be exchanged in real time as they are received. This information will be released by each side to the news media in accordance with its own obligations and practices," "ASTP Public Information Plan–Part II," ASTP 20 050 Part II, 13 Sept. 1974.

49. *Aviation Week & Space Technology,* 14 Oct. 1974, p. 11. Other comments included "Soviets Keep Soyuz Launch Site 'Off Limits,' "

Editor & Publisher, 30 Nov. 1974; "From Russia, with Live," *Broadcasting,* 14 Oct. 1974; "Soviet Refuses to Lift Its Ban on U.S. Newsmen at Launching," *New York Times,* 12 Oct. 1974; "Russians Insist on Bar on Press," *New York Times,* 24 Oct. 1974; "The Unwelcome Mat," *Huntsville Times,* 14 Oct. 1974; "Veil on Launch," *Washington Star-News,* 12 Oct. 1974; and "Russians To Provide Soyuz TV," *Houston Post,* 10 Oct. 1974.

50. Fletcher to Wes M. Gallagher, 22 Oct. 1974. This was also sent as a TWX.

51. Ibid.

52. "ASTP Minutes of Public Information Working Group Meeting, USSR Academy of Sciences and US National Aeronautics and Space Administration," 6-16 Apr. 1975; and "ASTP Public Information Plan–Part II," ASTP 20 050 Part II, 10 July 1975.

53. News release, issued by the office of Senator William Proxmire, Wisconsin, 2 July 1975.

54. Ibid.

55. Proxmire to Fletcher, 2 July 1975.

56. Fletcher to Proxmire, 3 July 1975.

57. Ibid.; and Thomas O'Toole, "U.S.-Soviet Flight Delay Is Rejected," *Washington Post,* 3 July 1975.

58. "Apollo-Soyuz Flight," *Congressional Record,* 11 July, S12417. Proxmire inserted Thomas O'Toole, "Apollo-Soyuz: Another 'Wheat Deal,' " *Washington Post,* 11 July 1975, and Tom Braden, "The Space Link-Up," *Washington Post,* 5 July 1975.

59. News release, issued by the office of Senator William Proxmire, Wisconsin, 14 July 1975.

60. Interview, Thomas P. Stafford-Ezell, 6 Apr. 1976.

61. NASA, OMSF, Apollo/Soyuz Test Program Office, "ASTP Flight Readiness Review," TM052-001-1A, June 1975.

62. Lightning had been a problem since Mercury and Gemini, with the first major incident occurring on 17 Aug. 1964 when a lightning strike at launch complex 19 interrupted testing of Gemini Launch Vehicle 2. This and other problems caused postponement of the GLV-2 flight until 19 Jan. 1965. Subsequently, GT-5 was postponed when a similar near miss affected the onboard computers. This resulted in a two-day delay. See James M. Grimwood, Barton C. Hacker, with Peter J. Vorzimmer, *Project Gemini*

Technology and Operations; A Chronology,
NASA SP-4002 (Washington, 1969), pp. 154 and
209. During the *Apollo 12* launch, the spacecraft
was affected by lightning-associated electrical
disturbances at 36.5 and 52 seconds after liftoff.
See NASA, "Analysis of Apollo 12 Lightning
Incidents," MSC-01540, Feb. 1970. Following
the *Apollo 12* flight, greater attention was given
to the lightning problem. For ASTP, see William
R. Durett to Donald D. Arabian, memo, "KSC
Lightning Research," 8 Aug. 1973; Lee to William H. Rock, memo, "Impact of Lightning
Strikes on the ASTP Mission," 30 Aug. 1973; Lee
to Rock, memo, "Impact of Lightning Strikes on
the ASTP Mission," 11 Dec. 1973; Rock to Lee,
memo, "Impact of Lightning Strikes on the
ASTP Mission," 20 Dec. 1973; Richard G. Smith
to Rock, memo, "Impact of Lightning Strikes on
the ASTP Mission," 16 Feb. 1974; Rock to
distribution, memo, "ASTP Lightning Review,"
27 Mar. 1974; Rock to Lee, memo, "Impact of
Lightning Strike on the ASTP Mission," 1 Apr.
1974; TWX, Lee to Lunney, Ellery B. May,
Rock, and Walter J. Kapryan, "Plan for Lightning
Investigation at KSC, Summer, 1974," 24 Apr.
1974; Lee to Lunney, May, and Rock, memo,
"ASTP Lightning Meeting," 20 May 1974; TWX,
Lee to Lunney, Rock, Isom A. Rigell, and May,
"Spacecraft Lightning Tests," 24 June 1974; May
to Lee, memo, "AS-210 Simulated In-Flight
Lightning Test," 27 June 1974; Rock to Lee,
memo, "ASTP In-Flight Lightning Tests," 2 July
1974; Lee to Lunney, May, and Rock, memo,
"ASTP Simulated In-Flight Lightning Tests," 25
July 1974; Lunney to Kapryan, memo, "Simu-
lated Lightning Test on CSM 119," 9 Aug. 1974;
TWX, Rock to Lee, "Simulated Lightning Tests
at KSC," 15 Aug. 1974; Rock to distribution,
memo, "Crew Safety Panel Approval, Concept of
Lightning Protection System for ASTP," 12 Aug.
1974; Lee to Lunney, May, and Rock, memo,
"Launch Mission Rules Changes for Lightning," 19
Aug. 1974; TWX, Lee to Lunney, May, and
Rock, memo, "ASTP Simulated In-Flight Light-
ning Tests," 6 Sept. 1974; TWX, Lee to Rock,
Lunney, and May, "Lightning Review," 3 Oct.
1974; Rock to distribution, memo, "ASTP
Launch Mission Rule," 4 Nov. 1974; John F.
Yardley to Low, memo, "ASTP Severe Weather
Launch Rules," 20 June 1975; Robert L. Blount,
Richard D. Gadbois, Dwight L. Suiter, and John
A. Zill, NASA, JSC, "ASTP Simulated Lightning
Test Report," JSC-09221, Nov. 1975; NASA
News Release, KSC-30-75, "Saturn/Apollo

Launch Tower Sports 'New Look,'" 2 Mar.
1975; and NASA, OMSF, Apollo/Soyuz Test
Program Office, "ASTP Flight Readiness Review," TM052-001-1A, June 1975.

63. Charles J. Newman, Weather Bureau,
Office of Meteorological Operations, Space Oper-
ations Support Division, *Frequency and Duration
of Thunderstorms at Cape Kennedy, Part 1,*
WBTM SOS-2 (Silver Spring, Md., June 1968);
and NASA, OMSF, Apollo/Soyuz Test Program
Office, "ASTP Flight Readiness Review,"
TM052-001-1A, June 1975.

64. NASA, OMSF, Apollo/Soyuz Test Pro-
gram Office, "ASTP Flight Readiness Review,"
TM052-001-1A, June 1975; NASA News Release,
MSFC, 75-119, "Apollo Soyuz Flight Readiness
Review Is Held," 12 June 1975; and NASA News
Release, KSC-104075, "ASTP Given Clearance
for Final Preparations," 12 June 1975. But Low
still had some questions; see Low to Lee, memo,
"ASTP Questions," 16 June 1975.

65. TWX, Lunney to Bushuyev, "Apollo
Status Report," 23 June 1975.

66. "Minutes of the ASTP Telephone Conversation [U.S. Minutes]," 26 June 1975; and
"Minutes of the June 26, 1975 Telephone Conversation between Lunney and Bushuyev [USSR
Minutes]," 26 June 1975.

67. NASA News Release, JSC, 75-59, "ASTP
Crew Begins Medical Isolation," 24 June 1975;
[NASA News Release, KSC (unnumbered)],
"ASTP Prime Crew Activities," 25-29 June 1975;
and data supplied by Mike S. Brzezinski.

68. NASA News Release, KSC-115-75,
"ASTP Dress Rehearsal Begins Wednesday," 25
June 1975; [Response to query, JSC], "Suit
Fitting Background" [n.d.]; and TWX, Lunney
to Bushuyev, "Apollo Status Report," 29 June
1975. See also NASA, JSC, Engineering and
Development, "Weekly Activity Report," 6-12
July 1975, p. 3:

> During CDDT, excessive leakage was dis-
> covered in the suit loop of the spacecraft
> while in the low pressure (vent) mode, al-
> though no leakage was evidenced at high
> pressure (3.75 psi). Troubleshooting con-
> ducted in the spacecraft isolated the problem
> to Astronaut Brand's suit. Following comple-
> tion of CDDT, Astronaut Brand's flight suits
> were returned to JSC where additional failure
> analysis established the source of this low
> pressure leak to be in the crotch radius area.
> The leak is induced when the crewman is
> tightly strapped into the couch position caus-
> ing the zipper sealing lips to unseat. After
> examining several possible corrective con-

cepts, a zipper insert [made of vinyl tubing] was designed which can be placed between the sealing zipper and the donning assist zipper thereby preventing the sealing lips from being unseated by the couch position condition. Certification tests of the zipper insert have been conducted to verify its acceptability for flight and to assure no adverse effects on use of the suits in the pressurized mode. Vance Brand had also fit checked his suits with the insert installed to verify acceptability.

69. [NASA News Release, KSC (unnumbered)], "ASTP Prime Crew Activities," 30 June-7 July 1975; TWX, Lunney to Bushuyev, "Apollo Status Report," 2 July 1975; [NASA News Release, KSC (unnumbered)], "Status Report ASTP-2," 4:30 p.m., 7 July 1975; [NASA News Release, KSC (unnumbered)], "ASTP Apollo Crew Activities," 7-8 July 1975; [ASTP News Release, KSC (unnumbered)], "ASTP Apollo Crew Activities," 8-15 July 1975; [NASA News Release, KSC (unnumbered)], Status Report ASTP 3," 8:00 a.m. EDT, 8 July 1975; [NASA News Release, KSC (unnumbered)], "Status Report–ASTP #4," 4:30 p.m., 8 July 1975; [NASA News Release, KSC (unnumbered)], "Status Report ASTP 5," 8:00 a.m., 9 July 1975; TWX, Lunney to Bushuyev, "Apollo Status Report," 9 July 1975; and data supplied by Brzezinski. See also TWX, Lunney to Bushuyev, "Apollo Status Report," 10 July 1975.

70. [NASA News Release, KSC (unnumbered)], "Status Report ASTP #12," 2:00 p.m., 11 July 1975; [NASA News Release, KSC (unnumbered)], "Status Report ASTP #14," 5:00 p.m., 11 July 1975; [U.S.S.R.], Press Bulletin No. 1, "ASTP Mission Events," 12 July 1975; [NASA News Release, KSC (unnumbered)], "Status Report #15," 9:30 a.m., 12 July 1975; [NASA News Release, KSC (unnumbered)], "Status Report ASTP #16," 2:30 p.m., 12 July 1975; [NASA News Release, KSC (unnumbered)], "Status Report #17," 5:00 p.m., 12 July 1975; [U.S.S.R.], Press Bulletin, "Two Days before the Liftoff," 13 July 1975; and A. I. Ostashev and N. I. Zelenshikov, "Soyuz 19 u poroga orbitiy" [Soyuz 19 on the threshold of space,] in Soyuz i Apollon, pp. 235-244.

71. TWX, Vladen S. Vereshchetin to Lunney, 7 July 1975; [NASA News Release, JSC (unnumbered)], "USSR Academy of Sciences ASTP Delegation for Mission" [n.d.]; [U.S.S.R.], Press Bulletin No. 3, "Two Days before the Liftoff," 13 July 1975; and [NASA, Apollo Program Office], "USA Specialists in Moscow during ASTP Mission" [n.d.].

XI

1. Except where indicated, the account of the flight is based upon notes taken by Edward C. Ezell and Linda N. Ezell during the mission.

2. ASTP mission commentary transcript, MC 8/2 [mission commentary tape #8, p. 2], 15 July 1975.

3. ASTP mission commentary transcript, MC 9/1-2 and MC 10/1, 15 July 1975.

4. ASTP mission commentary transcript, MC 12/1-2, 15 July 1975.

5. ASTP mission commentary transcript, MC 26/1 and MC 33/1, 15 July 1975.

6. ASTP mission commentary transcript, MC 34/1, MC 32/1, MC 36/1, and MC 37/1-3, 15 July 1975.

7. ASTP mission commentary transcript, MC 42/1, SR 6/1 [Soviet mission commentary tape #6, p. 1] and MC 51/1, 15 July 1975.

8. NASA, JSC, Test Division, Program Operations Office, "ASTP Technical Air-to-Ground Voice Transcription," JSC-09815, July 1975, pp. 3-4. In NASA, JSC, Crew Training and Procedures Division Training Office, "ASTP Technical Crew Debriefing," JSC-09823, 8 Aug. 1975, pp. 3-1 through 3-2, the crew noted once again considerable longitudinal oscillation during the operation of the first stage.

9. Program Operation Office, "ASTP Technical Air-to-Ground Voice Transcription," p. 10; interview, Thomas P. Stafford-Ezell, 6 Apr. 1976; and NASA, JSC, "Apollo Soyuz Mission Evaluation Report," JSC-1067, Dec. 1975, pp. 10-1 and 10-2.

10. Interview, Stafford-Ezell, 6 Apr. 1976.

11. NASA, JSC, "Apollo Soyuz Mission Evaluation Report," JSC-10607, Oct. 1975, p. 3-1.

12. Program Operations Office, "ASTP Technical Air-to-Ground Voice Transcription," pp. 18-19; and [NASA News Release], Apollo News Center, JSC, "ASTP Change-of-Shift Debriefing," 15 July 1975. For further information on ATS 6, see NASA, Applications Technology Satellite-6 (ATS-6) [NASA Facts Series] (Washington, 1975); NASA News Release, HQ, 75-194, "ATS-6 Now on Station in View of India–Anomaly Being Studied," 2 July 1975; "ATS-6: Extra Special Satellite," Rendezvous [Textron/Bell, Aerospace Division] (Summer, 1974), pp. 8-11; and NASA, JSC, "Apollo Soyuz Mission Evalua-

tion Report," JSC-10607, Oct. 1975, pp. 11-2 through 11-6.

13. Program Operations Office, "ASTP Technical Air-to-Ground Voice Transcription," pp. 46-47.

14. Ibid., p. 59.

15. Ibid,. pp. 60-63; and [NASA News Release], Apollo News Center, JSC, "ASTP Change-of-Shift Debriefing," 16 July 1975.

16. [NASA News Release], Apollo News Center, JSC, "ASTP Change-of-Shift Debriefing," 16 July 1975; and NASA, MSC, Mission Evaluation Team, "Apollo 14 Mission Anomaly Report No. 1, Failure to Achieve Docking Probe Capture Latch Engagement," MSC 05101, Oct. 1971.

17. Interview, John P. Donnelly and Robert J. Shafer-Ezell, 26 and 28 Jan. 1976; and ASTP mission commentary transcript, SR 13/1 and 13/2, 15 July 1975.

18. Program Operations Office, "ASTP Technical Air-to-Ground Voice Transcription," pp. 69-84; and Crew Training and Procedures Division Training Office, "ASTP Technical Crew Debriefing," p. 4-4. See also interview, Robert D. White-Ezell, 19 Aug. 1975:

> The technicians who had installed the pyros for the gas bottles which actuate the structural latches had misaligned one of the pyro caps so it blocked the hole into which the release handle shaft was to be inserted. That release handle and shaft had been redesigned after *Apollo 14*. Formerly the release handle was attached at all times, after *Apollo 14* it was stowed away from the shaft. This change was made to permit the discharge of the gas bottles when the capture latches had failed to engage. When the handle was in place, the gas bottles could not be fired, because the release shaft would bend upon retraction. The handle would jam against the hatch and bend the shaft. This alteration opened the way for the type of pyro misalignment which occurred in the ASTP mission. That misalignment showed up in the pre-mission buy-off photographs. Both the NASA and Rockwell inspectors missed this misalignment. . . . The crew had to remove the pyro cover and realign the pyro (which incidentally was the one fired for the structural latching) so the handle and shaft could be installed.

19. Program Operations Office, "ASTP Technical Air-to-Ground Voice Transcription," p. 85; and ASTP mission commentary transcript, MC 121/1, SR 31/1-3 and SR 32/1-3, 16 July 1975.

20. Program Operations Office, "ASTP Technical Air-to-Ground Voice Transcription," pp. 138-148; and ASTP mission commentary transcript, MC 158/1 and SR 38/1-2, 16 July 1975.

21. Program Operations Office, "ASTP Technical Air-to-Ground Voice Transcription," pp. 176, 191, and 198.

22. Ibid., pp. 199-200.

23. Ibid., p. 219.

24. Ibid., pp. 222-223.

25. Ibid., p. 231; and Crew Training and Procedures Division Training Office, "ASTP Technical Crew Debriefing," pp. 4-13 and 4-14.

26. ASTP mission commentary transcript, SR 72/1-2, 17 July 1975.

27. Program Operations Office, "ASTP Technical Air-to-Ground Voice Transcription," p. 257.

28. Ibid., pp. 261-264.

29. Ibid., p. 265.

30. Ibid., p. 267; and F. Dennis Williams to Arnold W. Frutkin, memo, "ASTP Symbolic Activities: Items Carried on the First International Flight" [n.d.]. The items included the FAI certificate of docking, U.S. and U.S.S.R. flags, one U.N. flag, commemorative placques, a commemorative medallion, six copies of the May 1972 Nixon-Kosygin Space Agreement in English and Russian, American and Russian tree seeds, silver medallions for presentation to Leonov and Kubasov, and copies of papers authored by K. E. Tsiolkovsky and Robert H. Goddard.

31. [NASA News Release], Apollo News Center, JSC, "Change-of-Shift Debriefing #7," 17 July 1975.

32. Program Operations Office, "ASTP Technical Air-to-Ground Voice Transcription," pp. 287, 294-301, and 303.

33. [NASA News Release], Apollo News Center, JSC, "Change-of-Shift Debriefing #10," 18 July 1975.

34. Program Operations Office, "ASTP Technical Air-to-Ground Voice Transcription," pp. 340-355; Hal Piper, "Moscow Shops Filled for Space Launchings," *Baltimore Sun,* 16 July 1975; Bill Anderson, "It's Just Another (yawn) Space Flight," *Chicago Tribune,* 17 July 1975; Gregg Kilday, "Hyping the Space Show," *Los Angeles Times,* 17 July 1975; "Millions Watch Awestruck; Soviet Leaders Hail Mission," *Christian Science Monitor,* 22 July 1975; and Lee Whinfrey, "Platitudes Dull in Space, Too," *Philadephia Inquirer,* 22 July 1975.

35. Program Operations Office, "ASTP Technical Air-to-Ground Voice Transcription," pp. 364-367.

36. Ibid., pp. 405-415; Thomas O'Toole, "Apollo Soyuz Crews Make Final Visit," *Wash-*

ington Post, 19 July 1975; Albert Sehlstedt, Jr., "Space Visit Ends," *Baltimore Sun,* 19 July 1975; and Peter Reich and James Pearre, "Apollo, Soyuz Crews End First Space Visit," *Chicago Tribune,* 19 July 1975.

37. Program Operations Office, "ASTP Technical Air-to-Ground Voice Transcription," pp. 419-420.

38. Crew Training and Procedures Division Training Office. "ASTP Technical Crew Debriefing," p. 4-27.

39. Program Operations Office, "ASTP Technical Air-to-Ground Voice Transcription," pp. 494-497; and Crew Training and Procedures Division Training Office, "ASTP Technical Crew Debriefing," p. 4-27.

40. Interview, White-Ezell, 19 Aug. 1975; Roscoe Lee to Task ASTP E-101 File, TRW memo, "Control System Post-flight Analysis for the Initial Mission Report," 30 July 1975; James O'Kackson, "Russ Bare Soyuz Strain in Linkup," *Chicago Tribune,* 21 July 1975; and Robert C. Toth, "Russians Upset by Hard Space Docking," *Los Angeles Times,* 21 July 1975.

41. Program Operations Office, "ASTP Technical Air-to-Ground Voice Transcription," pp. 539-540.

42. Interview, Roger A. Burke-Ezell, 14 Aug. 1975.

43. Interview, Donald K. Slayton-Ezell, 2 Mar. 1976.

44. Interview, Stafford-Ezell, 6 Apr. 1976.

45. ASTP mission commentary transcript, SR 132/1-2, 19 July 1975, SR 134/1-2, SR 143/1-2, SR 144/1, and SR 145/1-2, 20 July 1975.

46. Program Operations Office, "ASTP Technical Air-to-Ground Voice Transcription," p. 619.

47. ASTP mission commentary transcript, SR 156/1 and SR 159/1-2, 20 July 1975, and SR 171/2, 21 July 1975; "2 Cosmonauts, Soyuz Spacecraft Back in the USSR," *Washington Star,* 21 July 1975; Robert C. Toth, "Soyuz Touches Down Perfectly in Soviet Desert," *Los Angeles Times,* 22 July 1975; Hal Piper, "Soyuz Ends Its Half of Joint Flight," *Baltimore Sun,* 22 July 1975; and "Crowds Flock to Watch Rare Public Landing," *Washington Post,* 22 July 1975.

48. Program Operations Office, "ASTP Technical Air-to-Ground Voice Transcription," p. 679.

49. Ibid., pp. 925-940 and 989-990; and "Astronauts Look Forward to the Space Shuttle Era," *Washington Star,* 23 July 1975.

50. Program Operations Office, "ASTP Technical Air-to-Ground Voice Transcription," p. 1037.

51. Crew Training and Procedures Division Training Office, "ASTP Technical Crew Debriefing," pp. 5-4 and 5-5; NASA News Release PC-56, "ASTP Crew Status Briefing," 25 July 1975; NASA News Release PC-57, "ASTP Crew Medical Status Briefing," 26 July 1975; NASA News Release PC-58, "ASTP Crew Medical Status Briefing," 26 July 1975; NASA News Release PC-59, "ASTP Crew Status Briefing," 27 July 1975; NASA News Release PC-60, "ASTP Post Flight Systems Review," 28 July 1975; NASA, JSC, "Apollo Soyuz Mission Evaluation Report," JSC-10607, Oct. 1975, pp. 14-4 through 14-13; and Richard Saltus, "Apollo Leak: Crew Failed to Trip Switches," *Washington Post,* 29 July 1975.

52. Crew Training and Procedures Division Training Office, "ASTP Technical Crew Debriefing," p. 5-5.

53. Interview, Stafford-Ezell, 6 Apr. 1976; and NASA, JSC, "Apollo Soyuz Mission Anomaly Report No. 1, Toxic Gas Entered Cabin during Earth Landing Sequence," JSC-10638, Dec. 1975.

54. NASA News Release, JSC, 75-69 "Astronaut Slayton to Undergo Surgery," 19 Aug. 1975; NASA News Release PC-63, "D. K. Slayton Medical Briefing," 20 Aug. 1975; press release, M. D. Anderson Hospital, Houston, Tex., "D. K. Slayton," 26 Aug. 1975; and R. Z. Sagdeyev and V. A. Olshevskiy, "Sovmestnymi usiliyami" [Through joint efforts], in *Soyuz i Apollon, rasskazivayut sovetskie uchenie inzheneri i kosmonavti—ychastniki sovmestnikh rabot s amerikanskimi spetsialistami* [Soyuz and Apollo, related by Soviet scientists, engineers, and cosmonauts—participants of the joint work with American specialists], Konstantin D. Bushuyev, ed. (Moscow, 1976), pp. 180-198.

EPILOGUE

1. NASA, JSC, Announcement, "Establishment of Shuttle Payload Integration and Development Program Office," 4 Aug. 1975; and Edward C. Ezell, notes on telecon, 29 Oct. 1975.

2. "Minutes of Joint Meeting, USSR Academy of Sciences and US National Aeronautics and Space Administration," Nov. 1975.

3. Jonathan Spivak, "The First Space Handshake," *Wall Street Journal,* 22 July 1975; and

news release, issued by the office of Senator William Proxmire, Wisconsin, 16 July 1975. Also see *Rukopozhati v kosmose* [Handshake in space] (Moscow, 1975), a collection of Soviet news accounts describing the joint mission, published by *Izvestiya;* and Kostantin D. Bushuyev, ed., *Soyuz i Apollon, rasskazivayut sovetskie uchenie inzheneri i kosmonavti—ychastniki sovmestnikh rabot s amerikanskimi spetsialistami* [Soyuz and Apollo, related by Soviet scientists, engineers, and cosmonauts—participants of the joint work with American specialists] (Moscow, 1976).

4. Robert Hotz, "Techno-Politics in Space," *Aviation Week and Space Technology,* 28 July 1975, p. 30.

5. Interview, Christopher C. Kraft-Ezell, 12 Apr. 1976.

6. Interview, Thomas P. Stafford-Ezell, 6 Apr. 1976.

7. ASTP mission commentary transcript, PC 36/E1, 17 July 1975.

8. Interview, Caldwell C. Johnson-Ezell, 27 Mar. 1975.

9. Interview, Kraft-Ezell, 12 Apr. 1976; and interview, Stafford-Ezell, 6 Apr. 1976.

10. Interview, Kraft-Ezell, 28 Mar. 1976.

11. Interview, Kraft-Ezell, 12 Apr. 1976.

12. Low to Monte D. Wright, NASA History Office, 2 Aug. 1976.

13. NASA News Release, HQ, 76-2, "Study of Cosmos Experiments Is Under Way," 13 Jan. 1976.

14. Melvin Calvin and Oleg G. Gazenko, *Foundations of Space Biology and Medicine: Joint USA/USSR Publication in Three Volumes* (Washington and Moscow, 1975-1976).

Sources and Research Materials

This essay is intended to serve as a guide to the sources used in preparing this history. As such, it is not designed to be an inclusive catalogue. For those who are interested in how we researched this book, for those who would like at some future date to follow in our steps, or for those who would attempt a contemporary history of their own, we would offer this road map to the materials from which we have woven the story of the Apollo-Soyuz Test Project.

From the standpoint of sources, this book can be divided into two parts—chapters I through III; and the prologue and chapters IV through the epilogue. In the former, we used the traditional sources familiar to the researcher—books, periodical and newspaper articles, and occasional primary documents from within the National Aeronautics and Space Administration. Whenever possible we made an attempt to use both Russian and English language publications in an effort to present a balanced view of the "Years Before," "Dryden and Blagonravov," and "Routes to Space Flight." A number of books were used over and over again in writing these background chapters:

Astashenkov, P. T. *Akademik S. P. Korolev.* Moscow, 1969. (Available in English as *Academician S. P. Korolev, Biography.* Air Force Foreign Technology Division-HC-23-542-70.)

Daniloff, Nicholas. *The Kremlin and the Cosmos.* New York, 1972.

Frutkin, Arnold W. *International Cooperation in Space.* Englewood Cliffs, N.J., 1965.

Green, Constance McLaughlin, and Lomask, Milton. *Vanguard: A History:* NASA SP-4202, Washington, 1970.

Harvey, Dodd L., and Ciccoritti, Linda C. *U.S.-Soviet Cooperation in Space.* Coral Gables, Fla., 1974.

Krieger, F. J. *Behind the Sputniks: A Survey of Soviet Space Sciences.* Washington, 1958.

Logsdon, John M. *The Decision to Go to the Moon: Project Apollo and the National Interest.* Cambridge, Mass., and London, 1970.

Narimanov, G. S., ed. *The Conquest of Space in the USSR.* NASA TTF-15,678, Washington, 1974. (Translation of *Osvoyeniye kosmicheskogo prostranstva v SSSR* [1972]. Moscow, 1974.)

Petrov, G. I., ed. *Osvoenie kosmicheskogo prostva v SSSR: ofitsial'nye soobscheniya TASS i materialy tsentral'noi pechati Oktyabr', 1967-1970 gg.* Moscow, 1971. (Available in English as *Conquest of Outer Space in the USSR: Official Announcements by TASS and Material Published in the National Press from October 1967 to 1970.* NASA TTF-725, New Delhi, 1973.)

Riabchikov, Evgeny. *Russians in Space.* Translated by Guy V. Daniels. Garden City, N.Y., 1971. (An official view of the Soviet space program prepared under the direction of Novosti Press and published in the United States by Doubleday and Co.)

Rosholt, Robert L. *An Administrative History of NASA, 1958-1963.* NASA SP-4101, Washington, 1966.

Skuridin, G. A., ed. *Mastery of Outer Space in the USSR, 1957-1967.* NASA TTF-773, 1975. (Translation of *Osvoyeniye kosmicheskogo prestranstva v SSSR, 1957-1967 gg.* Moscow, 1971.)

Smolders, Peter. *Soviets in Space: The Story of the Salyut and the Soviet Approach to Present and Future Space Travel.* Translated by Marian Powell. Guildford and London, 1973.

Swenson, Loyd S., Jr., Grimwood, James M., and Alexander, Charles C. *This New Ocean: A History of Project Mercury.* NASA SP-4201, Washington, 1966.

Umansky, S. P. *Chelovek na komisheskoy orbite.* Moscow, 1975. (Available in English as *Man in Space Orbit.* NASA TTF-15973.)

U.S., Congress, Senate, Committee on Aeronautical and Space Sciences. *Documents on International Aspects of the Exploration and Use of Outer Space, 1954-1962.* 88th Cong., 1st sess., 1963.

———. *International Cooperation and Organization for Outer Space.* 89th Cong., 1st sess., 1965.

———. *Soviet Space Programs, 1962-1965: Goals and Purposes, Achievements, Plans, and International Implications.* 89th Cong., 2d sess., 1966.

———. *Soviet Space Programs, 1966-70: Goals and Purposes, Organization, Resources, Facilities and Hardware, Manned and Unmanned Flight Programs, Bioastronautics, Civil and Military Applications, Projections of Future Plans, Attitudes toward International Cooperation and Space Law.* 92d Cong., 1st sess., 1971.

———. *Soviet Space Programs, 1971: A Supplement to the Corresponding Report Covering the Period 1966-70.* 92d Cong., 2d sess., 1971.

———. *Soviet Space Programs, 1971-75 Overview, Facilities and Hardware, Manned and Unmanned Flight Programs, Bioastronautics,*

Civil and Military Applications, Projections of Future Plans, Vols. I and II. 94th Cong., 2d sess., 1976.

————. *Soviet Space Programs: Organization, Plans, Goals, and International Implications.* 87th Cong., 2d Sess., 1962.

Vladimirov, Leonid. *The Russian Space Bluff.* Translated by David Floyd. London, 1971.

We should also call attention to the two major Soviet publications on the joint mission: *Rukopozhatie v kosmose* [Handshake in space] (Moscow, 1975), a collection of Soviet news accounts describing the joint mission published by *Izvestiya* (also available as "Handshake in Space," NASA TTF 17045); and Konstantin D. Bushuyev, ed., *Soyuz i Apollon, rasskazivayut sovetskie uchenie, inzheneriy i kosmonavtiy—uchastniki sovmestnikh rabot s amerikanskimi spetsialistami* [Soyuz and Apollo, related by Soviet scientists, engineers, and cosmonauts—participants of the joint work with American specialists] (Moscow, 1976), a collection of essays written by the Soviet Working Group chairmen and other leading participants in the mission published on the first anniversary of the 15 July 1975 launch.

The following periodicals and newspapers were used repeatedly:

Aviation Week and Space Technology
Baltimore Sun
Department of State Bulletin
Houston Chronicle
Izvestiya [The (latest) news]
Krasnaya Zvezda [Red star]
Missiles and Rockets
Nauka i Zhin' [Science and life]
New York Times
Pravda [Truth]
Space Business Daily
Trud [Labor]
Wall Street Journal
Washington Post
Washington Star

In the prologue, chapters IV through XI, and the epilogue, we have relied upon two types of primary sources—official NASA documents and oral history materials. An examination of the variety of primary materials will give the reader a better understanding of the documentation we have left behind at the History Office of the Johnson Space Center (JSC).

When we arrived in Houston in the spring of 1974, work was in full swing on ASTP. Since we planned to observe the negotiations and testing activities as time permitted, James M. Grimwood, JSC Historian, recommended that we keep a record of the events we witnessed. We quickly evolved a scheme for log notes numerically filed from 1 to 65, which were, in effect, after-action reports. These notes, which ranged from transcribed transactions of Glynn Lunney's tag-up meetings to reports on meetings to ephemera collected during the joint sessions (e.g., agenda, lists of delegates, and invitations to leisure activities) to transcriptions of interviews, served as an *aide mémoire* for the time when we began to write about the events we witnessed. Simultaneously, we began to collect documentation that would be necessary to write the history.

Over the years since the establishment of the History Office at JSC, Grimwood and his able assistant, archivist, and editor, Sally D. Gates, have cultivated a sense of history at the center. Houston participants who keep their own "desk archives" relating to a particular project have been encouraged to send these non-official copies of documents to the History Office, which maintains unofficial but valuable working archives relating to the history of manned space flight. Unlike the official record copies that are retired to the Federal Records Center at Forth Worth, Texas, these items, mainly photocopies, are in effect pre-screened for historical value and are readily at hand to the official historians. If we had been trying to write the same history from documents at the Federal Records Center, it would have taken years. When we arrived at JSC, Gates had already sorted out a large number of ASTP documents as part of an ongoing effort to segregate materials according to project as time is available for her to do so. While these materials were not arranged in any fashion, this group of letters, memoranda, telexes, and minutes of meetings formed a basis for our files.

Among the materials Gates had collected were a group of documents covering the period October 1970 to May 1972, which had been sent to the History Office by René A. Berglund prior to his retirement in early 1974. Thus, we had a large body of documents waiting for us, all of which were considered by those who had been working on ASTP to be of primary importance. The first question that faced us was how best to organize these materials. We decided to separate Working Group documents (minutes, reports, test activities, and data) from correspondence. From Hugh M. Scott of NASA and Jerry Siemers and Harry Hall of Boeing,* we learned that Boeing was maintaining a data file of all materials generated for or at meetings—agenda, briefings, technical documents exchanged, photographs and drawings exchanged, joint communiques, ASTP documents, Interacting

---

*Boeing had a contract to manage ASTP technical documentation.

SOURCES AND RESEARCH MATERIALS

Equipment Documents, and Interacting Equipment Revision Notices. Using their data file, we established our own files for each meeting including the earlier meetings (before July 1972) for which Boeing had not been responsible. Since the joint meetings were the major feature in the organization and functioning of this project and since all other activities were organized with these meetings in mind, this filing arrangement gave us a systematic method of keeping track of Working Group activities. When the project was completed, we had a compilation of the major documents prepared by the two sides. Whenever possible, we acquired copies of draft documents as well. A complete list of the numbered ASTP project documents is given in attachment 1. Attachment 2 provides an example of a data file.

Our second task was the organization of the correspondence created by ASTP. After considerable trial and error, we reverted to a simple chronological arrangement for these communications. However, we did create three correspondence files that are not included in this general chronological arrangement. First, the correspondence between the Apollo Spacecraft Program Office (ASPO) and the spacecraft contractor, Rockwell International, is highly technical, dealing mainly with design changes, production progress, component availability, and other such information related to preparing the command and service module (CSM) and the docking module (DM). Correspondence relating to public affairs activities is organized separately, as well. This material was collected with the help of Robert J. Shafer and his secretary, Patsy Respess. We selected these data from John P. Donnelly's and Shafer's ASTP files at Headquarters, and Respess and Evelyn L. Taylor assisted by copying nearly 1500 pages of correspondence and items relating to the negotiation of the Public Affairs Plan. Bennet James and John Riley at JSC provided additional materials to help us complete our collection of pertinent public affairs documents. The third group of materials we segregated from the general ASTP correspondence related to the scientific payload.

The documents that make up the correspondence files were obtained from a variety of sources, such as retired reading files. Throughout NASA there exist unofficial and official copies of most correspondence. At JSC, copies of correspondence are distributed to the appropriate offices at the center concerned with its contents. (A memorandum, for example, addressed to "Distribution" would be circulated among various individuals and/or offices at the discretion of the author.) The Center Director would receive all policy and much top level management correspondence from within the agency, and his staff systematically would gather, copy, and circulate these items to the Director and his staff in the Director's Daily Reading File. Once these files had served their function of keeping management informed, they were routinely retired to the History Office.

For our ASTP archives, more detailed files were obtained from the Apollo Spacecraft Program Office, where Betty Cornett, Mary F. Crocker, and Betty Sue Fedderson of Glynn Lunney's staff kept track of the day-to-day aspects of the project. Lunney's files were the most useful to us since much of their contents was written from the vantage point of the Project Manager. Key materials were brought together in these files, representing the documents that Lunney and his staff thought to be the most important. More routine materials and all official record copies were filed in the ASPO Correspondence and Records Office. We owe a debt of thanks to Virginia Trotter who guided us through that maze of documents, permitting us to borrow armloads of folders at a time.

Another boon to our document collection came when we were placed on the distribution list for all correspondence sent from the Apollo Program Office. In turn, this office also distributed the correspondence it received. Starting early in April 1974, we had a reading file of our own with which we could keep abreast of current project activities. After the mission, the Working Group chairmen and other participants sent us boxes of their personal reading files and other working materials. We sorted through these, weeding out the duplicates and the materials that were too detailed. Exercising historical judgment, we tried to preserve copies of all items that we used in our source notes and any additional materials that might help future historians who would wish to pursue a particular point in more detail.

These JSC materials were supplemented by documents acquired from NASA Headquarters. George Low and his secretary, Shirley Malloy, were very gracious in sending us copies of such pertinent materials as trip reports from his files. Secretary Donna Skidmore and other members of Chester M. Lee's staff were also very helpful when it came to providing copies of documents from Captain Lee's reading files. After the splashdown, Lee's reading files were retired to the Headquarters History Office. Archivist Lee D. Saegesser promptly sent us these folders so we could check them against our holdings. In addition, throughout our work on ASTP, Saegesser has inundated us weekly with news clippings (Soviet and American), translations of Soviet journal articles, and numerous other items related to the joint project. Saegesser's presence in Washington and his complete enthusiasm for helping researchers saved us many hours of searching for specific documents and several trips to Headquarters.

When pulled together and arranged in chronological order, these letters, memoranda, trip reports, and minutes of telephone conversations formed the backbone of the ASTP history skeleton. The most formal of these items were the letters. The agency usually employs letters to communicate with the outside world and formally within NASA. We found letters to the Soviets, to members of Congress, to Defense Department personnel, and to the scientific community. During ASTP, all project-related letters between the United States and the Soviet Union were channeled through the respective ASTP Technical Directors, Lunney and Bushuyev. Letters from NASA were sent over Lunney's signature. From JSC, the letter went to Chet Lee's office, which in turn sent it to Arnold Frutkin's International Affairs office. It would then be delivered to the State Department for dispatch to the Soviet Union via diplomatic pouch. The American Embassy in Moscow delivered the letter to Bushuyev at the Soviet Academy of Sciences. All letters to the Soviets were sent in English. All letters from the U.S.S.R. were received in Russian. Each side translated the correspondence it received.

Telexes were generally used for high priority communications. These TWXs included messages to other NASA centers, to other government agencies, to contractors, and to the Soviets. Like telephone conversations, the telex gave Lunney and Bushuyev much quicker and more direct communication. It became a very valuable management tool in the course of preparing for the joint mission.

By far the most common form of communication we encountered was the memorandum, which NASA uses for most internal correspondence. We have chosen to cite these by the names of the author and addressee rather than by the mail codes generally used by NASA. Memos cover a variety of subjects. For example, trip reports from NASA engineers, negotiators, and astronauts were distributed in memo form so Lunney and others could get a better idea of what happened on the working trips to the Soviet Union.

Because we live in the era of the telephone, many actions and decisions were not recorded in formal documents. Therefore, we found it particularly helpful to interview the participants frequently, either by telephone or in person. Interviewing is both the strength and the weakness of contemporary history. With interviews, we obtained explanations of cryptic or confusing documents or gathered insights not recorded in the official records. The joint minutes of the Working Group meetings were distilled; interviews often gave us opportunities to discover what lay behind certain diplomatically phrased passages. Or when we encountered briefing charts that gave us only a clue to an important story, the interview supplied the details.

But interviews are also a potential hazard. Individuals can selectively remember some facts and as conveniently forget others, so we always made an effort to confirm any one version of an event with documentation and other interviews. A more common problem was the failure to remember at all. Engineers and technical managers at NASA shared a common tendency to forget about a project or event once it had passed. Always looking at today's technical problems or concerns, they often do not remember earlier crises because they were resolved. Therefore, it was frequently necessary for us to have a document or photograph in hand with which to jog memories. A common response was, "I'd forgotten all about that until now." Clearly, problems ceased to be problems—sometimes ceased to exist—once they were solved. However, one of the values of writing a history so close to the events is to preserve elements of the past that might otherwise not be recorded or might simply disappear from memory. Many readers of our comment edition reacted the same way—"Did all that really happen?"

The interviews we collected vary in length and detail. Some were lengthy conversations that were tape recorded and transcribed. Others were short 5- to 15-minute discussions about specific topics; for example, the reaction control system (RCS) impingement problem discussed in chapter IX. Still others were conducted over the telephone with only notes for a record. At all times, we received only the fullest cooperation. The following persons aided us through interviews:

Anderson, Oscar E., Jr.	Hawkins, W. Royce
Biggs, Charles A., Sr.	Holmes, Tamara
Brand, Vance D.*	Jaax, James R.
Brzezinski, M. S.	James, Bennett W.
Burke, Roger A.	Johnson, Caldwell C.*
Cernan, Eugene A.*	Jones, James C.
Cheatham, Donald C.	King, John W.
Covington, Clarke*	Kraft, Christopher C.*
Creasy, William K.*	Latter, Natalie
Culbertson, Philip E.	Lee, Chester M.
Cundieff, Lonnie D.	Lee, Roscoe
Dietz, R. H.*	Low, George M.
Donnelly, John P.*	Lunney, Glynn S.*
Epstein, Donalyn*	Nicholson, Leonard S.*
Frutkin, Arnold W.*	Overmyer, Robert F.*
Gilruth, Robert R.	Paine, Thomas O.†
Guy, Walter W.*	Pollock, S. T.
Haken, Richard L.	Riley, John E.*
Handler, Philip†	Roberts, James Leroy
Hardy, George B.	Ross, Thomas O.

*More than one interview.
†By letter only.

410

Scott, David R.
Shafer, Robert J.*
Slayton, Donald K.*
Smith, Herbert E.
Smylie, Robert E.*
Stafford, Thomas P.
Syromyatnikov, V. S.
Tatistcheff, Alex

Taub, Willard M.*
Taylor, Ada
Timacheff, Nicholas*
Travis, A. Don
Waite, J. C.
Webb, James E.[†]
White, Robert D.

Three other individuals that deserve our thanks are Mary Kerber, who helped us with the typing and retyping; Robert V. Gordon from the JSC Public Affairs Office, who always made sure that we knew about ASTP briefings for the press, news releases, and other pertinent activities; and Andrew R. Patnesky, JSC Photographer, who supplied us with so many excellent photographs of ASTP activities.

Another category of source material deserves special mention—the responses we received on the comment edition of this history. Early in December 1975, we distributed 135 copies of our draft. The comments this early version brought varied considerably in scope, format, and value, but a number were very useful in completing the final manuscript and in saving us from embarrassing mistakes. Those who commented were:

Anderson, Oscar E.
Brand, Vance D.
Brieseth, Christopher
Compton, W. David
Covert, Elizabeth R.
Covington, Clarke
Creasy, William K.
Dietz, R. H.
Donnelly, John P.
Emme, Eugene M.
Epstein, Donalyn
Forostenko, Anatole
Frutkin, Arnold W.
Gates, Sally D.
Giuli, R. Thomas
Grimwood, James M.

Guy, Walter W.
Hall, R. Cargill
Hecht, Kenneth F.
Holley, I. B.
Huss, Carl R.
Jaax, James R.
King, John W.
Kraft, Christopher C.
Kranzberg, Melvin
Larson, Ray
Lavroff, Ross
Lee, Chester M.
Lockyer, William
Low, George M.
Lunney, Glynn S.
Maines, Howard G.

*More than one interview.
[†]Interview by someone other than the authors.

411

Morton, Louis

Nicholson, Leonard S.

Riley, John E.

Roberts, James Leroy

Roland, Alex

Shafer, Robert J.

Slayton, Donald K.

Smith, Herbert E.

Smylie, Robert E.

Stafford, Thomas P.

Taub, Willard M.

Taylor, Ada

Timacheff, Nicholas

Underwood, Richard W.

White, Robert D.

Wright, Monte D.

Young, Kenneth A.

Zavoico, Irene

In addition to the correspondence and project documentation, we used NASA news releases, transcripts of press conferences, mission-related briefings, and the air-to-ground transcripts to add life and human interest to the text. Equally useful at times were technical reports prepared by the contractors. Unlike earlier programs, ASTP did not generate a large number of press kits or handbooks for the mission, but we did find those generated during the lunar flights to be quite useful regarding the Apollo spacecraft and the Saturn launch vehicle. ASTP produced many public affairs firsts. Among these news materials, one unique pair of documents was developed for the joint mission—the bilingual editions of NASA, "Apollo Soyuz Test Project Kit" [July 1975], and Soviet Academy of Sciences, "Apollo-Soyuz Test Project Information for Press" [July 1975].

The reader is invited to peruse the chapter notes for other source materials not mentioned.

ATTACHMENT 1—COMPLETE LIST OF IDENTIFIED ASTP DOCUMENTS

| ASTP 10 000 | Project Technical Proposal |

Planning Documents

ASTP 20 000, Part I	Organization Plan for Apollo Soyuz
ASTP 20 000, Part II	Organization Plan Part II
ASTP 20 010	Transportation of Equipment between the USSR and the USA
ASTP 20 020	ASTP Glossary
ASTP 20 021	ASTP Acronyms and Abbreviations
ASTP 20 022	Representative Crew Communications
ASTP 20 050, Part I	ASTP Public Information Plan—Part I
ASTP 20 050, Part II	ASTP Public Information Plan—Part II

Safety Assessment Reports

ASTP 20 101	Safety Assessment Report for the Apollo Structural Ring Latches
ASTP 20 102	Safety Assessment Report for Apollo Propulsion and Control Systems
ASTP 20 103	Safety Assessment Report for Apollo Fire Safety and Flammability
ASTP 20 104	Safety Assessment Report for Apollo Pyrotechnic Devices
ASTP 20 105	Safety Assessment Report for Apollo Cabin Pressure
ASTP 20 106	Safety Assessment Report for Apollo Manufacturing, Test and Checkout
ASTP 20 107	Safety Assessment Report for Apollo Radio Command Systems
ASTP 20 201	Safety Assessment Report for the Soyuz Structural Ring Latches
ASTP 20 202	Safety Assessment Report for Soyuz Propulsion and Control Systems
ASTP 20 203	Safety Assessment Report for Soyuz Fire Safety and Flammability
ASTP 20 204	Safety Assessment Report for Soyuz Pyrotechnic Devices
ASTP 20 205	Safety Assessment Report for Soyuz Cabin Pressure
ASTP 20 206	Safety Assessment Report for Soyuz Manufacturing, Test and Checkout
ASTP 20 207	Safety Assessment Report for Soyuz Radio Command System

Scheduling Documents

ASTP 30 000	Project Schedule Documents

Mission Documents

ASTP 40 000	Mission Requirements
ASTP 40 001	Design Characteristics for Soyuz and Apollo
ASTP 40 010	Onboard Television and Photography Plan

ASTP 40 012	Activity Plan for Apollo and Soyuz Mockups
ASTP 40 100	Launch Window Plan
ASTP 40 200	Trajectory Plan
ASTP 40 201	Trajectory Computation Model
ASTP 40 300	Flight Plan Guidelines
ASTP 40 301	Joint Crew Activities Plan
ASTP 40 400	Mission Operations Plan
ASTP 40 401	Control Centers Interaction Plan
ASTP 40 402	Prelaunch Preparation Plan
ASTP 40 500	Contingency Plan
ASTP 40 600	Onboard Joint Operations Instructions
ASTP 40 700	Crew and Ground Personnel Training Plan
ASTP 40 701	Summary Training Plan for USA/USSR Flight Controllers
ASTP 40 702	Checkout Program of Ground Personnel Interaction Procedures in December 1974
ASTP 40 703	Control Centers Training Plan with Crew Participation in March 1975
ASTP 40 704	Control Centers Training Plan with Crew Participation for May 1975
ASTP 40 705	Control Centers Training Plan with Crew Participation for June 1975
ASTP 40 706	Plan for Specialists and Flight Crew Activities with Flight Spacecraft at the American and Soviet Launch Complexes
ASTP 40 800	Post Mission Report

Interacting Equipment Documents

IED 50 001	Technical Requirements for Compatible Docking Systems for the Apollo Soyuz Test Project
IED 50 002	Apollo Soyuz Joint Development Plan, Docking Systems
IED 50 003	Test Plan for Scale Models of Apollo Soyuz Docking System
IED 50 004	Apollo Soyuz Physical Interface Requirements
IED 50 005	Apollo Soyuz Docking System Load Requirements
IED 50 006	Apollo Soyuz Docking System Thermal Interface

IED 50 007	USSR Ground Support Equipment/USA Docking System Equipment, Mechanical and Electrical Interface Requirements
IED 50 008	USA Ground Support Equipment/USSR Docking System Equipment, Mechanical and Electrical Interface Requirements
IED 50 009	Apollo Soyuz Joint Development Test Plan, Docking Systems
IED 50 010	Apollo Soyuz Joint Qualification Test Plan, Docking Systems
IED 50 011	Apollo Soyuz Preflight Compatibility Verification Test Plan, Docking Systems
IED 50 012	Results of Apollo Soyuz Docking Systems Scale Model Tests
IED 50 013	Results of Apollo Soyuz Docking Systems Development Tests
IED 50 014	Results of Apollo Soyuz Docking Systems Qualification Tests
IED 50 015	Results of Apollo Soyuz Docking Systems Preflight Compatibility Verification Test
IED 50 016	Apollo Soyuz Docking System Sequence of Docking and Undocking
IED 50 101	Technical Requirements for the Radio Communications and Ranging System
IED 50 102	Interface Signal Characteristics for the Radio Communications and Ranging System
IED 50 103	Compatibility Test Plan for the Communications Systems
IED 50 104, Part I	Compatibility Test Procedures for the Radio Communications and Ranging System
IED 50 104, Part II	Compatibility Test Results for the Radio Communications and Ranging System
IED 50 105	Soyuz Test System/Compatibility Test Laboratory Interface Requirements
IED 50 106	Inflight VHF Coverage Analysis
IED 50 107	Circuit Margins for the Radio Communication and Ranging System
IED 50 108	Definition of Terms and Abbreviations for the Radio Communications and Ranging System
IED 50 109	Development Plan for the Radio Communications and Ranging System

IED 50 110   Test Procedure and Results for the Implementation of the Apollo VHF Transceiver and Range Tone Transfer Assembly into the Soyuz Spacecraft

IED 50 112   Report on Investigation of Radio Frequency Effects of Apollo Soyuz and Ground Transmitter on Spacecraft Receivers

IED 50 113   Radio Frequency Interference Compatibility Data

IED 50 114   Apollo VHF Equipment Management Requirements

IED 50 115   Plan for Implementation of the Apollo VHF Equipment into the Soyuz Spacecraft

IED 50 116   Preflight Compatibility Verification Test Plan, Apollo VHF Equipment

IED 50 117   Apollo VHF/AM Equipment Preflight Verification Procedures and Results

IED 50 118   Preflight Compatibility Verification Test Plan, VHF/FM Equipment

IED 50 119   Preflight Test Procedures and Results, VHF/FM Equipment

IED 50 120   VHF/AM Ground Test Equipment Calibration Procedures

IED 50 121   VHF/AM Flight Equipment and Ground Test Equipment/Test Procedures

IED 50 201   Technical Requirements for the Docking Alinement Targets

IED 50 202   Verification Test Plan for Docking Alinement Targets

IED 50 203   Results of Verification Testing of Docked Alinement Targets

IED 50 205   Development Plan for Implementation of the Docked Alinement Target

IED 50 301   Technical Requirements for External Lights

IED 50 401   Technical Requirements for Stabilization and Control System

IED 50 402   Verification Test Plans of Stabilization and Control Systems

IED 50 403   Verification Test Results of Stabilization and Control Systems

IED 50 404   Docking Initial Contact Condition Criteria

IED 50 405	Nominal and Contingency Control Procedures for Docking, Docked and Undocking Operations
IED 50 406	Axis Convention To Be Used for Spacecraft Maneuver Definitions
IED 50 501	Technical Requirements, Inter-Control Center Communications System
IED 50 502	Development Plan, Inter-Control Center Communications System
IED 50 503	Test Plan, Inter-Control Center Communications System
IED 50 504	Test Procedures and Test Results, Inter-Control Center Communications System
IED 50 505	Inter-Control Center Communications Maintenance and Operations Procedures
IED 50 506	Message Format Conventions Inter-Control Center Communications System
IED 50 507	Television Line Transmission Schedule
IED 50 601	Cable Communications Requirements
IED 50 602	Development Plan for Cable Communications
IED 50 603	Cable Communications Preflight Compatibility Verification Test Plan
IED 50 604	Cable Communications Preflight Compatibility Verification Test Procedures and Results
IED 50 605	Plan for Preflight Tests for Electromagnetic Compatibility of Cable Communications Terminal Devices
IED 50 606	Procedures for Preflight Tests for Electromagnetic Compatibility of Cable Communications Terminal Devices
IED 50 607	Test Plan and Procedures for TV Lighting and Fit Checks of USA Terminal Devices in the Soyuz Mockup in Moscow
IED 50 608	Report of Joint Test on the Apollo Flight Spacecraft at Kennedy Space Center
IED 50 609	Report on the Joint Tests on the Soyuz Flight Spacecraft at Baikonur Launch Site
IED 50 701	Command Module Environment Definition
IED 50 702	Docking Module Environment Definition
IED 50 703	Soyuz Environment Definition

IED 50 704	Soyuz/DM Environment Interface Definition
IED 50 706	General Operational Description of the Systems for Environmental Control and Crew Transfer in the Docking Module
IED 50 715	Fire Safety Control Requirements for Materials Transferred from Apollo to Soyuz
IED 50 716	Apollo Atmosphere Toxicological Requirements
IED 50 717	Soyuz Atmosphere Toxicological Requirements
IED 50 719	Crew Transfer Operations Definition
IED 50 720	Materials Fire Safety Certification for USA Equipment Transferred to Soyuz
IED 50 721	Materials Fire Safety Certification for USSR Equipment Transferred to Apollo
IED 50 722	USA Radio Equipment/Soyuz Structure Thermal Interaction
IED 50 723	Functional Description of the Provisions for Transfer and Mixed Crew Presence in Soyuz Spacecraft
IED 50 724	Analysis of Non-Nominal Situations Involving the Soyuz Life Support Systems and Apollo Environmental Control Systems
IED 50 725	General Operational Description of Command Module Environmental Control System
IED 50 726	Report on Results of Soyuz Life Support System and Transfer Provision Tests
IED 50 727	Environmental Control System Test and Operational Verification of the Apollo Spacecraft
IED 50 728	Assessment of the Joint Operation of the Soyuz Life Support System and Apollo Environmental Control System Based on Independent Testing and Flight Experience
IED 50 729	Report of Flight Readiness of the Environmental Control and Life Support Systems and Transfer Provisions for the Soyuz and Apollo Spacecraft for Joint Operations in ASTP
IED 50 803	Solar Eclipse
IED 50 804	Zone-Forming Fungi
IED 50 805	Microbial Exchange

| IED 50 806 | Furnace System Experiment |
| IED 50 807 | Ultra-Violet Absorption Experiment |

ATTACHMENT 2—"DATA FILE OF SOVIET MEETING JULY 6-18, 1972"

1.0 Signed Documents and Meeting Minutes

The ASTP/Interacting Equipment Documents (IED) documents and summary minutes signed on 17 July 1972 were placed in the vaults the week of 7 August 1972.

The masters are available for reproduction on an as required basis. These documents include:

1	ASTP 10 000
2	ASTP 20 000, Part I
3	ASTP 30 000
4	MSC 05 887
5	IED 50 001
6	IED 50 002
7	IED 50 004 Reproducibles
	April 3, 1972
	July 17, 1972
8	IED 50 101
9	IED 50 201
10	IED 50 205
11	IED 50 301
12	IED 50 401
13	IED 50 402
14	IED 50 404
15	IED 50 601
16	IED 50 602
17	IED 50 701
18	IED 50 702
19	IED 50 703
20	Signed Summary Minutes—All Working Groups

2.0 Data Provided by USA

2.1 Preliminary Documents

| 2.1.1 | ASTP 40 100 | Launch Window |
| 2.1.2 | ASTP 40 200 | Trajectory Plan |

2.1.3	ASTP 40 300	Crew Activities Plan
2.1.4	ASTP 40 400	Mission Operations Plan
2.1.5	ASTP 40 500	Contingency Plan
2.1.6	ASTP 40 600	Detailed Operational Procedures
2.1.7	ASTP 40 700	Training Plan
2.1.8	IED 50 005	Functional Requirement Loads & Bending Moment
2.1.9	IED 50 006	Functional Requirement, Docked Thermal Interface
2.1.10	IED 50 102 (BUILD)	Performance and Interface Signal Characteristics for RF Communications & Ranging System (BUILD)
2.1.11	IED 50 102 (EXCHANGE)	Performance and Interface Signal Characteristics for RF Communications & Ranging System (EXCHANGE)
2.1.12	IED 50 103 (BUILD)	Compatibility Test Plan for RF Communications and Ranging System (BUILD)
2.1.13	IED 50 103 (EXCHANGE)	Compatibility Test Plan for RF Communications and Ranging System (EXCHANGE)
2.1.14	IED 50 109	Development Plan, Apollo VHF Transceiver and Range Tone Transfer to Soyuz Spacecraft
2.1.15	IED 50 110 (EXCHANGE)	Development Plan, Soyuz VHF-FM Transfer to Apollo Spacecraft (EXCHANGE)
2.1.16	IED 50 113*	RFI Compatibility Data Requirements
2.1.17	IED 50 202	Verification Test Plan, Installation of Docking Alinement Target
2.1.18	IED 50 203	Results, Verification Test Plan, Installation of Docking Alinement Target
2.1.19	IED 50 302	Verification Test Plan, External Light System for Rendezvous and Docking

*This number was formerly IED 50 507.

2.1.20	IED 50 303	Results, Verification Testing, External Light System for Rendezvous and Docking
2.1.21	IED 50 709	Liquid Cooled Garment Definition
2.1.22	IED 50 711	Pressure Garment Assembly Definition
2.1.23	IED 50 712	Extravehicular Visor Assembly Definition
2.2.24	IED 50 713	Emergency Oxygen Purge System Definition
2.1.25	IED 50 715	Flammability Control Requirements for Transferred Materials
2.1.26	IED 50 716	Toxicological Considerations

2.2 Other Data Provided by USA

2.2.1	WG 1	
	Item 1	Presentation material
	Item 2	Mission model
	Item 3	Proposed agenda
2.2.2	WG 3	Outline of Apollo/Soyuz Docking System Dynamics Testing
2.2.3	WG 4	
	2.2.3.1	Preliminary Apollo VHF Transceiver Control Panel Interconnections for Installation in Soyuz
	2.2.3.2	Range Tone Transfer Assembly Specification Control Drawing LSC 380-00080
	2.2.3.3*	Outline and Mounting—Transceiver Assembly, VHF-RCA Drawing 8359401
	2.2.3.4*	Interconnections Diagram—Transceiver Assembly, VHF-RCA Drawing 8359363
	2.2.3.5	Operating Description of the Apollo/Skylab Television Camera

*2.2.3.3 and 2.2.3.4 were Xerox copies reproduced from the blueprints and given to the Soviets; 8½- by 11-inch masters have been made of both documents; copies of these form part of the data stored.

2.2.3.6	Color photographs (4) of television camera equipment*
2.2.3.7	System configuration—Westinghouse Drawing 2 RD 2600—10 sheets
2.2.3.8	Color television outline—RCA Drawing 2265870
2.2.3.9	Ground Commanded Television Assembly Operation and Checkout Manual
2.2.3.10	Ground Commanded Television Assembly Interim Final Report

2.2.4   WG 5

Data package	Information provided by Working Group 5 for the purpose of preparing material for future meetings
Item 2	Issues
Item 3	First Transfer Sequence
Item 5	Technical Comparison
Item 6	Docking Module Failure Conditions

3.0 Data Provided by USSR

3.1 Preliminary Documents

3.1.1	ASTP 10 000	Draft of Technical Proposals for the Experimental Soyuz/Apollo Flight, dated June, 1972. USA State Department Translation attached.
3.1.2	ASTP 10 000	Technical Proposals for Experimental Flight Soyuz/Apollo (Plan), dated July 5, 1972. NASA translated copy attached.
3.1.3	ASTP 20 000	Organization Plan (Draft). NASA translation attached.
3.1.4	ASTP 20 000	Remarks and Additions to the U.S.A. Document—"Proposed Organization Plan for Apollo/Salyut Mission." NASA translation attached.

---

*Two color photographs of Westinghouse camera (color Vugraphs of same in EE files [Room 220, building 440]); RCA photo 72-1-61c of camera (copy in EE files); and fourth photo of different view of RCA camera (copy available in vendor file [RCA Astrionics Division]).

3.1.5	ASTP 30 000	Project Schedule Document. NASA translation attached.
3.1.6	IED 50 101	General Requirements for the Soviet Communications System and American Communications and Ranging System for the Experimental Flight "Apollo-Soyuz." NASA translated copy attached.
3.1.7	IED 50 102	Agreement on Signal Characteristics on the Soviet and American Working Frequencies to Ensure the First Apollo-Soyuz Test Flight. NASA translated copy attached.
3.1.8	IED 50 103	Volume and Order of Testing for Compatibility the Radio Equipment for Communications and Range for the First Experimental Flight "Apollo-Soyuz." NASA translated copy attached.
3.1.9	IED 50 104	Determination of Terminology and Abbreviations Used in the Document on Radio Communications and Ranging during Preparation for the First Experimental Flight. NASA translated copy attached.
3.1.10	IED 50 301	General Technical Requirements for External Lights in the USSR Orbital Spacecraft "Soyuz" Performing Rendezvous and Docking with the USA Spacecraft "Apollo." NASA translation attached.
3.1.11	IED 50 305	Test Plans and Procedures for Verification Testing of the USSR Orbital Spacecraft "Soyuz," Performing Rendezvous and Docking with the USA Spacecraft "Apollo." NASA translation attached.
3.1.12	IED 50 306	Verification Test Results for External Lights on the USSR Orbital Spacecraft "Soyuz." Performing Rendezvous and Docking with the USA Spacecraft "Apollo." NASA translation attached.

3.2 Other Data

3.2.1	Russian report	Investigation of Docking Targets. State Department translation only.

3.2.2	Russian report	Systems of Differential Equations of Motion of a Spacecraft with Consideration of the Movement of the Liquid Charge in the Tanks during the Operational Mode of a Correcting Engine. State Department translation only.
3.2.3		Description of the VHF Transceiver for Spacecraft to Spacecraft Communications During the Joint Apollo Soyuz Flight and Rendezvous—"Vetka" Preliminary Project. NASA translation attached.
3.2.4		Voice cable communications diagram
3.2.5		Television diagram
3.2.6		Materials for the agreed upon Parameters of the Docking System Providing a Compatible USSR & USA Design

4.0 Agendas

4.1 Telegram
4.2 Proposed Agenda
4.3 Overall Agenda
4.4 Detail Plan
4.5 Social Agenda

5.0 Photographs

S-72-43527	S-72-43542	S-72-44301	S-72-44316
S-72-43528	S-72-43543	S-72-44302	S-72-44317
S-72-43529	S-72-43544	S-72-44303	S-72-44318
S-72-43530	S-72-43545	S-72-44304	S-72-44319
S-72-43531		S-72-44305	S-72-44320
S-72-43532	S-72-43757	S-72-44306	S-72-44321
S-72-43533	S-72-43758	S-72-44307	S-72-44322
S-72-43534	S-72-44143	S-72-44308	S-72-44323
S-72-43535	S-72-44144	S-72-44309	S-72-44324
S-72-43536		S-72-43310	
S-72-43537	S-72-44165	S-72-44311	S-72-44326
S-72-43538	S-72-44166	S-72-44312	S-72-44327
S-72-43539	S-72-44298	S-72-44313	S-72-44328
S-72-43540	S-72-44299	S-72-44314	S-72-44329
S-72-43541	S-72-44300	S-72-44315	S-72-44330

S-72-44331	S-72-44346	S-72-44383	S-72-44851
S-72-44332	S-72-44347	S-72-44398	S-72-44852
S-72-44333	S-72-44348	S-72-44399	
S-72-44334	S-72-44349		S-72-44871
S-72-44335	S-72-44350	S-72-44832	S-72-44872
S-72-44336	S-72-44351	S-72-44833	S-72-44873
S-72-44337	S-72-44352	S-72-44834	S-72-44874
S-72-44338	S-72-44353	S-72-44835	S-72-44875
S-72-44339	S-72-44354	S-72-44836	S-72-44876
S-72-44340	S-72-44355	S-72-44837	S-72-44877
S-72-44341	S-72-44356	S-72-44846	S-72-44878
S-72-44342		S-72-44847	S-72-44879
S-72-44343	S-72-44380	S-72-44848	S-72-44880
S-72-44344	S-72-44381	S-72-44849	S-72-44881
S-72-44345	S-72-44382	S-72-44850	S-72-44882

6.0 Debriefing Memos

6.1 Working Group 1
6.2 Working Group 2
6.3 Working Group 3—none made
6.4 Working Group 4
6.5 Working Group 5

7.0 Results Presentation

7.1 Communique on Results of Apollo-Soyuz Test Project Meetings

Note on Photography

Photography played a major role in the experimental documentation and study of the earth's surface as called for by the Earth Observations and Photography Experiment (MA-136). Unfortunately, the quality of the earth-looking photography was considerably below that of the photography obtained on previous manned space flights. Of the 1916 earth-looking photographs, 39 percent were considered good photographs. Of the remaining, about 29 percent were overexposed, 21 percent were out of focus, 5 percent were underexposed, and 6 percent had other deficiencies.

Two distinctly different Hasselblad 500 EL 70-millimeter cameras were used. One had a built-in precision 1-centimeter reseau grid and used either a 60- or 100-millimeter lens. Its main use was with the observations experiment. The other Hasselblad was used with either an 80- or 250-millimeter lens, retaining the capability of single lens reflex operation. The films used included Kodak Ektachrome High Definition Aerial Type SO-242, Kodak Ektachrome MS Type QX-807, and Kodak Ektachrome Infrared Aerochrome Type 2443.

Photography of *Soyuz 19* at rendezvous was accomplished using the Hasselblad Model 500 EL 70-millimeter reflex camera. Since the actual docking would have been difficult to record using hand-held equipment, the data acquisition camera (DAC), a 16-millimeter variable frame rate motion-picture camera, was employed.

Interior photography was taken with a Nikon 35-millimeter single lens reflex camera with a 35-millimeter lens, a Honeywell electronic flash, and Kodak Ektachrome EF Type SO-168 film. The Nikon was also used with a 300-millimeter lens for earth-looking purposes. Image motion from a hand-held camera taking photographs from a vehicle moving at nearly 8 kilometers per second usually results in a degraded photograph.

The selection of earth-looking photographs in this section shows a number of examples of the diverse subject matter available on our little planet. This information and the photographs and captions were made available by Richard W. Underwood, Technical Assistant to the Chief, Photographic Technology Division, Johnson Space Center.

Sun glitter highlights the division of the Nile into the western Rashud (Rosetta) and eastern Dumyat (Damietta) branches, which formed the great triangular delta fanning out to the Mediterranean Sea. In the lower left the valley of the Nile is quite narrow (10 to 20 kilometers) and entirely under cultivation. At lower left center the blue-green El Faiyum Depression is similarly cultivated. The lake is about 40 meters below sea level. In the lower right the Suez Canal can be seen from Suez northward past the Bitter Lakes to Ismailiya. 19 July 1975. (AST-9-556)

Egypt's second city Alexandria is loc in the upper center coast to the le the curved bay, which was the site o Battle of the Nile. The Rosetta Nile be traced from the lower center t mouth. The buff-color Western D shows areas of reclamation under d opment. The thin dark area in the l center left is Wadi el Natrun, a seri ten salt lakes below sea level. 20 1975. (AST-16-1257)

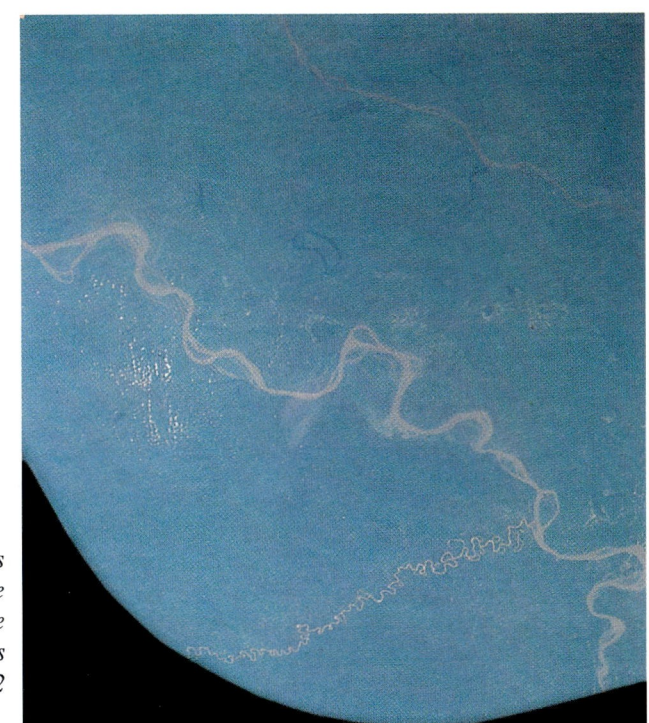

...ilt-laden Amazon River meanders ... the Amazon Basin of Brazil. The ... straighter Rio Japura is to the ... right. The smaller Jurura comes ... e Amazon from the lower left. 22 ...75. (AST-21-1682)

A most unique land form and drainage pattern is seen in the area of Angola known as Moxico and Cuando Cubango. The Cuando River crosses the photograph from upper left to lower right. The rivers flow southeast in straight parallel valleys 10 to 30 kilometers apart. South of the Cuando River, they flow parallel and straight east but 5 to 10 kilometers apart. In the lower left part the Cuito River system is seen. 18 July 1975. (AST-14-890)

Aitutaki Atoll in the Cook Islands clearly shows the contrast between the deep Pacific Ocean, the barrier reef, the lagoon, and the tree-covered main island. Aitutaki is about 1300 kilometers southeast of Samoa. 16 July 1975. (AST-1-39)

The island of Heirro in the Canary group is some 400 kilometers west of the Atlantic coast of Morocco. At one time this volcanic island was much larger. Half the old crater at left center is still seen but at some time in the distant past a massive eruption destroyed the other half of the island, crater and all. 22 July 1975. (AST-23-1939)

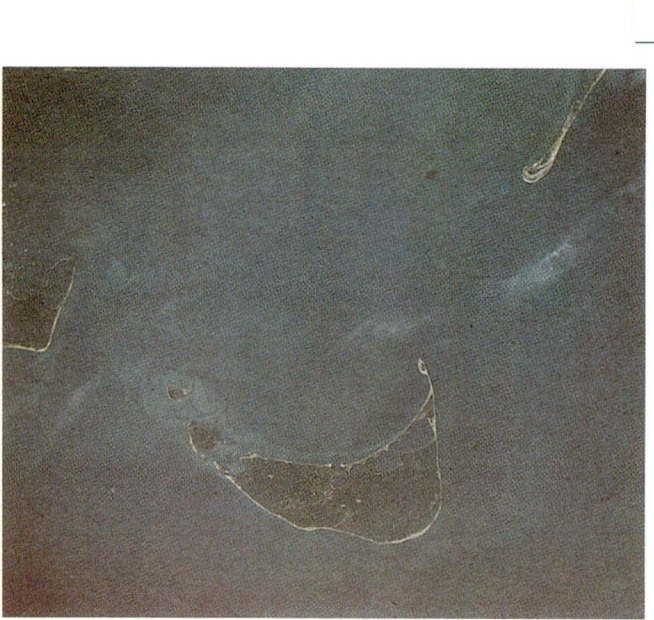

Nantucket Island, Massachusetts, is at the center of the photograph. At the upper right, a long sandy point, Monomo extends southward from Cape Cod. the left center is Chappaquiddick Isla and part of Martha's Vineyard. The islands represent the remnants of t terminal moraine of the great continen glaciers of the past. 23 July 1975. (AS 1-63)

In Chile, snow-covered Andean peaks penetrate the lower level clouds to make a striking photograph. The mountains seen are south of Santiago. (AST-23-1908)

Washington's Cascade Mountains are still snow covered in late July. In the lower left, Interstate 90 passes Cle Elum, Kachess, and Keechelus Lakes and loops over Snoqualmie Pass. The Stevens Pass Highway and Cascade Tunnel are in the upper right. Areas of controversial timber "clear cutting" are easily delineated. 20 July 1975. (AST-30-2601)

The Canadian Rockies of British Columbia and Alberta show extensive snow fields and glaciers in the area of Banff, Jasper, and Glacier (Canada) National Parks. The North Saskatchewan River is in the lower center. The Columbia River is to the left. Mount Columbia—Alberta's highest—the Athabasca Glacier, and the Clemenceau Icefield are clearly seen near the center. 23 July 1975. (AST-19-1570)

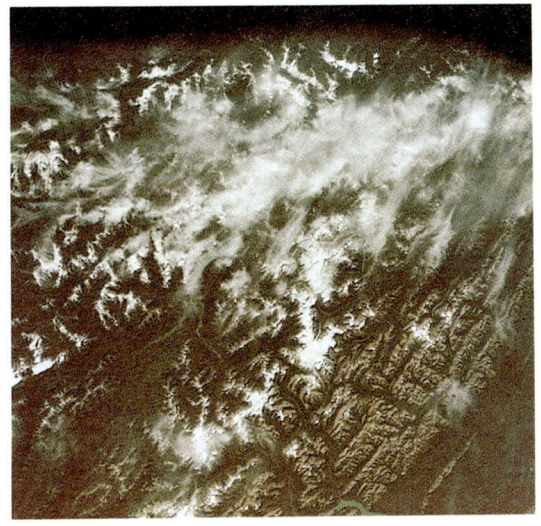

The astronaut's view of the approach to Gibraltar is well demonstrated in this photograph. Spain and Portugal are to the left. Morocco and Algeria are to the right. The Mediterranean Sea is in the background. 20 July 1975. (AST-27-2363)

An excellent vertical view of the Strait of Gibraltar was made about 1 minute after AST-27-2363. Morocco and Tangier are at the lower center. Spain and Cadiz are to the right. The Huelva-Palos area, Columbus' point of departure for the new world, is at the upper center. A small portion of Portugal is along the upper edge. Cape Trafalgar, the location of the great sea battle, guards the Atlantic side of the important strait. 20 July 1975. (AST-27-2366)

A unique view looking south at portions of Israel (lower center), Saudi Arabia (upper left), Egypt (upper right), and Jordan (lower left) clearly shows the great rifts that have developed into the Red Sea, Gulf of Suez, Gulf of Aqaba, and Dead Sea (lower center). The fertile Nile Valley is clearly traced in the upper right, Sinai is at the lower right, and the harsh deserts of the Al Hijaz are to the left. One can trace a portion of the ancient pilgrim route of the followers of Muhammad to Al Madinah and Mecca from the Levant. 19 July 1975. (AST-9-560)

A very striking view to the southwest of the Levant. Visible is all of Lebanon and portions of Syria, Jordan, Egypt, and Israel. Notable coastal places from lower right to upper left include Latakia, Tripoli, Beirut, Tyre, Haifa, Tel-Aviv, and Gaza. In the interior, one can locate such places of importance as Damascus, the Sea of Galilee, the Jordan River, Nazareth, Amman, Jerusalem, the Dead Sea, and Beersheba. 19 July 1975. (AST-9-564)

e giant swing-arm irrigation systems *nmon to the American desert invade *e Sahara at Libya's Al Kufra Oasis *wer center). It is said that the water *noved here fell to the earth in this *ssive basin many thousands of years *o. 20 July 1975. (AST-16-1244)

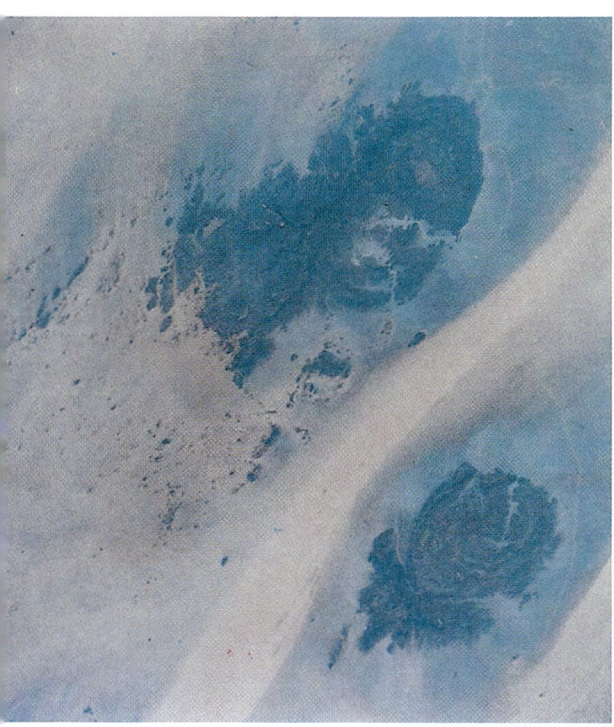

The area shown is where Libya, Egypt, and Sudan join in the east central Sahara. The circular dark basaltic intrusions are Jebel Uweinat (upper right) and Jebel Arkenu (lower right). The harsh Sahara winds move sand along paths that resemble rivers. 20 July 1975. (AST-2-130)

East of Queensland, Australia, the Great Barrier Reef extends for over 2000 kilometers. The bottom detail and reefs in the area known as the Cumberland Islands are clearly seen. 20 July 1975. (AST-2-104)

The sun glint reflecting off the Coral Sea east of Queensland, Australia, produces a unique effect along a portion of the Great Barrier Reef in the area of Capricorn Channel and Northumberland Islands. Sun glitter will accentuate the visibility of the surface state and the surface current situation. This is in complete contrast to AST-2-104, taken 70 hours later without sun glitter, which clearly shows subsurface detail. 17 July 1975. (AST-1-46)

Usually Lake Eyre North is seen from space as a large white salt pan in South Australia's Lake Eyre Basin. The very rare, very heavy rains of 1973 and 1974 created a lake 70 by 120 kilometers in size and some 10 meters below sea level. North is to the upper left. 23 July 1975. (AST-21-1726)

A large portion of the Los Angeles Basin is clearly seen in the lower portion of the photograph. The San Fernando Valley is at the upper center. The San Gabriel Mountains cross from upper center to lower right. The light lines that crisscross the photograph are in some cases freeways and in others paved rivers. The Pacific Coast is seen from Point Dume southward to Seal Beach. 16 July 1975. (AST-14-881)

A portion of the Los Angeles urban complex is seen in the left and lower portion of the photograph. The San Gabriel Mountains cross from the upper left corner to the lower right. The very active San Andreas Fault can be clearly traced as a straight line from the upper center edge to the lower right edge. The high San Gabriel Mountains block moisture reaching the Mojave Desert (right side), where irrigated fields clearly stand out. The dark fan at the right center is the result of a massive mud slide that originated near Wrightwood. 16 July 1975. (AST-14-882)

Appendix A
NASA Organization Charts

NATIONAL AERONAUTICS AND SPACE ADMINISTRATION
[29 January 1959]

EXECUTIVE DIRECTION

- NATIONAL AERONAUTICS AND SPACE COUNCIL
- CIVILIAN-MILITARY LIAISON COMMITTEE
- OFFICE OF THE ADMINISTRATOR
  - ADMINISTRATOR
  - DEPUTY ADMINISTRATOR
  - ASSOCIATE ADMINISTRATOR
  - ASSISTANTS
- ASSISTANT ADMINISTRATOR FOR CONGRESSIONAL RELATIONS
- OFFICE OF PUBLIC INFORMATION
- OFFICE OF INTERNATIONAL PROGRAMS
- OFFICE OF PROGRAM PLANNING AND EVALUATION
- GENERAL COUNSEL

- OFFICE OF SPACE FLIGHT DEVELOPMENT DIRECTOR
  - ASSISTANTS
  - SPACE SCIENCE ASSISTANT DIRECTOR
  - SPACE FLIGHT OPERATIONS ASSISTANT DIRECTOR
  - ADVANCED TECHNOLOGY ASSISTANT DIRECTOR
  - PROPULSION ASSISTANT DIRECTOR

- OFFICE OF AERONAUTICAL AND SPACE RESEARCH DIRECTOR
  - ASSISTANTS
  - STRUCTURES AND MATERIALS, AND AIRCRAFT OPERATING PROBLEMS ASSISTANT DIRECTOR
  - OFFICE OF UNIVERSITY RESEARCH
  - AERODYNAMICS AND FLIGHT MECHANICS ASSISTANT DIRECTOR
  - POWER PLANTS ASSISTANT DIRECTOR

PROGRAMING OPERATIONS

- OFFICE OF BUSINESS ADMINISTRATION DIRECTOR
  - ASSISTANTS
  - AUDIT DIVISION
  - PROCUREMENT AND SUPPLY DIVISION
  - SECURITY DIVISION
  - ADMINISTRATIVE SERVICES DIVISION
  - BUDGET AND FISCAL DIVISION
  - PERSONNEL DIVISION
  - TECHNICAL INFORMATION DIVISION

SPACE PROJECT CENTERS

- JET PROPULSION LABORATORY (CAL. TECH. CONTRACT) PASADENA, CALIF.
- BELTSVILLE SPACE CENTER BELTSVILLE, MD.
- WALLOPS STATION CAPE CANAVERAL

RESEARCH CENTERS

- AMES MOFFETT FIELD, CALIF.
- LANGLEY LANGLEY FIELD, VA.
- HIGH SPEED FLIGHT STATION
- LEWIS CLEVELAND, OHIO
- LIAISON OFFICES LOS ANGELES, CALIF.

NATIONAL AERONAUTICS AND SPACE ADMINISTRATION
OFFICE OF MANNED SPACE FLIGHT
[17 June 1969]

NASA WORKING ORGANIZATIONS FOR U.S.A./U.S.S.R. DOCKING SYSTEM MEETINGS [1971]

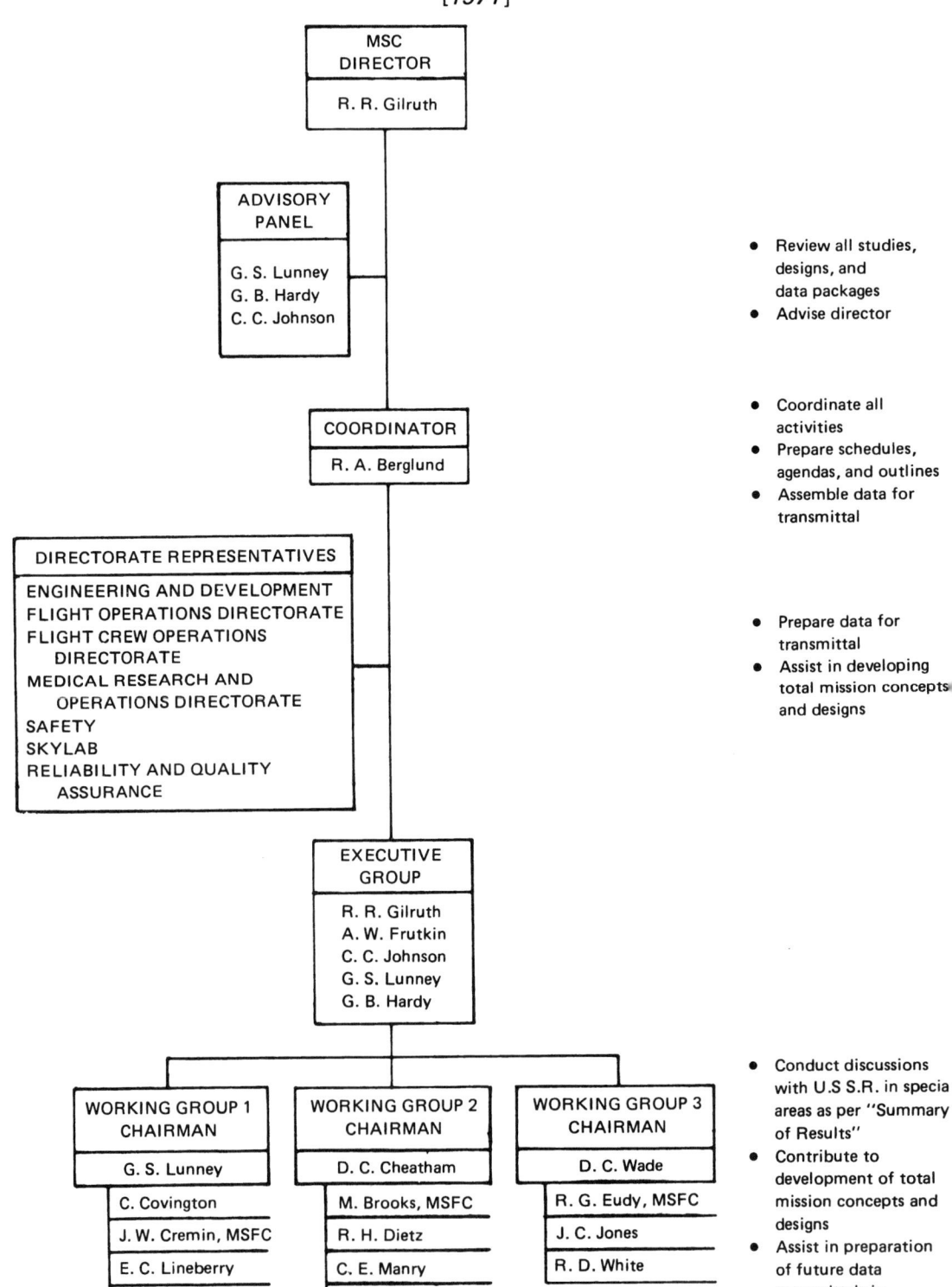

INTERNATIONAL RENDEZVOUS AND DOCKING MISSION STUDY TASK TEAM
[1971]

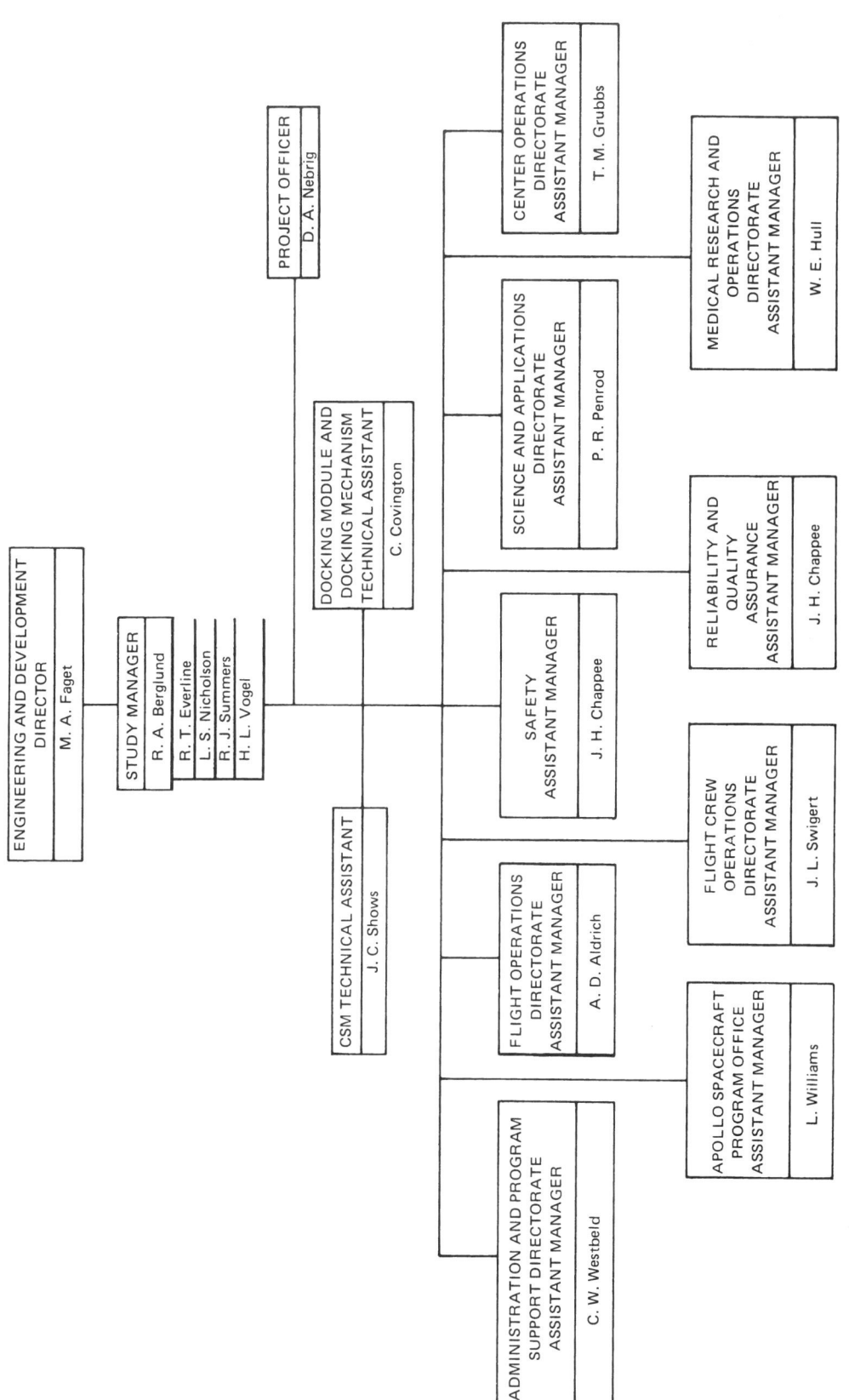

NATIONAL AERONAUTICS AND SPACE ADMINISTRATION MANNED SPACECRAFT CENTER [1972]

U.S./U.S.S.R. ORGANIZATION FOR ASTP
[28 May 1974]

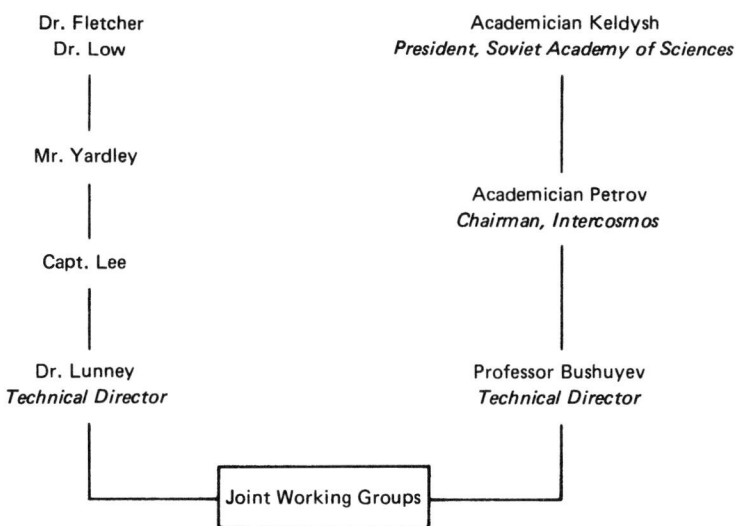

U.S.

U.S.S.R.

Dr. Fletcher
Dr. Low

Academician Keldysh
President, Soviet Academy of Sciences

Mr. Yardley

Academician Petrov
Chairman, Intercosmos

Capt. Lee

Dr. Lunney
Technical Director

Professor Bushuyev
Technical Director

Joint Working Groups

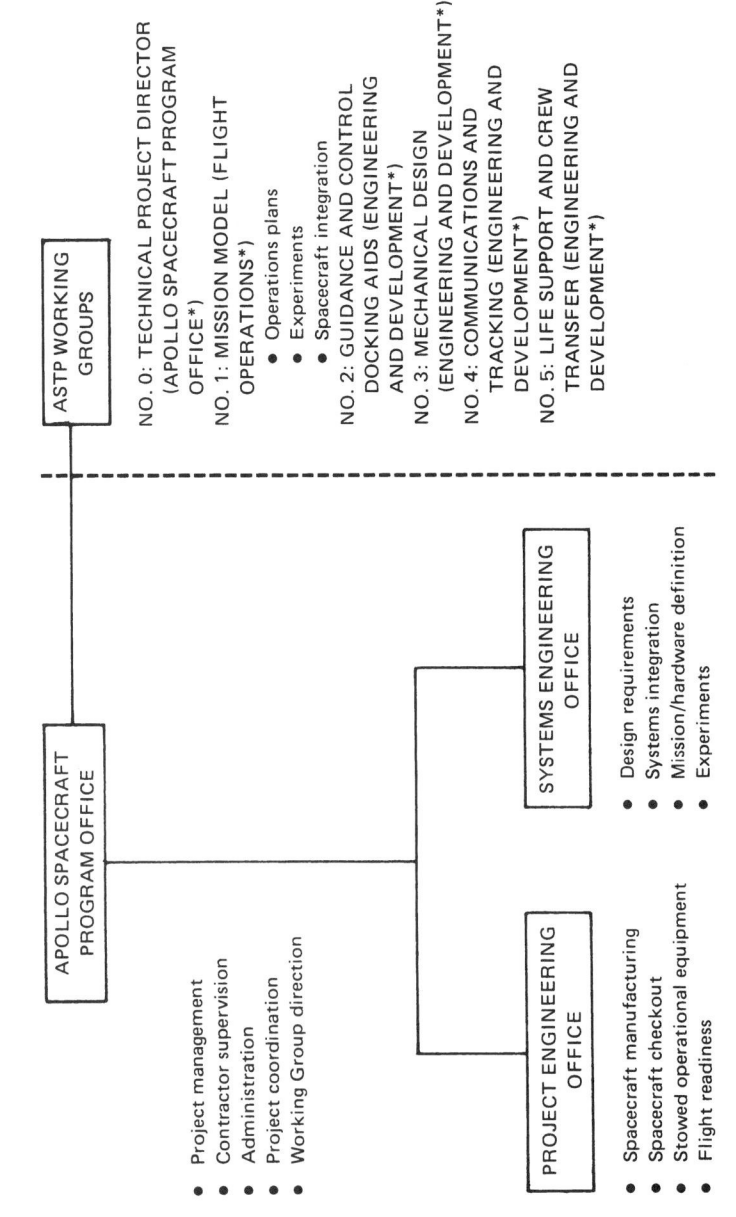

ASTP ORGANIZATION
[1974]

APOLLO SPACECRAFT PROGRAM OFFICE

- Project management
- Contractor supervision
- Administration
- Project coordination
- Working Group direction

PROJECT ENGINEERING OFFICE

- Spacecraft manufacturing
- Spacecraft checkout
- Stowed operational equipment
- Flight readiness

SYSTEMS ENGINEERING OFFICE

- Design requirements
- Systems integration
- Mission/hardware definition
- Experiments

ASTP WORKING GROUPS

NO. 0: TECHNICAL PROJECT DIRECTOR (APOLLO SPACECRAFT PROGRAM OFFICE*)

NO. 1: MISSION MODEL (FLIGHT OPERATIONS*)
- Operations plans
- Experiments
- Spacecraft integration

NO. 2: GUIDANCE AND CONTROL DOCKING AIDS (ENGINEERING AND DEVELOPMENT*)

NO. 3: MECHANICAL DESIGN (ENGINEERING AND DEVELOPMENT*)

NO. 4: COMMUNICATIONS AND TRACKING (ENGINEERING AND DEVELOPMENT*)

NO. 5: LIFE SUPPORT AND CREW TRANSFER (ENGINEERING AND DEVELOPMENT*)

*Office or directorate providing chairman of Working Group

JOHNSON SPACE CENTER
ASTP ACTIVITIES DELEGATED TO FUNCTIONAL DIRECTORATES
[1974]

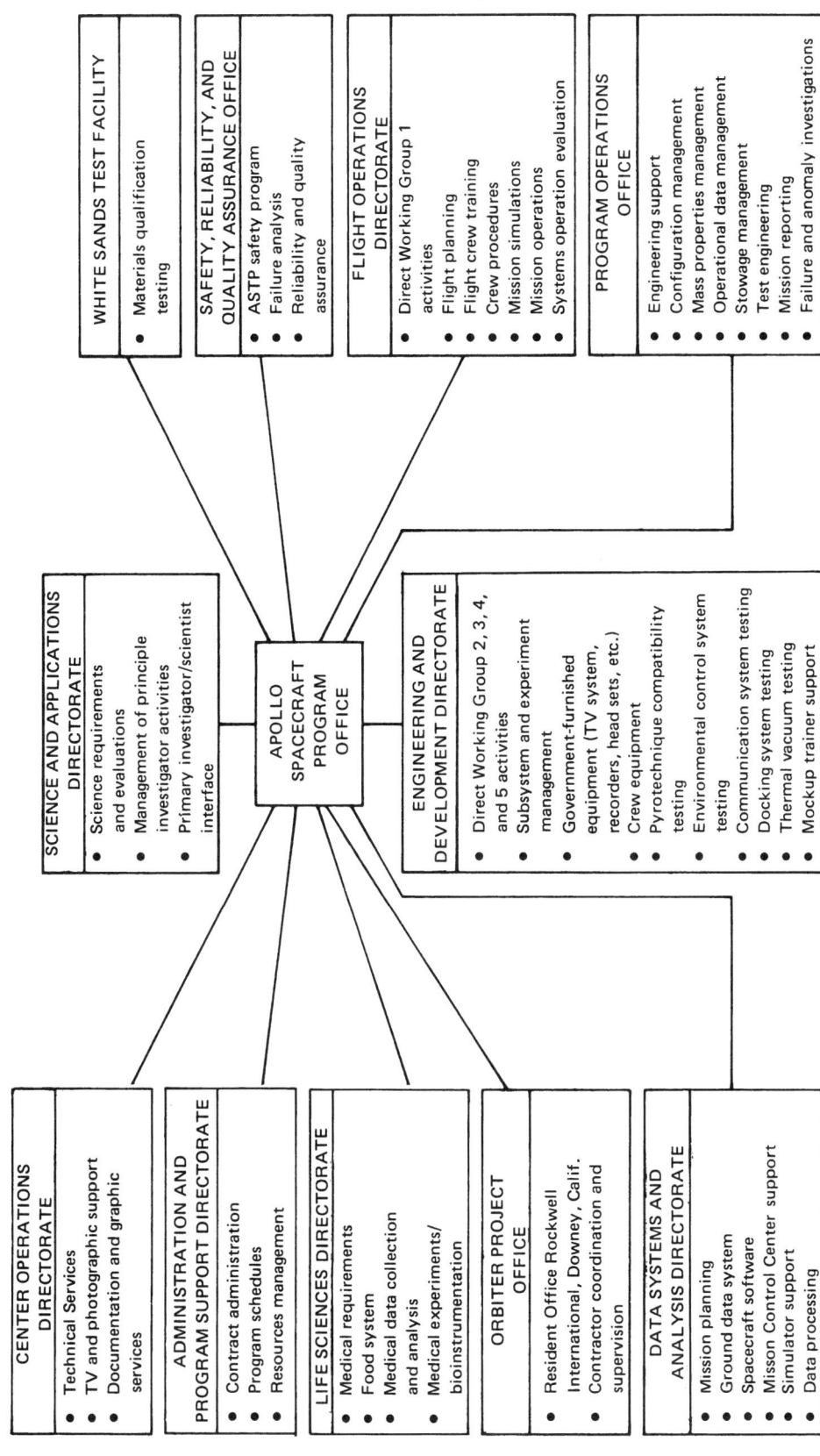

WHITE SANDS TEST FACILITY
- Materials qualification testing

SAFETY, RELIABILITY, AND QUALITY ASSURANCE OFFICE
- ASTP safety program
- Failure analysis
- Reliability and quality assurance

FLIGHT OPERATIONS DIRECTORATE
- Direct Working Group 1 activities
- Flight planning
- Flight crew training
- Crew procedures
- Mission simulations
- Mission operations
- Systems operation evaluation

PROGRAM OPERATIONS OFFICE
- Engineering support
- Configuration management
- Mass properties management
- Operational data management
- Stowage management
- Test engineering
- Mission reporting
- Failure and anomaly investigations

SCIENCE AND APPLICATIONS DIRECTORATE
- Science requirements and evaluations
- Management of principle investigator activities
- Primary investigator/scientist interface

APOLLO SPACECRAFT PROGRAM OFFICE

ENGINEERING AND DEVELOPMENT DIRECTORATE
- Direct Working Group 2, 3, 4, and 5 activities
- Subsystem and experiment management
- Government-furnished equipment (TV system, recorders, head sets, etc.)
- Crew equipment
- Pyrotechnique compatibility testing
- Environmental control system testing
- Communication system testing
- Docking system testing
- Thermal vacuum testing
- Mockup trainer support

CENTER OPERATIONS DIRECTORATE
- Technical Services
- TV and photographic support
- Documentation and graphic services

ADMINISTRATION AND PROGRAM SUPPORT DIRECTORATE
- Contract administration
- Program schedules
- Resources management

LIFE SCIENCES DIRECTORATE
- Medical requirements
- Food system
- Medical data collection and analysis
- Medical experiments/bioinstrumentation

ORBITER PROJECT OFFICE
- Resident Office Rockwell International, Downey, Calif.
- Contractor coordination and supervision

DATA SYSTEMS AND ANALYSIS DIRECTORATE
- Mission planning
- Ground data system
- Spacecraft software
- Misson Control Center support
- Simulator support
- Data processing

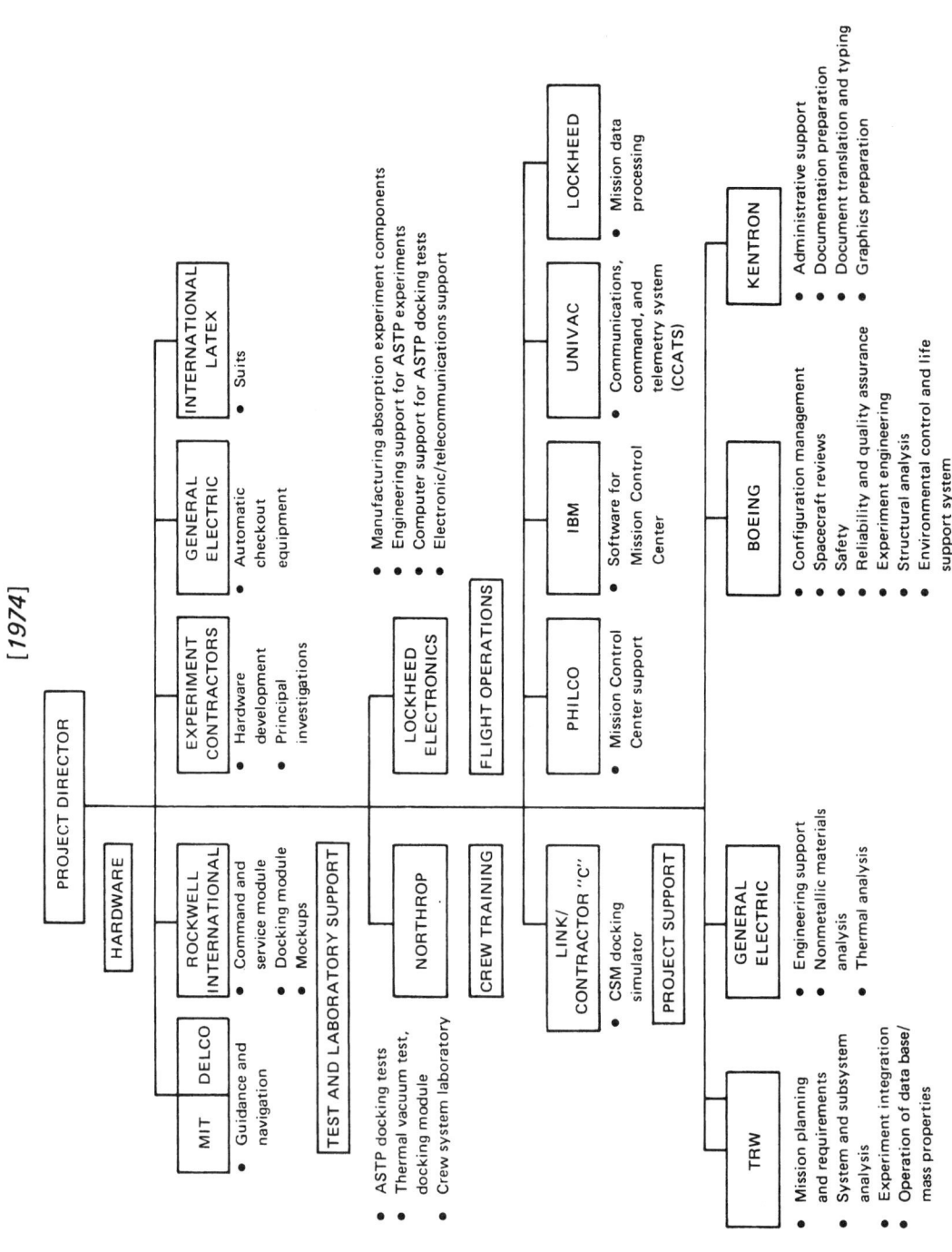

ASTP CONTRACTING TEAM
[1974]

PROJECT DIRECTOR

HARDWARE

ROCKWELL INTERNATIONAL
- Command and service module
- Docking module
- Mockups

MIT | DELCO
- Guidance and navigation

EXPERIMENT CONTRACTORS
- Hardware development
- Principal investigations

GENERAL ELECTRIC
- Automatic checkout equipment

INTERNATIONAL LATEX
- Suits

- Manufacturing absorption experiment components
- Engineering support for ASTP experiments
- Computer support for ASTP docking tests
- Electronic/telecommunications support

TEST AND LABORATORY SUPPORT

NORTHROP

LOCKHEED ELECTRONICS

- ASTP docking tests
- Thermal vacuum test, docking module
- Crew system laboratory

CREW TRAINING

FLIGHT OPERATIONS

LINK/ CONTRACTOR "C"
- CSM docking simulator

PHILCO
- Mission Control Center support

IBM
- Software for Mission Control Center

UNIVAC
- Communications, command, and telemetry system (CCATS)

LOCKHEED
- Mission data processing

PROJECT SUPPORT

TRW
- Mission planning and requirements
- System and subsystem analysis
- Experiment integration
- Operation of data base/ mass properties

GENERAL ELECTRIC
- Engineering support
- Nonmetallic materials analysis
- Thermal analysis

BOEING
- Configuration management
- Spacecraft reviews
- Safety
- Reliability and quality assurance
- Experiment engineering
- Structural analysis
- Environmental control and life support system

KENTRON
- Administrative support
- Documentation preparation
- Document translation and typing
- Graphics preparation

Appendix B
Development of Manned Space Flight, American and Soviet

Name of spacecraft	Launch date	Occupant	Payload weight (kg)	Flight time (hr:min:sec)	Basic flight objectives	Summary of results
	1959					
LJ-1	21 Aug.	Unmanned	1 007	00:00:20	Maximum dynamic pressure abort test; evaluation of launch escape and recovery systems.	Objectives not met; abort initiated during countdown.
Big Joe	9 Sept.	Unmanned	1 159	00:13:00	Ballistic flight; evaluation of heat-protection concept, aerodynamic shape, and recovery system.	Successful.
LJ-6	4 Oct.	Unmanned	1 134[1]	00:05:10	Ballistic flight; qualification of launch vehicle structure; evaluation of command system.	Successful.
LJ-1A	4 Nov.	Unmanned	1 007	00:08:11	Maximum dynamic pressure abort test; same as LJ-1.	Primary objective not met: escape motor ignition was late.
LJ-2	4 Dec.	Rhesus monkey "Sam"	1 007[1]	00:11:06	High-altitude abort test; evaluation of launch, abort, and reentry dynamics; recovery.	Successful.
	1960					
LJ-1B	21 Jan.	Rhesus monkey "Miss Sam"	1 007[1]	00:08:35	Maximum dynamic pressure abort test; same as LJ-1A; evaluation of launch and abort.	Successful.

452

	Date	Crew	Weight	Time	Objectives	Results
Beach abort	9 May	Unmanned	1 154	00:01:16	Off-the-pad abort test; qualification of structure and launch escape system for simulated pad abort.	Successful.
Korabl Sputnik I[2]	15 May	Simulated man	4 540	—	Place space cabin into orbit; test life support systems; recover cabin from orbit.	Put in near circular orbit; recovery operation malfunction, 19 May; burned up after 844 days.
MA-1	29 July	Unmanned	1 154	00:03:18	Ballistic flight; test of spacecraft/launch vehicle compatibility; thermal loads in critical abort.	Objectives not met; mission failed at about 60 sec after lift-off.
Korabl Sputnik II[2]	19 Aug.	Animals	4 600	25:00:00	Test of capsule and recovery system; evaluation of effects of space travel on biological payload (2 dogs, Strelka [Little Arrow] and Belka [Squirrel]; 12 mice; 2 rats; 15 flasks of fruit flies; plants).	Successful; 17+-orbit flight and recovery.
LJ-5	8 Nov.	Unmanned	1 141[1]	00:02:22	Maximum dynamic pressure abort test; qualification of launch escape system and structure.	Objectives not met; spacecraft did not separate from launch vehicle.
MR-1	21 Nov.	Simulated man	1 211	00:00:00	Suborbital flight; qualification of spacecraft/launch vehicle compatibility; posigrades.	Test objectives not met; launch vehicle shut down at lift-off.

NOTE—footnotes at end of table.

453

Name of spacecraft	Launch date	Occupant	Payload weight (kg)	Flight time (hr:min:sec)	Basic flight objectives	Summary of results
Korabl Sputnik III[2]	1 Dec.	Animals	4 563	24:00:00[1]	Test of equipment for manned flights; evaluation of effects of space travel on biological payload (2 dogs, Pchelka [Bee] and Mushka [Little Fly]; mice; insects; plants).	Spacecraft burned up on unprogrammed reentry, 2 Dec.
MR-1A	19 Dec.	Unmanned	1 211[1]	00:15:45	Suborbital flight; same as MR-1.	Successful; cutoff overspeed caused overshoot of recovery area.
	1961					
MR-2	31 Jan.	Chimpanzee "Ham"	1 203	00:16:39	Suborbital flight; acquisition of physiological and performance data on a primate in flight; systems qualification tests.	Successful; early depletion of liquid oxygen triggered escape rocket, which yanked spacecraft 209 km beyond recovery area.
MA-2	21 Feb.	Unmanned	1 154[1]	00:17:56	Ballistic flight; same as MA-1.	Successful.
Korabl Sputnik IV[2]	9 Mar.	Simulated man and animals	4 700	01:30:00[1]	Testing of structure and systems for manned flight; observation of effects of biological payload (1 dog, Chernushka [Blackie]; mice; guinea pigs).	Successful 1-orbit flight; upon recovery animal passengers were reported to be alive and well.

LJ-5A	18 Mar.	Unmanned	1 141[1]	00:23:48	Maximum dynamic pressure abort test; same as LJ-5.	Objectives not met; escape rocket ignited early.
MR-BD	24 Mar.	Unmanned	1 141[1]	00:08:23	Suborbital flight; evaluation of modifications to correct MR-1A and MR-2 malfunctions.	Successful.
Korabl Sputnik V[2]	25 Mar.	Simulated man and dog	4 695	01:30:00[1]	Further test of structure and systems, including recovery of biological payload and the simulated cosmonaut "Ivan Ivanovich" (1 dog, Zvezdochka [Little Star]).	Successful.
Vostok 1 Kedr Cedar	12 Apr.	Gagarin	4 725	01:48:00	Placement of manned spacecraft into orbit; safe recovery of cosmonaut and craft.	Successful 1-orbit flight; first manned orbital flight.
MA-3	25 Apr.	Unmanned	1 179[1]	00:07:19	1-pass orbital flight; evaluation of all systems, network, and recovery forces.	Objectives not met; launch vehicle failed to follow roll program; escape system operated.
LJ-5B	28 Apr.	Unmanned	1 141	00:05:25	Maximum dynamic pressure abort test; same as LJ-5 and LJ-5A.	Successful.
MR-3 Freedom 7	5 May	Shepard	1 290	00:15:22	Suborbital flight; familiarization of man with space flight; evaluation of response and control of craft.	Successful; first American in space.

Name of spacecraft	Launch date	Occupant	Payload weight (kg)	Flight time (hr:min:sec)	Basic flight objectives	Summary of results
MR-4 Liberty Bell 7	21 July	Grissom	1 286	00:15:37	Suborbital flight; same as MR-3.	Successful; after water landing, premature hatch release caused craft to sink; astronaut recovered.
Vostok 2 Oryel Eagle	6 Aug.	Titov	4 731	25:18:00	Study effects of prolonged weightlessness on cosmonaut after 17 orbits of flight.	Successful 17½-orbit flight.
MA-4	13 Sept.	Unmanned	1 179[1]	01:49:20	1-pass orbital flight; same as MA-3.	Successful; circuit anomaly in control system caused landing 120 km uprange.
MA-5	29 Nov.	Chimpanzee "Enos"	1 331	03:20:59	3-pass orbital flight; qualification of all systems and network for orbital flight recovery.	Successful; control system malfunction terminated flight after 2 passes.
	1962					
MA-6 Friendship 7	20 Feb.	Glenn	1 355	04:55:23	3-pass orbital flight; evaluation of effects on and performance of astronaut in space.	Successful; first American to orbit earth; control system malfunction required manual retro-fire and reentry.
MA-7 Aurora 7	24 May	Carpenter	1 349	04:56:05	3-pass orbital flight; same as MA-6; evaluation of spacecraft modifications and network.	Successful; horizon scanner circuit malfunction required manual retro-fire; yaw error caused

Spacecraft	Date	Pilot	Weight	Duration	Purpose	Results
						landing 402 km downrange; recovery in 3 hr.
Vostok 3 Sokol Falcon	11 Aug.	Nikolayev	4 723	94:09:59	Orbit of manned craft and precision recovery; orbit in close conjunction with Vostok 4.	Successful; transmitted live TV to Soviet ground stations; radio communication with Vostok 4; 64 orbits.
Vostok 4 Berkut Golden Eagle	12 Aug.	Popovich	4 729	70:43:48	Orbit in close conjunction with Vostok 3; same as Vostok 3.	Successful; first tandem flight; 48 orbits.
MA-8 Sigma 7	3 Oct.	Schirra	1 373	09:13:11	Six-pass orbital flight; same as MA-6 and MA-7 except for extended time.	Successful; partially blocked ECS coolant valve delayed stabilizing suit temperature until 2d pass; on target recovery.
1963						
MA-9 Faith 7	15 May	Cooper	1 376	34:19:49	22-pass orbital flight; evaluation of effects of a lengthier stay in space on man; verification of man as primary spacecraft system.	Successful; manual retro-fire and reentry; on target landing.
Vostok 5 Yastreb Hawk	14 June	Bykovsky	4 720	119:06:00	Study of effect of prolonged flight on human organism; tandem flight with Vostok 6; improvement of spacecraft equipment and pilotage.	Successful; completed record 81 orbits; 48 tandem orbits.

Name of spacecraft	Launch date	Occupant	Payload weight (kg)	Flight time (hr:min:sec)	Basic flight objectives	Summary of results
Vostok 6 Chaika Seagull	16 June	Tereshkova	4 713	70:50:00	Study of effects of space flight on a woman; tandem flight with Vostok 5.	Successful; first woman to fly in space; completed 48-orbit flight in tandem with Vostok 5.
PA-1	7 Nov.	Unmanned	7 177[3]	00:02:45	Evaluation of Apollo escape configuration during pad abort.	Successful.
	1964					
GT-1	8 Apr.	Unmanned	3 187	04:50:00	Test of structural integrity of Gemini spacecraft and compatibility of Gemini and Titan II launch vehicle.	Successful; recovery not planned; reentered after 64th orbital pass and disintegrated.
A-001	13 May	Unmanned	26 284[3]	00:05:50	Determination of aerodynamic characteristics of launch escape system.	Successful.
AS-101	28 May	Unmanned	SM: 1 892 CM: 4 218	00:10:24	Demonstration of compatibility of spacecraft with launch vehicle in launch and exit trajectory and environment for Apollo earth orbital flights.	Successful; no recovery planned; craft reentered on 54th orbital pass and disintegrated.
AS-102	18 Sept.	Unmanned	SM: 1 882 CM: 4 299	00:10:21	Demonstration of compatibility of spacecraft	Successful; no recovery planned; craft reentered

Name	Date	Crew	(kg)	Duration	Objectives	Results
					with launch vehicle in launch and exit trajectory; determination of launch and exit environmental parameters; demonstration of alternate mode of escape-tower jettison.	on 59th orbital pass and disintegrated.
Cosmos 47	6 Oct.	Unmanned	5 320[1]	24:00:00[1]	Unmanned precursor to Voskhod I.	Orbital data nearly identical to 12 Oct. Voskhod I flight.
Voskhod I Rubin Ruby	12 Oct.	Komarov Yegorov Feoktistov	5 320	24:17:03	Test of multiseat spacecraft; check of interaction of 3 cosmonauts; evaluation of medical findings on prolonged flight; test of soft-landing apparatus and shirtsleeve cabin environment.	Successful; 16-orbit flight.
A-002	8 Dec.	Unmanned	42 788[3]	00:07:23	Demonstration of launch vehicle performance and verification of abort capability in maximum dynamic pressure region.	Successful; CM recovered.
1965						
GT-2	19 Jan.	Unmanned	3 122	00:18:16	Demonstration of adequacy of reentry heat protection equipment, structural integrity, and capability of craft and systems; evaluation of	Successful; except for fuel cell results as fuel cell was deactivated before lift-off because of malfunction.

Name of spacecraft	Launch date	Occupant	Payload weight (kg)	Flight time (hr:min:sec)	Basic flight objectives	Summary of results
Cosmos 57	22 Feb.	Unmanned	5 683	NA	backup guidance steering signals. Unmanned precursor to Voskhod II.	Orbital data somewhat similar to 18 Mar. Voskhod II flight; exploded in orbit.
Voskhod II Almaz Diamond	18 Mar.	Belyayev Leonov	5 683	26:02:17	Performance and evaluation of EVA; use of manual reentry system.	EVA successful; overshot landing area and landed in snow-covered forest area; crew was rescued.
GT-3 Molly Brown	23 Mar.	Grissom Young	3 225	04:52:31	Evaluation of manned flight in Gemini and its 2-man design; evaluation of worldwide tracking network; maneuvering in orbit; control of reentry flight-path and recovery area; evaluation of systems: prelaunch, launch, and recovery.	Successful; performed 2 out of 3 experiments; Grissom became 1st man to make 2d space flight.
A-003	19 May	Unmanned	80 372[3]	00:05:03	Demonstration of performance of launch escape system in high-altitude region; demonstration of orientation of	Partially successful; required altitude not achieved; low-altitude conditions observed successfully.

Mission	Date	Crew	Weight	Time	Objective	Results
					CM to a main-heatshield-forward attitude after abort.	
GT-4	3 June	McDivitt White	3 574	97:40:01	Evaluation of effects of long exposure; demonstration of EVA; stationkeeping; rendezvous with detached 2d stage.	Successful; rendezvous attempt abandoned because of excessive fuel consumption; 22-min EVA; computer malfunction necessitated ballistic reentry.
PA-2	29 June	Unmanned	CM: 4 940	00:01:52	Demonstration of capability of launch escape system to abort from launch pad and recover.	Successful.
GT-5	21 Aug.	Cooper Conrad	3 605	190:55:14	Evaluation of performance of rendezvous guidance and navigation systems using a rendezvous evaluation pod; confirmation of physiological endurance for lunar mission.	Successful except for rendezvous failure because of fuel-cell heater problem; longest manned space flight in time and distance (8 days); performed 16 of 17 experiments.
GT-7	4 Dec.	Borman Lovell	3 663	330:35:01	Demonstration of 14-day mission capability; rendezvous target for GT-6A; stationkeeping; evaluation of lightweight pressure suit.	Successful; accurate controlled reentry; performed 20 experiments.
GT-6A	15 Dec.	Schirra Stafford	3 546	25:51:24	Rendezvous with GT-7; stationkeeping; spacecraft systems tests.	Successful; first rendezvous in space.

Name of spacecraft	Launch date	Occupant	Payload weight (kg)	Flight time (hr:min:sec)	Basic flight objectives	Summary of results
	1966					
A-004	20 Jan.	Unmanned	63 446[3]	00:06:50	Demonstration of satisfactory launch escape system performance for an abort in the power-on tumbling boundary region; demonstration of structural integrity of launch escape system airframe structure during same period.	Successful.
Cosmos 110	22 Feb.	Animals	5 200[1]	528:00:00[1] (22 days)[1]	Evaluation of prolonged effects of radiation in space travel on the biological payload (2 dogs, Veterok [Breeze] and Ugolek [Blackie]).	Successful; reportedly a Soyuz-type reentry vehicle.
AS-201	26 Feb.	Unmanned	SM: 3 654 CM: 4 990	00:37:20	Demonstration of compatibility and structural integrity of spacecraft and Saturn IB configuration; evaluation of heatshield performance.	Successful; 1st flight separation of launch vehicle and production spacecraft; 1st SM burn and restart at altitude; 1st full flight test of Block I spacecraft.
Cosmos 111	1 Mar.	Unmanned	NA	48:00:00	Orbital data similar to manned flight.	Possible precursor flight to 23 Apr. 1967 Soyuz 1 flight.

	Date	Crew	Weight	Time	Objectives	Results
GT-8	16 Mar.	Armstrong Scott	3 788	10:41:26	Rendezvous and docking with target vehicle; EVA; docking practice.	Rendezvous and docking successful; unexpected roll and yaw motion prevented further activity and necessitated early return; EVA not performed; safe recovery.
GT-9A	3 June	Stafford Cernan	3 750	72:20:50	Rendezvous and docking with target vehicle; EVA; docking practice.	Target vehicle had not been separated from its launch shroud, making docking impossible; 2-hr and 7-min EVA; rendezvous as planned; controlled reentry; landed 0.7 km from target.
GT-10	18 July	Young Collins	3 763	70:46:39	Rendezvous and docking with target vehicle; use of target vehicle propulsion system to rendezvous with 2d target vehicle; EVA.	Successful; 2 EVA periods.
AS-202	25 Aug.	Unmanned	SM: 4 466 CM: 5 471	01:33:00	Flight test of production Apollo Block I type spacecraft using Saturn IB launch vehicle; demonstration of structural integrity and compatibility; evaluation of heatshield performance.	Successful.
GT-11	12 Sept.	Conrad Gordon	3 798	71:17:08	1st revolution rendezvous and docking; EVA; docking practice; maneu-	Successful; reached record altitude of about 1400 km above earth.

Name of spacecraft	Launch date	Occupant	Payload weight (kg)	Flight time (hr:min:sec)	Basic flight objectives	Summary of results
					vers in docked configuration to high altitude.	
GT-12	11 Nov.	Lovell Aldrin	3 763	94:34:31	Rendezvous and docking with target vehicle; 3 EVAs; tethered station-keeping exercise; maneuvers.	EVA and rendezvous/docking exercises successful; trouble with target vehicle prevented maneuvers; last Gemini flight.
Cosmos 133	28 Nov.	Unmanned	6 575[1]	48:00:00[1]	Orbital data similar to manned flight.	Possible precursor flight to 23 Apr. 1967 Soyuz 1 flight.
	1967					
Apollo 1	27 Jan.	Grissom White Chaffee	20 412[1]	—	Not a flight, but a plugs-out test at Kennedy Space Center Launch Complex 34.	The 3 crewmembers were killed in a flash fire that swept through the spacecraft.
Cosmos 140	7 Feb.	Unmanned	6 575[1]	48:00:00[1]	Orbital data similar to manned flight.	Possible precursor flight to 23 Apr. flight of Soyuz 1.
Cosmos 146	10 Mar.	Unmanned	22 720[1]	192:00:00[1]	NA.	Alternatively identified as a precursor flight to either Zond or Soyuz.
Cosmos 154	8 Apr.	Unmanned	22 720[1]	48:00:00[1]	Orbital data similar to manned flight.	Possible precursor to a later Soyuz flight.

Name	Date	Crew	Weight	Duration	Purpose	Remarks
Soyuz 1 Rubin Ruby	23 Apr.	Komarov	6 451	26:40:00	First manned test of the Soyuz spacecraft.	Because the spacecraft was experiencing movement around its axis upon reentry, the craft crash-landed when its parachute lines became tangled, killing the cosmonaut upon impact.
Cosmos 186	27 Oct.	Unmanned	6 575[1]	84:40:00[1]	Demonstration of automatic docking of 2 Soyuz-type spacecraft (with Cosmos 188).	Successful 3½-hr docking period.
Cosmos 188	30 Oct.	Unmanned	6 575[1]	72:00:00[1]	Demonstration of automatic docking of 2 Soyuz-type spacecraft (with Cosmos 186).	Successful 3½-hr docking period.
Apollo 4 AS 501	9 Nov.	Unmanned	CSM: 23 401 LTA: 13 381	08:37:09	Demonstration of integrity and compatibility of launch vehicle and Apollo spacecraft, stage separation, heatshield performance, support facilities, and recovery.	Successful; 1st flight of Saturn V launch vehicle.
	1968					
Apollo 5 AS 204	22 Jan.	Unmanned	LM: 14 300	07:52:03	Verification of operation of LM and ascent propulsion systems and structure; evaluation of staging and launch vehicle performance.	Successful; first orbital test of LM; CSM replaced by dummy nose cone.

Name of spacecraft	Launch date	Occupant	Payload weight (kg)	Flight time (hr:min:sec)	Basic flight objectives	Summary of results
Zond 4	2 Mar.	Unmanned	22 720[1]	161:00:00[1]	Flight check of Zond spacecraft.	Reentry vehicle similar to that of Soyuz reentry vehicle; Zond series identified by Soviets as fully capable of carrying a human crew around the moon.
Apollo 6 AS 502	4 Apr.	Unmanned	CSM: 25 138 LTA: 11 794	09:49:45	Demonstration of structural and thermal integrity and compatibility of launch vehicle and spacecraft and stage separation; verification of operation of propulsion, guidance and control, and electrical systems; demonstration of mission support facilities.	Stage separation not achieved as planned; other objectives met.
Cosmos 212	14 Apr.	Unmanned	6 575[1]	120:00:00[1]	Further demonstration of automatic rendezvous and docking (with Cosmos 213).	Successful.
Cosmos 213	15 Apr.	Unmanned	6 575[1]	120:00:00[1]	Further demonstration of automatic rendezvous and docking (with Cosmos 212).	Successful.
Cosmos 238	28 Aug.	Unmanned	6 575[1]	72:00:00[1]	Orbital data similar to manned flight.	Possible precursor flight to later Soyuz flights.

Mission	Date	Crew/Payload	Weight	Duration	Objectives	Results
Zond 5	15 Sept.	Biological payload (?)	22 720[1]	162:24:00	Evaluation of effects of circumlunar flight on biological payload; demonstration of circumlunar flight and recovery of automatic spacecraft.	Successful; first Soviet water landing and recovery.
Apollo 7 AS 205	11 Oct.	Schirra Eisele Cunningham	CSM: 14 781	260:09:08	Demonstration of crew performance, rendezvous capability, and crew/vehicle/support facilities performance; qualification of heatshield.	Successful; rendezvous with Saturn IVB stage; 1st manned Apollo flight.
Soyuz 2	25 Oct.	Unmanned	6 350[1]	72:30:00	Target for Soyuz 3 rendezvous.	Successful.
Soyuz 3 Argon Argon	26 Oct.	Beregovoy	6 576	94:51:00	Perfection of rendezvous techniques in orbit; experiments with unmanned Soyuz 2.	Successful; performed automatic and manual rendezvous maneuvers.
Zond 6	10 Nov.	Biological payload (?)	22 720[1]	162:46:30	Cosmic ray experiment; determination of effects of flight on biological payload; circumlunar flight; lunar photography.	Successful.
Apollo 8 AS 503	21 Dec.	Borman Lovell Anders	CSM: 28 817 LTA: 9 026	147:00:42	Demonstration of performance in cislunar and lunar orbit environment; evaluation of crew performance in lunar orbit mission; demonstration of communications and tracking; high-resolution photography.	Successful; 1st manned lunar orbit; 1st manned Saturn V launch.

Name of spacecraft	Launch date	Occupant	Payload weight (kg)	Flight time (hr:min:sec)	Basic flight objectives	Summary of results
	1969					
Soyuz 4 Amur Amur	14 Jan.	Shatalov	6 626	71:23:00	Demonstration of rendezvous and manual docking of 2 manned spacecraft (with Soyuz 5).	Successful; performed docked maneuvers; overall volume = 18 cu. m.
Soyuz 5 Baykal Baikal	15 Jan.	Volynov Khrunov Yeliseyev	6 586	72:56:00	Demonstration of rendezvous and manual docking of 2 manned spacecraft; crew transfer by EVA to Soyuz 4.	Successful.
Apollo 9 AS 504 Gumdrop and Spider	3 Mar.	McDivitt Scott Schweickart	CSM: 26 801 LM: 14 575	241:00:53	Demonstration of LM performance (propulsion, rendezvous, and docking capabilities); EVA and intervehicular crew transfer.	Successful.
Apollo 10 AS 505 Charlie Brown and Snoopy	18 May	Stafford Young Cernan	CSM: 28 834 LM: 13 941	192:03:23	Qualification of combined spacecraft in lunar environment; LM rendezvous and CM docking in lunar gravitational field; evaluation of lunar navigation.	Successful; first lunar orbital mission with complete Apollo spacecraft.
Apollo 11 AS 506 Columbia and Eagle	16 July	Armstrong Collins Aldrin	CSM: 28 807 LM: 15 103	195:18:35	Demonstration of lunar landing with manned Apollo spacecraft;	Successful; first manned lunar landing; time on the moon = 21:36:20.9.

	Date	Crew	Mass (kg)	Duration	Purpose	Result
					"moon walk"; lunar photography.	
Zond 7	7 Aug.	Unmanned	22 720[1]	144:00:00[1]	Further demonstration of circumlunar flight; lunar photography.	Successful.
Soyuz 6 Antey Antasus	11 Oct.	Shonin Kubasov	6 578	118:42:00	Welding experiments in vacuum; biomedical experiments; stellar observations; photography; maneuvers with the Soyuz 7 and 8 spacecraft.	Successful.
Soyuz 7 Buran Snowstorm	12 Oct.	Filipchenko Volkov Gorbatko	6 571	118:41:00	Same as Soyuz 6 (no welding planned); maneuvers with Soyuz 6 and 8 spacecraft.	Successful.
Soyuz 8 Granit Granite	13 Oct.	Shatalov Yeliseyev	6 647	118:50:00	Demonstration of service as a command ship for 3-craft maneuvers, illustrating capability for future space station construction; demonstration of ground control capability in multicraft situation.	Successful.
Apollo 12 AS 507 Yankee Clipper and Intrepid	14 Nov.	Conrad Gordon Bean	CSM: 28 838 LM: 15 235	244:36:25	Precise lunar landing; lunar exploration; scientific experiments in Ocean of Storms area.	Successful; time on the moon = 31:31:12.

Name of spacecraft	Launch date	Occupant	Payload weight (kg)	Flight time (hr:min:sec)	Basic flight objectives	Summary of results
	1970					
Apollo 13 AS 508 Odyssey and Aquarius	11 Apr.	Lovell Swigert Haise	CSM: 28 945 LM: 15 196	142:54:41	Performance of lunar mission at Fra Mauro; selenological inspection; experiments; development of working capability in lunar environment; lunar photography.	Nominal mission aborted because of an abrupt loss of SM cryogenic oxygen associated with a fire in 1 of the 2 tanks at about 56 hr GET; LM provided power and life support until transfer to CM for reentry.
Soyuz 9 Sokol Falcon	1 June	Nikolayev Sevastyanov	6 501	424:59:00	Acquisition of extensive observations on the effects of prolonged flight on both the crew and the spacecraft; performance of numerous course correction exercises.	Successful; longest manned flight in time to date; cosmonauts experienced abnormal blood pressure and color perception and fatigue upon landing but survived the long flight satisfactorily.
Zond 8	20 Oct.	Unmanned	22 720¹	161:00:00	Further demonstration of circumlunar flight; lunar and planetary photography.	Successful; Soviets' second water recovery.
Cosmos 379	24 Nov.	Unmanned	6 575¹	NA	Initial orbital data similar to Soyuz-type flight data but with a later lunar or interplanetary orbital launch platform.	Possible test of new lunar-type engine system.

Name	Date	Crew	Weight	Duration	Purpose	Remarks
Cosmos 382	2 Dec.	Unmanned	22 720[1]	NA	Orbital data. (Speculation has been made that this flight may have been a precursor to manned lunar flight.)	NA.
	1971					
Apollo 14 AS 509 Kitty Hawk and Antares	31 Jan.	Shepard Roosa Mitchell	CSM: 29 240 LM: 15 264	216:01:58	Selenological inspection and sampling; development of capability to work in lunar environment; photographs of candidate exploration sights; experiments.	Basically successful; several minor problems prevented the lunar crew's performance of all objectives; time on the moon = 33:30:31.
Cosmos 398	26 Feb.	Unmanned	6 575[1]	NA	Initial orbital data similar to Soyuz-type flight data but with a later lunar or interplanetary orbital launch platform.	Possible test of new lunar-type engine system.
Salyut 1	19 Apr.	Unmanned	18 597[1]	—	Service as a space station for experiments and observation (to be manned by visiting Soyuz flight).	Successful; Soyuz 10 (Apr. 1971) and 11 (June 1971) crews visited Salyut 1; reentered atmosphere 11 Oct. 1971.
Soyuz 10 Granit Granite	23 Apr.	Shatalov Yeliseyev Rukavishnikov	6 577[1]	47:46:00	Test of new rendezvous and docking techniques (equipped with a docking collar).	Remained in docked position with Salyut 1 5 hr and 30 min; some speculation that crew had intended to enter station but could not because of malfunction.

Name of spacecraft	Launch date	Occupant	Payload weight (kg)	Flight time (hr:min:sec)	Basic flight objectives	Summary of results
Soyuz 11 Yantar Amber	6 June	Dobrovolskiy Patsayev Volkov	6 577[1]	570:22:00 (23+ days)	Occupation of the Salyut station and performance of extensive experiments, biological and astronomical; observation of spacecraft systems.	All flight objectives met successfully during 24-day stay on Salyut; crew was killed during landing procedures when a seal in the spacecraft failed before the ship entered earth's atmosphere; all scientific logs and films were recovered.
Apollo 15 AS 510 Endeavor and Falcon	26 July	Scott Worden Irwin	CSM: 30 370 LM: 16 430	295:11:53	Lunar inspection, survey, and sampling of materials in Hadley-Apennine region; experiments; evaluation of capability of lunar equipment during extended lunar surface stay.	Successful; used lunar rover during moon exploration; total lunar stay = 66:54:53; conducted 3 lunar EVA periods, plus 1 transearth coast EVA for a total of 19:47:00; 1 parachute failed on landing but the recovery was successful and the landing precise.
Cosmos 434	12 Aug.	Unmanned	6 575[1]	NA	Initial orbital data similar to Soyuz-type flight data but with later lunar or interplanetary orbital launch platform.	Possible test of new lunar-type engine system.

Spacecraft	Date	Crew	Weight (kg)	Duration	Purpose	Results
	1972					
Apollo 16 AS 511 Casper and Orion	16 Apr.	Young Mattingly Duke	CSM: 30 395 LM: 16 445	265:51:05	Lunar inspection, survey, and materials sampling in Descartes region; experiments on lunar surface and in flight; photography.	Successful; total lunar stay = 71:02:13; lunar rover driven 26.7 km; lunar EVA periods totaled 20 hr and 14 min.
Cosmos 496	26 June	Unmanned	6 575[1]	NA	Performance of test flight to verify safety modifications to Soyuz breathing ventilation valve.	Successful; reentered 2 July.
Apollo 17 AS 512 America and Challenger	7 Dec.	Cernan Evans Schmitt	CSM: 30 369 LM: 16 456	301:51:59	Lunar survey and sampling of materials in Taurus-Littrow region; surface experiments and photography; in-flight experiments and photography.	Successful; last Apollo flight to the moon; lunar stay = 74:59:38; 3 lunar EVA periods totaled 22:05:04; traveled 35 km in lunar rover.
	1973					
Salyut 2	3 Apr.	Unmanned	18 597[1]	—	Service as a space station for experiments and observations; perfection of design and onboard systems.	Not successful; suspected thruster problem caused craft to tumble out of control, resulting in a solar panel malfunction; reentered 28 May 1973.
Cosmos 557	11 May	Unmanned	18 500[1]	96:00:00[1]	Orbital data. (Observers believed craft was Salyut failure; U.S.S.R. denies any relationship to manned program.)	Mission apparently completed in 4 days; craft reentered 22 May 1973.

473

Name of spacecraft	Launch date	Occupant	Payload weight (kg)	Flight time (hr:min:sec)	Basic flight objectives	Summary of results
Skylab 1 AS 513 and 212	14 May	Unmanned	74 783	–	Service as an orbiting space station in which to conduct experiments for lengthy periods of time by visiting crews.	Successful; manned by visiting crews from Skylab Mission 2 (May 1973), 3 (July 1973), and 4 (Nov. 1973); still in orbit.
Skylab 2 CSM 116 AS 206	25 May	Conrad Kerwin Weitz	CM: 6 076	672:49:49 (28+ days)	Operation of orbital workshop as a habitable space structure for up to 28 days; acquisition of medical data on crew for use in extending duration of manned flight; in-flight experiments.	Successful; crew made repairs to workshop damaged during launch (Skylab parasol); some difficulty in performing the docking procedure.
Cosmos 573	15 June	Unmanned	6 575[1]	NA	Performance of test flight to verify safety modifications to Soyuz breathing ventilation valve.	Successful; reentered 17 June.
Skylab 3 CSM 117 AS 207	28 July	Bean Garriott Lousma	CM: 6 085	1427:09:04 (59+ days)	Acquisition of data for evaluating performance of unmanned workshop; reactivation of workshop; acquisition of medical data on the crew; in-flight experiments.	Successful; all scientific objectives met; crew met with some motion sickness during first 3 visit days.

	Date	Crew/Payload	Mass	Duration	Objective	Result
Soyuz 12 / Uraliy / Urals	27 Sept.	Lazarev Makarov	6 577[1]	47:16:00	Test of improved flight conditions; acquisition of spectrographic data of separate sections of the earth; evaluation of new space suit design.	Successful.
Cosmos 605	31 Oct.	Biological payload	5 500[1]	NA	Study of effects of space on living organisms (white rats, steppe turtles, insects, fungi); test of life-sustaining systems.	Successful; recovered 22 Nov.
Skylab 4 / CSM 118 / AS 208	16 Nov.	Carr Gibson Pogue	CM: 6 104	2017:15:31 (84+ days)	Evaluation of performance of unmanned workshop and its reactivation; acquisition of extensive medical data from crewmembers; in-flight experiments.	Successful; last visit to Skylab; total mission visit time = 4117:14:24; total mission flight time = 12 351:43:12; some initial problem with motion sickness; made extensive observations of Comet Kohoutek.
Cosmos 613	30 Nov.	Unmanned	6 570[1]	NA	Systems test of Soyuz to determine if craft could remain inactive in orbit and then be reactivated.	Successful; soft-landed 30 Jan. 1974.
Soyuz 13 / Kavkaz / Caucasus	18 Dec.	Klimuk Lebedev	6 577[1]	188:32:00	Observation of stars in the ultraviolet range using a special system of telescopes; survey of separate sections of earth's surface and acquisition of data; continuation of comprehensive verifica-	Successful.

Name of spacecraft	Launch date	Occupant	Payload weight (kg)	Flight time (hr:min:sec)	Basic flight objectives	Summary of results
	1974				tion of onboard systems; test of manual and automatic control and methods of autonomous navigation in various flight conditions.	
Cosmos 638	3 Apr.	Unmanned	6 570[1]	NA	Test of docking systems and rendezvous for ASTP.	Successful; reentered 13 Apr.
Cosmos 656	27 May	Unmanned	6 570[1]	NA	ASTP-related tests of Soyuz.	Successful; reentered 29 May.
Salyut 3	25 June	Unmanned	18 144[1]	—	Service as an unmanned orbital workshop to be visited and manned by Soyuz crews who will perform experiments and observations.	Successful; manned by Soyuz 14 (July 1974) crew; programmed reentry and disintegration on 24 Jan. 1975.
Soyuz 14 Berkut Golden Eagle	3 July	Popovich Artyukhin	6 577[1]	336:00:00[1]	Test of Salyut's engineering system and energy supply; in-flight experiments.	Successful.
Cosmos 672	12 Aug.	Unmanned	6 570[1]	NA	ASTP-related tests of Soyuz.	Successful; reentered 18 Aug.

Name	Date	Crew	Mass	Duration	Description	Result
Soyuz 15 Dunay Danube	26 Aug.	Sarafanov Demin	6 577[1]	48:12:00	Docking with Salyut 3; onboard tests and in-flight experiments.	Unsuccessful; night landing in adverse weather conditions; crew recovered.
Cosmos 690	22 Oct.	Biological payload	NA	NA	Tests of effects of radiation on animals (white rats); cesium 137 gamma ray source on board.	Successful; 21-day flight.
Soyuz 16 Buran Snowstorm	2 Dec.	Filipchenko Rukavishnikov	6 577[1]	142:24:00	Simulation of ASTP-type mission; U.S.-Soviet joint tracking exercise.	Successful.
Salyut 4	26 Dec.	Unmanned	18 144[1]	—	Unmanned orbiting space workshop to further test design, onboard systems, and equipment; scientific-technical studies conducted by visiting Soyuz crews.	Successful; manned by Soyuz 17 (Jan. 1975) crew.
1975						
Soyuz 17 Zenit Zenith	11 Jan.	Grechko Gubarev	6 577[1]	720:00:00[1] (30 days)[1]	Extensive series of scientific and medical experiments onboard Salyut 4; observation of effects of prolonged weightlessness on man.	Successful; set Soviet record for time in space.
Soyuz anomaly Uraliy Urals	5 Apr.	Lazarev Makarov	6 577[1]	00:04:45[1]	Visit to Salyut 4 space station.	Aborted shortly after launch because of launch vehicle malfunction; crew recovered safely near

Name of spacecraft	Launch date	Occupant	Payload weight (kg)	Flight time (hr:min:sec)	Basic flight objectives	Summary of results
						Gorno Altaisk (in Siberia near the People's Republic of China border); first manned mission abort.
Soyuz 18	24 May	Klimuk Sevastyanov	6 577[1]	1512:00:00[1] (63 days)	Visit to Salyut 4 space station.	Successful; set Soviet record for time in space; landed 26 July.
Soyuz 19	15 July	Leonov Kubasov	7 000[1]	142:30:54 (5+ days)	Rendezvous and docking with U.S. Apollo crew; joint activities including crew transfer.	Successful; landed 21 July.
Apollo	15 July	Stafford Brand Slayton	CSM: 12 905 DM: 2 006	217:28:24 (9 days)	Rendezvous and docking with U.S.S.R. Soyuz crew; joint activities including crew transfer.	Successful; landed 24 July.

A   Apollo
AS   Apollo-Saturn
ASTP   Apollo-Soyuz Test Project
BD   Booster development
CM   Command module
CSM   Command and service module
DM   Docking module
ECS   Environmental control system

EVA   Extravehicular activity
GET   Ground-elapsed time
GT   Gemini-Titan
LJ   Little Joe
LM   Lunar module
LTA   Lunar test article
MA   Mercury-Atlas
MR   Mercury-Redstone

NA   Not available
PA   Pad abort
SM   Service module

[1] Approximate.
[2] Sometimes called "spacecraft."
[3] Test vehicle at launch.

Appendix C
Signatories of Joint Meeting Minutes, Working Groups 1 to 5

Working Group 1 Meeting Minutes—Signatories

Member	U.S.A.	U.S.S.R.	1971 21-25 June	1972 26 Nov.-6 Dec.	1973 6-18 July	1973 9-19 Oct.	1973 15-30 Mar.	1973 18-29 June	1973 9-20 July	1973 1-12 Oct.	1973 19-30 Nov.	1974 14 Jan.-1 Feb.	1974 8 Apr.-3 May	1974 17-28 June	1974 24 June-11 July	1974 26 Aug.-20 Sept.	1974 9-27 Sept.	1974 16-27 Sept.	1974 24 Oct.-6 Nov.	1974 25 Nov.-21 Dec.	1975 20 Jan.-13 Feb.	1975 26 Jan.-13 Mar.	1975 7-28 Feb.	1975 6-16 Apr.	1975 14-25 Apr.	1975 9-13 June
Akoyev, I. G.		X					X					X				X								X		
Alexander, J. D.	X						X																			
Alibrando, A. P.	X																							X	X	
Artemov, B. P.		X		X	X	X		X																		
Babkov, O. I.		X																								X
Blackmer, A. M.	X						X		X	X		X	X								X					X
Blagov, V. D.		X				X	X		X	X		X	X			X					X					X
Bland, D. A.	X						X		X	X		X	X						X		X	X	X		X	X
Bobkov, V. N.		X	X	X		X	X				X	X	X	X		X			X	X	X	X				
Brooks, D. R.	X						X												X		X					
Brzezinski, M. S.	X						X																			
Busch, C. W.	X									X									X		X					
Bushuyev, K. D.		X	X	X																						
Bykovskiy, V. F.		X											X								X					
Cernan, E. A.	X										X															
Chernyshev, G. A.		X													X										X	X
Chistov, Yu. I.		X						X																		
Cockrell, B. F.	X						X																			
Covington, C.	X		X	X																						
Cremin, J. W.	X		X																							
Denisenko, V. A.		X																					X			
Denisov, Yu. S.		X			X	X	X	X	X	X		X	X			X							X		X	
Dennett, A. D.	X											X	X	X		X			X	X		X			X	

479

Working Group 1 Meeting Minutes—Signatories—Continued

Member	Country		1971	1972	1973							1974									1975					
	U.S.A.	U.S.S.R.	21-25 June	26 Nov.-6 Dec.	6-18 July	9-19 Oct.	15-30 Mar.	18-29 June	9-20 July	1-12 Oct.	19-30 Nov.	14 Jan.-1 Feb.	8 Apr.-3 May	17-28 June	24 June-11 July	26 Aug.-20 Sept.	9-27 Sept.	16-27 Sept.	24 Oct.-6 Nov.	25 Nov.-21 Dec.	20 Jan.-13 Feb.	26 Jan.-13 Mar.	7-28 Feb.	6-16 Apr.	14-25 Apr.	9-13 June
Dubiago, Yu.		X																								X
Dulnev, L. I.		X				X																				
Duruda		X											X													
Edmiston, C. R.	X												X								X					
Ferguson, G. M.	X						X		X																	
Fokin, Y. V.		X																						X		
Frank, M. P.	X				X	X	X		X	X		X				X		X	X	X					X	
Frolov, S. G.		X				X																				
Golovanov, B. A.		X				X																				
Gorshkov, L. A.		X		X	X																					
Gorulko, Yu. D.		X								X																
Graham, O. L.	X																			X						
Gringauz, K. I.		X						X																		
Grishin, S. D.		X																							X	
Haken, R. L.	X																			X						
Honeycutt, J. F.	X												X			X									X	
Ivanov, L. I.		X						X																		
Ivanov, V. B.		X																		X						
James, B. W.	X												X							X					X	
Jaschke, P. S.	X						X	X	X	X		X	X							X	X				X	
Jenness, M. D.	X									X		X	X													
Johnson, C. C.	X		X																							
Kain, R. R.	X														X											
Kalenkov, O. B.		X							X	X		X	X			X				X	X				X	
Khodarev, Yu. K.		X						X	X	X		X				X										
King, J. W.	X																			X						
Kondratkov, N. F.		X																			X	X				
Korolev, A. S.		X								X	X															
Kravets, V. G.	X															X									X	
Kubasov, V. N.		X											X													
Kunashev, B. S.		X										X									X					
Kuz'michev, Yu. P.		X										X									X					
Land, C. K.	X													X												
Larsen, A. M.	X											X	X													
Lavrov, I. V.		X	X	X																						
Lee, R.	X																			X						
Leonidov, N. D.		X																							X	
Leonov, A. A.		X								X	X	X		X												
Lineberry, E. G.	X		X	X	X	X																				
Littleton, F. C.	X											X	X			X				X	X				X	

Working Group 1 Meeting Minutes—Signatories—Continued

Member	U.S.A.	U.S.S.R.	21-25 June (1971)	26 Nov.-6 Dec. (1971)	6-18 July (1972)	9-19 Oct. (1972)	15-30 Mar. (1973)	18-29 June (1973)	9-20 July (1973)	1-12 Oct. (1973)	19-30 Nov. (1973)	14 Jan.-1 Feb. (1974)	8 Apr.-3 May (1974)	17-28 June (1974)	24 June-11 July (1974)	26 Aug.-20 Sept. (1974)	9-27 Sept. (1974)	16-27 Sept. (1974)	24 Oct.-6 Nov. (1974)	25 Nov.-21 Dec. (1974)	20 Jan.-13 Feb. (1975)	26 Jan.-13 Mar. (1975)	7-28 Feb. (1975)	6-16 Apr. (1975)	14-25 Apr. (1975)	9-13 June (1975)
Lunney, G. S.	X		X	X																						
Mager, J. E.	X									X			X													
Makarov, Ye. S.		X				X				X																
Maksimenko, A. H.		X			X																					
Maksimov, S. N.		X							X																	
Managadze, G. G.		X						X																		
Maslennikov, Ya. P.		X														X					X					
Matveev		X														X										
Merriam, R. S.	X						X		X																	
Metcalf, G. E.	X																								X	
Mezenov, L. F.		X														X				X	X				X	
Milyukov, Ya. P.		X																			X					X
Mosel, D. K.	X						X																			
Nesterenko, A. A.		X																		X					X	
Netherton, C. B.	X															X					X					
Niemeyer	X													X												
Nikolayev, A. G.		X				X																				
Nikolsky, G. M.		X						X																		
Novikov, N. S.		X																			X					
Novikov, S. I.		X								X																
Novikov, Yu. V.		X						X																		
Ol'shyevsky, V. A.		X														X									X	
Ostashyev, A. I.		X														X										
Overmyer, R. F.	X						X		X	X		X	X								X					
Patrushev, V. S.		X											X			X										
Pavelka, E. L.	X								X	X		X	X													
Permitin, V. Yu.		X												X							X				X	X
Petrov, B. N.		X											X							X						
Pippert, E. B.	X						X		X	X		X	X			X					X	X	X		X	
Pochitalin, I. G.		X	X																							
Pollock, S. T.	X																				X					
Putiatin		X												X							X					
Ragan, J. H.	X														X											
Reiten, L. A.	X																				X	X				
Reshetin, A. G.		X																			X					
Rivers, J. V.	X						X																			
Roditelev, N. D.		X														X					X					
Rogers, R.	X																									X
Rumyantsev, I. P.		X	X	X																				X		
Savitsky, Yu. I.		X										X				X					X				X	

Working Group 1 Meeting Minutes—Signatories—Concluded

Member	U.S.A.	U.S.S.R.	1971 21-25 June	1972 26 Nov.-6 Dec.	1973 6-18 July	9-19 Oct.	15-30 Mar.	18-29 June	9-20 July	1-12 Oct.	19-30 Nov.	1974 14 Jan.-1 Feb.	8 Apr.-3 May	17-28 June	24 June-11 July	26 Aug.-20 Sept.	9-27 Sept.	16-27 Sept.	24 Oct.-6 Nov.	25 Nov.-21 Dec.	1975 20 Jan.-13 Feb.	26 Jan.-13 Mar.	7-28 Feb.	6-16 Apr.	14-25 Apr.	9-13 June
Shafer, R. J.	X																							X		
Shatalov, V. A.		X									X						X							X		
Sheviakov, A.		X																								X
Shibanov, L. S.		X											X		X				X		X				X	
Shinkle, G. L.	X																		X							
Shustin, K. S.		X		X																						
Sivkov, V. I.		X											X	X	X				X		X					
Skotnikov, B. P.		X																	X							
Smith, H. E.	X																		X							
Smylie, R. E.	X		X	X	X																					
Stafford, T. P.	X					X	X	X	X	X	X	X	X	X										X		
Starodubtsev, G. V.		X							X																	
Stetsyura, V. P.		X										X	X	X												
Stupnikov, Yu. N.		X					X		X				X		X						X					
Suschev, G. A.		X												X					X							
Svirin, V. A.		X				X						X	X	X	X				X		X	X			X	X
Sytin, O. G.		X		X	X	X	X		X			X	X		X				X							
Temple, J. H.	X					X		X	X		X			X					X					X		
Timchenko, V. A.		X				X	X		X		X			X					X					X		
Tsybin, S. P.		X												X										X		
Varshavsky, V. P.		X				X	X		X	X	X	X		X				X		X	X			X	X	
Vaskevich, Ye. A.		X							X			X														
Vereshchetin, V. S.		X																					X			
Vice, J. R.	X													X												
Vlasov, G.		X											X						X							
Voronin, A. A.		X				X			X	X		X	X		X				X							
Ward, R. J.	X				X	X	X																			
Whitely, J. F.	X																									X
Yastrebov, V. D.		X																			X					
Yeats, H. D.	X																		X							
Yegorov, A. D.		X											X		X						X					
Yeliseyev, A. S.		X				X	X		X	X			X					X	X	X					X	
Yermakov, V. G.		X										X									X					
Yesipov, Ye. V.		X																					X			
Young, K. A.	X					X		X	X		X	X		X							X					
Zaloguyev, S. N.		X				X																				
Zelenshchikov, N. I.		X				X									X						X					

Note—This list includes meetings of Crew Training, Onboard Documentation, and Public Information Sub-groups. After July 1972, some of these individuals became members of Working Group 4 or 5.

Working Group 1 Meeting Minutes—Translation Verifiers

Translation verifier	U.S.A.	U.S.S.R.	21-25 June 1971	26 Nov.-6 Dec. 1972	6-18 July 1972	9-19 Oct. 1972	15-30 Mar. 1973	18-29 June 1973	9-20 July 1973	1-12 Oct. 1973	19-30 Nov. 1973	14 Jan.-1 Feb. 1974	8 Apr.-3 May 1974	17-28 June 1974	24 June-11 July 1974	26 Aug.-20 Sept. 1974	9-27 Sept. 1974	16-27 Sept. 1974	24 Oct.-6 Nov. 1974	25 Nov.-21 Dec. 1974	20 Jan.-13 Feb. 1975	26 Jan.-13 Mar. 1975	7-28 Feb. 1975	6-16 Apr. 1975	14-25 Apr. 1975	9-13 June 1975
Artemov, B. P.		X																X	X	X						
Forostenko, A.	X											X											X		
Garipova, R. I.		X											X	X										X	
Gnovtcheff, M. A.	X																			X					
Habarin, N. V.		X																		X					
Karakulko, N.	X													X					X	X					X
Klochkovskaya, T. V.		X									X														
Krukov, A. B.		X					X																		
Latter, N. K.	X						X	X	X		X	X					X			X				X	
Lavrov, R. N.	X																		X						
Levitin, L. E.		X						X																	
Mamantov, I. A.	X								X																
Milukov, I. V.		X											X												
Pepov, F.		X								X															
Preobrazhensky, Yu. V.		X							X																
Rodzianko, A.	X																						X		
Samofal, K. S.		X														X							X		
Savel'yeva, I. S.		X																		X					
Sementovsky, A.	X													X			X								
Sharonova, Yu. A.		X																							X
Sokolov, S. A.		X																	X			X		X	
Taylor, A. G.	X																							X	
Timacheff, N.	X								X																
Yatsenko, V. A.		X						X		X	X													X	

Note—This list includes meetings of Crew Training, Onboard Documentation, and Public Information Subgroups. After July 1972, some of these individuals became members of Working Group 4 or 5.

Working Group 2 Meeting Minutes—Signatories

Member	U.S.A.	U.S.S.R.	21-25 June 1971	26 Nov.-6 Dec. 1972	10-17 May 1972	6-18 July 1972	9-19 Oct. 1972	24 Nov.-2 Dec. 1972	15-30 Mar. 1973	18-29 June 1973	9-20 July 1973	8 Apr.-3 May 1974	26 Aug.-20 Sept. 1974	20 Jan.-13 Feb. 1975	5-22 May 1975
Ageyev, A. A.		X							X	X		X	X		X
Alekseyev, B. I.		X								X		X	X		X
Babkov, O. I.		X	X	X	X	X			X	X	X	X	X	X	
Bennett, D.	X											X		X	
Berry, R. L.	X						X		X	X					
Bobkov, V. N.		X									X				X
Brooks, M. F.	X		X												
Cheatham, D. C.	X		X	X	X	X									
Chernov, Z. S.		X		X											
Cox, K. J.	X						X		X	X					
Dietz, R. H.	X			X	X	X									
Dobrosel'sky, V. K.		X	X												
Haken, R. L.	X								X	X	X	X	X		
Hardee, J. H.	X														X
Helms, C. W.	X										X				
Ispolatov, O. G.		X	X	X	X	X									
Keller, T.	X												X		
Khomenko, Yu. P.		X	X			X									
Kozyrev, V. I.		X			X	X									
Kubiak, E. T.	X							X	X						
Kupriyanov, I. K.		X	X	X	X										
Lattier, E. E.	X					X									
Lee, R.	X											X	X	X	X
Legostayev, V. P.		X	X	X	X	X	X	X	X	X	X	X	X	X	X
Lindsay, K.	X							X	X	X	X	X			
Manry, C. E.	X			X	X	X		X	X						
Minayev, V.		X													X
Morgulev, A. S.		X	X	X	X										
Podelyakin, V. A.		X	X	X	X		X	X	X	X					
Pollock, S. T.	X												X	X	
Prostov, L.		X											X		
Raspletin, V. A.		X	X	X	X	X									
Reid, R.	X		X	X	X	X		X	X						
Rountree, R. W.	X								X	X					
Shmyglevskiy, I. P.		X	X	X	X	X	X	X	X	X	X	X		X	
Shores, P. W.	X					X									
Skotnikov, B. P.		X									X		X	X	X
Smith, H. E.	X						X	X	X	X	X	X	X	X	X
Snipes, S. E.	X												X	X	
Travis, A. D.	X					X									
Turnbull, J.	X												X		
Zakomorny, V.		X									X				

Note—After July 1972, some of these individuals became members of Working Group 4 or 5.

Working Group 2 Meeting Minutes—Translation Verifiers

Translation verifier	U.S.A.	U.S.S.R.	21-25 June (1971)	26 Nov.-6 Dec. (1972)	10-17 May	6-18 July	9-19 Oct.	24 Nov.-2 Dec.	15-30 Mar. (1973)	18-29 June	9-20 July	8 Apr.-3 May (1974)	26 Aug.-20 Sept.	20 Jan.-13 Feb. (1975)	5-22 May
Artemov, B. P.		X									X				
Bourov, V. A.		X								X					
Klemanov, S. A.		X								X					X
Latter, N. K.	X									X					
Lavroff, R. N.	X								X			X			
Mamantov, I. A.	X												X		
Preobrazhensky, Yu. V.		X								X					
Safonova, E. S.		X										X	X		
Sementovsky, A.	X														X
Sergeyeva, Ye. N.		X												X	X
Taylor, A. G.	X												X		X
Timacheff, N.	X										X				
Zhuravlev, V. D.		X							X						

Note—After July 1972, some of these individuals became members of Working Group 4 or 5.

Working Group 3 Meeting Minutes—Signatories

Member	Country		Date of meeting															
			1971		1972				1973				1974				1975	
	U.S.A.	U.S.S.R.	21-25 June	26 Nov.-6 Dec.	27 Mar.-3 Apr.	6-18 July	9-19 Oct.	6-16 Dec.	15-30 Mar.	27 June-11 July	13-23 Nov.	3-24 Dec.	8 Apr.-3 May	1 July-5 Sept.	26 Aug.-20 Sept.	23 Oct.-11 Dec.	20 Jan.-13 Feb.	5-22 May
Bagno, V. I.		X			X													
Belikov, E. M.		X								X		X		X	X	X		
Bloom, K. A.	X					X	X	X	X	X			X		X	X		
Bobrov, Ye. G.		X			X	X	X	X	X	X	X		X	X	X	X	X	X
Brown, R. G.	X							X										
Chizhikov, B. S.		X			X			X	X	X	X			X	X	X		
Craig, M. K.	X								X									
Creasy, W. K.	X					X	X	X	X	X	X		X	X	X	X	X	X
Duhovsky, Ye. A.		X								X	X							
Eudy, G.	X		X															
Hardee, J. H.	X												X		X	X		
Holder, B. W.	X								X	X		X	X	X				
Jivoglotov, V. N.		X						X	X									
Johnson, C. C.	X			X	X	X	X											
Jones, J. C.	X		X	X	X													
Kirkpatrick, C. A.	X								X									
Klimenko, Ye. I.		X						X	X	X								
Kudryavtzev, V. V.		X						X	X	X	X		X	X	X	X		
Lebedev, Ye. F.		X						X	X	X	X				X	X		
Maksimenko, A. H.		X			X													
Nikiforov, A V.		X	X	X														
Pavlov, V. H.		X								X								
Polyakov, V. G.		X								X								
Rosenberg, O. M.	X							X										
Schliesing, J. A.	X						X		X				X	X	X	X	X	
Shkliayev, P. N.		X						X		X								
Sokolovsky, A. A.		X								X	X							
Syromyatnikov, V. S.		X	X	X	X	X	X	X	X	X	X	X	X	X	X	X	X	X
Taylor, J. T.	X								X	X								
Temnov, S. S.		X											X		X	X		
Vibbart, C. M.	X								X				X	X				
Wade, D. C.	X		X	X	X	X	X	X										
White, R D.	X		X	X	X	X	X	X	X	X	X	X	X	X	X	X	X	X
Williams, L. G.	X					X	X		X	X								
Zajac, R. H.	X								X									
Zhivoglotov, V.		X	X	X														

Working Group 3 Meeting Minutes—Translation Verifiers

Translation verifier	U.S.A.	U.S.S.R.	1971 21-25 June	1972 26 Nov.-6 Dec.	1972 27 Mar.-3 Apr.	1972 6-18 July	1972 9-19 Oct.	1972 6-16 Dec.	1973 15-30 Mar.	1973 27 June-11 July	1973 13-23 Nov.	1973 3-24 Dec.	1974 8 Apr.-3 May	1974 1 July-5 Sept.	1974 26 Aug.-20 Sept.	1974 23 Oct.-11 Dec.	1975 20 Jan.-13 Feb.	1975 5-22 May
Artemov, B. P.		X								X								
Glikman, J. O.	X															X		
Harrin, E. N.	X								X	X	X	X	X	X	X	X	X	X
Kolesnikov, L. I.		X													X			
Levitin, L. E.		X								X								
Pershikov, O. B.		X										X						
Sergeyeva, H. N.		X															X	
Sharonova, Yu. A.		X														X		
Skakhova, N. I.		X																X
Zhuravlev, V. D.		X								X			X		X			

Working Group 4 Meeting Minutes—Signatories

Member	Country		Date of meeting										
			1972		1973			1974				1975	
	U.S.A.	U.S.S.R.	6-18 July	24 Nov.-2 Dec.	15-30 Mar.	18-29 June	1-12 Oct.	14 Jan.-1 Feb.	8 Apr.-3 May	3-20 June	26 Aug.-20 Sept.	20 Jan.-13 Feb.	5-22 May
Babkov, O. I.		X		X	X								
Bagno, V. I.		X	X	X									
Bobylev, V. A		X						X	X	X	X		
Bohlmann, R. R.	X										X		
Brunjes, J.	X											X	
Busch, C. W.	X					X							
Dickerson, E. T.	X											X	
Dietz, R. H.	X		X	X	X	X	X	X	X		X	X	X
Eisenhauer, D.	X											X	
Forsander, R. A.	X										X	X	
Galin, Ye. N.		X				X	X	X	X		X	X	X
Gilson, K.	X					X	X						
Gorlin, Ye. I.		X			X	X	X	X	X	X	X	X	X
Graham, O. L.	X							X	X			X	
Graves, T. J.	X											X	
Hamilton, M. W.	X							X	X			X	
Heidler, H. F.	X										X		
Ispravnikov, Yu. V.		X				X	X	X	X	X	X	X	X
Ivanov, V. B.		X				X	X	X			X	X	
Johnson, G. W.	X				X	X		X	X			X	
Kas'yanov, D. K.		X			X	X	X	X					
Kelley, J. S.	X					X			X			X	
Khomenko, Yu. P.		X				X							
Khristolyubov, P. I.		X						X	X	X	X	X	
Kryukov, A. B.		X			X								
Kurbatov, V. V.		X						X	X		X	X	X
Lattier, E. E.	X		X	X	X	X	X	X	X		X	X	X
Lindsay, O. C.	X							X	X			X	
Mager, J. E.	X				X	X	X						
Morgulev, A. S.		X	X	X	X	X			X	X			
Nekrasov, G. A.		X				X	X		X	X			X
Nikitin, B. V.		X			X	X	X	X	X	X	X	X	X
Olenev, B.		X				X							
Panter, W. C.	X							X	X	X		X	X
Porter, J. A.	X					X	X		X				
Raspletin, V. A.		X				X							
Robinson, R. L.	X								X			X	
Roditelev, N. D.		X					X		X		X	X	X
Ryadinsky, B. F.		X					X	X	X	X	X	X	X

APPENDIX C

Working Group 4 Meeting Minutes—Signatories
—Concluded

Member	Country		Date of meeting										
			1972		1973			1974				1975	
	U.S.A.	U.S.S.R.	6-18 July	24 Nov.-2 Dec.	15-30 Mar.	18-29 June	1-12 Oct.	14 Jan.-1 Feb.	8 Apr.-3 May	3-20 June	26 Aug.-20 Sept.	20 Jan.-13 Feb.	5-22 May
Savitsky, Yu. I.		X			X	X							
Shack, P. E.	X											X	
Shirokov, V. S.		X						X	X		X	X	X
Shores, P. W.	X		X	X	X	X	X	X	X		X	X	
Snitkin, S. S.		X					X			X	X		X
Sosulin, G. K.		X					X		X		X		X
Souslennikov, V. V.		X					X						
Speier, W. M.	X			X	X	X	X	X	X		X	X	
Stelter, L.	X						X						
Sushchev, G. A.		X					X	X	X		X	X	
Sustannikov, V.		X					X						
Tkachenko, Yu. P.		X			X								
Travis, A. D.	X		X	X	X	X	X	X	X		X	X	X
Vasiliev, Yu. N.		X					X						
Vermillion, B. K.	X						X	X	X		X	X	
Vlasov, G.		X									X		
Whitson, R.	X										X		
Wooten, C. M.	X								X		X	X	
Zelenkov, S. V.		X					X	X		X	X	X	
Zelenshichikov, N. I.		X										X	X

Working Group 4 Meeting Minutes—Translation Verifiers

Translation verifier	Country		Date of meeting										
			1972	1973				1974				1975	
	U.S.A.	U.S.S.R.	6-18 July	24 Nov.-2 Dec.	15-30 Mar.	18-29 June	1-12 Oct.	14 Jan.-1 Feb.	8 Apr.-3 May	3-20 June	26 Aug.-20 Sept.	20 Jan.-13 Feb.	5-22 May
Afanansenko, P.	X											X	
Glikman, J. O.	X										X		
Holmes, T.	X					X							
Krukov, A. B.		X			X			X	X		X	X	X
Levitin, L. E.		X								X			
Rykov, G. A.		X								X			
Savel'yeva, I. S.		X					X						
Sementovsky, A	X				X			X	X		X		X
Zhuravlev, V. D.		X				X							

APPENDIX C

Working Group 5 Meeting Minutes—Signatories

Member	Country U.S.A.	U.S.S.R.	1972 6-18 July	9-19 Oct.	1973 15-30 Mar.	18-29 June	1-12 Oct.	1974 14 Jan.-1 Feb.	11-22 Mar.	8 Apr.-3 May	15-29 July	26 Aug.-20 Sept.	18-29 Nov.	1975 20 Jan.-13 Feb.	5-22 May
Dolgopolov, Yu. S.		X	X	X			X	X	X	X	X	X		X	X
Ellis, W. E.	X							X	X		X			X	
Grafe, R. L.	X							X	X	X	X	X		X	
Guy, W. W.	X		X	X	X		X	X	X	X	X	X		X	X
Harris, R. S.	X												X	X	X
Hawkins, W. R.	X		X	X											
Hughes, D. F.	X							X	X	X	X	X		X	
Jaax, J. R.	X						X	X		X	X	X		X	X
Joslyn, A. W.	X						X						X		
Kalenkov, O. B.		X			X										
Kholodkov, V. N.		X						X	X	X					
Kireev, V. I.		X							X						
Klimenko, I. V.		X			X	X		X	X	X					
Makarov, A. A.		X									X				
Mayo, R. E.	X								X						
Merhoff, P. C.	X						X								
Monahov, B.		X					X								
Novikov, V. V.		X					X	X	X		X	X		X	
Pravetsky, V. N.		X			X					X					
Shkliayev, P. N.		X					X								
Smylie, R. E.	X				X	X	X								
Stavitsky, A. K.		X										X	X		X
Taylor, J. T.	X					X	X	X		X		X	X		
Trepov, Y. Y.		X					X	X		X					
Ustenko, V. F.		X			X	X		X	X						
Zajac, R. H.	X													X	
Zakolov, V. A.		X									X			X	
Zaytsev, Ye. N.		X					X			X	X		X	X	X
Zedekar, R. G.	X				X	X		X	X		X		X	X	

Note—This list includes meetings of the Thermal Subgroup.

Working Group 5 Meeting Minutes—Translation Verifiers

Translation verifier	U.S.A.	U.S.S.R.	6-18 July	9-19 Oct.	15-30 Mar.	18-29 June	1-12 Oct.	14 Jan.-1 Feb.	11-22 Mar.	8 Apr.-3 May	15-29 July	26 Aug.-20 Sept.	18-29 Nov.	20 Jan.-13 Feb.	5-22 May
			1972		1973			1974						1975	
Barsukov, G. P.		X							X						
Diomina, G. S.		X					X		X				X		
Erohina, T.	X														X
Harrin, E. N.	X					X			X				X		
Holmes, T.	X				X		X	X		X					
Krukov, A. B.		X						X							
Lasareff-Mironoff, A.	X											X	X	X	
Mamantov, I. A.	X						X								
Yavorskaya, O. S.		X										X	X	X	X
Yevdokimov, V.		X			X	X	X								

Note—This list includes meetings of the Thermal Subgroup.

492

Appendix D
Summary of U.S./U.S.S.R. Meetings

Date	Place	American delegation leader	Soviet delegation leader	Participating groups	Summary of results
1970					
16-28 Oct.	Moscow	R. R. Gilruth	B. N. Petrov	Pre-ASTP	Discussed in chapter IV.
1971					
16-21 Jan.	Moscow	G. M. Low	M. V. Keldysh	Pre-ASTP	Discussed in chapter V.
21-25 June	Houston	R. R. Gilruth	B. N. Petrov	Working Groups (WGs) 1, 2, 3	Discussed in chapter V.
26 Nov.-6 Dec.	Moscow	G. S. Lunney	K. D. Bushuyev	WGs 1, 2, 3	Discussed in chapter VI.
1972					
27 Mar.-3 Apr.	Houston	D. C. Wade	V. S. Syromyatnikov	WG 3	Discussed in chapter VI.
4-6 Apr.	Moscow	G. M. Low	V. A. Kotelnikov	Executive	Discussed in chapter VI.
10-17 May	Moscow	D. C. Cheatham	V. P. Legostayev	WG 2	Discussed in chapter VI.
24 May	Moscow	R. M. Nixon	A. N. Kosygin	Summit	Discussed in chapter VI.
6-18 July	Houston	G. S. Lunney	K. D. Bushuyev	WGs 0,1, 2, 3, 4, 5	Discussed in chapter VII.
9-19 Oct.	Moscow	G. S. Lunney	K. D. Bushuyev	WGs 0, 1, 2, 3, 5	Discussed in chapter VII.
24 Nov.-6 Dec.	Houston	H. E. Smith R. H. Dietz	V. P. Legostayev	WGs 2, 4	Discussed in chapter VII.

6-16 Dec. 1973	Moscow	D. C. Wade	V. S. Syromyatnikov	WG 3	Discussed in chapter VII.
15-30 Mar.	Houston	G. S. Lunney	K. D. Bushuyev	WGs 0, 1, 2, 3, 4, 5	Technical directors (TD) reviewed documentation and milestones contained in ASTP 30 000. Special attention was given to crew preparation, and a crew activities plan was discussed. Lunney and Bushuyev agreed to have technical specialists from each side in the other country's control center during flight, and plans were made for visits to each nation's control center. The TDs decided that the flight directors (FD) would have prime responsibility for implementing flight-related decisions during the mission, with the TDs acting as advisers and consultants. They also discussed scientific experiments, training schedules, and flammability safety studies, as well as reviewed the results of a meeting (Feb. 1973) between U.S.S.R. and U.S. medical doctors. The TDs reaffirmed that all medical discussions relating to ASTP would be handled as part of the WG meetings and agreed that the return of a mixed crew should be studied further with a decision on this issue being planned for the next meeting. The increased tempo of activities led to the decision to begin regularly scheduled telephone conversations every 2 weeks, beginning 24 Apr. 1973.[1] WG 1 confirmed the following guidelines for flight: rendezvous on 29th orbit; Soyuz maneuvers on 4th and 17th orbits; circular docking orbit at 225 km; orbit inclination = 51.8°; Apollo to launch 7.5 hr after Soyuz. Both sides agreed to ASTP 40 201, "Trajectory Computation Model," which contained coordinate systems, atmospheric, and gravitational models. Onboard

Date	Place	American delegation leader	Soviet delegation leader	Participating groups	Summary of results
					documents were discussed further, as was development of the "Control Centers Interaction Plan," ASTP 40 401.[2] WG 2 continued discussion of Nov.-Dec. 1972 topics: tracking requirements for Apollo and the orientation requirements for Soyuz during rendezvous, exchange of spacecraft surface materials, external lights, and control systems. A Preliminary Systems Review (PSR) was conducted for the docking mechanism on Soyuz and for the NASA test fixture that would assure the target's proper alignment. The PSR data were reviewed by the TDs and approved.[3] WG 3 worked on the interface seals for the docking system and decided on the types of seal samples that would be exchanged for testing. IED 50 004, the engineering drawing of the docking system, was updated to reflect changes in the design. The U.S.S.R. and U.S. safety assessment reports on the inadvertent release of the structural latches were reviewed, and additional analysis of this topic was scheduled.[4] WG 4 also continued discussion of earlier topics: interface signal characteristics for radio communications (intercom), compatibility tests for those systems, inter control center communications, VHF/FM Preliminary Design Review (PDR), etc. Training sessions for U.S.S.R. specialists in the testing of U.S. VHF/AM equipment were scheduled for 30 Apr.-11 May at the Grumman Aerospace Corporation factory at Bethpage, NY.[5] WG 5 considered questions of life support system compatibility, transfer procedures, and fire safety. The Soviets pre-

		D. R. Scott	V. P. Legostayev	WGs 1, 2, 4, 5
18-29 June	Moscow			

sented a description of the changes that were being made in the Soyuz atmosphere regeneration system, presented in a preliminary version of IED 50 723. Because of flammability concerns, all equipment transferred to Apollo would have to be certified. This included 2 cameras (still and TV) and cosmonaut flight dress. The Soviets indicated that practically all their electrical equipment was being tested to determine its safety for use in a 50-percent O_2 environment.[6]

WG 1, represented by an Experiments Subgroup, discussed the scientific activities to be conducted during the mission and the form and contents of documents dealing with them. P. S. Jaschke and Yu. S. Denisov acted as co-chairmen.[7] WG 2 conducted a Design Acceptance Review (DAR) of the Soyuz docking target, using a high fidelity model. Soyuz control requirements were discussed, and although previous documentation specified the docked attitude, control, maneuver, and translation requirements for Soyuz, the Soviets were "unprepared to and reluctant" to provide the data and level of detail necessary to fulfill the agreed requirements. The necessary information was made available later when Legostayev made a special trip to Houston in July. Safety Assessment Reports (SAR) were also considered, but Bushuyev was "surprised at the level of detail of the US version," especially the fact that Ed Smith had included material on the control system problems encountered in Gemini VIII and Apollo 13.[8] WG 4's progress was particularly unsatisfactory. Scott in his report to Lunney commented, "The lack of agreed documentation, the late documentation, and the

Date	Place	American delegation leader	Soviet delegation leader	Participating groups	Summary of results
					occasional unscheduled absence of Soviet delegates (Nikit[in], Savitiski, Morgulev) made an already overburdened agenda extremely difficult to complete, which understandably it was not. Professor Bushuyev was made aware of the problems early in the session and stated that positive action would be taken . . . but it was not adequate to completely resolve all of the deficiencies. . . ."[9] This experience reinforced Lunney's desire to iron out the documentation problems and subsequently led to the October Mid-Term Review. The WG 5 Thermal Subgroup met and discussed a number of temperature effects on materials relating to the docking system, communications equipment, and life support systems. The Americans agreed to give the Soviets data on the possible impingement of the Apollo reaction control system (RCS) on the Soyuz docking seals.[10]
27 June–11 July	Moscow	R. D. White	V. S. Syromyatnikov	WG 3	SARs dealing with inadvertent opening of structural ring latches were updated, and the U.S. SAR was signed. Drafts of development test procedures were discussed. Syromyatnikov agreed to provide a Soviet seal to permit testing of both seals at Rockwell. Before WG 2 departed, a meeting was held to discuss the effects that Apollo maneuvering during the docked phase might have on the Soyuz solar panels; this issue was resolved. DAR, Phase I, review of U.S. drawings, was completed.[11]

9-20 July	Houston	G. S. Lunney	K. D. Bushuyev	WGs 0, 1, 2	
					The purpose of the "meeting was to discuss specific technical problems, continue development of trajectories and flight plans, tentatively coordinate . . . the scientific experiment program, and to familiarize the cosmonauts with the design and operation of the Apollo spacecraft systems." The TDs agreed that prime responsibility for the mission will shift from the TDs to FDs at the beginning of the joint pre-launch phase as defined in ASTP 40 500. Test flights of Soyuz equipped with ASTP systems were planned by the U.S.S.R. The two sides discussed possibilities of joint participation in or observation of test activities and flight preparations of compatible equipment. Six SARs were discussed (control systems, fire and cabin pressure, pyrotechnics, manufacturing test and checkout, structural latches, and ground command capability). As SARs were completed, they were to be turned over to WG 1 so that appropriate procedures could be written into flight plans. The Moscow meeting was reviewed, and TDs agreed that more data were needed for Soyuz control systems. Bushuyev also said that steps would be taken to eliminate WG 4 problems. TDs discussed transferring responsibility for inter control center communications from WG 4 to WG 1 and the draft of Part I of the Public Information Plan, which had been presented by the U.S. in Mar. 1973. The U.S.S.R. wanted to add some specific language and agreed to provide comments by late Aug.[12] A general review was held on the status of the joint experiments; Lunney explained that the internal NASA review of proposed experiments was still in progress.[13] WG 1 continued work on flight related documentation; e.g.,

Date	Place	American delegation leader	Soviet delegation leader	Participating groups	Summary of results
					"Flight Plan Guidelines," ASTP 40 300; "Joint Crew Activities Plan," ASTP 40 301; "Control Centers Interaction Plan," ASTP 40 401; "Contingency Plan," ASTP 40 500; and "Onboard Joint Operations Instructions," 40 600. Alternative launch trajectories were discussed at length. Flight controllers training was planned–U.S.S.R. controllers to visit Houston in Oct. 1974; U.S. controllers to visit the U.S.S.R. in Dec. 1974. Two joint control center training sessions were scheduled for Apr. 1975 and for within 10 days of launch. (Crew training activities discussed in chap. VIII.) WG 2 continued its work on a full agenda of control system items that related to the spacecraft in a docked configuration.[14]
1-20 Oct.	Moscow	G. S. Lunney	K. D. Bushuyev	WGs 0, 1, 4, 5	The results of this meeting are discussed in chapter VIII. In addition to mission related activities, the U.S. team raised the question of "Joint Requirements for Compatibility of Future Space Systems. C. Covington and the Soviet WG chairmen resumed this discussion, during which the Americans made a number of proposals that would alter the ASTP agreements to accommodate the next generation of spacecraft, e.g., the suggestion to increase the transfer tunnel diameter from 0.8 to 0.92 m and to increase the number of structural latches to permit the docking of larger spacecraft. The changes to the size of the docking gear would necessitate its complete re-design. Although these talks on future systems were to have been continued in subsequent

Date	Location				
15-18 Oct.	Moscow	G. M. Low	B. N. Petrov	Mid-Term Review	plenary sessions, the Soviets did not wish to discuss this during the preparations for the test flight, and it was not discussed again until the post-mission meeting in Nov. 1975.[15] Discussed in chapter VIII.
13-23 Nov.	Houston Downey, Calif.	R. D. White	V. S. Syromyatnikov	WG 3	Seal verification tests observed by the Soviets were held at Rockwell on 12-19 Nov. This was a subgroup effort designed to resolve some difficulties that had plagued the development and testing of the docking seals. Tests were a success and solved the problems.[16]
19-30 Nov.	Star City	T. P. Stafford	V. A. Shatalov	Crews	Discussed in chapter IX.
3-24 Dec.	Houston	R. D. White	V. S. Syromyatnikov	WG 3	Actually, the minutes dated 3-24 Dec. relate to a series of discussions that were ancillary to the development tests for the docking system, spanning the period 16 Sept.-24 Dec.[17]
1974					
14 Jan.-1 Feb.	Houston	G. S. Lunney	K. D. Bushuyev	WGs 1, 2, 4	35 Soviet specialists engaged in meetings with Johnson Space Center (JSC) counterparts, conducting additional discussions of mission plans and experiments, communications and tracking, and life support and transfer.[18]
4 Feb.-5 Apr.	Houston	R. H. Dietz	B. V. Nikitin	WG 4	Soviet members of WG 4 stayed in Houston to participate in communications system compatibility tests. These tests were designed to verify that all communications and tracking equipment that interfaced between Apollo and Soyuz would work satisfactorily. During this evaluation, the performance of the American and Soviet VHF/

Date	Place	American delegation leader	Soviet delegation leader	Participating groups	Summary of results
					AM equipment was studied separately and as installed in the electrical equivalents of Apollo and Soyuz communications systems. The Soviet and American cable intercommunications system was also tested. The test program identified several compatibility discrepancies in the cable communications, audio circuits, and other areas, which were subsequently corrected.[19]
11–22 Mar.	Air Force base near Moscow	W. W. Guy	I. V. Lavrov	WG 5	Discussed in chapter IX.
8 Apr.– 3 May	Houston	G. S. Lunney	K. D. Bushuyev	WGs 0, 1, 2, 3, 4, 5	In addition to the crew training discussed in chapter IX and the WG 5 compatibility tests and environmental control system (ECS) tests previously described, the TDs received reports on a number of other activities. They reviewed the Mar. 1974 meeting of the Public Affairs specialists and the subsequent talks on Part II of the Public Information Plan, which concerned the release of information during the flight. A Soviet proposal was discussed, the U.S. prepared a revised version of Part II, and further talks were scheduled. The TDs also looked at the results of the discussions on joint experiments; the experiment Interacting Equipment Documents (IEDs) were signed, and the experiment schedules were incorporated into ASTP 30 000. They also agreed to further tests of the Soyuz pyrotechnics, using a full scale mockup of the front of that craft.

502

Date	Location			WG	
3-20 June	Moscow	W. C. Panter	E. I. Gorlin	WG 4	Dates for the visit of specialists and crewmembers to the launch sites were discussed, and a tentative schedule agreed upon. The TDs visited the communication system compatibility test laboratory, the ECS breadboard test facility, and the thermal vacuum test facility.[20] WG 1 personnel studied onboard motion pictures, still photography, and television, and tested the mechanical compatibility of U.S.S.R. mockup TV cameras with the structure of the command module (CM) and the docking module (DM). In addition to experiments, trajectory, flight control, and ground control subjects, onboard documentation was a major topic of discussion.[21] WG 2 continued its work on spacecraft tracking, safety assessment problems (pyrotechnics), spacecraft external coating reflectivity characteristic measurement results, control system functioning, and docking targets.[22] WG 3 worked on further aspects of the design and dynamics of the docking system. They also prepared for the joint qualification tests by agreeing to IED 50 010, "Apollo Soyuz Joint Qualification Test Plan, Docking Systems."[23] WGs 4 and 5 conducted additional work relating to their tests. WG 5 evaluated flight uniform material from the U.S.S.R. and found it satisfied the non-flammability requirements for use in the Apollo pure O_2 environment.[24] Tests of 3 sets of U.S. VHF/AM hardware [transceiver and range tone transfer assembly (RTTA)] to prove its fitness as flight hardware to be installed in Soyuz were completed successfully.[25]
17-28 June	Star City	A. D. Dennett	V. N. Bobkov	WG 1 subgroup	The TV and Photo Subgroup evaluated lighting and facilities available for television and photog-

503

Date	Place	American delegation leader	Soviet delegation leader	Participating groups	Summary of results
					raphy onboard Soyuz in a high fidelity mockup. Tests were performed with crews in flight clothing to simulate the actual mission.[26]
24 June–11 July	Star City	T. P. Stafford	A. A. Leonov	Crews	Discussed in chapter IX.
1 July–5 Sept.	Houston	R. D. White	V. S. Syromyatnikov	WG 3	Tests were held on the dynamic docking test system to qualify the flight docking systems as prescribed in the "Apollo Soyuz Joint Qualification Test Plan," IED 50 010; they were completed successfully.[27]
15–29 July	Houston	W. W. Guy	Yu. S. Dolgopolov	WG 5	Soviet specialists witnessed manned and unmanned thermal vacuum tests of flight article docking module 1; tests were successful.[28]
26 Aug.–20 Sept.	Moscow	G. S. Lunney	K. D. Bushuyev	WGs 0, 1, 2, 3, 4, 5	TDs reviewed post-flight scientific data exchange from experiments, docking system qualification test results, and spacecraft communications during various phases of joint activities; they discussed other aspects of test and checkout activities down to the time of launch. As part of WG 1's effort, the TDs agreed that NASA was to track the upcoming precursor flight of Soyuz.[29] (See chap.. IX.) WG 1, in addition to continuing work on flight documentation, discussed flight photography and TV, joint experiments, and crew training. Major attention was given to the preparations of the control centers and personnel for the mission.[30] WG 2 neared completion of all its topics and discussed the RCS impingement question.[31]

Date	Place				Description
					(See chap. IX.) WG 3 reviewed the Institute of Space Research facilities for performing the matecheck of the flight docking systems to begin on 23 Oct. WG 4 continued discussions of in-flight communications, and their agreements were approved by the TDs. The effects of U.S. VHF radiation on Soyuz were discussed, and agreement was reached on testing the Soyuz pyros.[32] WG 5 conducted analysis and review of contingency situations that might affect Soyuz and Apollo ECS; they revised "Materials Fire Safety Certification of USSR Equipment Transferred to Apollo," IED 50 721, and "Materials Fire Safety Certification for USA Equipment Transferred to Soyuz," IED 50 720; they also continued completion of other documentation.[33]
9-27 Sept.	Houston	T. P. Stafford	A. A. Leonov	Crews	Continued preparation of the crews for flight with emphasis on rendezvous and transfer phases. Medical examinations of cosmonauts for medical experiments during the flight were completed.[34]
16-27 Sept.	Moscow Kaliningrad	M. P. Frank	A. S. Yeliseyev	WG 1 subgroup	The first training session for flight controllers was held in the U.S.S.R. From 16-20 Sept., the controllers heard lectures at the Space Research Institute relating to the Soyuz and its subsystems; from 23-25 Sept., they were given familiarization sessions at the mission control center in Kaliningrad.[35]
23 Oct.-11 Dec.	Moscow	R. D. White	V. S. Syromyatnikov	WG 3	Conducted matechecks with flight docking equipment, more formally called "Preflight Compatibility Verification Test of Docking Systems," involving a complete examination of the gear to establish its flight readiness. The tests were successful, and the results were recorded in IED

Date	Place	American delegation leader	Soviet delegation leader	Participating groups	Summary of results
					50 015. At this meeting, a subgroup session was conducted on 15-22 Nov. to ascertain the status of modifications made to the guide pins and sockets.[36]
24 Oct.-11 Dec.	Houston	M. P. Frank	A. S. Yeliseyev	WG 1 subgroup	Training sessions for Soviet flight controllers in the U.S. were begun; from 24-31 Oct., lectures were given on Apollo systems; from 31 Oct.-6 Nov., the U.S.S.R. personnel were acquainted with the JSC control center and selected aspects of mission management.[37]
18-29 Nov.	Moscow	A. W. Joslyn	Ye. I. Klimenko	WG 5 subgroup	The Thermal Subgroup met to discuss effects of space temperatures on docking seals and the requirements for cooling the U.S. VHF transceiver mounted in Soyuz.[38]
25 Nov.-21 Dec.	Houston	M. P. Frank	V. A. Timchenko	WG 1	Continued work on the following topics and related documents: spacecraft design characteristics, joint scientific experiments, onboard documents, control center joint pre-flight practice simulations, inter control center communications systems, and photo and TV work in Apollo mockups.[39]
1975					
20 Jan.-13 Feb.	Houston	G. S. Lunney	K. D. Bushuyev	WGs 0, 1, 2, 3, 4, 5	TDs reviewed the various test and documentation activities conducted since the Aug.-Sept. 1974 plenary meeting. Bushuyev reported on Soyuz

Date	Location			Group	
26 Jan.–13 Mar.	Houston	E. B. Pippert	V. P. Varshavsky	WG 1 subgroup	16. (See chap. IX.) This was the last plenary session in which all the WGs were represented. The next full scale meeting was to be the Flight Readiness Review (FRR). The WG activities mainly involved completing documentation. WG 1 was the only group that had large scale tasks to complete since it was the group charged with conducting the joint flight.[40] The Onboard Documentation Subgroup met to complete work on the "Onboard Joint Operations Instructions," ASTP 40 600, and the "Joint Crew Activities Plan," ASTP 40 301. When signed off by the WG chairmen and the TDs, these documents became the basic statement of how the joint phases of the mission would be conducted.[41]
7–28 Feb.	Kennedy Space Center Houston	T. P. Stafford	V. A. Shatalov	Crews	U.S. and U.S.S.R. crews visited Kennedy Space Center (KSC) on 8 February for orientation visits to the vehicle assembly building, the launch pads, the firing room, and the crew quarters at the launch site. At JSC, beginning on the 10th, the crews received briefings on the 5 joint experiments and the rules and procedures governing crew actions in various emergency situations; they continued to work on transfer and communications training. The Soviets flew the Apollo command and service module (CSM) simulator to review rendezvous and docking as seen from the U.S. side. They also had 2 run-throughs for each crew with the DM mockup to review transfer and contingency procedures. And the Soviet prime crew tasted samples of the food they would eat aboard Apollo.[42]

Date	Place	American delegation leader	Soviet delegation leader	Participating groups	Summary of results
20-21, 24-25, 27-28 Mar.	Houston Kaliningrad	L. A. Reitan	L. F. Mezenov	WG 1 subgroup	On these three 2-day sessions, the Mission Control Center (MCC) Houston and MCC Moscow personnel participated in training sessions that involved the simulation of selected aspects of the flight.
6-16 Apr.	Moscow	A. P. Alibrando	V. S. Vereschetin	WG 1 subgroup	The Public Information (PI) specialists discussed the symbolic activities of the crews during the flight—the exchange of TV transmission, the in-flight press conference, exchange of still and motion pictures, and the participation of the PI representatives in the May joint simulations.[43]
14-25 Apr.	Moscow	M. P. Frank	V. A. Timchenko	WG 1	Final preparations for flight and the Flight Readiness Review were continued. The delegates discussed the initial meeting of the crews, a TV tour from space of the U.S.S.R. narrated by Kubasov, and a number of other topics related to ground and flight aspects of the mission. The joint On-board Documentation Subgroup met from 14-30 Apr. to complete revisions and updates to ASTP 40 600 and 40 301.[44]
14-30 Apr.	Star City Baykonur Cosmodrome	T. P. Stafford	A. A. Leonov	Crews	The final crew training session was completed. In addition to review of joint activity phases of the flight plan and additional time in the Soyuz mockup, the U.S. crews practiced contingency procedures in Soyuz. On 28 Apr., the crews visited Baykonur Cosmodrome. (See chap. IX.) This work completed all the training as outlined in ASTP 40 700.[45]

5-22 May	Moscow	G. S. Lunney	K. D. Bushuyev	WGs 0, 2, 3, 4, 5	These sessions concluded the preparations for the flight and the FRR. The TDs received a tour of Baykonur to review the status of spacecraft readiness for the launch. The TDs also agreed to the "Outline for the Initial Mission Report," which would summarize the flight results of ASTP. WG 2 discussed their FRR presentation and the docking target alignment tests. WG 3 also worked on their FRR presentation. WG 4, in addition to FRR preparation, reviewed the pre-flight equipment checks that were conducted at the cosmodrome on 12-17 May. WG 5 completed work on the FRR.[46]
23 May	Moscow	G. M. Low	V. A. Kotelnikov	Flight Readiness Review	Discussed in chapter X.
13-20 May	Houston Kaliningrad	L. A. Reitan	L. F. Mezenov	WG 1 subgroup	Second simulation exercise for the control center personnel completed.
9-13 June	Moscow	D. A. Bland	V. P. Varshavsky	WG 1 subgroup	Onboard Documentation Subgroup completed its work.
29 June-1 July	Houston Kaliningrad	L. A. Reitan	L. F. Mezenov	WG 1 subgroup	Final full scale dress rehearsal of control center operations prior to the mission.
17-19 July	In earth orbit	T. P. Stafford	A. A Leonov	Crews	Joint activities of ASTP flight. (Discussed in chap. XI.)

[1] "Summary of Results of the March 1973 Meeting of Specialists of the USSR Academy of Sciences and the USA National Aeronautics and Space Administration on Compatible Systems for Rendezvous and Docking of Manned Spacecraft and Stations," in "Apollo Soyuz Test Project, Minutes of Joint Meeting, USSR Academy of Sciences and US National Aeronautics and Space Administration," 15-30 Mar. 1973.

[2] "Minutes of the Joint Meeting of Working Group I," in "Minutes of Joint Meeting," 15-30 Mar. 1975.

[3] "Working Group No. 2, Minutes of Meeting on Apollo Soyuz Test Project," in "Minutes of Joint Meeting," 15-30 Mar. 1973.

[4] "Working Group #3, Minutes of Meeting on Assuring Compatibility, Docking Systems," in "Minutes of Joint Meeting," 15-30 Mar. 1973.

[5] "Working Group No. 4, Minutes of Meeting on Apollo Soyuz/Test Project," in "Minutes of Joint Meeting," 15-30 Mar. 1973.

[6] "Minutes, Working Group 5," in "Minutes of Joint Meeting," 15-30 Mar. 1973.

[7] D. R. Scott to G. S. Lunney, memo, "ASTP Mission to Moscow, June-July, 1973," 31 July 1973.

[8] Ibid.

[9] Ibid.; and R. H. Dietz to G. S. Lunney, memo, "Miscellaneous Observations/Recommendations Precipitated by the June 1973 Meeting for Working Group No. 4," 13 July 1973.

[10] D. R. Scott to G. S. Lunney, memo, "ASTP Mission to Moscow," 31 July 1973.

[11] Ibid.; and "Minutes, Working Group #3," 27 June-11 July 1973.

[12] "Summary of Results of the July 1973 Meeting of Specialists of the USSR Academy of Sciences and the USA National Aeronautics and Space Administration on Compatible Systems for Rendezvous and Docking of Manned Spacecraft and Stations," 20 July 1973, in "Apollo Soyuz Test Project, Minutes of Joint Meeting, USSR Academy of Sciences and US National Aeronautics and Space Administration," 9-20 July 1973.

[13] Ibid.; and "Proposals of a Program for Scientific Experiments," USSR-WG1-005 [n.d.].

[14] "Working Group No. 2," in "Minutes of Joint Meeting," 9-13 July 1973.

[15] "Minutes of Meeting," 8-12 Oct. 1973, in "Apollo Soyuz Test Project, Minutes of Joint Meeting, USSR Academy of Sciences and US National Aeronautics and Space Administration," 1-12 Oct. 1973.

[16] R. D. White to G. S. Lunney, memo, "Working Group #3 Joint Meeting on November 13-23, 1973," 28 Nov. 1973; and interview, R. D. White-Ezell, 30 Sept. 1975.

[17] "Working Group 3, Minutes of the Meeting Assuring Joint Compatibility of the Docking Systems," 13-23 Nov. 1973; and "Apollo Soyuz Test Project, Results of Apollo Soyuz Docking Systems Development Tests," IED 50 013, 25 Dec. 1973.

[18] NASA News Release 74-9, "ASTP Working Groups to Meet," 10 Jan. 1974.

[19] R. H. Dietz to E. C. Ezell, memo, 1 Oct. 1975; and "Apollo Soyuz Test Project, Working Group 4, Minutes of Joint Meeting in Houston, USSR Academy of Sciences and US National Aeronautics and Space Administration," 8-26 Apr. 1974.

[20] "Apollo Soyuz Test Project, Minutes of Joint Meeting, USSR Academy of Sciences and US National Aeronautics and Space Administration," 8 Apr.-3 May 1974; and NASA News Release [unnumbered], "ASTP Press Conference," 3 May 1974.

[21] "Apollo Soyuz Test Project, Working Group 1, Minutes of Joint Meeting in Houston, USSR Academy of Sciences and US National Aeronautics and Space Administration," 15 Apr.-3 May 1974.

[22] "Apollo Soyuz Test Project, Working Group 2 Minutes of Joint Meeting in Houston, USSR Academy of Sciences and US National Aeronautics and Space Administration," 15 Apr.-3 May 1974.

[23] "Apollo Soyuz Test Project, Working Group 3 Minutes of Joint Meeting in Houston, USSR Academy of Sciences and US National Aeronautics and Space Administration," 15 Apr.-3 May 1974.

[24] "Apollo Soyuz Test Project, Working Group 5 Minutes of Joint Meeting in Houston, USSR Academy of Sciences and US National Aeronautics and Space Administration," 15 Apr.-3 May 1974.

[25] "Apollo Soyuz Test Project, Minutes, Joint Tests Conducted on the USA Supplied VHF/AM Equipment in Moscow," 3-20 June 1974.

[26] "Apollo Soyuz Test Project, Minutes of Joint Meeting and Testing on the Soyuz Mockup in Gagarin Cosmonaut Training Center," 28 June 1974.

[27] "Apollo Soyuz Test Project, Minutes of Joint Meeting, Working Group 3, USSR Academy of Sciences and US National Aeronautics and Space Administration," 1 July-5 Sept. 1974.

[28] "Apollo Soyuz Test Project, Minutes of Working Group 5," 15-29 July 1974; and "USA Docking Module Thermal/Vacuum Testing Quick Look Report" [n.d.].

[29] "Summary of Results of the August-September 1974 Meeting of Specialists of the USA NASA and USSR Academy of Sciences on the Preparations for Conduct of the Joint Test Flight of Apollo and Soyuz," 20 Sept. 1974, in "Apollo Soyuz Test Project, Minutes of Joint Meeting, USSR Academy of Sciences and US National Aeronautics and Space Administration," 26 Aug.-20 Sept. 1974.

[30] "Minutes of the WG-1 Joint Meeting," in "Minutes of Joint Meeting," 26 Aug.-13 Sept. 1974.

[31] "Working Group 2 Minutes of the Meeting in Moscow," in "Minutes of Joint Meeting," 9-20 Sept. 1974.

[32] "Working Group 4 Minutes," in "Minutes of Joint Meeting," 9-20 Sept. 1974.

[33] "Minutes, Working Group 5," in "Minutes of Joint Meeting," 9-20 Sept. 1974.

[34] "Apollo Soyuz Test Project, Minutes of the Astronaut and Cosmonaut Joint Training Visit at the Johnson Space Center," 9-27 Sept. 1974.

[35] "Apollo Soyuz Test Project, Results of the Joint US/USSR Flight Controller Training," 16-27 Sept. 1974.

[36] "Apollo Soyuz Test Project, Minutes of Joint WG3 Meeting, Docking Systems Compatibility Assurance Meeting," 23 Oct.-11 Dec. 1974; and "Apollo Soyuz Test Project, Working Group 3, Minutes on the Question of Resolving the Problem of Binding of the Docking System Guide Pin and Socket," 22 Nov. 1974.

[37] "Apollo Soyuz Test Project, Results of the Joint USA/USSR Flight Controller Training," 24 Oct.-6 Nov. 1974.

[38] "Apollo Soyuz Test Project, Minutes, Working Group 5 on Thermal Problems," 18-29 Nov. 1974.

[39] "Apollo Soyuz Test Project, Meeting Minutes, Joint Meeting of Working Group 1," 25 Nov.-20 Dec. 1974.

[40] "Apollo Soyuz Test Project, Minutes of Joint Meeting, USSR Academy of Sciences and US National Aeronautics and Space Administration," 20 Jan.-13 Feb. 1975.

[41] "Apollo Soyuz Test Project, Minutes of Joint WG1 Onboard Documentation Subgroup Meeting," 26 Jan.-13 Mar. 1975.

[42] "Apollo Soyuz Test Project, Minutes of the Astronaut and Cosmonaut Joint Training Visit at the Johnson Space Center," 7-28 Feb. 1975.

[43] "Apollo Soyuz Test Project, Minutes of Public Information Working Group Meeting, USSR Academy of Sciences and US National Aeronautics and Space Administration," 6-16 Apr. 1975.

[44] "Apollo Soyuz Test Project, Minutes of Working Group 1 Joint Meeting, USSR Academy of Sciences and US National Aeronautics and Space Administration," 14-25 Apr. 1975; and "Apollo Soyuz Test Project, Minutes of Joint WG 1 Onboard Documentation Subgroup Meeting," 14-30 Apr. 1975.

[45] "Minutes of the Joint Training of Soviet and American ASTP Crews at Gagarin CTC," 14-30 Apr. 1975.

[46] "Apollo Soyuz Test Project, Minutes of Joint Meeting, USSR Academy of Sciences and US National Aeronautics and Space Administration," 5-22 May 1975; and "Apollo Soyuz Test Project, Report on the Joint Tests on the Soyuz Spacecraft at 'Baikonur' Launch Site," IED 50 609, 21 May 1975.

Appendix E
ASTP Scientific Experiments

Since the joint flight with the Soviets grew in part from studies on how to utilize excess Apollo hardware, the mission planners in Houston naturally gave considerable attention to the scientific experiments that would be flown. As early as mid-1971, René Berglund received a proposal from Paul R. Penrod, who was working with the Advanced Programs Office representing the Science and Applications Directorate, suggesting scientific activities for an International Rendezvous and Docking Mission (IRDM). Penrod stressed maximum use of existing hardware, maximum crew involvement, use of the docking module (DM) as an experiment station, minimum use of extravehicular activity (EVA), and a schedule leading to either a mid-1974 or mid-1975 launch. While none of the actual experiments proposed by Penrod at this time were flown on ASTP, his suggestions became leading criteria in choosing experiments for the joint flight.[1]

In mid-October 1971, Penrod recommended to Berglund that one of the exciting aspects of an American-Soviet mission was the possiblity of conducting joint experiments during the docked phase of the operation. Such exercises would not affect the feasibility of an international mission, and certainly it would provide meaningful activities for the docked portion of the venture.[2] In December 1971, a letter from Penrod was sent to selected potential experimenters informing them of NASA's "interest in directly involving the user community in the payload planning for the International Rendezvous and Docking Mission."[3] Implicit in this early work were some basic assumptions that would shape subsequent efforts to select ASTP experiments. There would be two categories of scientific investigations—NASA (unilateral) and joint (bilateral). Crewmembers would be active participants in the experiments, which would fall into three groups—stellar phenomena, materials processing, and earth observations. Another key feature of these early discussions was the "austere funding climate" that dictated the use of CSM 111, which did not have the scientific instrument module bay, plus a $10-million ceiling on the cost of the total experiment package.[4]

Formalization of the experiment effort came in the fall of 1972. On 4 October, an initial meeting was held in Washington, during which Houston personnel explained to NASA Headquarters staff the engineering and operational constraints on the planning effort.[5] To simplify the experiment planning, a NASA Working Group was given internal responsibility for overseeing this work. Further, to prevent confusion in the negotiations with the Soviets, M. Pete Frank's Working Group 1 was given sole responsibility for coordinating efforts on bilateral experiments.[6] Through the first six months of 1973, NASA examined candidate experiments.

As this work progressed, Representative Olin Teague urged the agency to make alternative plans for the mission in the event that the Soviets failed for either political or technical reasons to rendezvous with Apollo. Teague believed that it was "essential that the NASA portion of the mission be capable of making a justifiable, independent, scientific and technological contribution without reliance on a Soviet rendezvous."[7] As indicated in chapter VII ("Creating a Test Project"), George Low and the Headquarters staff decided to rely upon the Soviets and not exceed the $10-million budget for experiments.

On 29 June 1973, Administrator Fletcher issued in letter form an "Announcement of Flight Opportunity" for the ASTP mission. Fletcher said, "In addition to developing mutual space rescue capability, the U.S. spacecraft will be able to carry about 400 pounds [181 kg] to conduct other space experiments of high importance." He also emphasized that "investigations that capitalize on the unique nature of this flight and are of common interest to both the U.S. and the U.S.S.R. are, of course, of interest." Enclosed with the letter was a schedule for experiment planning, development, and implementation:

1. *Proposals Due at NASA*   July 23, 1973
   (If appropriate, a prior proposal with a memo updating it will be acceptable.)
2. *Experiment Selection*   Week of July 30, 1973
3. *Selected Experimenter Notification*   August 20, 1973
4. *Interface Control Documentation Complete*   October 1, 1973
5. *Mockup Complete*   March 1, 1974
6. *Training Simulator* (Plus thermal model,   April 1, 1974
   if required)
7. *Definitive Training Unit*   August 1, 1974
8. *Qual[ification] Test Complete*   October 15, 1974
9. *Flight Unit Delivery:*
   Experiments requiring installation in docking
   module or require penetration of docking
   module wall   August 1, 1974

   CSM Installation:

      Complex type   December 1, 1974
      Stowage type   April 15, 1975

10. *Roll Out to Launch Pad* (Only limited access   March 1, 1975
    to experiments after this date)
11. *Launch*[8]   July 15, 1975

NASA sponsored a seminar "with outstanding experts in space science and in the conduct of applications programs in space" at Woods Hole, Massachusetts, on 7 July 1973. "The seminar members were asked to debate possible investigations and guidelines for the final selection of investigations."[9] At the seminar, Homer E. Newell, NASA's Associate Administrator, explained that the Announcement of Flight Opportunity had been issued because of outside dissatisfaction with the earlier efforts within the agency to select experiments for ASTP. NASA's preliminary payload proposal "had been presented to the Space Science Board [of the National Academy of Sciences] and the Physical Sciences Committee [of NASA] and it was received less than enthusiastically. Consequently, it was decided to issue a general Announcement . . . and to convene a special panel to aid in the evaluation process."[10] Following this seminar, Newell, in a letter to Fletcher, reported that a special ad hoc committee would be created, consisting of all but one of the Woods Hole attendees and five other specialists from the scientific and technical community.

> During the week of 30 July, a formally chartered, *closed to the public,* Ad Hoc Evaluation Committee will assemble at the Johnson Space Center to evaluate all proposals including those evaluated as unacceptable for technical and merit reasons in the preliminary review . . . and to categorize them according to suitability for the mission. The proposers will be asked to make presentations and otherwise explain and expand upon their proposals as an expedient to the evaluation process. . . .
>
> Following the activities of the ASTP Program Office, the Ad Hoc Evaluation Committee and costing studies, the Manned Space Flight Experiments Board will review the categorized list developed by the Ad Hoc Committee. This list will include the life science experiments which will have undergone a separate review by the American Institute of Biological Sciences and specific members of the Space Medicine and Biology Committee, Space Science Board of the NAS [National Academy of Sciences]. The MSFEB reviews will be attended by Science, Applications, Technology, and Life Sciences personnel. . . .
>
> We then plan that a presentation will be made to you and Dr. Low on the results of the evaluation and on the integration and cost aspects of the proposed experiments.[11]

On 16 August 1973, Fletcher approved an experiments payload, as presented by Chet Lee, the Program Director. This payload had been approved by the Manned Space Flight Experiments Board (MSFEB) on 10 August from a recommended list provided by the Science, Applications and Technology Ad Hoc Committee, the Life Sciences Ad Hoc Panel, and the American Institute of Biological Sciences Ad Hoc Panel. A total of 146 proposals were received: 24 in the life sciences, 75 in applications and technology, and 47 in the physical sciences. The 18 experiments approved on 16 August were the following:[12]

MA no.	Experiment	Principal investigator	Institution
083	Extreme ultraviolet survey	Bowyer	University of California
088	Helium glow	Bowyer	University of California
059	Ultraviolet atmospheric absorption	Donahue	University of Pittsburgh
048	Soft X-ray	Friedman	Naval Research Laboratory
089	Doppler tracking	Weiffenbach	Smithsonian Institution
010	Furnace	Boese	Marshall Space Flight Center
060	Interface marking in crystals	Gatos	Massachusetts Institute of Technology
070	Zero-g processing of magnets	Larson	Grumman Aerospace Corporation
085	Crystal growth from the vapor phase	Wiedemeier	Rensselaer Polytechnic Institute
041	Surface-tension-induced convection	Reed	Oak Ridge National Laboratory
131	Sodium chloride/lithium eutectic	Yue	University of California
044	Monotectic and synthetic alloys	Cho-Yi Ang	Northrop Corporation
014	Electrophoresis	Hannig	Max Planck Institut für Biochemie
107	Biostack III	Bücker	University of Frankfurt
031	Cellular immune response	Criswell	Baylor College of Medicine
032	Polymorphonuclear leukocyte response	Martin	Baylor College of Medicine
AR-002	Microbial exchange	Taylor	Johnson Space Center
106	Light flash	Tobias	University of California

Glynn Lunney kept the Soviets informed of the status of experiment proposals through his regular telephone conferences with Professor Bushuyev. During their conversation of 23 August, Lunney advised the Professor that the following bilateral experiments had been approved by Administrator Fletcher: artificial solar eclipse, microbial exchange, multipurpose furnace, ultraviolet absorption, and doppler tracking.[13]

As work on the experiments progressed, Chet Lee's office became concerned over their rising costs. Since this increase was largely caused by the amount of documentation required to qualify them for the flight, Lee recommended a relaxation of the procedures:

> The latest cost estimates for experiments hardware indicate that a substantial part of the cost growth we have seen is attributable to implementation of the necessary tasks and effort to meet the Apollo quality and reliability standards which were established to provide the highest assurance that hardware was reliable and safe. The application of these standards to the Apollo and Skylab experiment package was a major factor in their success. The high costs, resulting from the implementation of these standards for high reliability, was warranted because of the high initial investment in the lunar flights, whose primary objective became science following the initial lunar landing. Since science is a secondary objective for ASTP, the capital investment in experiments should be much lower. Therefore, in order to reduce costs we should not require the same degree of documentation, formal reviews, etc. that provided the highest assurance that the reliability and quality standards are being met. Therefore, it is necessary that for the ASTP experiment hardware, the Apollo reliability and quality guidelines be relaxed except where safety of the crew is involved.[14]

Lunney agreed with this evaluation and advised Lee that his office had reviewed the situation and had selected an approach that would minimize costs but still provide high quality hardware. A cost reduction effort was initiated in December 1973 to reduce the cost of the ASTP experiments and to serve as a pilot project for evaluating experiment cost reduction in future programs.[15]

As the Johnson Space Center (JSC) prepared for the flight, new experiments were added and others were deleted or altered. At an MSFEB meeting on 7 January 1974, six more experiments were approved for ASTP subject to the availability of funds and payload capability. Concurrently, the experiment cost ceiling was raised to $16 million. Earth observations and photography (MA-136) was expanded and given full experiment status. Stratospheric aerosol measurement (MA-007) and crystal growth (MA-028) were conditionally approved pending a review by the Ad Hoc Committees. Gas release (MA-043) was also approved tentatively, contingent upon low impact on the docking module design and on the spectrometer used for ultraviolet absorption (MA-059). The other three new experiments were electrophoresis technology (MA-011), geodynamics (MA-128), and barium cloud (MA-017). The barium cloud and gas release investigations were dropped from consideration during the summer of 1974 because of technical and expense difficulties.[16]

During the next year, the experimenters were busy with preparations for the flight.[17] On 26 June 1974, while the principal investigators and contractors worked on their hardware, NASA officially appointed R. Thomas Giuli, of the JSC Science and Applications Directorate, to be the ASTP Program Scientist. His responsibilities included coordinating all scientific aspects of the mission.[18] Subsequently, Giuli summarized in the *Apollo-Soyuz Test Project Preliminary Science Report** the programmatic aspects of the experiments performed unilaterally by the U.S. and jointly with the U.S.S.R.:

> The Apollo-Soyuz Test Project ... experiments package comprised 28 separate experiments. Twenty-one were unilateral U.S. experiments, five were joint U.S.-U.S.S.R. experiments ... and two were unilateral West German experiments (i.e., funded by the Federal Republic of Germany). Together, these experiments formed a well-integrated program of complementary scientific objectives. In several cases, related experiments used different techniques in pursuit of the same or similar scientific objectives. A comparison of the scientific results from these experiments may be useful in defining the best technique to pursue in future space missions.
>
> The individual experiments are grouped in this report according to category and topic. The space sciences experiments are presented in order of the distance away from the center of the Earth that the objectives of study lie. The soft X-ray objects lie deep in our galaxy and even beyond our galaxy. The extreme ultraviolet (EUV) objects lie within a few hundred light-years from the solar system, whereas the portion of the interstellar medium investigated by the helium glow experiment lies within a few astronomical units. The corona photographed during the artificial solar eclipse lay within approximately 50 solar radii from the Sun. Two crystal detectors that have potential application for future gamma-ray astronomy payloads were carried onboard the Apollo spacecraft to measure their susceptibility to radioactivation by cosmic particle bombardment. The tenuous Earth atmosphere at the spacecraft altitude was investigated by ultraviolet absorption, and the aerosol

---

*Published in Feb. 1976 as NASA TM X-58173, this 529-page report provided a detailed synopsis of scientific results as analyzed through 1975. This document is available through the National Technical Information Service, Springfield, Va. 22161. Vol. I of a Summary Science Report was published in 1977 as NASA SP-412 and vol. II is in preparation.

component of the atmosphere below the spacecraft was investigated by stratospheric measurements. Features of the Earth surface were observed and photographed by the Apollo crew, and the structure of the Earth below the surface was investigated by two spacecraft-spacecraft doppler techniques.

The life sciences experiment addressed three primary topics. One was the effects of cosmic particle bombardment on live cells: the human eye retina (light flash), dormant eggs and seeds (biostack), and growing fungi (zone-forming fungi). (The fungi experiment also studied the effects of space-flight factors on biorhythm.) The second topic was the effects of space flight on the human immune system from the aspect of microbial transfer and ability to cause infection and from the aspect of the ability of the immune system to resist infection. The third topic was the effects of reduced gravity on the calcium metabolism of the killifish vestibular system. The purpose was to assess the feasibility of using the killifish vestibular system as a model for future investigation of space-flight effects on human calcium metabolism.

The materials processing effort addressed two topics: the separation of live cells and the improvement of physical properties of solid materials. The live cell separation was performed by each of two electrophoresis methods in which an electric field was applied through a buffer solution containing a mixture of cells with different biological functions (and hence with different negative surface charges). The cells separated into groups of cells with like biological function, each group being characterized by a unique value of cell surface charge. Each group thus acquired a unique speed through the buffer solution. The solid materials were processed by a high-temperature (melting) technique and an ambient-temperature (crystal growth from solution) technique.

The subsequent sections in this report describe in detail the conceptual, instrumental, and operational aspects of each experiment and include a preliminary assessment of scientific results. This section describes the major preliminary results of a few of the experiments (astronomy, Earth atmosphere, Earth observations, biological materials processing, and solid materials processing) as known in December 1975.[19]

ASTRONOMY

MA-048: Soft X-Ray Observation

Objectives

The objectives of this experiment were to study the spectra of a large number of known celestial X-ray sources in the range from 0.1 to 10 kiloelectron volts, search for periodicities and other variability in these sources, and more precisely map the soft X-ray diffuse background.

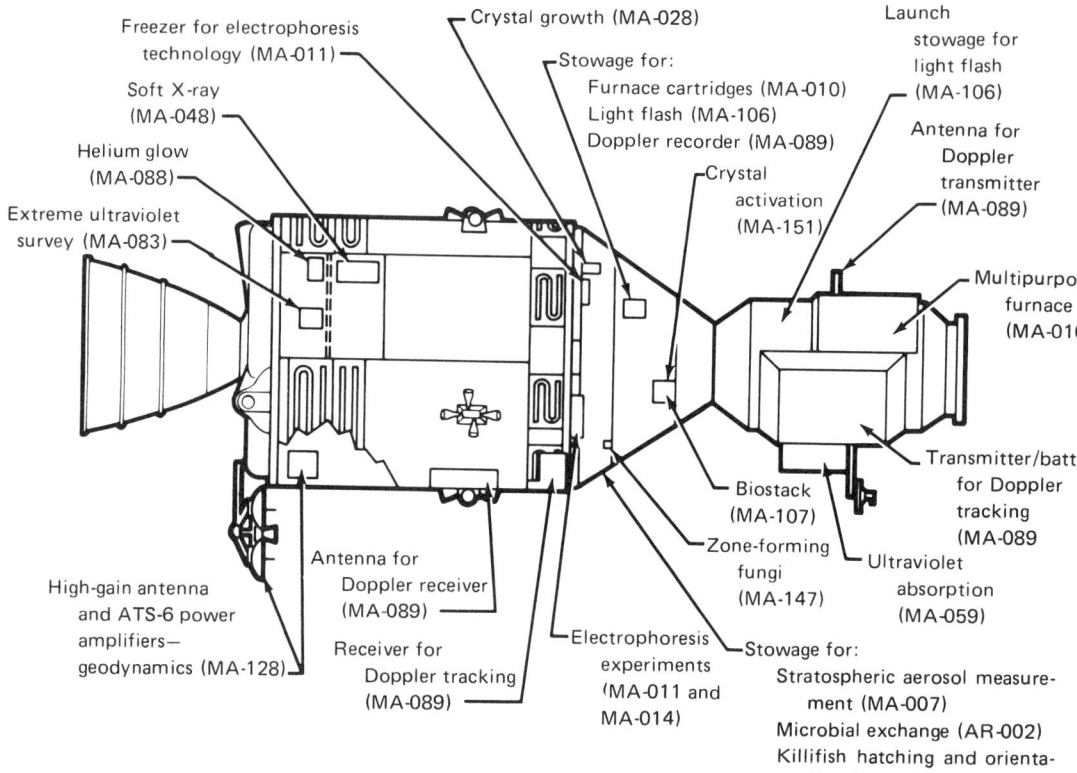

Freezer for electrophoresis technology (MA-011)

Soft X-ray (MA-048)

Helium glow (MA-088)

Extreme ultraviolet survey (MA-083)

Crystal growth (MA-028)

Stowage for:
Furnace cartridges (MA-010)
Light flash (MA-106)
Doppler recorder (MA-089)

Crystal activation (MA-151)

Launch stowage for light flash (MA-106)

Antenna for Doppler transmitter (MA-089)

Multipurpose furnace (MA-01

High-gain antenna and ATS-6 power amplifiers— geodynamics (MA-128)

Antenna for Doppler receiver (MA-089)

Receiver for Doppler tracking (MA-089)

Electrophoresis experiments (MA-011 and MA-014)

Biostack (MA-107)

Zone-forming fungi (MA-147)

Stowage for:
Stratospheric aerosol measure-ment (MA-007)
Microbial exchange (AR-002)
Killifish hatching and orienta-tion (MA-161)

Ultraviolet absorption (MA-059)

Transmitter/batt for Doppler tracking (MA-089

Experiment equipment locations.

Principal Investigator

H. Friedman
Naval Research Laboratory

Results

Data were obtained on approximately twelve sources. Unexpectedly, the instrument developed an intermittent high-voltage discharge problem that resulted in the loss of about 75 percent of the anticipated data. Among the results that were obtained was the discovery of the first known pulsar (star whose radiation pulsates very rapidly) outside our galaxy.

MA-083: Extreme Ultraviolet Survey

Objective

The objective of this experiment was to conduct the first sensitive search for extreme ultraviolet (EUV) radiation from non-solar sources.

Principal Investigator

> S. Bowyer
> University of California at Berkeley

Results

The EUV telescope functioned perfectly during the entire mission. All the primary goals of the experiment were achieved. EUV radiation was detected from four of the thirty stars investigated, which were selected from a variety of classes of stars. Extensive data on the EUV background radiation were also acquired.

MA-088: Interstellar Helium Glow

Objective

The objective of this experiment was to study the motion of helium in the local intestellar medium, as that medium passes through the solar system, to determine several poorly known properties of the local interstellar gas.

Principal Investigator

> S. Bowyer
> University of California at Berkeley

Results

The instrument used was a photometer sensitive to two solar extreme ultraviolet spectral lines that are resonantly scattered by helium gas. The instrument surveyed the entire celestial sphere during a series of slow rolling maneuvers by the Apollo spacecraft. The equipment operated properly; usable data were obtained and are being evaluated.

MA-148: Artificial Solar Eclipse (Joint U.S.S.R.-U.S. Experiment)

Objective

In this U.S.S.R.-proposed exercise, one of five joint experiments, the Apollo crew was responsible for performing the required spacecraft maneuvers and for photographing the eclipse shadow on the Soyuz vehicle, and the Soyuz crew was responsible for photographing the solar corona.

Principal Investigator

> G. M. Nikolsky
> Institute of Terrestrial Magnetism Ionosphere and Radio Wave Propagation

U.S. Point of Contact

> R. T. Giuili
> Johnson Space Center

Results

The U.S.S.R. investigator responsible for the scientific analysis of the experiment reports detection of the solar corona. The results were published in the *ASTP Summary Science Report,* a special publication by NASA.

MA-151: Crystal Activation

Objective

The objective of this experiment was to fly two gamma ray detectors in the command module for post-flight analysis of the radioactivity induced in them by cosmic rays during the flight. The purpose was to measure the instrument background caused by detector activation that interferes with detection of gamma radiation in the 0.02- to 10-megaelectron-volt range from earth orbit. These measurements will be used to estimate this background and thus assist in the development of gamma ray instrumentation and detectors for future experiments in this relatively new field of gamma ray astronomy.

Principal Investigator

> J. I. Trombka
> Goddard Space Flight Center

APPENDIX E

Results

Good data were obtained, which also could be correlated with results of a similar experiment carried on *Apollo 17.*

EARTH ENVIRONMENT

MA-059: Ultraviolet Absorption (Joint U.S.-U.S.S.R. Experiment)

Objectives

The objective of this experiment was to apply optical absorption spectroscopy to the investigation of neutral atomic oxygen and nitrogen (as low as 2 million atoms per cubic centimeter) and their temperatures in the earth's atmosphere at the spacecraft altitude (220 kilometers). The technique was to send monochromatic light beams, the wavelengths of which correspond to neutral atomic oxygen and nitrogen resonance lines (1304 and 1200 Å, respectively), from the Apollo to the Soyuz. The beams were bounced back to the spectrometer aboard the Apollo by a Soyuz-mounted retroreflector.

U.S. Co-Principal Investigators

> T. M. Donahue
> University of Michigan
>
> R. D. Hudson
> Johnson Space Center

Soviet Principal Investigator

> V. G. Kurt
> Institute of Space Research

Results

The O and N densities obtained with this experiment were consistent with the best previous determinations from space experiments employing different techniques and from theoretical models, thus opening the way for a broader application of this technique for atmospheric research.

MA-007: Stratospheric Aerosol Measurement

Objective

This experiment was designed to demonstrate the feasibility of long-term remote sensing of aerosols in the stratosphere from a manned or unmanned spacecraft. Increasing interest in the stratosphere has led to the investigation of methods for remote sensing from earth-orbiting satellites. Data gained from this experiment will be housed in the design of subsequent satellite equipment.

Principal Investigator

T. J. Pepin
University of Wyoming

Results

Excellent aerosol data were obtained in the stratosphere; pollution measurements were obtained down into the troposphere.

MA-136: Earth Observations and Photography

Objective

Astronaut visual observations and photography of surface features (of the moon with Apollo, of earth with Skylab) have demonstrated the usefulness of the large scale view as an aid to interpretation of surface features and phenomena. The human eye's large dynamic range and sensitivity to color and texture have enhanced the perspective of the photographic results. This experiment (a combination of investigations) was designed to permit the crew to perform a number of observations which, based upon Skylab experience, would yield the greatest return of information. The topics of interest were geology, deserts, oceanography, hydrology, and meteorology. A large team of outside scientists constituted the investigator team for this experiment.

Principal Investigator

F. El-Baz
Smithsonian Institution

Results

The data returned were discussed in the *Preliminary Science Report* (p. 10-16). "The astronauts are enthusiastic about their contributions, and the participating scientists have a considerable amount of new data to be interpreted and analyzed. This analysis will further our vistas in numerous fields of Earth science."

MA-089: Doppler Tracking

Objective

This experiment was designed to test the feasibility of improved mapping of earth gravity field anomalies by means of the low-low satellite-to-satellite tracking method. In this case, the low satellites were the command and service module (CSM) and the DM, which were separated to a distance of about 300 kilometers. The CSM received radio signals transmitted from the DM. Such investigations of the earth's gravity field are expected to provide new information on the structure of the earth, with application to continental drift theories.

Principal Investigator

G. C. Weiffenbach
Smithsonian Astrophysical Observatory

Results

When the data are fully analyzed, the investigators anticipate that mass anomalies of approximately 200 to 350 kilometers in size affecting the gravity field will be resolved.

MA-128: Geodynamics

Objective

This experiment was designed to test the feasibility of improved mapping of earth gravity field anomalies by means of the low-high satellite-to-satellite tracking method. In this case, the low satellite was the CSM, and ATS 6 was the high satellite.

Principal Investigator

    F. O. Vonbun
    Goddard Space Flight Center

Results

Early results indicate this method of satellite-satellite tracking yields high quality data for investigations of the gravity field.

LIFE SCIENCES–RADIATION EFFECTS

Interest has developed in studying the effect of high charge and high energy (HZE) particles on human tissue during prolonged space flight. Of particular interest are the effects on non-generative cells, such as the tissue composing the central nervous system. The HZE particles (generally the heavier and energetic cosmic rays) may have destructive effects on human cells under some circumstances. Experiments MA-106, MA-107, and MA-147 were designed to investigate how cosmic rays affect live cells.

MA-106: Quantitative Observation of Light Flashing Sensations

Objective

Light flashes caused by the interaction of cosmic particles and the eyes have been observed by astronauts on all space missions since *Apollo 11*. This experiment compared measurements of the observer's visual sensitivity with measurements of the radiation environment.

Principal Investigator

    T. F. Budinger
    Lawrence Berkeley Laboratory
    University of California at Berkeley

Results

The light flash sensations recorded by the astronauts were well correlated to the detection of HZE particles and protons by onboard electronic and emulsion detectors. The sensations were 25 times more numerous when the spacecraft traveled in the high latitude regions than when it traveled the latitudes between 30° N and 30° S. Ground-based experiments are proceeding to verify the conclusions drawn from the flight data concerning the efficiency of the eye as a detector for various types of particles.

MA-107: Biostack III (German Experiment)

Objective

The objective of this experiment was to continue and extend the research carried out in *Apollos 16* and *17* (Biostacks I and II) to study the biological effects of individual heavy cosmic particles of high-energy loss not available on earth, to study additional space-flight factors, to obtain knowledge on the mechanism by which HZE particles damage biological materials, to get information on the spectrum of charge and energy of the cosmic ions in the spacecraft, and to estimate the radiation hazards to man in space.

Principal Investigator

H. Bücker
University of Frankfurt

Results

Very high resolution impact data were obtained. The consequent effects on the biological specimens are being studied by growing specimens and observing the associated mutations.

MA-147: Zone-Forming Fungi (Joint U.S.S.R.-U.S. Experiment)

Objective

Where MA-107 involved dormant cells that were later cultured or nurtured into growing systems (e.g., seeds of plants and eggs of brine shrimp), this experiment employed growing cells to determine the effect of HZE upon them. Mutations of both types of cells were the objective in both cases to determine the possible effects on humans. Both experiments were planned to examine long-term effects by growing second generation systems from the mutated systems, which would be compared to cells that were not impacted by the HZE particles. Effects of zero gravity were to be analyzed by comparison of flight materials with similar organisms that were not flown.

Soviet Principal Investigator

I. G. Akoyev
Institute of Biological Physics

U.S. Principal Investigator

G. R. Taylor
Johnson Space Center

Results

Differences were detected in growth rates and spore formation between flight samples and ground control samples. The factors causing these differences are currently under study.

LIFE SCIENCES–IMMUNE SYSTEM

Experiments performed by the U.S.S.R. and the U.S. on their space flights have shown that (1) microbes transfer between crewmembers and from crewmembers to the spacecraft; (2) numbers of types of microbes reduce significantly, whereas numbers of microbes of a given (surviving) type increase significantly; and (3) immunological resistance of crewmembers may change during flight. AR-002, complemented by laboratory analysis of blood samples to be performed by MA-031 and MA-032, investigated separately questions of how space flight alters the ability of microbes to infect humans and how space flight alters the ability of humans to resist infection.

AR-002: Microbial Exchange (Joint U.S.-U.S.S.R. Experiment)

Objective

Monitoring two separate crews, which differed microbiologically and immunologically, provided an opportunity to study in-flight cross-contamination patterns. Microbe investigation was accomplished by analyzing spacecraft and crewmember skin swab samples before, during, and after flight.

U.S. Principal Investigator

> G. R. Taylor
> Johnson Space Center

Soviet Principal Investigator

> F. N. Zaloguyev
> Institute of Biomedical Problems
> Ministry of Public Health

Results

The major portion of the planned post-flight laboratory analysis continues. Analysis of the specimen collection and distribution activities indicates that most of the experiment objectives will be satisfied. Analyses of the medically important micro-organisms from U.S. crewmen have shown in-flight inter-crew transfer of potential pathogens but no other changes of medical significance.

MA-031: Cellular Immune Response

Objective

The cellular immune response of the three astronauts was studied before and after the nine days of flight.

Principal Investigator

> B. S. Criswell
> Baylor College of Medicine

Results

Significant changes in the phytohemagglutinin (PHA) lymphocytic responsiveness occurred in the cellular immune response of the astronauts. Parameters studied were white blood cell concentrations, lymphocyte numbers, B- and T-lymphocyte distributions in peripheral blood, and lymphocyte responsiveness of PHA, pokeweed mitogen, Concanaval in A, and influenza virus antigen.

529

MA-032: Effects of Space Flight on Polymorphonuclear Leukocyte Response

Objective

A series of blood samples taken from the astronauts at intervals from thirty days before flight to thirty days after recovery was used to determine the effects of space flight on polymorphonuclear leukocytes (PMN).

Principal Investigator

R. R. Martin
Baylor College of Medicine

Results

Analysis continues but this experiment successfully documented that no consistent, potentially serious abnormalities in the PMN function were produced in the ASTP crewmembers. A broader experience, including similar studies on future space-flight missions, will be required before definite conclusions can be drawn.

LIFE SCIENCES–VESTIBULAR SYSTEM

MA-161: Killifish Hatching and Orientation

Objective

The objective of this experiment was to determine the effect of the zero gravity environment on the sense of balance in living organisms. The killifish *Fundulus heteroclitus* was used to study embryonic development and vestibular adaptation in orbital flight. A series of staged embryos in five individual compartments of a polyethylene bag and a series of preconditioned juvenile fish in a similar bag were mounted on the wall of the service module (SM) and photographed periodically during the mission to record the swimming activity of the fish and the condition of the eggs. At splashdown, vestibular sensitivity of the juvenile fish and of the hatchlings from the eggs was tested in a rotating, striped drum. Subsequently, additional vestibular orientation tests during parabolic trajectory flight, light orientation tests, and geotaxis tests were performed.

Principal Investigator

> H. W. Scheld
> Baylor College of Medicine

Results

Juvenile fish in a null-gravity environment exhibited looping swimming activity similar to that observed during *Skylab 3*. Hatchlings from the 336-hour egg stage were also observed to loop. At splashdown, both juveniles and hatchlings exhibited a typical diving response suggesting relatively normal vestibular function. Juveniles exhibited swimming patterns suggestive of abnormal swim bladders. Rotating drum tests confirmed that no radical changes in vestibular function had occurred, but more subtle changes may be apparent after analysis of motion pictures. Other analyses continue.

APPLICATIONS

Biological Materials

For various types of biological research and medical application, it is necessary to separate pure samples of live cells from a mixture of different types of live cells. The separation process is often not amenable to centrifuge or filter techniques because the different types of cells are not sufficiently dissimilar in size, shape, or mass. Electrophoresis is a separation method that utilizes the fact that live cells have a negative surface charge, and the quantity of this charge is as unique to each type of cell as the cell's biological function. Thus, if a mixture of different types of cells is placed in an electrolytic buffer solution (the composition of which is chosen to preserve the biological vitality of the cells), and if an electrical field is applied, the different types of cells should separate into individual zones according to their individual electrophoretic mobilities. In ground-based laboratories, the performance of this process is limited by effects that are the result of the 1-g environment; for example, the density difference between sample zones and buffer solution causes sedimentation, and heating of the electrophoretic column by the electric field causes destabilizing currents. Both effects are counterproductive. On ASTP, two methods of electrophoresis were tested to determine if better results could be obtained from processing materials in zero gravity.

MA-011: Electrophoresis Technology

Objective

Using the static column of buffer solution with the electrical field aligned along the column, a given amount of sample mixture was introduced at one end and the developing sample zones traveled individually (at different rates) down the column. This was a complete experiment in that it addressed both the major issues for future application: how to process the samples and how to preserve the samples.

Principal Investigator

R. E. Allen
Marshall Space Flight Center

Results

The hardware functioned as planned. Frozen live cells were successfully transported into space; electrophoretic processing was performed; and viable cells were returned to earth. This experiment provided a significant step forward in the development of a biological processing facility in space.

MA-014: Electrophoresis Experiment (German Experiment)

Objective

Using the free-flow method in which a buffer solution flows along a tube with the electrical field aligned perpendicularly to the tube, the sample mixture was inserted continuously at one end and the individual substances separated laterally from each other into multiple streams, which were collected continuously at the other end of the tube. This method is conceptually capable of producing larger quantities, whereas the static column method is most applicable for producing "starter" quantities, which then can be cultured into larger quantities in the laboratory. This experiment addressed only the problem of sample processing and did not involve sample preservation.

Principal Investigator

K. Hannig
Max Planck Institut für Biochemie
Munich

Results

The feasibility of separating living cells under zero gravity conditions was demonstrated.

MA-010: Multipurpose Electric Furnace

Objectives

Based upon a similar furnace (M-518) flown on Skylab, this furnace was used to heat and cool material samples in space, thereby taking advantage of the lack of thermal convection and sedimentation during the liquid or gaseous phase of the material being processed. Seven experiments were performed. The guiding design requirement for the multipurpose electric furnace system was to produce an apparatus that provided the widest possible flexibility in applying predetermined temperature distributions and temperature/time sequences within the constraints imposed by existing interfaces. Although the Skylab multipurpose furnace met all expectations of performance and reliability, it was apparent that improvement in function could be obtained with some specific modifications for ASTP. The system consisted of three essential parts: the furnace, a programmable electronic temperature controller that provided the desired temperatures, and a helium rapid cooldown system.

Principal Investigator

A. Boese
Marshall Space Flight Center

Results

The entire multipurpose furnace system performed perfectly. Final results on all the experiments are pending.

MA-041: Surface-Tension-Induced Convection

Principal Investigator

R. E. Reed
Oak Ridge National Laboratory

MA-044: Monotectic and Synthetic Alloys

Principal Investigators

> C. Y. Ang and L. L. Lacey
> Marshall Space Flight Center

MA-060: Interface Markings in Crystals

Principal Investigator

> H. C. Gatos
> Massachusetts Institute of Technology

MA-070: Zero-g Processing of Magnets

Principal Investigator

> D. J. Larson, Jr.
> Grumman Aerospace Corporation
> Bethpage, New York

MA-085: Crystal Growth from the Vapor Phase

Principal Investigator

> H. Wiedemeier
> Rensselaer Polytechnic Institute

MA-131: Halide Eutectic Growth

Principal Investigator

> A. S. Yue
> University of California

MA-150: Multiple Material Melting (Joint U.S.S.R.-U.S. Experiment)

Soviet Principal Investigator

> I. Ivanov
> Institute of Metallurgy

MA-028: Crystal Growth

Objective

The objective of this experiment was to assess a novel process for growing single crystals of insoluble substances by allowing two or more reactant solutions to diffuse toward each other through a region of pure solvent in zero gravity. This experiment was designed to produce superior crystals and to improve our understanding of the theory of crystal growth.

Principal Investigator

M. D. Lind
Rockwell International Science Center

Results

The experiment was entirely successful and yielded crystals of about expected size, quality, and growth.

NOTES

1. Paul R. Penrod to René A. Berglund, memo, "International Rendezvous and Docking Mission (IRDM) Experiment Requirements," 17 Sept. 1971.

2. Penrod to Berglund, memo, "Joint USA-USSR Experiments during the Docked Phase of IRDM," 15 Oct. 1972.

3. For example, Penrod to William O. Davis, 22 Dec. 1971. See "Post Skylab Missions Familiarization Briefing," 28 Dec. 1971.

4. William O. Armstrong to Berglund, memo, "Payload Planning for Post Skylab CSM Missions," 10 Mar. 1972.

5. Richard J. Allenby to distribution, memo, "ASTP Investigations," 20 Oct. 1972.

6. John E. Naugle to Dale D. Myers, memo, "Joint NASA/USSR Experiments," 2 Nov. 1972; and Myers to distribution, memo, "Joint NASA/USSR Experiments," 19 Dec. 1972.

7. Olin E. Teague to James C. Fletcher, 1 May 1973.

8. Fletcher to distribution, 29 June 1973, with enclosure.

9. Homer E. Newell to Fletcher, 27 July 1973.

10. "Summary Minutes Apollo-Soyuz Test Project (ASTP) Seminar," Houston House, Woods Hole, Mass., 7 July 1973.

11. Newell to Fletcher, 27 July 1973.

12. Chester M. Lee to Glynn S. Lunney, memo, "ASTP Experiment Payload," 27 Aug. 1973.

13. Lunney to Konstantin Davydovich Bushuyev [28 Aug. 1973].

14. Lee to Lunney, Ellery B. May, and William H. Rock, memo, "ASTP Experiments Payload," 30 Nov. 1973.

15. Lunney to Lee, memo, "ASTP Experiments Payload," 14 Dec. 1973; and Lawrence G. Williams, "ASTP Experiment Development Evaluation Report," 21 Aug. 1975.

16. Lee to Lunney, memo, "ASTP Experiments Payload Addition," 1 Feb. 1974; William C. Schneider to Lee, memo, "Approval of Experiments MA-007, Stratosphere Aerosol Measure-

ment and MA-028, Crystal Growth for Apollo/Soyuz Test Project," 19 Mar. 1974; Lee to Naugle, memo, "ASTP Barium Cloud Experiment–MA-017," 9 Apr. 1974; TWX, Lee to Lunney et al., "ASTP Barium Cloud Experiment–MA-017," 24 Apr. 1974; Lee to Lunney, memo, "ASTP Barium Cloud Experiment–MA-017," 24 Apr. 1974; Lee to Armstrong, memo, "ASTP Barium Cloud Experiment–MA-017," 25 Apr. 1974; and Lee to John F. Yardley, memo, "ASTP Barium Cloud Experiment, MA-017," 10 July 1974.

17. Robert O. Aller to attendees, memo, "ASTP Joint Experiments Meeting, NASA Headquarters on June 26, 1974," 1 July 1974.

18. JSC Announcement, "Key Personnel Assignment," 26 June 1974.

19. NASA, *Apollo-Soyuz Test Project Preliminary Science Report,* TM X-58173 (Springfield, Va., 1976), pp. 1-1 and 1-2.

Appendix F
ASTP Launch Vehicles

As part of the joint agreement to use existing hardware, the Soviet and American launch vehicles employed in ASTP were standard boosters with proven records of performance. The Soviets utilized a modernized version of their Soyuz launch vehicle (*Rakyeta nosityel soyuz*), and the Americans used the Saturn IB. This appendix summarizes the information available concerning the performance characteristics of those boosters and the pre-flight preparations of the ASTP vehicles.

RAKYETA NOSITYEL SOYUZ—BACKGROUND

The Soyuz launch vehicle has a design lineage that can be traced to the boosters that placed the first Sputniks into orbit. In the early 1950s, the Soviets developed a kerosene and liquid-oxygen-fueled rocket motor for use in their first intercontinental ballistic missiles (ICBMs). When four of these motors were clustered together with two steerable vernier motors, the Soviets called the combination the RD 107 engine; when four motors were combined with four steerable motors the designation was RD 108. The initial ICBM, the SS-6 (Sapwood in NATO terminology), had four RD 107 units attached as strap-ons to a central core, which was powered by a RD 108. There was a total of twenty main rocket motors and twelve steering motors.

In Soviet practice, the four strap-on units (each 19 meters long and 3 meters in diameter at its base) constituted the first stage of the launch vehicle, while the central core (28 meters by 2.95 meters) was the second stage. Together these stages had been the workhorses of the Soviet space program since 1957. Starting with this basic combination, the Soviets had adapted their launch vehicle to different roles by varying the upper stages attached to it. Early satellites were launched using just the first two stages. Later *Luna 1* through *3* and the manned Vostok series were launched using the Lunik third stage. Planetary probes and Voskhod were lifted into space by the SS-6 and the more powerful Venik third stage. Soyuz and Salyut were orbited by the SS-6 and third stages of respectively greater power. In the

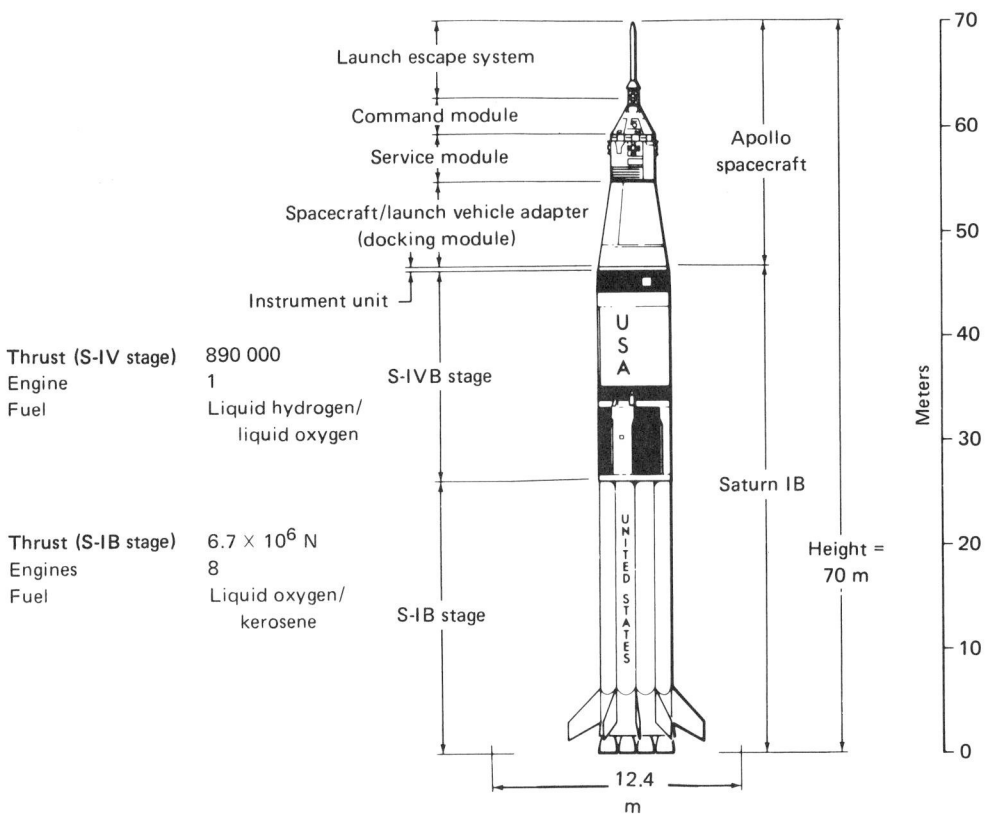

Thrust (S-IV stage)	890 000	
Engine	1	
Fuel	Liquid hydrogen/ liquid oxygen	

Thrust (S-IB stage)	6.7×10^6 N	
Engines	8	
Fuel	Liquid oxygen/ kerosene	

Apollo launch configuration

Apollo and Soyuz

case of the joint project, the *Soyuz 16* and *19* spacecraft were boosted by the latest version of the SS-6 and the Soyuz third stage. The aborted 5 April 1975 flight utilized an older version of the standard Soyuz launch vehicle.

As employed in ASTP, the Soyuz launch vehicle had the following characteristics: each RD 107 produced approximately 845 000 newtons (190 000 pounds) of thrust, and the RD 108 produced about the same, for a total of 4.7 million newtons (950 000 pounds) at sea level; the Soyuz third stage (8 meters long and 2.6 meters in diameter) generated a vacuum thrust of approximately 294 000 newtons (66 000 pounds).*

At launch the engines of the first and second stages were ignited simultaneously. After 120 seconds of flight, the strap-on units were

*These thrust figures are calculated from data made available by the Soviets.

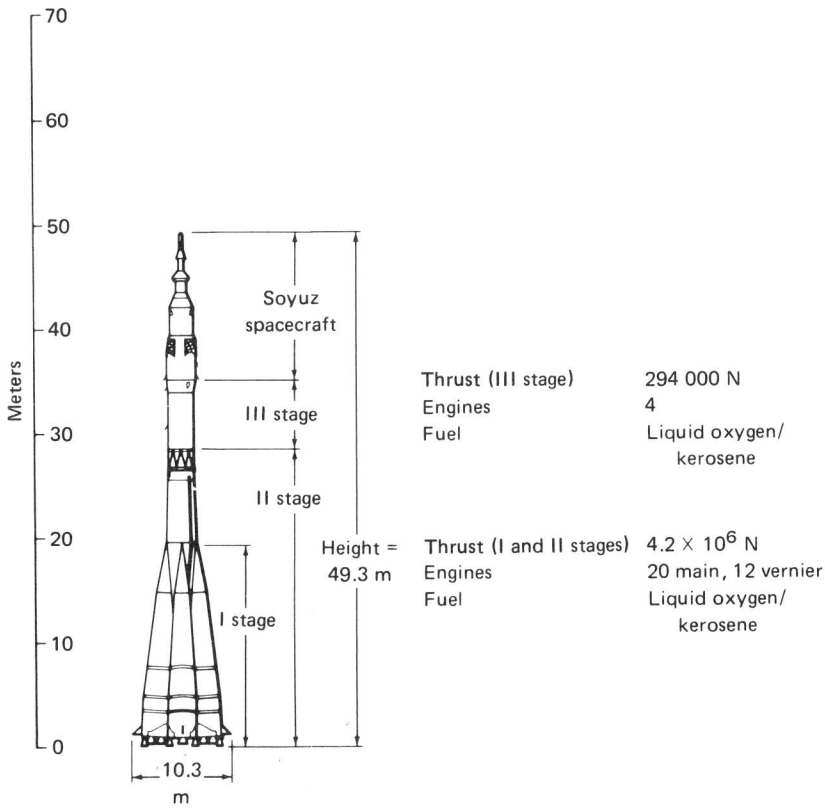

Thrust (III stage)	294 000 N
Engines	4
Fuel	Liquid oxygen/ kerosene

Thrust (I and II stages)	4.2×10^6 N
Engines	20 main, 12 vernier
Fuel	Liquid oxygen/ kerosene

Soyuz launch configuration

space vehicles.

jettisoned. The central core continued to burn until 270 seconds after lift-off, thus accounting for the core being called a sustainer engine. At 270 seconds, the engines of the third stage were ignited, and the second stage was jettisoned. The spacecraft continued on its powered flight until 530 seconds when the third-stage engines were shut down and the spacecraft began its orbital flight around the earth.[1]

SATURN IB–BACKGROUND

A member of the Saturn launch vehicle family, the Saturn IB was conceived in 1962 as a more powerful (uprated) version of the Saturn I launch vehicle. The newer booster was capable of lifting larger payloads than its predecessor and was put to use during the Apollo earth orbital test missions and the command and service module (CSM) flights to Skylab and ASTP. All the lunar voyages of Apollo used the much more powerful Saturn V.* As employed in the ASTP mission, the Saturn IB's first stage produced 6.7 million newtons (1.5 million pounds) of total thrust from its eight kerosene and liquid-oxygen-powered H-1 engines. Its second stage, the S-IVB, used a single J-2 engine fueled by liquid hydrogen and liquid oxygen to produce 890 000 newtons (200 000 pounds).

The first Saturn IB launch (AS 201) took place in February 1966. Nine years later SA 210 lifted the ASTP CSM into orbit. Key dates in the life history of SA 210 are given in table F-1, and key dates in the life history of CSM 111, docking module (DM) 2, and descent stage (DS) 5 are given in table F-2. Once the launch vehicle and the spacecraft were received at KSC,

*Both the Saturn IB and its predecessor helped to lay the foundation for the Saturn V program. The Saturn V had three stages—the S-IC, the S-II, and the S-IVB. The Saturn IC had five kerosene and liquid oxygen F-1 engines producing 33.4 million newtons (7.5 million pounds), while the Saturn II stage produced 4.5 million newtons (1 million pounds) with five J-2 engines. The Saturn IVB was the same stage as used on the S-IB launch vehicle.

Table F-1.—*Summary of Life History of SA 210*

Phase	S-IB	S-IVB	Instrument unit
Start of fabrication	6 Sept. 1966	15 Feb. 1966	Mar. 1967
Completion of fabrication	4 Jan. 1967	3 Jan. 1967	June 1968
Completion of testing	7 Mar. 1967	21 Mar. 1967	June-Aug. 1968
Static firing tests	9-22 May 1967	NA	NA
Start of plant storage	30 Aug. 1967	23 Apr. 1967	June 1969
Termination of plant storage	30 Oct. 1972	12 Jan. 1971	May 1974
Shipment to Kennedy Space Center (KSC)	17 Apr. 1974	7 Nov. 1972	May 1974
Start of KSC storage	26 Apr. 1974	8 Nov. 1972	May 1974
Termination of KSC storage	4 Dec. 1974	8 Mar. 1974	Dec. 1974
Stacked on mobile launcher	13 Jan. 1975	14 Jan. 1975	Jan. 1975

Table F-2.—*Summary of Life History of U.S. ASTP Modules*

Phase	CSM 111	DM 2	DS 5
Start of fabrication	Mar. 1967	Apr. 1973	Sept. 1973
Completion of fabrication	Mar. 1970	June 1974	Sept. 1974
Plant storage completed	July 1972	NA	NA
Initiation of ASTP modifications	Aug. 1972	NA	NA
Completion of ASTP modifications	Mar. 1973	NA	NA
Completion of ASTP experiments/ATS 6 modifications	Aug. 1974	NA	NA
Arrival at KSC	7 Sept. 1974	29 Oct. 1974	3 Jan. 1974

NA = not applicable.

the launch site team began to run a number of tests and began final flight preparations:

S-IB and S-IVB stacked on mobile launcher	14 Jan.
Manned altitude chamber tests with prime crew	14 Jan.
Instrument unit and boilerplate CSM erection	16 Jan.
Manned altitude chamber tests with backup crew	16 Jan.
Mating of DM and CSM	27 Jan.
Crew compartment fit and function test involving American and Soviet crews	10 Feb.
Mating of DM to spacecraft lunar module adapter (SLA)	24 Feb.
Mating of CSM to SLA	5 Mar.
Replacement of cracked fins	11-19 Mar.
Spacecraft delivery to Vehicle Assembly Building and erection on the stacked launch vehicle	17 Mar.
Installation of launch escape system	22 Mar.
Tests of lightning mast	23 Mar.
Rollout to launch pad[2]	24 Mar.

SATURN IB STRESS CORROSION PROBLEM

According to a 1972 "Apollo Experience Report," stress corrosion cracking had been the most common cause of structural-materials failures in the Apollo program. "The frequency of stress-corrosion cracking has been high and the magnitude of the problem, in terms of hardware lost and time and money expended, has been significant."[*][3] Since some of the alloys used in the construction of the Saturn IB launch vehicle were known to be susceptible to stress corrosion, routine inspections had long been a standard procedure. After the discovery in late 1973 of cracks in eight stabilizing fins of the S-IB stage used to launch *Skylab 4,* the SA 210 fins were given special attention. A crack was first noted on a test fin undergoing a stress corrosion check at the Michoud Assembly Facility, New Orleans. A subsequent, more detailed investigation of all eight fins of SA 210 at KSC on 19 February 1975 discovered cracks in the hold-down fittings in two of the fins.[4] In a telex to Professor Bushuyev, Glynn Lunney explained that "this fitting serves no purpose in flight, but supports the launch vehicle on the hold-down

*When certain metal alloys are exposed to a corrosive environment while at the same time they are subjected to an appreciable, continuously maintained, tensile stress, rapid structural failure can occur as a result of stress corrosion. This is known as stress corrosion cracking and is characterized by a brittle-type failure in a material that is otherwise ductile.

arms of the mobile launcher. The critical load on this fitting would occur during 'rebound' if the launch were to be aborted after engines were started and before hold-down arms are released. Fins without cracks have been modified to reduce the stress in the area where cracks initiated. Portions of the fittings were also treated to provide compressive stresses in the surface which also prevents cracking. A fin with these fixes was tested to 142 percent of the design rebound load. Modified fins are now being installed and there is no delay in launch schedule."[5] After the replacement of all eight fins, which solved the stress corrosion problem, this issue was certified to have been corrected during the Headquarters Flight Readiness Review, 12 June 1975.[6]

LAUNCH OPERATIONS

Table F-3 lists the schedule of events prior to the launch of both Soyuz and Apollo.

Table F-3.—*Launch Preparations*

Time[1]	Time to launch	Procedure
13 July		
10:30 a.m.	Apollo: $T - 42$ hr, 30 min	Start Apollo countdown.
11:00 a.m.	Soyuz: $T - 34$ hr, 30 min	First Soyuz launch vehicle readied for propellant loading; second Soyuz also on its launch pad.
5:00 p.m.	Apollo: $T - 35$ hr, 30 min	Cryogenic loading of Apollo fuel cells begun; completed at 11:00 p.m.
15 July		
2:20 a.m.	Soyuz: $T - 6$ hr	Soyuz launch vehicle batteries installed.
4:20 a.m.	Soyuz: $T - 4$ hr	First Soyuz launch vehicle propellant loaded.
6:20 a.m.	Soyuz: $T - 2$ hr	U.S.S.R. Soyuz crew enters first Soyuz spacecraft and U.S. mobile service structure is moved from the U.S. Apollo launch pad.
6:50 a.m.	Apollo: $T - 9$ hr	Begin clearing blast danger area for launch vehicle propellant loading.
7:35 a.m.	Soyuz: $T - 45$ min	The U.S. reports the last Apollo status "Go for Soyuz launch."
7:42 a.m.	Apollo: $T - 8$ hr, 8 min	Initial target update to the launch vehicle digital computer for rendezvous with Soyuz.
8:19:40 a.m.	Soyuz: $T - 20$ s	The U.S.S.R. Soyuz is launched and confirmation by voice follows immediately from the U.S.S.R. control center.
8:30 a.m.	Soyuz: $T + 10$ min	The U.S. starts the Apollo launch vehicle propellant loading: liquid oxygen in first stage and liquid oxygen and liquid hydrogen in second stage; continues for 4 hr and 22 min.

Table F-3.—*Launch Preparations—Concluded*

Time[1]	Time to launch	Procedure
10:35 a.m.	Apollo: $T - 5$ hr, 15 min	Flight crew alerted.
10:50 a.m.	Apollo: $T - 5$ hr	Crew medical examination.
11:20 a.m.	Apollo: $T - 4$ hr, 30 min	Brunch for crew.
12:20 p.m.	Apollo: $T - 3$ hr, 30 min	30-min built-in hold.
12:44 p.m.	Apollo: $T - 3$ hr, 6 min	Crew leaves manned spacecraft operations building for LC-39 via transfer van.
1:02 p.m.	Apollo: $T - 2$ hr, 48 min	Crew arrives at Pad B.
1:10 p.m.	Apollo: $T - 40$ min	Start flight crew ingress.
1:59 p.m.	Apollo: $T - 1$ hr, 51 min	Start space vehicle emergency detection system test.
2:29 p.m.	Apollo: $T - 1$ hr, 21 min	Target update to the launch vehicle digital computer for rendezvous with Soyuz.
2:52 p.m.	Apollo: $T - 58$ min	Launch vehicle power transfer test.
3:05 p.m.	Apollo: $T - 45$ min	Retract Apollo access arm to standby position (12°).
3:08 p.m.	Apollo: $T - 42$ min	Final launch vehicle range safety check (to 35 min).
3:15 p.m.	Apollo: $T - 35$ min	Final target update to launch vehicle digital computer for rendezvous with Soyuz.
3:20 p.m.	Apollo: $T - 30$ min	The U.S.S.R. provides a nominal Soyuz status "Go for Apollo launch."
3:35 p.m.	Apollo: $T - 15$ min	Maximum 2-min hold for adjusting lift-off time; spacecraft to full internal power.
3:42 p.m.	Apollo: $T - 8$ min	The U.S.S.R. reports the last Soyuz status "Go for Apollo launch."
3:44 p.m.	Apollo: $T - 6$ min	Space vehicle final status check.
3:45 p.m.	Apollo: $T - 5$ min	Apollo access arm fully rejected.
3:47 p.m.	Apollo: $T - 3$ min, 7 s	Firing command (automatic sequence).
3:49 p.m.	Apollo: $T - 50$ s	Launch vehicle transfer to internal power.
	Apollo: $T - 3$ s	Ignition sequence start.
	Apollo: $T - 1$ s	All engines running.
3:50 p.m.	Apollo: $T - 0$ s	*Lift-off.*

[1] EDT.

NOTES

1. Reliable data on Soviet launch vehicles are hard to find. This summary is based on the following sources: [Soviet Academy of Sciences], "Apollo-Soyuz Test Project; Information for Press," 1975, pp. 76-78; Charles S. Sheldon II, "The Soviet Space Program Revisited," *TRW Space Log* (1974), pp. 2-19; Peter L. Smolders, *Soviets in Space* (Guildford and London, 1973), pp. 62-68; U.S. Congress, Senate, Committee on Aeronautical and Space Sciences, *Soviet Space Programs, 1966-70; Staff Report,* 92d Cong., 1st sess. (9 Dec. 1971), pp. 130-132 and 559-563; and ASTP mission commentary transcript, MC 9/1, 15 July 1975.

2. NASA, MSFC, KSC, et al., "Saturn IB News Reference," Dec. 1965 (changed Sept. 1968); and Ellery B. May to Edward C. Ezell, 24 Feb. 1976, with enclosed data on SA 210.

3. NASA, JSC, Robert E. Johnson, "Apollo Experience Report, the Problem of Stress-Corrosion Cracking," TN S-344 (MSC-07201), review copy, July 1972, p. 1.

4. NASA, MSFC, "Design Guidelines for Controlling Stress Corrosion Cracking," 15 June 1970; [Chrysler Corp.], C. C. Davis to R. J. Nuber, memo, "Submittal of CCSSD ECP's EP 12112 and EP 12112T–Additional Structural Components Requiring Stress Corrosion Inspec-

tion and Supplemental Test ECP," 10 Jan. 1974; NASA, MSFC, "ASTP Launch Vehicle Stress Corrosion Review," 11 Nov. 1974; R. J. Schwinghammer to Ellery B. May, memo, "Stress Corrosion Assessment of AS-210," 14 Nov. 1974; NASA, MSFC, "ASTP SA-210 Launch Vehicle Design Certification Review," 15 Nov. 1974; and NASA News Release, KSC-27-75, "Two ASTP Saturn IB Fins to Be Replaced," 25 Feb. 1975.

5. TWX, Glynn S. Lunney to Konstantin Davydovich Bushuyev, 17 Mar. 1975.

6. NASA News Release, MSFC, 75-43, "All Eight Saturn I-B Fins to Be Replaced," 28 Feb. 1975; and NASA, HQ, "Saturn IB Stress Corrosion," General Management Review Report, 17 Mar. 1975.

Index

INDEX

547

INDEX

INDEX

The Authors

The Partnership was written by a husband-wife team, working out of the History Office of the Johnson Space Center, Houston, Texas. Edward Clinton Ezell, born in Indianapolis, Indiana (1939), received his A.B. from Butler University, Indianapolis (1961); M.A. from the University of Delaware (1963), where he was a Hagley Fellow; and Ph.D. in the history of science and technology from Case Institue of Technology, Cleveland (1969). He taught at North Carolina State University, Raleigh, and Sangamon State University, Springfield, Illinois, before contracting with the National Aeronautics and Space Administration to write a history of the Apollo-Soyuz Test Project. Linda Neuman Ezell, from Fulton County, Illinois, was born in 1951. She graduated from Sangamon State University in 1974 and is currently doing graduate work at the University of Houston at Clear Lake City. Dr. Ezell has also published in the field of military technology: recent examples are "Science and Technology in the 19th Century," a chapter in *A Guide to the Sources of United States Military History* (Handen, Conn.: Archon Books, 1975), and *Small Arms of the World,* 11th ed. (Harrisburg, Pa.: Stackpole Books, 1977).